Beyond Constructivism

Models and Modeling Perspectives on Mathematics Problem Solving, Learning, and Teaching

Beyond Constructivism

Models and Modeling Perspectives on Mathematics Problem Solving, Learning, and Teaching

Edited by

Richard Lesh
Purdue University

Helen M. Doerr
Syracuse University

LAWRENCE ERLBAUM ASSOCIATES, PUBLISHERS
Mahwah, New Jersey   London

The final camera copy for this book was supplied by the editors.

Copyright © 2003 by Lawrence Erlbaum Associates, Inc.
All rights reserved. No part of this book may be reproduced in any form, by photostat, microform, retrieval system, or any other means, without prior written permission of the publisher.

Lawrence Erlbaum Associates, Inc., Publishers
10 Industrial Avenue
Mahwah, NJ 07430

Cover art by Pamela A. White
Cover design by Kathryn Houghtaling Lacey

Library of Congress Cataloging-in-Publication Data

Beyond constructivism : models and modeling perspectives on mathematics problem solving, learning, and teaching / edited by Richard Lesh, Helen M. Doerr.

    p. cm.

Includes bibliographical references and index.
ISBN 0-8058-3821-X (cloth : alk. paper)
ISBN 0-8058-3822-8 (pbk. : alk. paper)
1. Mathematics—Study and teaching—Psychological aspects.
    2. Constructivism (Education).   I. Lesh, Richard A.   II. Doerr, Helen M.
QA11.2 .B49   2002
510'.71—dc21   2001040919
                                                                       CIP

Books published by Lawrence Erlbaum Associates are printed on acid-free paper, and their bindings are chosen for strength and durability.

Printed in the United States of America
10 9 8 7 6 5 4 3 2

CONTENTS

Preface ix

PART I: INTRODUCTION TO A MODELS AND MODELING PERSPECTIVE

1. Foundations of a Models and Modeling Perspective on Mathematics Teaching, Learning, and Problem Solving
   Richard Lesh and Helen M. Doerr — 3

2. Model Development Sequences
   Richard Lesh, Kathleen Cramer, Helen M. Doerr, Thomas Post, and Judith S. Zawojewski — 35

3. Origins and Evolution of Model-Based Reasoning in Mathematics and Science
   Richard Lehrer and Leona Schauble — 59

4. Piagetian Conceptual Systems and Models for Mathematizing Everyday Experiences
   Richard Lesh and Guadalupe Carmona — 71

5. A Semiotic Look at Modeling Behavior
   Paul E. Kehle and Frank K. Lester, Jr. — 97

PART II: A MODELS AND MODELING PERSPECTIVE ON TEACHER DEVELOPMENT

6. A Modeling Perspective on Teacher Development
   Helen M. Doerr and Richard Lesh — 125

7. A Modeling Approach for Providing Teacher Development
   Roberta Y. Schorr and Richard Lesh — 141

8. A Modeling Approach to Describe Teacher Knowledge
   Karen Koellner Clark and Richard Lesh — 159

9. Task-Analysis Cycles as Tools for Supporting Students' Mathematical Development
   Kay McClain — 175

10. Explanations Why? The Role of Explanations in Answers to (Assessment) Problems
    Martin van Reeuwijk and Monica Wijers — 191

PART III: MODELS AND MODELING AS VIEWED BY HEAVY USERS OF MATHEMATICS

11	What Mathematical Abilities Are Needed for Success Beyond School in a Technology-Based Age of Information? *Richard Lesh, Judith S. Zawojewski, and Guadalupe Carmona*	205
12	The EPICS Model in Engineering Education: Perspectives on Problem-Solving Abilities Needed for Success Beyond Schools *William Oakes and Anthony G. Rud, Jr.*	223
13	The Case for Cases *Geza Kardos*	241
14	Introduction to an Economic Problem: A Models and Modeling Perspective *Charalambos D. Aliprantis and Guadalupe Carmona*	255
15	A Models and Modeling Perspective on Technology-Based Representational Media *Tristan Johnson and Richard Lesh*	265
16	A Models and Modeling Perspective on Skills for the High Performance Workplace *Melissa J. Dark*	279

PART IV: MODELS AND MODELING IN PROBLEM SOLVING AND LEARNING

17	Ends-in-View Problems *Lyn English and Richard Lesh*	297
18	A Models and Modeling Perspective on Problem Solving *Judith S. Zawojewski and Richard Lesh*	317
19	A Models and Modeling Perspective on the Role of Small Group Learning Activities *Judith S. Zawojewski, Richard Lesh, and Lyn English*	337
20	Local Conceptual Development of Proof Schemes in a Cooperative Learning Setting *Guershon Harel and Richard Lesh*	359
21	A Models and Modeling Perspective on Metacognitive Functioning in Everyday Situations Where Problem Solvers Develop Mathematical Constructs *Richard Lesh, Frank K. Lester, Jr., and Margret Hjalmarson*	383
22	Interest, Identity, and Social Functioning: Central Features of Modeling Activity *James A. Middleton, Richard Lesh, and Michelle Heger*	405

PART V: MODELS AND MODELING BEFORE AND AFTER MIDDLE SCHOOL

23 Beyond Constructivism: An Improved Fitness Metaphor for the Acquisition of Mathematical Knowledge
Susan J. Lamon
435

24 Using a Translation Model for Curriculum Development and Classroom Instruction
Kathleen Cramer
449

25 Integrating a Models and Modeling Perspective With Existing Research and Practice
Marilyn Carlson, Sean Larsen, and Richard Lesh
465

26 Models of Functions and Models of Situations: On the Design of Modeling-Based Learning Environments
Beba Shternberg and Michal Yerushalmy
479

PART VI: NEXT STEPS

27 From Problem Solving to Modeling: The Evolution of Thinking About Research on Complex Mathematical Activity
Frank K. Lester, Jr., and Paul E. Kehle
501

28 In What Ways Does a Models and Modeling Perspective Move Beyond Constructivism?
Richard Lesh and Helen M. Doerr
519

References	557
Author Index	583
Subject Index	589

PREFACE

This book evolved out of research that began more than 25 years ago, based on a National Science Foundation-supported project investigating the question: *what is needed, beyond having a mathematical idea that enables students to use it in everyday problem-solving situations*? (Lesh, Landau, & Hamilton, 1983). Large portions of this research involved creating mathematically rich situations that were similar to the old television program *Candid Camera*. That is, we videotaped students in "real life" problem-solving situations that were thought to be typical of those in which elementary-but-powerful mathematical constructs are useful in the everyday lives of students, or their friends or families. As we observed and analyzed students' mathematical thinking in such real life problem-solving situations, our research teams gradually developed far more sophisticated notions about (a) the nature of situations where mathematics is useful beyond school, and (b) the nature of the understandings and abilities that contribute to success in the preceding situations. We came to recognize the importance of a broader range of elementary-but-deep mathematical understandings and abilities, compared to those emphasized in standardized tests, textbooks, and teaching.

We also came to recognize a much broader range of students who are able to make extraordinary achievements in such situations. Furthermore, many of the students who emerged as being extraordinarily capable were from highly underprivileged backgrounds—and had long histories of poor achievement in school. Therefore, our research became increasingly more action-oriented, and was aimed at developing innovative ways to achieve the following goals:

- *Provide early democratic access to powerful conceptual tools* (constructs or capabilities) that enable all students to achieve extraordinary results (Lesh & Doerr, 2000).
- *Assess deeper and higher order understandings and abilities* that are likely to provide the foundations for success beyond schools in a technology-based *age of information* (Lesh & Lamon, 1993).
- *Identify and encourage a broader range of students* whose outstanding abilities and achievements often aren't apparent in past settings involving traditional tests, textbooks, and teaching (Lesh, Hoover, & Kelly, 1993).
- *Provide new ways for students to document their achievements and abilities* in ways that open doors for entry into desirable schools, professions, and jobs (Lesh, 2001).

For the past 3 to 5 years, the preceding goals have provided major unifying themes for collaborative research efforts within: (a) the *Models and Modeling*

Working Group that is associated with the University of Wisconsin National Center for Improving Student Learning and Achievement in Mathematics and Science Education, (b) the *Models and Modeling Working Group* that is associated with the North American chapter of the International Group for the Psychology of Mathematics Education (*PME-NA*), and (c) the multi-university research seminar series associated with Purdue and Indiana University's jointly sponsored *Distributed Doctoral Program in Mathematics Education.*[1] The doctoral program has used Internet-based multimedia communication technologies to facilitate both real time and asynchronous collaborations among graduate students and faculty members in a variety of leading research institutions throughout Australia, Canada, England, Mexico, and the United States.

Throughout all of the preceding working groups, collaborative seminars, and research or development projects, discussions have been organized around the same themes that are used to identify subsections of this book. That is, we have focused on clarifying the nature of a *models and modeling perspective* on: (a) teacher development, (b) problem solving beyond school, (c) influences of technology on teaching and learning, and (d) extensions of modeling perspectives beyond the middle school grades. For example, in mathematics education research and development, two of the important strands of the last three decades have been on the conceptual development of the individual (and we mean this to broadly include both psychological and social perspectives on that development) and on mathematical problem solving by individuals and (more recently) by groups of individuals. However, the focus on conceptual development has largely forefronted the individual's cognitive development often using problem-solving situations. But, in focusing on explaining conceptual development, such research has not explained how students become better problem solvers. The problem-solving research, on the other hand, has focused on describing and explaining what it is that good problem solvers do, but has left conceptual development in the background. The models and modeling perspective put forward in the chapters of this book brings these two research traditions together so that conceptual development and problem solving can be seen as codeveloping from a single perspective. This is because modeling is local conceptual development.

Throughout the book, each chapter is accompanied by Internet-based resources that include: a 15-minute slide show in which the author(s) give an overview of the main points addressed in the chapter, and digital appendixes that include downloadable problem-solving activities for students or teachers, videotaped and/or transcribed problem-solving sessions, and accompanying tools and resources for teachers and researchers. These resources can be downloaded from the Web site for Purdue's *Center for 21st Century Conceptual*

---

[1] Each of these projects was supported in part by the University of Wisconsin's *National Center for Improving Student Learning and Achievement* in Mathematics and Science Education. This publication and the research reported here are supported under the Educational Research and Development Centers Program (PR/Award Number R305A60007) administered by the Office of Educational Research and Improvement (OERI), U.S. Department of Education. The findings and opinions expressed here are those of the authors and do not necessarily reflect the views of the supporting agencies.

PREFACE   xi

Tools http://tcct.soe.purdue.edu/books_and_journals/models_and_modeling/. By appending Web-based resources to each chapter, it is possible to keep the chapters as focused as possible while making available supporting digital resources for the reader.

The final chapter of this book is organized in a table format that identifies many of the most significant ways that models and modeling (M&M) perspectives move beyond constructivism. In particular, the following questions are addressed:

1. What is the nature of *"real" experience*?
2. Can we know what is in the mind of another person?
3. What does it mean to be a *constructivist*?
4. What is different about *assembly-ists*, *constructivists* and *CONSTRUCT-ivists*?
5. Do constructs need to be *discovered* independently by students?
6. What is the relationship between the mastery of *basics skills* and the development of *deeper and higher order understandings and abilities*?
7. In what ways do M&M perspectives go beyond *Piaget's views* about students' developing mathematical constructs?
8. In what ways do M&M perspectives go beyond *cognitively guided instruction*?
9. In what ways do M&M perspectives go beyond descriptions of *cognitive development* that reduce to ladder-like stages?
10. In what ways do M&M perspectives go beyond research on *situated cognition*?
11. In what ways do M&M perspectives go beyond *social constructivism*?
12. In what ways do M&M perspectives go beyond research on students' *learning trajectories*?
13. In what ways do M&M perspectives go beyond research on Vygotsky's *zones of proximal development*?
14. In what ways do M&M perspectives go beyond research on *representation and semiotic functioning*?
15. In what ways do M&M perspectives go beyond research on *problem solving—and information processing*?
16. In what ways do M&M perspectives go beyond research on *attitudes, interests, affect, identity, and beliefs*?
17. In what ways do M&M perspectives go beyond research on *metacognition and higher order thinking*?
18. In what ways do M&M perspectives go beyond research on what is needed for success beyond school in a technology-based *age of information*?
19. In what ways do M&M perspectives go beyond research on *teacher development*?

Each section of the table refers back to the chapters in the book that address the issue at hand. Some readers may find it useful to briefly read relevant sections of the table *before* reading the main chapters of the book; others may wish to use the table to help summarize points as they are emphasized in various sections of the book. For example, any of the questions below might make useful organizing issues to consider for the sections indicated.

Part I: Introduction to a Models and Modeling Perspective
          Questions: 1, 2, 3, 4, 5, 6, 7, 8, 9, 10, 11, 12, 13
Part II: A Models and Modeling Perspective on Teacher Development
          Questions: 4, 5, 6, 7, 8, 9, 10, 11, 12, 13, 19
Part III: Models and Modeling as Viewed by Heavy Users of Mathematics
          Questions: 1, 5, 6, 10, 14, 15, 17
Part IV: Models and Modeling in Problem Solving and Learning
          Questions: 1, 4, 5, 11, 13, 14, 15, 17
Part V: Models and Modeling Before and After Middle School
          Questions: 14, 15, 16, 17, 18
Part VI: Next Steps
          Questions: 5, 6, 9, 15, 16, 17

There is no one way to read through this book. The reader is invited to cycle through the sections, the chapters, the digital overviews and appendixes, and the summarizing table that is in the closing chapter in ways that support revising and refining ideas about problem solving, and teaching and learning mathematics. This book is intended for practitioners (including teachers, researchers, curriculum developers, instructional designers, and school-based decision makers) who are interested in the development of powerful conceptual tools for addressing teaching and learning the mathematics and problem solving that is needed in the 21st century.

ACKNOWLEDGMENTS

We gratefully acknowledge the contributions made to the chapters of this book from the feedback, review, and critique generated by the mathematics education doctoral students and faculty at Purdue Calumet, Indiana University, Purdue University, Queensland University of Technology, SUNY–Buffalo, Arizona State University, Rutgers University, the University of Montreal, and Syracuse University. Appreciation is especially given to Guadalupe Carmona, Michelle Heger, Margret Hjalmarson, and Vickie Sanders, who assisted in the many details of assembling and producing the works in this volume.

– Richard Lesh
– Helen M. Doerr

I
INTRODUCTION TO A MODELS AND MODELING PERSPECTIVE

Chapter 1

Foundations of a Models and Modeling Perspective on Mathematics Teaching, Learning, and Problem Solving

Richard Lesh
Purdue University

Helen M. Doerr
Syracuse University

At the end of this chapter, Appendixes A, B, and C, are three examples of problem-solving activities that we refer to as *model-eliciting activities*—so called because the products that students produce go beyond short answers to narrowly specified questions—which involve sharable, manipulatable, modifiable, and reusable conceptual tools (e.g., models) for constructing, describing, explaining, manipulating, predicting, or controlling mathematically significant systems. Thus, these descriptions, explanations, and constructions are not simply processes that students use on the way to producing "the answer;" and, they are not simply postscripts that students give after "the answer" has been produced. They ARE the most important components of the responses that are needed. So, the process is the product! (See Fig. 1.1)

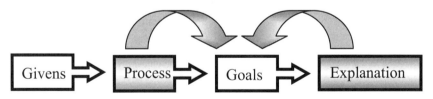

FIG. 1.1: Often, in model-eliciting activities, the process is the product.

Figure 1.2 illustrates a second way that model-eliciting activities differ from traditional textbook word problems. This difference results from the fact that, beyond the computational skills that most traditional textbook word problems are intended to emphasize, it is often the case that the main thing that is problematic for students to make meaning of symbolically described situations. Whereas, for the kind of problems that are emphasized in this book, almost exactly the opposite kind of processes are problematic. That is, model-eliciting activities are similar to many real life situations in which mathematics is useful. Students must try to make symbolic descriptions of meaningful situations.

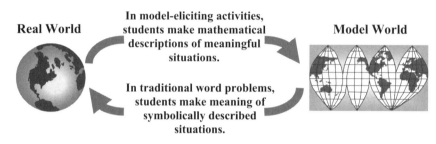

FIG. 1.2. Mathematizing versus decoding.

According to traditional views of learning and problem solving, learning to solve "real life" problems is assumed to be more difficult than solving their anesthetized counterparts in textbooks and tests. But, according to a *models and modeling perspective*, almost exactly the opposite assumptions are made.

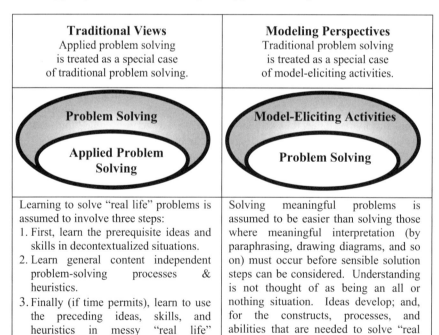

FIG. 1.3. Applied problem solving ≠ model-eliciting activities.

1. FOUNDATIONS OF MODELS AND MODELING PERSPECTIVE

A point emphasized throughout this book is that model-eliciting activities can be designed to lead to significant forms of learning. This is true because model-eliciting activities usually involve mathematizing—by quantifying, dimensionalizing, coordinatizing, categorizing, algebratizing, and systematizing relevant objects, relationships, actions, patterns, and regularities. Consequently, in cases where the conceptual systems that students develop are mathematically significant sense-making systems, the constructs that are extended, revised, or refined may involve situated versions of some of the most powerful elementary—but deep—constructs that provide the foundations for elementary mathematical reasoning. In particular, the three model-eliciting activities that follow were designed to focus on elementary functional reasoning that involves proportions, products, ratios, or linear combinations of variables. For each of the three problems, our research has shown that students often invent (or significantly extend, refine, or revise) constructs that are more powerful than anybody has dared to try to teach to them using traditional methods (Lesh, Hoover, & Kelly, 1993).

(a) The Big Foot Problem in Appendix A is a middle school version of a case study (simulation of a real life problem-solving situation) that we have seen used for instruction (or assessment) in a variety of graduate programs or professional schools at Purdue University, Syracuse University, and other institutions. In particular, the Big Foot Problem is similar to a paleontology problem that involved estimating how fast a particular species of dinosaur could run using photographs of footprints discovered at an archeological site in Utah. The central mathematical idea that students need to use in the Big Foot Problem involves some form of proportional reasoning—perhaps based on linear relationships such as $Y = S \bullet X$ (where S is the factor for scaling up or scaling down from one situation to another).

To prepare for the Big Foot Problem, students are introduced to the context and to relevant information by reading and discussing a newspaper article about the famous tracker, Tom Brown, who lives in New Jersey's pine barrens and who often works with police to help them find lost people or escaped criminals. Tom was taught tracking skills by his Apache grandfather who also showed him how to live in wild country with no tools or food except those he could make for himself using things he could find. So, Tom has become like the famous detective, Sherlock Holmes. He can see tracks where other people cannot see any, and, just by looking at footprints, he often is able to make amazingly accurate estimates about the people who made these footprints—how tall they are, how much they weigh, whether they are men or women, and how fast they are running or walking.

To start students working on the Big Foot Problems, after they have discussed the newspaper article about Tom Brown, and after they have read the statement of the problem that is given below, students are shown a flat box (2 x 2 x 2) that contains a huge footprint in a 1" layer of dried mud—as shown in Fig. 1.4.

Early this morning, the police discovered that, sometime late last night, some nice people rebuilt the old brick drinking fountain in the park where lots of neighborhood children like to play. The parents in the neighborhood would like to thank the people who did it. But, nobody saw who it was. All the police could find were lots of footprints. One of the footprints is shown here. The person who made this footprint seems to be very big. But, to find this person and his friends, it would help if we could figure out how big he is? ---- Your job is to make a "HOW TO" TOOL KIT that police can use to make good guesses about how big people are - just by looking at their footprints. Your tool kit should work for footprints like the one shown here. But, it also should work for other footprints.

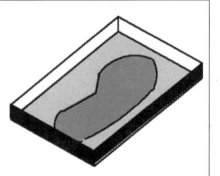

FIG. 1.4. The Big Foot Problem.[1]

(b) The Volleyball Problem that is given in Appendix B is an example of a middle school version of a case study that we first saw used at Northwestern University's Kellogg Graduate School of Management. But, it is also similar to problems that occur when teachers calculate grades for students by combining performance measures from quizzes, tests, lab projects, and other assignments; it resembles what teachers do when they devise "scoring rubrics" to asses students' work on complex tasks; and, it is also like what is done in *places-rated almanacs*, or *consumer guides*, or other publications that compare and assess complex systems. This is because the basic problem involves combining measurements involving several different types of qualitative and quantitative information. In particular, the Volleyball Problem involves formulating an "index of volleyball playing potential" by combining the following six types of information gathered during tryouts on the first day of a girls' summer volleyball camp: (1) height of the player (e.g., Gertrude is 6'1", whereas Beth is 5'2"); (2) vertical leap in inches (e.g., Amy jumped vertically 27 inches, whereas Robin jumped 15 inches); (3) running ability in a 40–meter dash in seconds (e.g., Beth ran 40 meters in 5.98 seconds, whereas Nikki took 8.18 seconds); (4) serve results (e.g., number of serves successfully completed out of 10); (5) Spike results (e.g., out of 5 attempts, Robin made 3 kills, 1 dink-unreturned, and 1 in the net); and (6) coaches comments (e.g., Kim is a great blocker.) After reading a newspaper article that tells about the summer volleyball camp in one small town, the students discuss problems that the league

---

[1] The footprint was made using a size 24 Reebok shoe that is exactly like the one Shaquille O'Neal wears to play basketball. The shoe is about 16 inches long from heel to toe, and it's about 5 1/2 inches wide near the ball of Shaquille's foot.

1. FOUNDATIONS OF MODELS AND MODELING PERSPECTIVE

had because they did not have a systematic procedure to divide girls into equal teams. To begin the problem, the students are given a sample of specific information about 18 girls who were in the tryout sessions. The students' job is to write a letter to the camp organizers describing a procedure for using information from tryouts to sort girls into equal teams during the coming summer (when at least 200 girls are expected to come to tryouts).

Solutions to the Volleyball Problem usually involve some type of weighted combination (e.g., weighted sums, weighted average, or some other linear functional relationship) of the information that is given. But, regardless of what approach is used, students must quantify qualitative information, and, they must devise ways to deal with the fact that (for example) several different types of qualitative and quantitative information must be taken into account—and, high scores are good for serve results, whereas, low scores are good for running.

(c) The Paper Airplane Problem that is given in Appendix C is an example of a middle school version of a case study that we first saw used in Purdue's graduate program for aeronautical engineering. To begin the paper airplane problem, students read an article about how to make a variety of different types of paper airplanes. Then, the students try to make several of these airplanes; and, they test their flight characteristics using several events that involve trying to hit a target on the three different kinds of flight paths shown below.

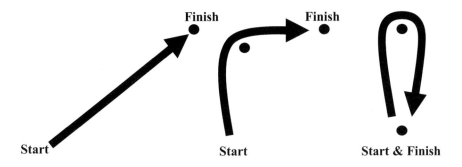

FIG. 1.5. Three flight paths for paper airplanes.

For each paper airplane, three test flights where made for each of the three flight paths; for each of these nine flights, three measurements were recorded: (a) the total distance flown, (b) the distance from the target, and (c) the time in flight. The goal is for students to write a letter to students in another class describing how such data can be used to assess paper airplanes for following four kinds of flight characteristics: (a) best floater (i.e., going slowly for a long time), (b) most accurate, (c) best boomerang, and (d) best overall.

WHAT IS A "MODELS AND MODELING PERSPECTIVE" ABOUT MATHEMATICS TEACHING, LEARNING, AND PROBLEM SOLVING?

This question has no single correct answer that can be taken as a generally accepted starting point for our discussions. This is because mathematics education researchers work at the intersection of education, psychology, and mathematics—where it is also important to communicate meaningfully with parents, policy makers, community leaders, teachers, administrators, teacher educators, curriculum designers, and others. Consequently, any language that is chosen to communicate nontrivial ideas tends to conjure up some unintended meanings from some people while failing to elicit some intended meanings from others. Yet, if any subgroup attempts to formulate more sophisticated and precise terminology, these attempts tend to sound like jargon mongering to nonspecialists in this group.

The aim of this book is to develop a "blue collar theory" that is based on assumptions that sound sensible and useful to ordinary people—but that also lead to implications that are powerful and nonobvious for decision-makers whose ways of thinking influence schools (or the design of productive learning and assessment experiences for K through 16 students). Thus, we have adopted the language of models and modeling not only because of its well-documented power in so many scientific endeavors, but also because it tends to sound simple, meaningful, and useful to people whose understanding and support we want to enlist—such as teachers, parents, and leaders in future-oriented science departments and professional schools at Purdue or Syracuse University.

As an example of an alternative approach, cognitive scientists might prefer to use terms that seem to them to be technically accurate and precise (such as "representations" or "schemata" or "situated cognitive structures") to describe the conceptual systems that we refer to simply as "models". But, although we recognize the importance of the meanings that these cognitive scientists are trying to convey, we have found that these meanings are seldom clear to nonspecialists in cognitive science. So, we have adopted simpler terminology that ordinary people consider to be productive and unpretentious—as long as these interpretations are close to those we intend, without carrying too much unintended conceptual baggage.

By striving for simplicity, readers should be forewarned that, as theoretical perspectives gradually unfold throughout this book, initial common-sense notions about models and modeling often need to be revised, extended, or refined beyond first-iteration ways of thinking that we have chosen as a starting points to begin discussions.

Another reason for adopting the language of models and modeling is because our experience working with teachers and other educators suggests that, when needs become apparent for greater precision or accuracy in our use of language, the discussions that arise usually focus on issues that should not have been avoided anyway—regardless what language we used. For example, in later chapters of this book, readers see that it turns out to be an advantage rather than

1. FOUNDATIONS OF MODELS AND MODELING PERSPECTIVE

a disadvantage that, when everyday language is used, the word "model" can be used as either a noun, or a verb, or an adjective. This payoff occurs because, when it becomes necessary to clear up the preceding kind of apparent ambiguities, the refinements in language that are needed focus attention on important relationships among *behavioral objectives* of instruction (e.g., facts and skills), *process objectives* of instruction (e.g., general problem solving strategies), and *cognitive objectives* of instruction (e.g., conceptual systems for making sense of experiences). More is said about this later.

HERE ARE TWO MISINTERPRETATIONS TO AVOID

In spite of the simplicity and generative power of a models and modeling perspective, needless confusions often arise if we do not explicitly disavow the following two common misinterpretations about our use of the term "model".

- When educators speak about model teaching or model classrooms, the term is being used as an adjective that is synonymous with the word "ideal". In this book, we do not want to invoke this meaning.
- Ever since mathematics educators first became aware of Piaget's research on children's mathematical reasoning (1970), they have been developing models to describe the conceptual systems that they believe students use to make sense of certain kinds of mathematical experiences. Sometimes, these researcher's models are described as if they were the actual conceptual systems that children employ (e.g., Carpenter, Fennema, & Lamon, 1988). But, for most of the cases considered in this book, it is important to emphasize the distinction between models that researchers develop to describe children's thinking, and models that children develop (explicitly, not just implicitly) to construct, describe, or explain mathematically significant systems that they encounter.

In general, the models we are talking about are the latter type, not the former. That is, in model-eliciting activities, students produce conceptual tools that include explicit descriptive or explanatory systems that function as models which reveal important aspects about how students are interpreting the problem-solving situations. Of course, it is not possible for researchers to be certain about many aspects of the conceptual systems that underlie the metaphors, diagrams, symbols, and other media that students use to make sense of their experiences. But, other aspects may be revealed explicitly—such as what kind of quantities the students are thinking about, what kind of relationships they believe are important, and what kind of rules do they believe govern operations on these quantities and quantitative relationships. Clearly, these visible components are part of the students' models and conceptual systems; they are not part of researchers' models. This distinction is important!

WHAT MEANINGS DO WE ASSOCIATE WITH MODELS AND MODELING?

In other recent publications (Lesh & Doerr, 2000), we have differentiated between models, representations, schemas, and other technical terms that are commonly used to describe students' modeling activities. But, for the purposes of this book, the following first-iteration definitions describe the most essential characteristics of the kind of models that we emphasize in this book.

> *Models* are conceptual systems (consisting of elements, relations, operations, and rules governing interactions) that are expressed using external notation systems, and that are used to construct, describe, or explain the behaviors of other system(s)—perhaps so that the other system can be manipulated or predicted intelligently.
>
> A *mathematical* model focuses on structural characteristics (rather than, for example, physical or musical characteristics) of the relevant systems.

In future-oriented fields ranging from aeronautical engineering to business administration to agricultural sciences, when modern scientists construct, describe, or explain complex systems, many of their most powerful conceptual tools implicitly or explicitly involve ways of thinking that are expressed in the form of mathematical and/or scientific models. What is the nature of these models and under what circumstances (and for what purposes) are they needed? Consider the following examples:

- In aeronautical engineering, scientists may construct model airplanes to guide the design and development of real airplanes (or other complex systems that do not occur naturally in nature). In such cases, the need for models may arise because it is too expensive or too dangerous to experiment using the real systems.
- In agricultural sciences, or in earth and atmospheric sciences, scientists may construct computer-based simulations (i.e., models) to investigate weather patterns or other complex phenomena that occur naturally. In this case, models may be needed to simplify the real systems. Or, they may be used to investigate (predict, avoid, manipulate or control) these systems in advance of difficulties that may arise.
- In fields such as economics or business management, statistical models may be embodied in equations and graphs, where the goal may be to provide simplified descriptions of complex patterns and regularities that are not directly accessible, or that cannot be simulated using exact deterministic tools.
- In cognitive psychology, psychologists may develop computer programs to simulate human behaviors in a given collection of problem solving situations. In these cases, the goal may be to explain possible ways of

thinking in forms that suggest how to induce changes in relevant understandings or abilities.
- Based on everyday experiences, children may use a variety of metaphors, analogies, diagrams, models, or stories to describe or explain phenomena involving electricity, shadows, objects that float, or other familiar situations. For example, the behavior of electrical circuits may be described as if it was like water flowing through tubes.

These examples suggest that many similarities exist between (a) the models that children develop to make sense of structurally interesting systems in their own everyday experiences, and (b) the models that psychologists (or other scientists) develop to describe and explain the behaviors of complex systems (such as children's conceptual systems) that they wish to understand or explain. But, this claim does not imply that the conceptual systems that children develop are nothing more than simple versions of the formal mathematical models that are developed by professional mathematicians and scientists. In fact, later in this book, we will explain why, even in the case of scientists and mathematicians, the conceptual systems that they use to develop formal models are not simply unsophisticated versions of the resulting formal systems. Yet, the language of models and modeling is robust enough to help ordinary people think in productive ways about how some of the most important types of mathematical scientific thinking evolve for children, scientists, teachers, and other adults.

WHERE DO MODELS RESIDE? IN THE MIND? IN REPRESENTATIONAL MEDIA?

Do the kind of models we are talking about reside inside the minds of learners or problem solvers? Or, are they embodied in the equations, diagrams, computer programs, or other representational media that are used by scientists, or other learners and problem solvers? Our answer is: Both! First, because they are conceptual systems, they are partly internal and are similar to (but usually more situated and contextualized than) the conceptual systems that cognitive scientists refer to as cognitive structures (or schemas for interpreting experiences). Second, in mathematics and science education, the conceptual systems that are most powerful and useful seldom function in sophisticated ways (or in nontrivial situations) unless they are expressed using spoken language, written symbols, concrete materials, diagrams or pictures, computer-programs, experience-based metaphors, or other representational media. Furthermore, in an age of information many of the most important things that influence peoples' daily lives are communication systems, social systems, economic systems, education systems, and other systems—which are created by humans as a direct result of conceptual systems that are used to structure the world at the same time they structure humans' interpretations of that world. Thus, (internal) conceptual systems are continually being projected into the (external) world.

Based on the preceding observations, the main point that we want to emphasize in this chapter is that, because different media emphasize (and de-emphasize) different aspects of the systems they were intended to describe: (a) meanings associated with a given conceptual system tend to be distributed across a variety of representational media, (b) representational fluency underlies some of the most important abilities associated with what it means to understand a given conceptual system, and (c) solution processes for model-development activities (or other types of problem solving experiences) often involve shifting back and forth among a variety of relevant representations (such as those depicted in Fig. 1.6).

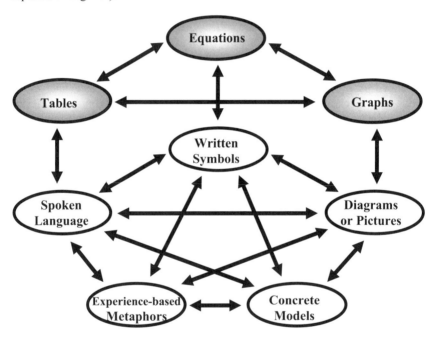

FIG. 1.6. Meanings of conceptual systems are distributed across a variety of representational media.

Even though a mathematician may imagine that all of his or her thinking is expressed in written symbolic notation systems, other media also tend to be involved that may include complex mixes of spoken language, diagrams, and metaphors, as well as written symbols. Similarly, for youngsters, Lesh (1996) illustrates these same points by giving examples from a research project that investigated the question: What's needed, beyond having an idea (as indicated by school performance and standardized tests) to be able to use it in everyday problem solving situations outside of school? To investigate this question, one of the approaches used was for researchers to create staged versions of real life settings (similar to those used in the old television show *Candid Camera*) in which students acted out situations similar to those described in word problems

1. FOUNDATIONS OF MODELS AND MODELING PERSPECTIVE 13

in traditional textbooks and tests. (e.g., Tom ate ¼ of a pizza yesterday, and ⅓ of a pizza today. How much did he eat altogether?)

This research showed that, whereas word problems generally present all of the relevant data in a homogeneous form (as written symbols), more realistic counterparts of these problems often involved information that occurs in mixed modes (such as those illustrated in Fig. 1.6). For example, in a real pizza parlor, the amount of pizza eaten yesterday might be given in the form of a spoken word (or a written symbol, or a picture or diagram), whereas, the amount that is to be eaten today might be given using real slices of pizza (which might be described using some other concrete or graphic media). Consequently, when students solved these types of multimedia problems, they needed to find ways to convert all of the relevant information to a homogeneous form (so relevant computations could be performed). In fact, even if all of the information was given in a homogeneous form, students' solution processes often involved the parallel and interacting use of a variety of different representational media.

In Fig. 1.6, the representation systems that are shaded (e.g., equations, tables, and graphs) correspond to "the big three" that researchers such as Kaput (1987) have emphasized in innovative curriculum materials for high school or college level mathematics, whereas, those that are not shaded have been emphasized in research and development activities aimed at elementary and middle school (Lesh, Post, & Behr, 1987); recently, new technology-based layers have been developed for each of the media depicted in Fig. 1.6.

All of the preceding kinds of representational systems will be discussed in greater detail later in this book (e.g., see Johnson & Lesh, chapter 15). For now, the central point to emphasize is that, regardless which (or how many) different representation systems students use to express their thinking, the underlying conceptual systems are similar to icebergs in the sense that a large portion of what is important is not visible in any single media. This is because different media emphasize and de-emphasize different aspects of the underlying conceptual system—because mathematical descriptions tend to focus on structural (rather than physical) characteristics of relationships, patterns, or regularities that lie beneath the surface of the objects that are most visible.

HOW LARGE (OR SMALL) ARE CONCEPTUAL MODELS IN ELEMENTARY MATHEMATICS OR SCIENCE?

Sometimes, in elementary mathematics or science, the models that provide exceptional utility and power may be as large, complex, and multifaceted as the wave or particle descriptions of light—which may be expressed using media ranging from simple diagrams, to spoken language, to experience-based metaphors, to complex systems of formal equations. Or, they may be as small as a single arithmetic sentence that is used to describe a particular problem-solving situation. All may be called models!

Figure 1.7 gives an example of an intermediate sized model that was created by a group of middle school students in one of our recent research projects focusing on the kind of model-eliciting activities that were described at the beginning of this chapter (Lesh, Hoover, Hole, Kelly, & Post, 2000). The problem involved the development of a conceptual tool to help a car dealer (and one of his or her potential customers) estimate the size of monthly payments for different sized loans and interest rates. The model the students developed was expressed in the form of a spreadsheet with accompanying graphs.

Year	Amount Owed	Interest Rate	Payment/Month	Total Paid
0	$10,000	7.0%	$300	$0
1	$7,100	7.0%	$300	$3,600
2	$3,997	7.0%	$300	$7,200
3	$677	7.0%	$300	$10,800
4	($2,876)	7.0%	$300	$14,400
5	($6,677)	7.0%	$300	$18,000

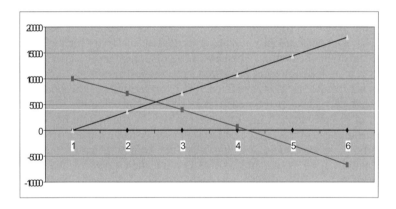

FIG. 1.7. A spreadsheet for determining interest payments for car loans.

In Fig. 1.7, the spreadsheet and graphs provide a system for describing and exploring the relationships among time, interest rates, monthly payments, and the amount of money remaining to be paid (or that has been paid) at any given time. In other words, they provide an intermediate sized model for describing relationships among many of the most important quantities related to this car loan situation.

Another way that constructs or conceptual tools may vary in size is that some focus on big ideas (or conceptual systems) that are especially powerful; whereas, others involve only elaborate chains of low-level ideas and skills. Also, some are easily modifiable (for other situations), reuseable (for other purposes), and sharable (with other people), whereas, others do not lead to significant (generalizable) forms of learning. Or, if students go beyond thinking with a conceptual tool to also think about it, then straightforward ways become apparent for dealing with important aspects of what people mean when they talk about deeper and higher order understandings of powerful elementary constructs.

WHY ARE TRADITIONAL CONCEPTIONS OF ELEMENTARY MATHEMATICS INADEQUATE?

Several answers to this question are apparent in media such as local newspapers, or national newspapers such as *USA Today*. In topic areas like editorials, sports, business, entertainment, advertisements, and weather, these newspapers often look more like computer displays than traditional printed prose. They are filled with tables, graphs, formulas, and charts that are intended to describe, explain, or predict patterns or regularities associated with complex and dynamically changing systems. Clearly, a revolution is occurring concerning what is basic in the core curriculum areas (reading, writing, and mathematics.) Technology-based tools are now used on a daily basis in fields ranging from the sciences to the arts and the humanities, as well as in professions from agriculture to business and engineering, and in employment positions from entry-level to the highest levels of leadership. And, these new conceptual tools are more than simply new ways to carry out old procedures; they are radically expanding the kind of problem solving and decision-making situations that should be emphasized in instruction and assessment.

The essence of an *age of information* is that, as soon as humans develop a conceptual tool for making sense of their experiences, they use this tool to create new realities and experiences. Consequently, the world is increasingly filled with complex systems—communication systems, information storage and retrieval systems, economic and finance systems, and systems for planning and monitoring behaviors of other complex systems. Therefore, people who can make (and make sense of) these complex systems enjoy many opportunities, whereas others risk being victimized by credit card plans, or other similar systems created by humans. To see why, consider the section of modern newspapers that gives advertisements for automobiles. Then, think about how these advertisements looked 20 years ago. They have changed dramatically! Today, it is often difficult to determine the actual price of cars that are shown. What is given are mind boggling varieties of loans, leases, and buy-back plans that may include many options about down payments, monthly payments, and billing periods. Why have these changes occurred? One answer is: They have changed because of the kind of graphing spreadsheets described in Fig. 1.7. By using the graphing and computational capabilities of modern spreadsheets (or calculators), it is easy for car sellers to work out sophisticated leasing and loan plans—using no more than a few basic ideas from elementary mathematics. Yet, the kind of mathematical understandings and abilities that are needed involve dynamic, iterative, and graphic ways of thinking that are quite different than those that have been emphasized in traditional schooling. For example:

- Beyond computing with numbers, thinking mathematically often involves describing situations mathematically. Furthermore, relevant descriptions often must go beyond simple closed-form algebraic equations, and, relevant mathematizing processes may range from

quantifying qualitative information, to dimensionalizing space, and coordinatizing locations.
- Systems that need to be described often go beyond those involving simple counts and measures to also include locations, dimensions, shapes, rules, qualities that must be quantified, quantities that are continuously changing and accumulating, quantities that have both size and direction, or quantities that can not be seen (and that must be measured indirectly). Also, relevant quantities go beyond dealing with "naked" numbers (such as 5, 12, or 126) to include information about both "how much" and "of what" (e.g., 6 apples, 3 apples-per-box)—so that considerations about measurement and units may become critically important.
- Rather than operating on pairs of numbers, students often need to work with whole lists (or sequences, or series, or arrays) of quantities; rather than simply adding, subtracting, multiplying, or dividing, they often need to investigate patterns, trends, and other regularities.
- Products often go beyond simple numeric answers (e.g., 12 apples) to include: descriptions (e.g., using texts, tables, or graphs), explanations (e.g., using culturally and socially embedded discourse to describe why something that appears to be true is not true), justifications (e.g., using persuasive discourse for recommending one procedure over another), or *constructions* (e.g., of shapes that satisfy certain specifications, or of procedures for accomplishing specified goals). Also, requested descriptions, explanations, and constructions usually need to be sharable, transportable, or reusable. So, generalizations and higher order thinking are required as problem solvers must go beyond thinking with the products to also think about them.
- The problem solvers often include more than individual students working in isolation. They may include teams of specialists who are engaging in socioculturally relevant discourse while using a variety of powerful conceptual tools. Furthermore, realistically complex problems often need to be broken into manageable pieces—so that results of subprojects need to be monitored and communicated in forms so that a variety of people can work collaboratively.

These observations suggest that thinking mathematically is about constructing, describing, and explaining at least as much as it is about computing; it is about quantities (and other mathematical objects) at least as much as it is about naked numbers; and, it is about making (and making sense of) patterns and regularities in complex systems at least as much as it is about pieces of data. Also, relevant representation systems include a variety of written, spoken, constructed, or drawn media; and, representational fluency is at the heart of what it means to understand most mathematical constructs. For more details about the evolving nature of modern elementary mathematics (and mathematical abilities), the next sections of this chapter shift attention beyond models (the products) toward modeling (the process).

1. FOUNDATIONS OF MODELS AND MODELING PERSPECTIVE

NONTRIVIAL PROBLEMS OFTEN INVOLVE MULTIPLE MODELING CYCLES

According to the examples that have been given in this chapter, models for describing situations mathematically sometimes must be constructed (or expressed) using sophisticated representational media (ranging from systems of equations, to sophisticated software). But, in other cases, useful models may involve nothing more than ordinary spoken language, drawings, or experience-based metaphors. That is, perfectly adequate models may be based on systems that already exist and that are familiar and accessible to the scientist (or the child). Or, in other cases, the model may consist of a simpler or more manageable ("toy") version of the more complex system. In any case, one simple type of problem solving involves a four-step modeling cycle of the type described in Fig. 1.8. The four steps include: (a) *description* that establishes a mapping to the model world from the real (or imagined) world, (b) *manipulation* of the model in order to generate predictions or actions related to the original problem solving situation, (c) *translation* (or prediction) carrying relevant results back into the real (or imagined) world, and (d) *verification* concerning the usefulness of actions and predictions.

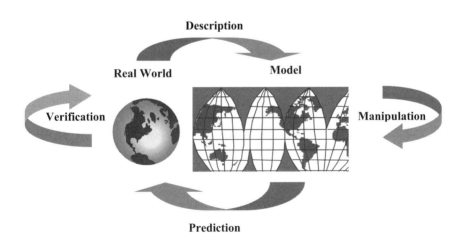

FIG. 1.8. Modeling cycles often involves four basic steps.

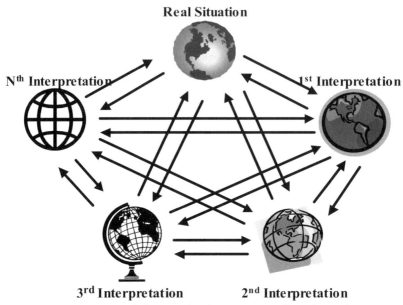

FIG. 1. 9. Modeling may involve sequences of iterative cycles.

Unfortunately, when problem solvers work to develop productive interpretations of complex problem solving situations, a single modeling cycle often is not enough. As Fig. 1.9 suggests, if the mathematical interpretation of a problem is not trivial, solutions typically involve several "modeling cycles" in which descriptions, explanations, and predictions are gradually refined, revised, or rejected—based on feedback from trial testing. Therefore, several levels and types of responses nearly always are possible (with the one that is best depending on purposes and circumstances), and students themselves must be able to judge the relative usefulness of alternative ways of thinking. Otherwise, the problem solvers have no way to know that they must go beyond their initial primitive ways of thinking; and, they also have no way of judging the strengths and weaknesses of alternative ways of thinking—so that productive characteristics of alternative ways of thinking can be sorted out and combined.

Figure 1.9 portrays different interpretations that problem solvers go through when developing a model. Nevertheless, it is somewhat difficult to illustrate how children's conceptual systems (or models) tend to be less like rigid and stable worlds than they are like loosely related and shifting collections of tectonic plates. These plates exist as relatively fuzzy, undifferentiated, and disintegrated fragments that yield piecemeal ways of thinking, which are often remarkably barren, distorted, and inconsistent from one moment to the next.

1. FOUNDATIONS OF MODELS AND MODELING PERSPECTIVE

MODELING OFTEN INVOLVES LOCAL CONCEPTUAL DEVELOPMENT

Following the Big Foot Problem in Appendix A, two more model-eliciting activities also are given that involve proportional reasoning (A/B = C/D); and, for each of these model-eliciting activities, transcripts are given of videotaped problem-solving sessions that show typical solution processes used by average ability middle school students. These transcripts are especially interesting because the development of proportional reasoning has been investigated extensively by both developmental psychologists (e.g., Piaget & Inhelder, 1958) and mathematics educators (e.g., Lesh, Behr, & Post, 1987)—and because there are striking similarities between the modeling cycles that the students went through to solve these problems, and the stages that children typically go through concerning the natural evolution of proportional reasoning.

Model-eliciting activities in which the conceptual systems that need to be developed involve basic cognitive structures that have been investigated by researchers in developmental psychology or mathematics education were selected. After observing hundreds of solutions to such problems, one conclusion we have formed is that, for the modeling cycles that students go through during sixty-minute problem solving sessions often appear to be local or situated versions of the stages that developmental psychologists have observed over time periods of several years. To see an example of this phenomenon, consider the following summary of the Big Foot transcript in Appendix A, and notice that similar modeling cycles are apparent in all of the transcripts in Appendix A—as well as other Appendixes throughout this book.

TABLE 1.1
A Comparison of Local Modeling Cycles and General (Piagetian) Stages of Development

Modeling Cycles That Occur During The Solution Of A Single Model-Eliciting Activity	Stages in the Development of Piagetian Conceptions of General Proportional Reasoning
The multiple-cycle solution described below is typical of those produced by many middle-school students during 1 hour problem-solving sessions.	Developmental psychologists (Piaget, 1964) have observed the following general stages in the evolution of children's capabilities to reason using multiplicative proportions.

Interpretation #1—based on qualitative reasoning: For the first 8 minutes of the session, the students used only qualitative judgments about the size of footprints for people of different size and sex—or wearing different types of shoes. e.g.,
     Wow! This guy's huge.
     You know any girls that big!
     Those're Nikes. The tread's just like mine.
	Stage #1: Reasoning based on qualitative relationships and on only a salient subset of relevant information: In their most primitive responses to Piaget's proportional reasoning activities, students tend to ignore part of the relevant data. For example, in a balance beam task, they may notice only the size of the weights, but ignore the distance of the weights from the fulcrum.

Interpretation #2—based on additive reasoning: A student puts his foot next to the footprint. Then, he uses two fingers to mark the distance between the toe of his shoe and the toe of the footprint. Finally, he moves his hand to imagine moving the distance between his fingers to the top of his head. So, if one footprint is 6" longer than another, then students guess that the difference in height is also 6". That is, instead of thinking in terms of multiplicative proportions (A/B = C/D), the students are thinking in terms of additive differences.

Note: At this point in the session, the students' thinking is quite unstable. For example, nobody notices that one student's estimate is quite different than another's, and predictions that don't make sense are ignored. Gradually, as predictions become more precise, differences among predictions begin to be noticed, and, attention focuses on answers that don't make sense. Still, "errors" generally are assumed to result from not doing the procedure carefully.

Third Interpretation—based on primitive multiplicative reasoning: Here, students' reasoning is based on the notion of being "twice as big". That is, if my shoe is twice as big as yours, then I'd be predicted to be twice as tall as you.

Fourth Interpretation—based on pattern recognition: The students focus on <u>trends</u> across a sequence of measurements; and, footprint-to-footprint comparisons are used to make estimates about height-to-height relationships. This way of thinking is based on the implicit assumption that the trends should be LINEAR—which means that the relevant relationships are automatically (but unconsciously) treated as being multiplicative.

Here, try this.—Line up at the wall.—Put your heels here against the wall.—Ben, stand here. Frank, stand here.—I'll stand here 'cause I'm about the same (size) as Ben . {She points to a point between Ben and Frank that's somewhat closer to Ben}—{pause}—Now, where should this guy be?—Hmmm. {She sweeps her arm to trace a line passing just in front of their toes.}—{pause}—

Students are able to solve problems requiring only qualitative reasoning, but they fail on problems in which quantitative reasoning also is needed. For example, for the two pairs of ratios below, it is easier to compare the first pair than to compare the second pair partly because: (i) if $a < b$ and $c > d$, then it is always true that $a/c < b/d$, but (ii) if $a < b$ and $c < d$, then the relative SIZES of the quantities must be taken into account to determine whether $a/b < c/d$ or $a/b > c/d$.

10 cookies to 2 children <?> 8 cookies to 3 children

10 cookies to 3 children <?> 8 cookies to 2 children

Stage #2: Reasoning based on additive relationships: Early attempts at quantifying often involve directly-perceivable additive differences (i.e., A - B = C - D) rather than multiplicative relationships (which are seldom directly perceivable). For example, if a student is shown a 2 X 3 rectangle and is asked to "enlarge it," a correct response often is given by "doubling" to make a 4 X 6 rectangle. But, if the request is then made to "enlarge it again for a 9" base, the same students often draw a 7 X 9 rectangle—adding 3 to both sides of the 4 X 6 rectangle.

Stage 3: Reasoning based on pattern recognition and replication: Students' earliest uses of multiplicative "pre-proportional reasoning" are often based on a sort of "pattern recognition & replication" strategy—which some have called a "build up" strategy (e.g., Hart, 1984; Karplus & Peterson, 1970, Piaget & Inhelder, 1958). For example, if youngsters are given a table of values, like the ones below, they may solve the problem by noticing a pattern which they can then apply to discover an unknown value. *A candy store sells 2 pieces of candy for 8 cents. How much do 6 pieces of candy cost? Solution:*
- *2 pieces for 8 cents*
- *4 pieces for 16 cents*
- *6 pieces for 24 cents*

According to Piaget, such reasoning doesn't represent "true" proportional reasoning because students don't need to be aware of the reversibility of the relevant operations. Reversibility is demonstrated if, when a change is made in one of the four variables in a proportion, the student is able to compensate by changing one of the other variables by an appropriate amount.

1. FOUNDATIONS OF MODELS AND MODELING PERSPECTIVE

Over there, I think.—{long pause}— Ok. So, where's this guy stand? —About here. {She points to a position where the toes of everyone's shoes would line up in a straight line.}

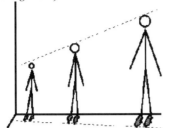

Note: Here, all three students are working together to measure heights, and the measurements are getting to be much more precise and accurate.

5 *Interpretation:* Now, the students are being *very* explicit about footprint-to-height comparisons. That is: A person's height is estimated to be about six times the size of the person's footprint.

*Everybody's a six footer!
(referring to six of their own feet.)*

Stage #4: *Reasoning based on multiplicative proportional reasoning:* According to Piaget (Piaget & Inhelder, 1958), the essential characteristic of proportional reasoning is that it must involve a *reversible relationship between two quantitative multiplicative relationships*. That is:

- It must involve more than qualitative reasoning.
- It must involve more than a relationship between two concrete objects (or two directly-perceivable quantitative differences); it must involve a "second-order" *relationship between two relationships*.
- It must be reversible in the sense that it involves a recognition of structural similarity between two elementary mathematical systems.

Overall, proportional reasoning develops from global conceptualizations (often additive in nature) which focus one-at-a-time on only the cues that are most salient (and often superficial), toward multiplicative conceptualizations which deal with SETS of information (or relationships) several-at-time within a well organized system of proportions *A is to B as C is to D*.

Among the many hundreds of the model-eliciting activities that we have videotaped and analyzed during the past 25 years, most of the solutions involve multiple modeling cycles (e.g., Lesh & Akerstrom, 1982; Lesh & Doerr, 2000; Lesh, Landau, & Hamilton, 1983;). Furthermore, as Table 1.1 illustrates, the modeling cycles that students go through often are strikingly similar to the stages of development that psychologists and educators have observed (over time periods of several years) concerning the natural development of the relevant constructs; and, the mechanisms that contribute to general cognitive development also are very similar to the mechanisms that contribute to local conceptual development in model-eliciting activities (Lesh & Doerr, 2000; Lesh & Kaput, 1988).

MODEL-ELICITING ACTIVITIES INVOLVE LOCAL CONCEPTUAL DEVELOPMENT

The transcripts in Appendixes A, B, and C all illustrate the facts that: (a) during the course of solving a single model-eliciting activity, students often make significant modifications (e.g., extensions, refinements, revisions, or rejections) to their own current ways of thinking, and (b) the underlying conceptual systems that students develop often involve constructs (such as those related to proportional reasoning, or other forms of multiplicative reasoning) that provide the foundations for some of the most powerful big ideas in elementary mathematics. In other words, model-eliciting activities can be designed so that they lead to significant forms of learning (Lesh et al., 2000). Furthermore, because significant forms of conceptual development occur during relatively brief periods of time, it often is possible to observe the processes that students use to extend, differentiate, integrate, refine, or revise the relevant constructs. Consequently, to investigate cognitive development, it is possible for researchers to go beyond descriptions of successive states of knowledge to observe the processes that promote development from one state to another (Lesh, 1983). Also, because it is possible to create conditions that optimize the chances that development will occur, without dictating the directions that this development must take, it is possible to go beyond investigating typical development in natural environments to also focus on induced development within carefully controlled and mathematically enriched environments (Kelly & Lesh, 2000).

Thinking of model-eliciting activities as local conceptual development sessions has many practical and theoretical implications. For example, mechanisms that contribute to local conceptual development can be used to help explain the situated development of students' general reasoning capabilities (Lesh, 1987), or mechanisms that developmental psychologists have shown contribute to general conceptual development can be used to help clarify the nature of problem solving processes (heuristics, strategies) that facilitate students' abilities to use mathematical ideas in everyday situations (Lesh & Zawojewski, 1987).

The application of developmental psychology to problem solving is a relatively new phenomenon in mathematics education research (Lesh, 1987). For example, traditionally, problem solving has been defined as getting from givens to goals when the path is not immediately obvious (or it is blocked); and heuristics have been conceived to be answers to the question: What can you do when you are stuck? But in model-eliciting activities, different modeling cycles are characterized by different ways of thinking about givens, goals, and possible solution steps, and, what is problematic is for students to develop more productive ways of thinking. In other words, in traditional problem solving, the goal is to process information using procedures associated with a fixed construct (that simply needs to be identified, retrieved, and executed correctly), whereas, in model-eliciting activities, it is the constructs themselves that need to be processed. Therefore, the kinds of heuristics and strategies that are most useful in local conceptual development sessions tend to be quite different than those that have been emphasized in traditional problems in which the solutions typically involve only a single interpretation cycle. For model-eliciting activities that involve a series of modeling cycles, the heuristics and strategies that are most useful tend to be aimed at helping students find productive ways to adapt, modify, and refine ideas that they do have, rather than helping them identify relevant ways of thinking when they have none.

1. FOUNDATIONS OF MODELS AND MODELING PERSPECTIVE

When multiple modeling cycles are needed to solve a given problem, two of the most important phenomena that need to be explained are: How is it possible that students recognize the need to develop beyond their first primitive conceptualizations of problem solving situations? And, how is it possible that students are able to develop toward interpretations that get progressively better which are not based on initial naïve concepts of best? These questions will be addressed from a variety of directions throughout this book. But, for now, what is clear just by looking at the transcripts in Appendixes A, B, and C is that: It is possible! One of the most encouraging aspects of research on model-eliciting activities is that, in session after session of the type shown in Appendixes A, B, and C, students who have been diagnosed as being below average in ability often prove that they are capable of inventing (or significantly modifying, refining, or adapting) powerful mathematical or scientific constructs that are much more sophisticated than anybody had dared to teach them—based on their past performance in situations involving traditional textbooks, tests, and teaching (Lesh et al., 2000).

IN MODEL-ELICITING ACTIVITIES, STUDENTS OFTEN INVENT (OR SIGNIFICANTLY EXTEND, REVISE, OR REFINE) POWERFUL MATHEMATICAL CONSTRUCTS

For the transcripts that are given in Appendixes A, B, and C, the three students whose work is shown were African-American seventh graders who were assigned to a remedial mathematics class because their past achievements had been far below average in mathematics classes and on standardized tests. Their classroom was in an inner city school in a large east coast urban school district whose students perform far below average on standardized tests given by the state.

The first problem solving session took place in November, and, prior to this session, the students had worked as a team on six other model-eliciting activities. Therefore, the three students had become accustomed to working on their own during problem solving sessions that often required the team at least 60 minutes to complete. Also, they had learned that they were expected to plan and assess their own work, without asking their teacher questions such as: What are we supposed to do? Are we done yet? Finally, in addition to their experiences participating in teams that solved model-eliciting activities, each of the three students also had played the role of "the client" who reviewed and assessed the results that other teams of students produced.

Research on model-eliciting activities has shown that the kind of problem solving performances that are recorded in Appendixes A, B, and C are by no means uncommon, even in classes of formerly under-achieving students (Graduate Record Examination Board, 1997). In fact, some of the main goals for research on model-eliciting activities is to recognize and reward a broader range of mathematical abilities than those traditionally emphasized in textbooks, tests, and teaching—and, as a result, to recognize and reward a broader range of students who have these abilities (Lesh, Hoover, & Kelly, 1993). Nonetheless, unless students have established appropriate expectations during at least half a dozen problem solving experiences similar to those described in Appendixes A, B, and C, they should not be expected to exhibit similar behaviors. For example, experience has shown that many under-achieving students seem to have "real heads" (that function "on the streets" outside of school) as well as "school heads" (that are used to respond to questions from textbooks, tests, and teachers), and, these two heads seldom appear to communicate. Yet, in order to observe the kind of behaviors that are described throughout this book, it is critical for students to engage their "real heads" (Lesh et al., 2000) in their school activities, and close the gap with their "school heads."

How is it possible for average ability students to invent powerful constructs? Notice, we are not talking about the development of clever language and notation systems that it took humans thousands of years to develop. We are talking about the development of models, metaphors, and other descriptive systems for making sense of familiar experiences in which special care has been taken to ensure that students will recognize the need for the relevant constructs. (Note: The history of science suggests that recognizing the need for a construct often is a large share or what it is needed for its development. One reason this is true is because the recognition of need often presupposes the availability of prerequisite conceptual tools.)

In elementary science, it is clear that, independently from the influence of schools, children often develop models (perhaps in the form of experience-based metaphors that may be expressed using natural language and child-generated diagrams and notations) for making sense of everyday experiences involving phenomena such as gravity, electricity, light, magnetism, or buoyancy. Furthermore, it is also obvious that, to help children develop beyond primitive ways of thinking about such phenomena, it is seldom enough to introduce new facts and skills. If no attempts are made to induce changes in the underlying models and metaphors for describing and explaining the relevant situations, then new facts and skills tend to be grafted onto old sense-making systems—with little impact on the student's thinking.

In elementary mathematics, exactly the opposite assumptions are made. That is, it is assumed that children are incapable of developing their own metaphors, models, and sense-making systems for thinking about situations in which the objects go beyond simple counts and measures—by involving objects such as ratios, rates, proportions, transformations, coordinates, sequences, patterns, quantified qualitative information, quantities that are continuously changing or accumulating, quantities that have both a size and a direction, or quantities that can not be seen (and consequently must be measured indirectly). And, it is assumed that, to teach the preceding ideas, all that is needed is to introduce a few basic facts and skills.

In mathematics, pessimism is based on the well-known facts that many of the constructs we want children to learn depend on clever ideas and notation systems that took many centuries for humans to develop, and many children grow to adulthood without developing beyond extremely primitive conceptual systems for dealing with ratios, rates, proportions, and other foundations for mathematical thinking (Behr, Lesh, & Post, 1987). But, similar statements are equally true in science. So, these facts alone do not justify pessimism about children's abilities to engage in productive model development activities.

In model classrooms where we have seen children make dramatic learning gains, a common characteristic of instruction is that children are continually put in situations where they must express, test, revise, and refine (or reject, or construct) their conceptual systems for making sense of math-rich or science-rich problem solving situations (e.g., Cramer, Post, Lesh, & Behr, 1998; Lehrer & Schauble, 2000; Lesh & Schorr, 2001; Lesh et al., 1993;). In these classrooms, when teachers talked about (and assessed) children's mathematical abilities, they went beyond asking "What arithmetic computations are the students able to do?" to also ask "What kind of situations are the students able to describe (explain, predict)?

Two of the reasons why hundreds of teachers have helped to develop model-eliciting activities are that: (a) they did not believe that standardized tests were assessing many of the most important ideas and abilities that they were trying to emphasize, and (b) they believed that many of their students had far greater understanding and ability than standardized tests implied (Lesh et al., 2000). When these teachers saw the kind of impressive results that many of their lowest achieving students were able to produce when they worked on model-eliciting activities that involved multiple modeling cycles, they recognized that the responses these students were giving on standardized tests often did not show what they could do; they only showed what they would do when they saw no reason (and did not have the time or inclination) to go beyond first-cycle interpretations.

Beyond these practical implications of emphasizing model-eliciting activities in instruction and assessment, the final sections of this chapter focus on implications for theory development.

MODEL DEVELOPMENT INVOLVES A VARIETY OF DIMENSIONS AND INITIAL WAYS OF THINKING TEND TO BE BARREN, DISTORTED, AND UNSTABLE

When solutions to model-eliciting activities involve a sequence of modeling cycles, as in the examples that are given in Appendix A, students' initial interpretations of givens, goals, and available solution steps tend to be quite barren, distorted, and unstable compared with later interpretations, and the development of progressively useful models tends to involve a variety of dimensions such as: simple–complex, particular–general, concrete–abstract, intuitive–formal, situated–decontextualized, external–internal, undifferentiated–analytic, crude–refined, or unstable–stable. Later in this book, we describe similarities and differences among these dimensions. But for now, a point to emphasize is that the right sides of these developmental pairs do not necessarily represent more advanced states of thinking. For example, in a given problem solving situation, the models (or ways of thinking) that prove to be most useful are not necessarily those that are most complex, most general, most abstract, most formal, or most de-contextualized. Yet, "survival of the stable" is one of the main principles that determines which conceptual tools are selected, revised, refined, or rejected, and gradually increasing stability provides some of the most important drivers that lead to transitions from primitive to more powerful ways of thinking.[2]

What are some of the most significant characteristics of unstable conceptual systems? In general, initial primitive interpretations tend to focus one-at-a-time on surface-level characteristics of situations, whereas, later and more sophisticated interpretations are more likely to emphasize deeper patterns and regularities. That is, early (unstable) ways of thinking tend to: (a) take into account only small amounts of the information that is relevant and available, (b) ignore seemingly obvious model-reality mismatches; and, (c) overlook mismatches among alternative ways of thinking. For example, inconsistencies in thinking may go unnoticed if they are expressed using different media (spoken language, written symbols, drawings, or experience-based metaphors), or when learners or problem solvers shift their attention from one aspect of the situation to another, previously noticed global characteristics may be ignored when attention focuses on details, or previously noticed details may be ignored when attention focuses on global characteristics (or on other types of details). So, facts and observations that are salient at one moment may be forgotten a moment later when somewhat different perspectives are adopted.

WHAT PROCESSES CONTRIBUTE TO MODEL DEVELOPMENT?

When problem solving involves several modeling cycles, and when initial interpretations tend to be barren and distorted, a primary question that researchers must answer is: How can students go beyond the limitations of their own initial ways of thinking? Most answers to this question derive from the fact that, for the most powerful constructs in elementary mathematics, understanding is not an all or nothing affair. For a given model-eliciting activity, the relevant constructs can be expected to be at intermediate stages of development, and development is possible along a variety of dimensions. However, by observing students who make progress in model-eliciting activities, more detailed answers also emerge.

There are straightforward ways to help students develop powerful sense-making systems in both mathematics and science. In general, the procedures involve (a) helping students coordinate relevant systems into stable conceptual systems-as-a-whole, and (b) putting students in situations where they are able to reveal, test, and revise/refine/reject alternative ways of thinking.

---

[2] The Piagetian notion of "cognitive conflict" is a special case where the gradually increasing stability of relatively unstable ways of thinking provides a driving force for development.

1. FOUNDATIONS OF MODELS AND MODELING PERSPECTIVE

In some ways, the processes that are involved in developing stable conceptual systems are similar to those required to coordinate stable procedural systems in fields ranging from sports, to cooking, to the performing arts, to mathematics and the sciences. For example, regardless whether a person is learning to play tennis, or golf, or a musical instrument, only learning a set of skills is not enough to make that person a good tennis, golf, or instrument player. Thus, construction (that is, the assembly of complex systems from simple systems) is not the only process that contributes to development. Progressively more powerful and useful systems are sorted out and differentiated at least as much as they are assembled, linked together, combined, or integrated; in general, the most important processes that contribute to the development of increasingly powerful and useful systems tend to be quite different than those that apply to the construction of a machine or a computer program. Productive processes tend to be more like those that contribute to the evolution of communities of living, adapting, and evolving biological systems. In particular, development is not likely to occur unless mechanisms are available to ensure: (a) *diversity*—so that a variety of ways of thinking are available; (b) *selection*—so that unproductive ways of thinking will be refined, revised, or rejected, (c) *propagation*—so that productive ways of thinking are spread and integrated throughout the conceptual landscape; and (d) *conservation*—so that productive ways of thinking are preserved over time (Lesh et al., 2000)

For both, conceptual systems and procedural systems, as increased levels of coordination (stability) is achieved, more details tend to be noticed, which in turn creates the need for increasing levels of coordination. Also, as higher levels of this coordination are achieved, the objects in later systems often consist of actions or patterns in earlier systems. Therefore, emergent phenomena often arise when thinking is reorganized to focus on patterns and regularities associated with higher level systems-as-a-whole.

IN MODEL-ELICITING ACTIVITIES, STUDENTS' ABILITIES VARY ACROSS TASKS AND ACROSS MODELING CYCLES WITHIN A GIVEN TASK

The transcripts that are given in Appendix A clearly reveal three facts.

First, the quality of the constructs that the students developed is impressive. This is especially true because the students involved were in a remedial class for "at risk" students in a school district representing a highly underprivileged urban community. Furthermore, all three students had failed virtually every item dealing with fractions, ratios, rates, or proportions on a recent standardized test.

Second, after appearing to move from primitive stages of development to more advanced stages of development for the Big Foot Problem, these students reverted back to primitive ways of thinking when they began to work on a Running Speed Problem (or a Sears Catalog Problem)—which psychologists, educators, or mathematicians would characterize as having the same underlying structure (i.e., proportional reasoning). Even though their modeling cycles evolved more rapidly for the second and third problems, their apparent Piagetian stage of development varied across modeling cycles (within any one of the problems) and it also varied from one problem to another that involved the use of the same basic ideas.

Third, even though each of the three model-eliciting activities required students to develop conceptual tools that were sharable and reusable, significant generalization across tasks only occurred after instructional activities were provided (see Appendix A for details). These activities helped students go beyond thinking with the relevant constructs to also think about these constructs—by introducing clever language, graphs, and notations for expressing their ideas.

Even though Piaget and other developmental psychologist often speak of proportional reasoning (and other mathematical constructs) as though they were manifestations of general cognitive structures, research on model-eliciting activities shows that the evolution of such constructs is characterized by a gradual increase in local competence more than by the acquisition of some general or all-purpose reasoning strategy. For example, conceptual systems related to proportional reasoning are first developed as situated models that apply to particular problem solving situations. Then, these models are gradually extended to larger classes of problems. It is only at relatively mature levels of thinking that knowledge begins to be organized around abstractions rather than experiences, and these mature levels of thinking are not likely to evolve unless students are challenged to develop models and conceptual tools that are sharable, reuseable, and transportable, introduced to powerful representation systems for expressing relevant constructs, and encouraged to go beyond thinking with these constructs to also think about them.

In addition to the preceding facts, earlier sections of this chapter also emphasized that conceptual evolution tends to involve a number of dimensions (e.g., concrete–abstract, global–analytic, undifferentiated–refined, simple–complex, intuitive–formal, situated–decontextualized, external–internal)—and that students' understanding of a given mathematical construct tends to be distributed across a variety of representation systems. Therefore, the sophistication of students' responses often varies considerably depending on which representation systems are emphasized in a given problem—and which dimensions of development are stressed in model development. Consequently, for a given collection of tasks, the predicted order of difficulty often changes significantly if attention shifts from one dimension of development to another, or from one representation system to another. For example, a child who appears to be functioning at level N for a given task may "fold back" to function a level N-n if the task is changed in ways that would seem superficial to a mathematician, or a psychologist, or a teacher (Kieren, 1994). Similarly, in research investigating the ways that mathematics is used in real life problem solving situations (e.g., Lave, 1988; Nunes, 1992; Saxe, 1991), it has become clear that performance often depends a great deal on the nature of the conceptual tools and resources that are familiar and available to a student in a given problem solving situation. Thus, a child's apparent stage of reasoning often varies significantly across constructs, across contexts, and across representations, as well as from moment to moment within a given model-eliciting activity. In other words, conceptual development appears to be far more situated, piecemeal, and unstable than ladder-like descriptions have suggested (diSessa, 1988; Greeno, 1991). Thus, the process of gradually sorting out and refining unstable conceptual systems tends to be quite different than the process of assembling (or constructing) stable conceptual systems.

Conceptual development in areas such as proportional reasoning is characterized by gradually increasing local competence more than by the acquisition of some general or all-purpose reasoning strategy. Does this variability in performance mean that Piaget's description of conceptual development is relatively useless as an explanatory theory in areas such as proportional reasoning? We believe that examples such as the one in this chapter suggest exactly the opposite. The most important aspects of Piaget's perspective have to do with: (a) the emphasis on the roles that conceptual models play in the interpretation of problem solving (or learning) situations, (b) the fact that the model refinement sequences have certain characteristics that are predictable from one problem situation to another, and (c) the mechanisms that drive the model-refinement process apply in local conceptual development situations.

In spite of the fact that Piaget-inspired analyses of conceptual development often result in ladder-like descriptions of development which generally presuppose that experience is organized around abstractions, and in spite of the fact that these ladder-like descriptions do not take into account the situated and unstable nature of early understandings, many of the most basic foundations of Piaget's theoretical perspectives should not be abandoned in mathematics education research. In particular, Piaget's emphasis on the complex *systemic* nature of mathematical constructs remains viable, especially at a time when complexity theories in mathematics are providing increasingly powerful ways to explain and predict the adaptive behaviors of complex systems (Kaufman, 1993). Also, the fact that model-eliciting activities involve local conceptual development suggests that Piaget's description of the mechanisms that drive development (e.g., cognitive conflict, reversibility, reflective abstraction) continue to be applicable to the evolution of situated conceptual systems.

Thus, modified Piagetian perspectives continue to be applicable to 1 to 2 hour model-eliciting activities. But, instead of focusing on the development of general Piagetian cognitive structures that are hypothesized to apply to a wide variety of specific constructs, contexts, and problems, a models and modeling perspective emphasizes the development of constructs that share the systemic characteristics of Piaget's general cognitive structures—but that are far more situated and dependent on available tools, representations, and resources.

In areas such as proportional reasoning, thinking of conceptual development as a gradually increasing local competence, rather than as a global manifestation of some general cognitive structure, appears to result in more (not less) importance for developmental theories—and to a clearer understanding on the part of educators of the close links that must exist between learning and problem solving. Problem solving should be a critical step on the way to learning; it is not simply an activity that should occur after a concept has been learned and "all or nothing" views of learning are completely inconsistent with the "gradually increasing local competence" view of learning (and problem solving) that is described in this chapter.

SUMMARY

According to this chapter's models and modeling perspectives about teaching, learning, and problem solving, one can envision a classroom as being filled with a teeming community of conceptual systems (or models for making sense of experience)—each of which is competing for survival in the minds of the resident community of students (Papert, 1991). In such an environment, the law survival of the fittest becomes survival of the shared, the useful, and the stable. So, what this book is about is how these conceptual systems evolve, and how development can be encouraged and assessed, during individual learning and problem solving experiences—and across a series of such experiences.

Figure 1. 10 suggests that, in mathematics education, just as in more mature modern sciences, it has become necessary to move beyond machine-based metaphors and factory-based models to account for patterns and regularities in the behaviors of students, groups of students, teachers, and other complex systems. Thus, it is also become necessary to move beyond the assumption that the behaviors of these systems can be explained using simple linear combinations of uni-directional cause-and-effect mechanisms that can be accurately characterized using models from elementary algebra and statistics (Lesh, 2000).

From an Industrial Age		Beyond an Age of Electronic Technologies		Toward an Age of Biotechnologies
using analogies based on hardware		using analogies based on computer software		using analogies based on wetware
where systems are considered to be no more than the sum of their parts, and where the interactions that are emphasized involve no more than simple one-way cause-and-effect relationships.	⇒	where silicone-based electronic circuits may involve layers of recursive interactions which often lead to emergent phenomena at higher levels which are not derived from characteristics of phenomena at lower levels	⇒	where neurochemical interactions may involve "logics" that are fuzzy, partly redundant, partly inconsistent, and unstable—as well as living systems that are complex, dynamic, and continually adapting.

FIG. 1.10. Recent transitions in models for making (or making sense of) complex systems

1. FOUNDATIONS OF MODELS AND MODELING PERSPECTIVE 31

As Table 1.2 suggests, a models and modeling perspective has important implications for teaching, learning, and problem solving.[3] Furthermore, in well designed construct-eliciting activities (Lesh et al., 2000), problem solving not only leads to significant forms of learning, but a byproduct of these learning episodes is that students automatically produce auditable trails of documentation that reveal the nature of what is being learned (Lesh & Lamon, 1992). So, model-eliciting activities often function as thought-revealing activities (Lesh et al., 2000) that provide powerful tools for teachers and researchers to use as performance assessments of students' achievements that are seldom addressed on standardized tests (Lesh & Lamon, 1992).

TABLE 1.2.
A Brief Traditional Views Versus Modeling Views
Of Mathematics, Problem Solving, Learning, and Instruction

	Traditional Perspectives	**Modeling Perspectives**
The Nature of Mathematics	Behavioral objectives of instruction are stated in he form: *Given ... the student will ...* Knowledge is described using a list of input-output condition-action rules (definitions, facts, skills) - some of which are higher order meta-cognitive rules for making decisions about: (i) which lower-level rules should be used in a particular situation, and (ii) how lower-level rules should be stored and retrieved when needed.	Knowledge is likened, not to a machine, but to a living organism. It is recognized that many of the most important *cognitive objectives* of mathematics instruction are conceptual systems (e.g., mathematical models) which are used to construct, describe, or explain situations in which mathematics is useful.
The Nature of Problem Solving	Problem solving is characterized as a process of *getting from givens to goals when the path is not obvious*. But, in mathematics classrooms, problem solving usually is restricted to answering questions that are posed by some people, within situations that are described by others, to get from givens to goals that are specified by a third party, using strings of facts and rules that are restricted in ways that are artificial and unrealistic.	Some of the most important aspects of real life problem solving involves developing useful ways to <u>interpret</u> the nature of givens, goals, possible solution paths, and patterns and regularities beneath the surface of things. Solutions typically involve several "modeling cycles" in which descriptions, explanations, and predictions are gradually refined and elaborated.

---

[3] According to ways of thinking borrowed from the industrial revolution, teachers have been led to believe that the construction of mathematical knowledge in a child's mind is similar to the process of assembling a machine, or programming a computer. That is, complex systems are thought of as being nothing more than the sums or their parts; each part is expected to be taught and tested one-at-a-time, in isolation, and out of context; and, the parts are assumed to be defined operationally using naive checklists of condition-action rules. In contrast to the preceding perspective, scientists today are investigating complexity theory where the processes governing the development of complex, dynamic, self-organizing, and continually adapting systems are quite different than those that apply to simple machines. Parts interact. Logic is fuzzy. Whole systems are more than the sums of their parts; and, when the relevant systems are acted on, they act back.

The Nature of Experts	Humans are characterized as information processors; and, outstanding students (teachers, experts) are those who flawlessly remember and execute factual and procedural rules ... and who are clever at assembling these facts and rules in ritualized settings.	Experts are people who develop powerful models for constructing, manipulating, and making sense of structurally interesting systems; and, they also are proficient at adapting, and extending, or refining these models to fit new and changing situations.
The Nature of Learning	Learning is viewed as a process of gradually adding, deleting, linking, uncoupling, and de-bugging mechanistic condition- action rules (definitions, facts, or skills). Therefore, if the precise state of knowledge is known for an expert (E) and for a given novice (N), then the difference between these two states is portrayed as the subtracted difference (E-N).	Experts not only know more, they also know differently. Learning involves model building; and, relevant constructs develop along dimensions such as concrete-to-abstract, particular-to-general, situated-to-decontextualized, intuitive-to-analytic--to-axiomatic, undifferentiated-to-refined, and fragmented-to-integrated. Evolution involves differentiating, integrating, and refining unstable systems – not simply linking together stable rules; and, it also involves discontinuities and conceptual reorganizations – such as when students go beyond thinking with a model to also think about it.
The Nature of Teaching	Teaching involves mainly: (i) demonstrating relevant facts, rules, skills, and processes, (ii) monitoring activities in which students repeat and practice the preceding items, and (iii) correcting errors that occur.	Teaching focuses on carefully structuring experiences in which students confront the need for mathematically significant constructs, and in which they repeatedly express, test, and refine or revise their current ways of thinking.

Finally, several chapters in this book describe how thought-revealing activities for students often provide the basis for thought-revealing activities for teachers (as well as parents, administrators, school board members, and other relevant leaders in the community). For example, when students work on model-eliciting activities, relevant thought-revealing activities for teachers often focus on developing tools for making sense of students' work. So, whereas students are developing models to make sense of mathematical problem-solving situations, teachers are developing models to make sense of students' work. Instances of these thought-revealing activities for teachers include: (a) developing *observation forms* that enable teachers to structure the notes that they take whereas students are working *on* model-eliciting activities, (b) developing *ways of thinking sheets* that enable teachers to recognize a variety of ways that students think about given model-eliciting activities, and (c) developing *quality assessment procedures* (or scoring rubrics) that enable them to judge the relative strengths and weaknesses of the products that students.

One purpose of the thought-revealing activities for teachers is to encourage them to explicitly reveal, test, revise, or refine important aspects of their own ways of thinking. Research on Cognitively Guided Instruction (Carpenter & Fennema, 1992) has shown that one of the most effective ways to help teachers improve the effectiveness of their teaching is to help them become more familiar with their students' evolving ways of thinking about important ideas and topics,

1. FOUNDATIONS OF MODELS AND MODELING PERSPECTIVE

and thought-revealing activities enable teachers to observe trails of documentation from the ways of thinking of their own students. Consequently, thought-revealing activities for teachers that are based on thought-revealing activities for their students provide the basis for effective and efficient *on-the-job classroom-based teacher development activities* that are based on the same instructional design principles as those that we advocate for children.

All of the preceding themes are revisited repeatedly throughout this book. Our purpose in this chapter has been to give an overview of underlying themes—together with a brief description of what lies ahead.

Chapter 2

Model Development Sequences

Richard Lesh
Purdue University

Kathleen Cramer
University of Wisconsin—River Falls

Helen M. Doerr
University of Syracuse

Thomas Post
University of Minnesota

Judith S. Zawojewski
Purdue University

This chapter describes instructional modules that are based on a models and modeling perspective, and that were designed to meet goals that are unusual compared with those driving the development of most commercially produced materials for instruction or assessment. First, the modules were designed to provide rich research sites for investigating the interacting development of students and teachers. Therefore, they are modularized so that components can be easily deleted, extended, modified, or re-sequenced to suit the needs of researchers (or teachers) representing a variety of theoretical perspectives, purposes, and student populations. Second, to make it possible to observe processes that influence the development of students' and teachers' ways of thinking, the modules were designed to be thought revealing (Lesh, Hoover, Hole, Kelly, & Post, 2000) and to be efficient for producing maximum results using minimum investments of time and other resources. Consequently, from the perspective of teachers, they have the unusual characteristic of seeming to be small-but-easy-to-extend rather than being large-and-difficult-to-reduce. Third, they were designed to emphasize important understandings and abilities that are needed for success beyond schools in a technology-based age of information.

Even though many of the big ideas that are especially powerful in everyday situations have long traditions of being treated as foundation-level ideas in elementary mathematics, it will be clear, in transcripts that are given throughout this book, that many others are not. Also, the activities that we'll be describing often give special attention to levels and types of understanding (and ability) that seldom have been emphasized in traditional textbooks, tests, or teaching. Consequently, by emphasizing an unusually broad range of understandings and abilities, a broader range of students typically emerge as being highly capable; and, many of these students are surprisingly young or were formerly labeled "below average" in ability or achievement. In short, a goal of these modules has been to help researchers investigate the nature of situations in which surprising students produce surprisingly sophisticated results at surprisingly young ages, in surprisingly brief periods of time, and with surprisingly little direct instruction.

Our experience suggests that one of the most effective ways to achieve the preceding goals is to put students in situations where their everyday knowledge and experience enables them to clearly recognize the need for the constructs that they are being challenged to produce—and where a lack of facility with esoteric facts and skills does not prevent them from using their knowledge and experience to develop the required conceptual tools.

RELEVANT ASSUMPTIONS ABOUT THE NATURE OF MATHEMATICS

For the *model development sequences* that are described in this chapter, an underlying assumption is that one of the most important characteristics that distinguish mathematical knowledge from other categories of constructs is that mathematics is the study of (pure)[1] operational/relational structures. Therefore, we focus on the underlying structural aspects of mathematical constructs; and, we assume that a large part of the meanings for objects (such as equations, number systems, formulas, and graphs) derive from their existence as parts of mathematical systems in which they function. From this point of view:

- Doing "pure" mathematics means investigating systems for their own sake—by constructing and transforming and exploring them in structurally interesting ways, and by studying their structural properties (Steen, 1987, 1988).[2]

- Doing applied mathematics means using the preceding systems as models (or conceptual tools) to construct, describe, explain, predict, manipulate, or control other systems. (Lesh & Doerr, 1998).

Jean Piaget (Piaget & Beth, 1966) was one of the most influential people to emphasize the holistic structural character of children's mathematical reasoning; and, Zoltan Dienes (1960) was one of the most creative mathematics educators to specify principles for designing instructional activities to help children develop these structure-based understandings and abilities. Dienes focused on concrete-to-abstract dimensions of conceptual development by emphasizing

---

[1] The word "pure" is enclosed in parentheses because these conceptual systems are never completely pure. To be useful beyond trivial situations, they are always expressed using (several) representational systems, and meanings are influenced by unstated experiences, assumptions, and purposes of the particular humans who use them.
[2] Musicians, for example, begin to step into the realm of mathematics when they go beyond playing sequences of individual notes toward investigating structural properties associated with whole patterns of notes.

activities with concrete manipulatable materials such as bundling sticks, a counting frame abacus, or arithmetic blocks (sometimes called *Dienes Blocks*).

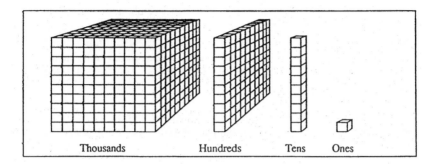

FIG. 2.1. Arithmetic blocks can be used for teaching place value arithmetic.

In previous publications, we have described productive ways that Dienes' principles can be extended to activities that involve computer-based graphics and animations rather than concrete materials (Lesh, Post, & Behr, 1987a). We also have described ways to apply Dienes' principles to teacher education programs whose aims are to teach teachers using techniques that we want them to use to teach their children (Bell, Fuson, & Lesh, 1976). This chapter briefly describes how Dienes' principles can be extended to include: (a) model development sequences that emphasize activities involving simulations of real life problem-solving situations (that are familiar and meaningful to the students) rather than artificial mathematics laboratory materials (that may be quite abstract in spite of the concrete material), and (b) problem-solving experiences that emphasize social dimensions of mathematical understanding—where students' interactions with peers or teachers are as significant interactions with concrete materials.

A BRIEF REVIEW OF DIENES' INSTRUCTIONAL PRINCIPLES

Dienes used the term *embodiment* to refer to concrete manipulatable materials (such as arithmetic blocks) that are useful props to help children develop elementary-but-powerful constructs that provide powerful foundations for elementary mathematical reasoning. He also emphasized the following four principles about how embodiments should be used in instruction:

1. *The Construction Principle*: Dienes, like Piaget, believed that many of the most important constructs in elementary mathematics must be abstracted, not from concrete objects, nor from isolated actions performed on concrete objects, but from *systems* of operations and relations that must be read into a given embodiment before the system itself can be read out. According to this point of

view, mental and physical actions only become elevated to the status of mathematical operations (or relations) when reflective abstraction treats them as being part of an operational-relational-organizational system-as-a-whole. Similarly, concrete materials only become embodiments of a given mathematical system after a child has coordinated the relevant actions to function as a system-as-a-whole in the context of these materials. Thus, concrete materials serve as supports for the student's conceptual activities; but, the abstraction is from the system of conceptual actions - not from the materials in which they operate.

2. *The Multiple Embodiment Principle*: To help a child go beyond thinking with a given construct (or conceptual system) to also think about it, several structurally similar embodiments are needed. Also, students need to focus on similarities and differences as the relevant system-as-a-whole functions in different contexts. Thus, students must go beyond investigating individual embodiments to investigate structure-related relationships among several alternative embodiments—perhaps by making translations or predictions from one embodiment to another.

3. *The Dynamic Principle*: In mathematics, components and characteristics of the relevant systems often refer to dynamic operations or transformations, rather than to static objects or states. Also, many of the most important characteristics involve invariance properties, or properties such as transitivity that apply to patterns or regularities rather than to isolated objects or actions. In particular, some of the most significant objects are variables (or operators, or transformations) rather than being objects in the usual sense of the term. Therefore, it is important for the relevant systems to be viewed as being dynamic rather than static, and it is important for attention to focus on patterns and regularities rather than on isolated pieces of information.

4. *The Perceptual Variability Principle*: Every embodiment of a mathematical system has some characteristics that the abstract system does not; and every mathematical system has some characteristics that the embodiment does not have. Therefore, when multiple embodiments are emphasized, it is important to use materials that have different perceptual characteristics. To select a small number of especially appropriate embodiments to focus on a given construct, irrelevant characteristics should vary from one embodiment to another so that these characteristics are "washed out" of the resulting abstraction. Collectively, the embodiments that are chosen should illustrate all of the most important structural characteristics of the modeled system.

In summary, Dienes's principles were designed to help students: (a) go beyond focusing on concrete materials, or on isolated actions, to focus on patterns and regularities that occur within systems of operations and relations that are imposed on the materials, (b) go beyond focusing on isolated embodiments to focus on similarities and differences among structurally similar systems that they come to embody, (c) go beyond static patterns and objects to focus on

2. MODEL DEVELOPMENT SEQUENCES

dynamic systems of operations, relations, and transformations, and (d) go beyond thinking with a given model to also think about it for a variety of problem-solving functions. For example, as Fig. 2.2 suggests, if Dienes' principles are used with elementary school children to focus on constructs that underlie our base-ten numeration system, a given child might investigate structural similarities among activities involving three different sets of concrete materials—such as arithmetic blocks, bundling sticks (popsicle sticks), and a counting frame abacus—and the relevant conceptual system is the system of operations and relations that is common to all three embodiments.

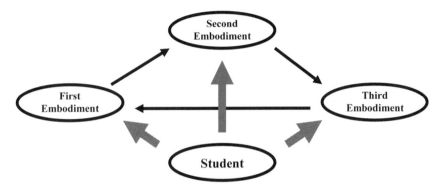

FIG. 2.2. Dienes' Multiple Embodiment Principle.

BEYOND THE PRIMARY SCHOOL, WHY DO SO FEW TEACHERS USE ACTIVITIES WITH CONCRETE MANIPULATABLE MATERIALS?

One answer to this question results from the commonly held belief that older children will be offended if "baby things" are used in instruction. But, activities with concrete materials often involve conceptual systems that are not at all simple, concrete, intuitive, or simple. In fact, in research with teachers and undergraduate students (Fuson, & Lesh, 1976; Carlson, 1998; Doerr & Tinto, 2000; Lesh, Post, & Behr, 1987b; Bell), deep conceptual weakness often become apparent if these adults are challenged to go beyond simply using memorized rules about the manipulation of written symbols—and are challenged to explain why these rules work using diagrams, concrete materials, experience-based metaphors, or other potential embodiments. Consequently, for teachers who experience such difficulties, fears often exist that, if these materials are used in instruction, the teacher's own misunderstandings would be exposed, and children would become confused in ways that the teachers could not explain. Beyond the preceding concerns, even when concrete manipulatible materials have been used in instruction, teachers and textbook authors generally use Dienes' materials without using his theory. For example:

- If concrete materials are used, they tend to be used to give demonstrations (or explanations) in which no construction activities are involved from students. Thus, both the construction principle and the dynamic principle are violated, and the relevant abstractions are expected to come from the materials themselves rather than from the (mental) actions that are performed on the materials.
- Because construction activities are omitted that would elevate the concrete materials to the status of being embodiments of the relevant conceptual system(s), the concrete materials often are used in ways that are no less abstract than using written symbolic symbol systems.
- Usually, only a single embodiment is used, and no attempt is made to explore structural similarities among related embodiments. Thus, the multiple embodiment principle is violated, and issues related to the perceptual variability principle never arise.

Because one of our goals is to extend Dienes principles to make them more useful to teachers beyond primary school, we have explored the possibility of replacing activities involving concrete materials with activities in which the contexts are grounded in the everyday experiences of students or their families. The result is depicted in Fig. 2.3 below.

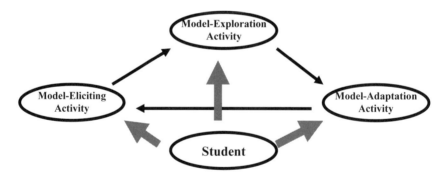

FIG. 2.3. Model development sequences.

Whereas Dienes' multiple embodiment principle recommends using a series of three or more embodiments for a given construct (see Fig. 2.2), we often replace the first and the third embodiments with experiences that we call model-eliciting activities. The result, illustrated in Fig. 2.3, is a version of Dienes' multiple embodiment principle that focuses on applied mathematics as much as pure mathematics—and on stories, diagrams, and experience-based metaphors as much as concrete materials.

MODEL DEVELOPMENT SEQUENCES INCLUDE SITUATIONS GROUNDED IN STUDENTS' EVERYDAY EXPERIENCES

As Fig. 2.3 suggests, the first activity in model development sequences is a model-eliciting activity. Details about model-eliciting activities have been described in several past publications (e.g., Lesh et. al, 2000; Lesh, Hoover, & Kelly, 1993). In general, they are similar to the case studies that are used for both instruction and assessment in future-oriented graduate programs and professional schools—in fields ranging from aeronautical engineering, to business management, to agricultural sciences—at universities such as Purdue, Syracuse, Minnesota, or Wisconsin. In these graduate programs and professional schools, many of the most important goals of instruction involve helping future leaders develop powerful models and conceptual tools for making (and making sense of) complex systems. That is, they are simulations of real life problem-solving situations; they require more than a few minutes to complete; teams of specialists often work together using powerful conceptual technologies; and, the central goal is to develop, test, revise and refine powerful, sharable, and re-useable conceptual tools that involve much more than simple answers to questions of the type emphasized in traditional textbooks and tests.

The Volleyball Problem that is given in Appendix A gives an example of a model-eliciting activity that is based on a case study that we observed used in Northwestern University's Kellogg School of Management. The original problem was designed to encourage graduate students to develop powerful ways of thinking about situations in which information about products (or places, or people) must be assigned "weights" and then aggregated in some way in order to assign "quality ratings" for some purpose. For example, in everyday situations, such weighting schemes are used (implicitly or explicitly) when automobiles are rated in a *consumer guide book*, cities are rated in a *places rated almanac*, sports teams are rated in newspapers, or students are rated based on scores on laboratory projects, weekly quizzes, unit tests, teachers' observations of classroom participation, plus performances on midterm and final examinations.

The Volleyball Problem gives information taken from tryout activities at a sports camp that specializes in volleyball instruction for young girls. The newspaper article that sets the context for this problem describes a situation where, on the first day of the camp, the following information was gathered for each girl who wanted to participate: height, vertical jumping ability (data from three jumps), serving ability (data from five serves), speed in running a 40-yard dash (results from three races), and comments from school coaches. The goal of the problem is to write a two-page letter to the camp director describing a scheme for using information from the tryouts to rank the girls—and to assign them to teams that are equivalent in overall ability.

The Volleyball Problem elicits a model in the sense that the product that the students produce is not simply a short numeric answer to a situation that has already been completely mathematized by others. Instead, the product is a description that expresses students' ways of thinking about issues such as: (a) how to quantify relevant qualitative information (e.g., coaches comments, weights that assign values to the relative importance of various factors), and (b) how to aggregate information in order to operationally define an "index of quality" for individual players and teams.

Appendix B gives a transcript that shows a typical solution process used by middle school students who have worked on the Volleyball Problem. The solution is typical in the following ways.

- The product that is produced goes through a series of iterative modeling cycles in which trial ways of thinking are tested and revised repeatedly.
- The mathematical understandings and abilities that are needed for success are quite different than those emphasized on standardized tests.
- Children who have been classified as below average (based on performance in situations involving traditional tests, textbooks, and teaching) often invent (or significantly revise or extend) constructs and ways of thinking that are far more sophisticated than anybody ever dared to try to teach to them (Lesh & Doerr, 2000). For example, in the Volleyball Problem, average ability middle school students often invent "weighted averages" (or "weighted sums")—in spite of the fact that their past performances on standardized tests suggested that they knew nothing about simple averages (or mean values).

In short, when middle school children work on well-designed model-eliciting activities, what we observe coincides with what professors express as "common knowledge" about graduate students (or on-the-job adults) who work on complex projects designed to help prepare them for leadership in future-oriented professions in business, engineering, or the sciences. That is:

- Relevant stages of problem solving include: problem posing, information gathering, mathematizing, planning, communicating, monitoring, and assessing intermediate results. Therefore, the levels and types of understandings and abilities are emphasized include many that involve many that go beyond traditional conceptions of content-related expertise.
- Basics from an industrial age are not sufficient to provide adequate foundations for future success in a technology-based age of information[3]; and, past conceptions of mathematics, science, reading, writing, and

---

[3] What is advocated here is *not* to abandon fundamentals. This would be as foolish in mathematics and science (or reading, writing, and communicating) as it is in basketball, cooking, or carpentry; and, greater accountability is what we support, not what we oppose. But, it is not necessary to master the names and skills associated with every item at Sears before students can begin to cook or to build things; and, great basketball teams are not likely to evolve by never allowing children to scrimmage until they had completed twelve years consisting of nothing but drills on skills. What is needed is a sensible mix of complexity and fundamentals; both must evolve in parallel; and, one does not come before (or without) the other.

communication are far too narrow, shallow, and restricted to be used as a basis for identifying students whose mathematical abilities should be recognized and encouraged.
- Students who are especially productive and capable in the preceding situations often are not those with records of high scores on standardized tests—or even high performance in traditional schooling.

To create simulations of real life problems that are especially designed to recognize and reward a broader range of students with exceptional potential in mathematics, hundreds of expert teachers have worked with us to develop the following six principles of instructional design. These principles were developed during a series of 10-week research studies investigating the development of teachers' assumptions about: (a) the nature of real life learning and problem-solving situations in which mathematical reasoning is useful in an age of information, and (b) the nature of the understandings and abilities that are needed for success in the preceding kinds of situations (Lesh et al., 2000).

1. *The Personal Meaningfulness Principle* (sometimes called *the "Reality" Principle*): Could this really happen in real life situations? Will students be encouraged to make sense of the situation based on extensions of their own personal knowledge and experiences? Will students' ideas be taken seriously, or will they be forced to conform to the teacher's (or author's) notion of the (only) correct way to think about the problem situation?

2. *The Model Construction Principle*: Does the task ensure that students clearly recognize the need for a model to be constructed, modified, extended, or refined? Does the task involve constructing, describing, explaining, manipulating, predicting, or controlling a structurally significant system? Is attention focused on underlying patterns and regularities rather than on surface-level information?

3. *The Self-Evaluation Principle*: Are the criteria clear to students for assessing the usefulness of alternative responses? Will students be able to judge for themselves when their responses are good enough? For what purposes are the results needed? By whom? When?

4. *The Model-Externalization Principle* (sometimes called the *Model-Documentation Principle*): Will the response require students to explicitly reveal how they are thinking about the situation (givens, goals, possible solution paths)? What kind of systems (mathematical objects, relations, operations, patterns, regularities) are they thinking about?

5. *The Simple Prototype Principle*: Is the situation as simple as possible, while still creating the need for a significant model? Will the solution provide a useful prototype for interpreting a variety of other structurally similar situations? Will the experience provide a story that will have explanatory power—or power for making sense of other structurally similar situations?

6. *The Model Generalization Principle*: Does the conceptual tool that is constructed apply to only a particular situation, or can it be modified and extended easily to apply to a broader range of situations? Students should be challenged to go beyond producing single-purpose ways to thinking to produce reusable, sharable, modifiable models.

These six principles are described in greater detail in Lesh, Hole, Hoover, Kelly, & Post (2000). For the purposes of this chapter, it is important to emphasize that, even though these principles appear to be sensible and straightforward to use, the teachers who developed them also used them to assess (or improve) activities that they found in textbooks, tests, and programs for performance assessment; and, what these teachers discovered was that, even when they started by selecting activities that seemed to be promising, nearly every task violated nearly every one of their principles—and these shortcomings were far from easy fix (Lesh et al., 2000).[4]

Whereas activities with concrete materials typically rely on carefully sequenced "guided questioning techniques" designed to lead students to think the way the author wants them to think, model-eliciting activities encourage students to repeatedly express, test, and refine or revise their own ways of thinking. Whereas activities with concrete materials typically assume that students know almost nothing about the construct the author wants them to develop, model-eliciting activities are more closely aligned with Jerome Bruner's famous claim that "the foundations of any subject can be taught to anybody, at any age, in some form" (Bruner, 1960, p. 12)—because, long before most constructs are understood as formal abstractions, they are used intuitively in meaningful situations. Thus, to encourage the development of such constructs, the first trick is to put students in familiar situations in which they clearly understand the need for the desired construct; the second is to ensure that responses can be based on extensions of students' everyday knowledge and experiences; and the third is to provide meaningful "design specs" involving constraints that enable students to "weed out" inadequate ways of thinking.

MODEL DEVELOPMENT SEQUENCES GO BEYOND RELYING ON ISOLATED PROBLEM-SOLVING ACTIVITIES

Isolated problem-solving activities are seldom enough to produce the kinds of results we seek. Sequences of structurally related activities are needed, and discussions and explorations are needed to focus on structural similarities among related activities.

---

[4] Novices often try to fix flawed problems by adding the phrase "explain your answer"—or by using scoring rubrics that focus on processes and underlying ways of thinking. But, such tactics are seldom effective for reasons described in detail in Lesh, Hoover, Hole, Kelly, & Post (2000).

2. MODEL DEVELOPMENT SEQUENCES

For many of the same reasons why the Dienes' multiple embodiment principle is needed to get the most instructional value out of embodiments of mathematical constructs and conceptual systems, model development sequences are needed to get the most instructional value out of model-eliciting activities. Also, because model-eliciting activities tend to rely much less heavily on narrowly guided questioning than most activities with concrete materials, discussions about sequences of several related activities are needed to ensure that the sequence-as-a-whole will have a clear sense of direction—even though the individual activities are relatively open-ended.

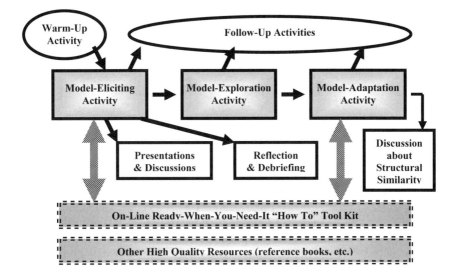

FIG. 2.4. A standard organizational scheme for model development sequences.

In *model-eliciting activities*, authors determine what kind of quantities and quantitative relationships (and operations) students must take into account, and they also determine which design specs provide constraints to weed out inadequate ways of thinking. Consequently, in order to get students to develop the desired conceptual tools, authors do not need to rely exclusively on guided questioning techniques. Furthermore, if discussions and explorations encourage students to investigate similarities and differences among structurally related tasks, then this provides another powerful tool to focus attention on the conceptual system that is common to all of the activities. Thus, model development sequences involve the following standard organizational scheme.

Warm-up activities usually are given the day before students are expected to begin work on the model-eliciting activity. Often, they are based on a math-rich newspaper article, or on a math-rich web site, which is followed by a half dozen questions aimed at:

- helping students read with a mathematical eye while also familiarizing them with the context of the model-eliciting activity—so that solutions are based on extensions of students real life knowledge and experiences, and so that time is saved that otherwise is wasted on "getting acclimated" during the model-eliciting activity.
- answering teachers' questions about "minimum prerequisites" for students to begin working on the model-eliciting activity.
- informing parents and other interested individuals about the practical importance of conceptual tools students are developing.

Model-eliciting activities usually require at least one or two full class periods to complete, and students usually are encouraged to work in teams with three students in each group. Often teachers use model-eliciting activities at the beginning of a unit that deals with the same big idea that underlies the main construct emphasized in the model-eliciting activity. Because model-eliciting activities require students to express their ways of thinking in forms that are visible to teachers, a goal is to identify students' conceptual strengths and weaknesses—to inform instructional decisions in much the same way as if the teacher had time to interview every student before teaching the unit.

Model-exploration activities are similar to Dienes' embodiment activities. However, rather than relying exclusively on concrete materials, model-exploration activities often involve computer graphics, diagrams, or animations (Lesh, Post, & Behr, 1987a). But, regardless what kind of "embodiments" are used, the goal is for students to develop a powerful representation system (and language) for making sense of the targeted conceptual system. For example, for the model-exploration activity that goes with the Volleyball Problem in Appendix A, the embodiment emphasizes the following three important processes.

- mapping scores for running, or jumping, or serving onto number lines.

    Example: If three people have average vertical jumps of 15', 17', and 26', they might be mapped to the numbers 3, 2, and 1 respectively—because they ranked 1st, 2nd, and 3rd in this event,.

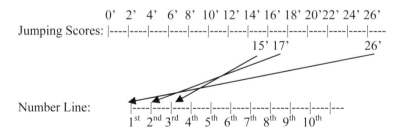

2. MODEL DEVELOPMENT SEQUENCES

- assigning weights to different quantities by stretching or shrinking the related number lines.
- investigating sums by combining lengths on several number lines.

Example: Suppose that Jen ranked 5th in jumping and 8th in running, whereas Trish ranked 8th in jumping and 5th in running; and, also suppose that the coaches consider jumping ability to be two times more important than jumping ability. Then, the arrows below show how Jen compares with Trish in overall ability—if only running and jumping abilities are considered.

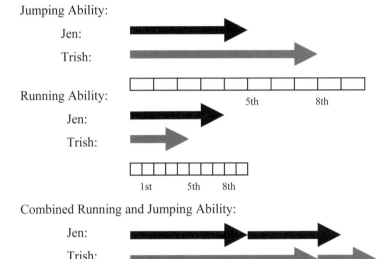

Note: According to the preceding way of measuring and combining scores, Jen is predicted to be a slightly better volleyball player than Trish because short arrows correspond to high rankings.

For the model-exploration activity that is given in Appendix A, other ways also are shown for mapping running, jumping, and serving scores to number lines; and, these same number lines also can be used to score qualitative comments from coaches. For example:

- Jumping scores can be mapped to a number line that goes from zero to ten, or they can be mapped to a number line that goes from zero to one hundred. Furthermore, in either case, the scores can be rounded off and restricted to whole numbers—or they can be assigned decimal values or fractional values.
- For either scores or rankings, weights can be assigned to different quantities by multiplying the numbers, or they can be assigned by

stretching or shrinking the number lines. In any case, good scores may correspond to either high numbers or low numbers; and, regardless what kind of mapping is used, consideration must be given to the fact that short time intervals are good for running, whereas long length intervals are good for jumping.

Using any of the preceding mappings, it is possible (metaphorically speaking) to add apples (jumping distances) and oranges (running times). That is, two fundamentally different kinds of quantities can be mapped to number lines that have homogeneous units which may be interpreted as ranks (ordinal numbers) or as point scores (cardinal numbers) or as directed distances (vectors). In any case, the mapping and weighting procedures that are used tend to distort the original information in different ways that often change the overall quality rating associated with combined scores. In particular, different volleyball players may appear to be best overall depending on the specific mapping and weighting procedures that are used. Should the difference between running scores of 4.0 seconds and 5.0 seconds be treated as being the same as the difference between running scores of 5.0 seconds and 6.0 seconds? Should a 15" jump be treated as half as good as a 30" jump? There are no value free answers to such questions. Every system of measuring makes some assumptions about the underlying phenomena being measured; and, this includes raw data that records running speeds using time intervals—and that records jumping ability using distance intervals.

For the model-exploration activity that is given in Appendix A, one of the big ideas is for students to become more aware of assumptions that are made any time real phenomena are mapped to number line models—and, in particular, any time weights (or lengths) are assigned to the underlying quantities, and any time scores are combined by adding, subtracting, multiplying, or dividing.

Appendix C gives a sample of different kinds of responses that middle school students often generate for the Volleyball Problem. Sometimes they may begin by finding a way to combine the scores for each volleyball player (perhaps using some sort of weighted sum, or weighted average), and then, they may rank the volleyball players using these combined scores. Or, they may begin by ranking volleyball players within each category of information; and then, they may devise a way to combine these rankings. In any case, mappings to number lines provide productive ways to think about (and visualize, and talk about) the procedures that are used—and the assumptions that are being made. This is why mappings to (and operations with) number lines were chosen to embody the conceptual systems that we wanted to highlight in the model development sequence related to the Volleyball Problem.

Model-exploration activities are intended to help students develop powerful language and representation systems that are useful to go beyond thinking with the relevant construct (or conceptual system) to also think about it. In particular, students often develop powerful conceptual tools that can be used to crush the problem they were given in the follow-up model-adaptation activity—which is the third major activity in model-development sequences.

2. MODEL DEVELOPMENT SEQUENCES

Model-adaptation activities sometimes have been called *model-application activities or model-extension activity*. This is because the goal of the activity often focuses on using the conceptual tool that was developed in the model-eliciting activity (and refined in the model-exploration) to deal with a problem that probably would have been too difficult to handle before the tool was developed. However, even though the tool was developed in a form that was intended to be easily sharable, and modifiable, and transportable, it seldom can be used in a new situation without making some significant adaptations. Furthermore, even though many model-adaptation problems are essentially more difficult versions of model-eliciting activities, many focus on problem *posing* at least as much as on problem solving—and on information gathering at least as much as on information processing.

Teachers often think of both model-eliciting activities and model-adaptation activities as being performance assessment activities—because both are similar to real life situations in which mathematics is used outside of school, both contribute to both learning and assessment, and both emphasize deeper and higher-order understandings and abilities that are seldom addressed on standardized tests. Consequently, teachers often think of model-eliciting activities often as pretests, and model-adaptation activities as post-tests for a given unit—where both focus on the main big ideas emphasized in the unit.

Another difference between model-eliciting activities and model-adaptation *activities* is that, for the former, students typically work in teams consisting of three to four people; whereas, for the latter, students often work alone. This is because most teachers are concerned about both students' abilities to work in groups, and their abilities to work independently.

In Appendix A, the model-adaptation activity that goes with the Volleyball Problem is about developing a scheme for evaluating "pocket radio recorders" for a consumer guidebook for kids.

Discussions about structural similarity are teacher-led activities that involve the whole class and that focus on structural similarities (and differences) among the constructs and conceptual systems emphasized throughout the model development sequence. A primary goal of these discussions is to provide experiences in which students go beyond thinking with these constructs and conceptual systems toward making the constructs explicit objects of thought. One way to do this is to ask questions that challenge students to identify corresponding parts of the three embodiments—and to make predictions from one embodiment to another.

Presentations and discussions are whole-class activities in which students make formal presentations about the results of work produced during either model-eliciting activities or model-adaptation activities. These results typically consist of two-page letters (or executive summaries) that provide conceptual tools that a client (who is identified in the statement of the problem) needs for a specific purpose (which also is identified in the statement of the problem). However, students also may express their results using brief multimedia presentations. In either case, the goal of the presentation and discussion session is for students to practice explaining their work, to see examples of a variety of ways of thinking, to discuss strengths and weaknesses of alternative approaches, and to identify directions for improvement in their own work—or the work of others. In such sessions, students often get immediate feedback about the quality of their work, even without getting detailed feedback from the teacher.

As an alternative to sessions in which students make presentations about their work, teams of students also can play the role of clients who must make decisions about the strengths and weaknesses of products that various groups of students produce. More details about presentation and discussion sessions are described in chapter 9, Kay McClain in this book.

Reflection and debriefing activities often consist of brief questionnaires in which students think back about their experiences during model-eliciting activities or model-adaptation activities. Sometimes these questionnaires focus on group dynamics; and, sometimes they focus on the roles that the individual student played during different stages of solution processes. Often, the purpose of these reflection and debriefing activities is for students to express, examine, assess, and (ultimately) control their own feelings, attitudes, preferences (values), beliefs, and behaviors—without necessarily exposing themselves to their teachers or peers. Consequently, we sometimes have found it useful to program relevant questionnaires into calculators (or computers) that are able to provide instant graphic summaries that play back to students what they appear to have said—without anybody else seeing this information. In these summaries, the goal is not to label students—as though their characteristics did not vary across time, across contexts, and even across different stages during the solution of an individual problem. The goal is simply to provide reflection tools that help students assume increasingly productive personae for learning and problem solving. More details about reflection and debriefing tools are described in chapter 22 by Jim Middleton in this book.

Follow-up activities often consist of problem sets that teachers generate to help students recognize connections between their experiences during model development sequences and their experiences based on more traditional kinds of classroom activities—centered around textbooks and tests. Quite often, in modern standards-based curriculum reform projects, these follow-up activities provide a concrete way to map the goals of model development sequences to goals specified in the school districts adopted curriculum guides.

The On-Line "How To" Toolkit is a computer-based version of the kind of *Schaum Outlines* (workbooks) that college students use to help them survive their required courses in topic areas such as physics, chemistry algebra, calculus, and statistics (e.g., Spiegel & Stephens, 1999). That is, the *"how to" tool kit* is a web-based archive of "canned demonstrations"—plus problem sets and examples of solved problems that give brief explanations of facts and skills that are most frequently needed in designated topic areas. At Purdue, undergraduate students develop materials for the *"how to" tool kit* as part of the coursework that focuses on uses of technology in education.

2. MODEL DEVELOPMENT SEQUENCES 51

Appendix D gives an example of a typical component of the *"how to" tool kit* that focuses on the topic of adding fractions. *Other high quality resources and references* include textbooks or other materials that teachers use as the basis for instruction in their classes. For example, in our own work in middle school mathematics, teachers participating in our projects often have been encouraged to use materials such as *Math in Context* (Romberg, 1997) or the *Rational Number Project: Fraction Lessons for Middle Grades* (Cramer, Post, Lesh & Behr, 1998). Sometimes teachers prefer to use these later types of materials as the foundation for their curriculum (and to use model development sequences as supplementary materials that focus on project-based learning). Or, sometimes they use project-based learning activities to provide the core of their curriculum (and use textbooks as supplementary resources). But, in either case, most teachers who have worked with us have found it useful to draw on both traditional and non-traditional resources.

COGNITIVE AND SOCIAL DIMENSIONS OF CONCEPTUAL DEVELOPMENT BOTH ARE MOST IMPORTANT IN INSTRUCTION

During model development sequences, the constructs and conceptual schemes that students develop are molded and shaped not only through interactions with concrete materials but also through interactions with other people. This is one reason why, for many of the activities within *model development sequences*, students work in teams with three to four students in each group.

Vygotsky's theories (Wertsch, 1991) and social constructivist perspectives (Cobb & Yackel, 1995) often are invoked to explain the importance of student-to-student and student-to-teacher interactions in instruction. For example, in this book, chapter 9 by McClain emphasizes the importance of communication and consensus building in the development of language and other social conventions, chapter 17 by English and Lesh emphasizes similar themes related to the identification and formulation of problems, approaches, and norms for assessing solution attempts, and chapter 21 by Lesh, Lester, and Hjalmarson (2001) emphasizes external-to-internal dimensions of conceptual development. For example, this latter chapter describes how external processes of explanation and justification gradually become internalized in the sense that, when children must go beyond blind thinking to also think about thinking, they are not likely to become proficient at carrying on (internal) dialogues with themselves if they lack experience carrying on (external) dialogues with others. Similarly, they are not likely to become proficient at monitoring and assessing their own behaviors if they lack experience monitoring and assessing the behaviors of others.

For the preceding reasons, model development sequences provide experiences in which students externalize processes that otherwise would have remained internal—and internalize processes that occur first in external forms.

But, in model development sequences, justifications for small group activities derive from Piaget-influenced cognitive perspectives at least as much as from Vygotsky-influenced social perspectives. Consequently, the remaining sections of this chapter focus on cognitive justifications for social interactions—beginning the notion of "societies of mind" and Piaget's notions of cognitive centering and cognitive egocentrism.

CENTERING, EGOCENTRISM AND "SOCIETIES OF MIND"

In the development of mathematical constructs and conceptual systems, the personally constructed nature of constructs is no less significant than the invented nature of social conventions—and the shared nature of knowledge. But, the construction of mathematical constructs involves not only the gradual coordination of increasingly sophisticated systems of operations and relations (which emphasize interactions with concrete materials), it also involves the coordination of perspectives (which emphasize interactions involving multiple people, or multiple perceptions by a single individual). This latter fact is true because, according to external-to-internal views of conceptual development, inter-personal interactions tend to precede intra-personal interactions—such as those that occur when a person needs to coordinate multiple ways of thinking about a single experience.

To see why the coordination of multiple perceptions becomes important in the development of mathematics constructs, it is useful to consider one of the most fundamental observations underlying modern cognitive science. That is, humans interpret learning and problem-solving situations using internal constructs and conceptual systems. Consequently, their interpretations of experiences are influenced, not just by external forces and events, but also by internal models and conceptual schemes, and when learners or problem solvers attempt to match their experiences with existing conceptual schemes, some relevant information always is ignored, de-emphasized, or distorted—whereas other meanings and information are projected into the situation that are not objectively given. For example, consider the following familiar observations about eye witnesses to everyday events. In descriptions given by eye witnesses to newsworthy events (e.g., fires, traffic accidents, robberies, or questionable judgments by referees to sporting events), different witnesses often report seeing very different "facts;" and, some observations are amazingly barren, distorted, and internally inconsistent, depending on the sophistication and stability of the sense making schemes (e.g., models and conceptual systems) that they use to interpret their experiences.

Similar conceptual characteristics also apply to auditory experiences, and to other domains of experiences, as well as to visual experiences. For example, when college students listen to lectures, or when they participate in discussions with professors or peers, a great deal of what is said is likely to go unnoticed unless they have developed a stable frame of reference for interpreting what is

2. MODEL DEVELOPMENT SEQUENCES

said, written, or shown. What one person says does not necessarily dictate what another person hears. Students may lose cognizance of the big picture when attention focuses on details; they may lose cognizance of details when attention focuses on the "big picture;" or, they may lose cognizance of one type of detail when attention focuses on other types of detail. Using terminology from Piagetian psychology, these conceptual characteristics are referred to as: (a) *centering*: noticing only the most salient information in the given situation while ignoring other relevant information, and (b) *egocentrism*: distorting interpretations to fit prior conceptions—thus attributing characteristics to the situation that are not objectively given.

Conceptual egocentrism and centering are especially apparent when unstable conceptual systems are used to make sense of experiences. But, to some extent, they occur whenever any model is used to interpret another system. This is because all models (or interpretation systems) are useful oversimplifications of the systems they are intended to describe. They simplify (or filter out) some aspects or reality in order to clarify (or highlight) others. Furthermore, these facts are especially apparent when models are used to describe situations in which:

- an overwhelming amount of information is relevant which must be filtered, simplified, and/or interpreted in some way in order to avoid exceeding human processing capabilities.
- some of the most important information has to do with patterns (trends, relationships, and regularities) in the available information—not just with isolated and unorganized bits of information.

In the preceding kinds of problem-solving situations, models are used so that meaningful patterns or relationships can be used to:

- base decisions on a minimum set of cues—because the model embodies an explanation of how the facts are related to one another.
- fill holes, or go beyond the filtered set of information—because the model gives a holistic interpretation of the entire situation, including hypotheses about objects or events that are not obviously given (and that need to be generated or sought out; Shulman, 1986).

Hallmarks of students early (and often *unstable)* conceptual systems include the facts that: (a) within a given interpretation, it often is difficult to keep forest-level and tree-level perspectives in mind at the same time, and (b) interpretations tend to shift unconsciously from one tree-level perspective to another, or from one forest-level perspective to a corresponding tree-level perspective—and then back to another forest-level perspective. Consequently, regardless of whether the problem solver is an individual or a group, a community of loosely related conceptual schemes (or fragments of conceptual schemes) tends to be available to interpret nearly any given situation; and, the result is a "society of mind" (Minsky, 1987) in which meaningful communication is needed among participants in a team—or among competing interpretations within a given individual.

Later in this book, chapter 19 by Zawojewski, Lesh, and English gives more details about cognitive perspectives concerning why (and what kind of) group interactions may be productive in model development sequences. In the meantime, the main points to emphasize here are that: (a) the preceding "society of mind" characteristic of real life modeling processes leads to some of the most interesting social dimensions of understanding that are relevant to model development sequences, and (b) one of the main instructional design principles that we use to address these social dimensions of understanding is called the *multiple perspective principle*.

THE MULTIPLE PERSPECTIVE PRINCIPLE

According to the Dienes' original version of the multiple embodiment principle, a student should investigate a series of structurally similar embodiments (as shown in Fig. 2.2), and the goal is to go beyond thinking with the relevant organizational/relational/operational system to also think about the system. But, when relevant conceptual systems are not yet functioning as well coordinated systems-as-a-whole, students tend to focus on surface-level characteristics of the concrete materials rather than on underlying patterns and regularities of the conceptual systems applied to these materials, and their thinking tends to be characterized by centering and conceptual egocentrism; that is, (a) the meanings that students associate with their experiences often are remarkably barren and distorted, (b) the student often has difficulty noticing more than a small number of surface characteristics of relevant experiences, and (c) the things students see in a given activity often varies a great deal from one moment to another—as attention shifts from one perspective to another, from the big picture to details, or from one type of detail to another.

Students with the preceding conceptual characteristics generally have difficulty recognizing structure-related similarities and differences among potentially related embodiments, and they also have difficulties coordinating related perspectives of a single problem-solving experience. Consequently, to help such students, the multiple perspective principle suggests that it often is useful to put students in situations where several alternative perspectives (or "windows") are juxtaposed for a single problem-solving experience (see Fig. 2.5a), and/or each student interacts with other students who are centering and distorting in ways that differ from their own (see Fig. 2.5b).

2. MODEL DEVELOPMENT SEQUENCES

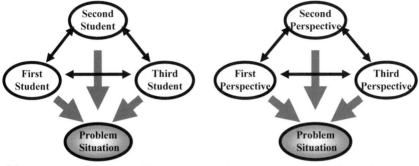

(a) Students view individual problem-solving episodes through several "windows."

(b) Students interact with other students during a single problem-solving experience.

FIG. 2.5. The (two-part) multiple perspective principle.

By closely juxtaposing multiple perspectives of a single situation, a goal is to encourage students to notice more and to distort less within any given perspective—and to think more deeply about experiences. For example:

- Teams of students may work together who have different technical capabilities, different cognitive styles, or different prior experiences.
- Within group problem-solving episodes, students may be encouraged to play different roles—such as manager, monitor, recorder, data gatherer, or tool operator.
- After problem-solving episodes are completed, students may use reflection tools to think about group functioning, or about roles played by different individuals, or about ideas and strategies that were and were not productive.
- After problem-solving episodes have been completed, students may serve on editorial boards that assess strengths and weaknesses of results that other groups produce, or they may play the role of the client who needs the result that the problem solvers were asked to produce.

Other techniques for juxtaposing multiple perspectives are discussed in greater detail in chapter 19 of this book, on *a models and modeling perspective about social dimensions of conceptual development* (Zawojewski, Lesh, & English, chap. 19, this volume). In the meantime, a point to emphasize is that simply juxtaposing multiple perspectives does not guarantee that these perspectives interact—nor that conflicting perspectives are recognized, nor that mismatches lead to cognitive conflicts of the type that Piaget believed to be important mechanisms to propel conceptual growth (Piaget & Beth, 1966). In the same way that a gradual process is required for students to coordinate their ways of thinking into coherent conceptual systems-as-a-whole, gradual processes also are needed to coordinate multiple perspectives of a single experience. Nonetheless, it often is possible to facilitate these processes by externalizing processes that otherwise would have been internal—by noticing, for example, that students may be able to monitor the behaviors of others before they can monitor their own behaviors. In general, this approach is based on Vygotsky's notion of a *zone of proximal development* which focuses on the gradual internalization of external processes (Wertsch, 1985).

An ultimate goal of the multiple perspective principle is to help individual students behave, within themselves, as though they were three people working together around a table—so that it is easier for them to overcome the barren and distorted inadequacies that are inherent in their own early interpretations of learning and problem-solving situations.

SUMMARY

The kind of model-development sequences described in this chapter are designed to be used in research, as well as in assessment or instruction. Furthermore, they are designed to focus on deeper and higher-order understandings of the conceptual schemes that underlie a small number of especially powerful constructs in elementary mathematics—rather than on trying to cover a large number of small facts and skills. Yet, skill-level abilities are not neglected because it is relatively easy to link model development sequences to instructional materials that deal with these latter types of ideas and abilities. In particular, model development sequences are modularized to make it easy for researchers or teachers to add, delete, modify, or re-sequence their components.

For example:

- It is possible to use model-eliciting activities (i.e., model-construction activities) as stand-alone problem-solving experiences—perhaps being preceded by a warm-up activity and followed by student presentations or discussions focusing on response assessment.

- It is possible to use model-eliciting activities (or model-adaptation activities) as performances assessments—and to use them somewhat like pre-tests (or post-tests) preceding (or following) a traditional instructional unit (or a chapter in a book) in which the relevant construct is emphasized. In this case, warm-up and follow-up activities might not be used.

(Pre-test)   (Post-test)

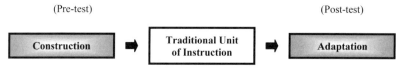

2. MODEL DEVELOPMENT SEQUENCES

- It is possible to have students engage in a complete model-development sequence—and to use traditional paper-based or computer-based materials as supplementary resources, where needed, for specific students to focus on specific facts and skills.

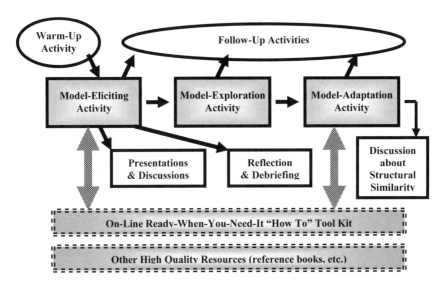

In general, in the preceding sequences, model-eliciting activities and model-exploration activities are designed for students to work together in teams consisting of three to four students each; other activities, such as the presentations and discussions, also are intended to emphasize student-to-student or student-to-teacher interactions as much as interactions with concrete materials. In fact, even in model-adaptation activities, where students often work individually, rather than in teams, the goal is to develop conceptual tools that are sharable, transportable, and reuseable for a variety of purposes in a variety of situations. Therefore, important social dimensions of conceptual development are not neglected.

Unlike constructivist teaching materials in which carefully guided sequences of questions provide the only means of leading students to assemble and adopt conceptual systems of the type the author has in mind, model development sequences put students in situations where they must express, test, and modify, revise, and refine their own ways of thinking during the process of designing powerful conceptual tools that embody constructs that students are intended to develop. In short, students adapt their own ways of thinking rather than adopting the author's (or teacher's) ways of thinking, and the adaptation (modification, extension, and revision) of existing conceptual systems is given as much attention as the construction (or assembly) of conceptual systems that are assumed to be completely new to the student(s).

Sometimes, it is useful for students to invent their own language, diagrams, metaphors, or notation systems that express their ways of thinking—in the presentations and discussions that follow model-eliciting activities, as well as in other activities that occur in model development sequences. But, in other cases, and especially in model-exploration activities (where a primary goal is to introduce students to powerful language, diagrams, metaphors, or notation systems), it often is not necessary to expect students to invent conventions that took many years to develop in the history of mathematics and science.

Concerning the artificial introduction of socially accepted language, symbols, and other representational media, dangers that arise result from the facts that it is easy to introduce language and symbols whose meanings presuppose the existence of conceptual schemes that students have not yet developed. The result is that students often sound like they understand constructs that they in fact do not. But, this reason for being conservative about the introduction of standard language, notations, and conventions is a very different than the notion that the value of these media depends mainly on votes from some group of people—rather than depending on their power and utility that they provide for the underlying conceptual systems they are intended to embody.

In addition to focusing on powerful constructs and conceptual systems, the kind of activities that are emphasized in Model development sequences are intended to go beyond isolated problem-solving experiences that are intended mainly as vehicles for emphasizing problem-solving processes.

Chapter 3

Origins and Evolution of Model-Based Reasoning in Mathematics and Science

Richard Lehrer
University of Wisconsin–Madison

Leona Schauble
University of Wisconsin–Madison

The purpose of our research is to construct an account of the early emergence and subsequent development of model-based practices and associated forms of reasoning in mathematics and science classrooms. Because this form of reasoning is contingent on particular forms of instruction sustained over time, our account of children's development relies on collaboration with teachers to design classrooms in which modeling is central. Science and mathematics educators have suggested that a central goal of instruction should be to introduce students to the "modeling game" (Hestenes, 1992), an enterprise that professional scientists and some mathematicians consider fundamental to their everyday work. These professionals do not regard mathematics and science as being primarily about the manipulation of formulas or other forms of computation, about learning facts and concepts, or even about problem solving—but rather, as a process of constructing, investigating, applying, interpreting, and evaluating models (Giere, 1992; Stewart & Golubitsky, 1992).

Modeling emphasizes the connections between mathematics and science, which are usually regarded as separate subjects. It also underscores the epistemological emphasis in both disciplines on structure (Steen, 1990). We are particularly interested in modeling that stretches toward mathematization, an emphasis increasingly aligned with trends in science since the time of Newton (Kline, 1980). We assume that to mathematize the world effectively, students require mathematical resources beyond arithmetic, including spatial visualization, data structure, measure, and uncertainty. Although these forms of mathematics are not typically taught in elementary grades, the National Council of Teachers of Mathematics has urgently suggested that they should be, and mathematics educators are working actively to develop a research base and curricular materials that can support this change (NCTM, 2000). Of course, studying the development of model-based reasoning requires having access to a context where it is likely to occur. However, modeling is not typically emphasized—or even included—in school science and mathematics, especially at the elementary grades. However, if one agrees that mathematics and science need to be centered around the on-going construction and revision of models, rather than the acquisition of finished bodies of knowledge, then modeling should be given a central place from the earliest years of instruction, not postponed until high school or beyond. Accordingly, we have been working closely with elementary and middle school conditions under which students routinely build and revise models and over the years of their instruction, develop an understanding of important classes of models. By important classes, we mean models that have proven fruitful over the course of time in their respective disciplines. Among the conditions necessary for participation in modeling are a wider base of mathematics, an emphasis on acquiring disciplinary forms of argument, the development of supporting classroom norms, and a teacher community organized around inquiry about the long-term development of student thinking. Together, we are working to build our cumulative knowledge about students' participation in modeling practices and the associated development of *model-based reasoning* over the first 6 years of schooling.

SIGNATURES OF MODELING

We are advocating here an unusually stringent criterion for what we consider "modeling," not to exclude young students from the enterprise, but to keep in sight the kinds of benchmarks and goals that instruction should continually be pushing toward. We agree with those who argue that even young children have foundational competencies on which model-based reasoning can be erected. For example, it is certainly the case that even preschoolers regard one thing as representing another. DeLoache's (1989) research suggests that before the age of 3, children understand that pictures and scale models represent the "real thing," and others (e.g., Leslie, 1987) have pointed out that pretend play, which occupies hours of a typical preschooler's time, is based on using objects to stand in for others. A child who pretends that a banana is a telephone, knows full well that it is actually just a banana. It is, however, an impressive cognitive achievement to hold both the real and pretend worlds in mind simultaneously, maintaining their separation while elaborating the symbolic world of pretense. There is also ample evidence (Brown, 1990; Gentner & Toupin, 1986) that young children routinely use analogy, both in language and in problem solving.

These early capabilities to regard one thing as representing another are obviously involved in the origins of model-based reasoning. For example, when preschoolers and first graders use counters for direct modeling to solve simple early number problems involving grouping and separating, they are recruiting their knowledge about the consequences of direct actions on materials to make inferences about quantities. Children are selecting some aspects of the world (counters) to represent others (quantities), and they are developing consistent mappings between representations and states of the world, real or imagined. There are some successful instructional programs that are based on these

3. MODEL-BASED REASONING IN MATH AND SCIENCE   61

features of young children's thinking and they collectively demonstrate the impressive capabilities of harnessing representation in service of cognitive development (Carpenter & Fennema, 1992; Rational number project).

However, important and impressive as it is, representational mapping is not all there is to modeling, so in many of these cases we would probably not say that these young children are engaged in modeling. First, by models, we refer not to "mental models" or private representations (e.g., Norman, 1993; Young, 1983), but to forms of representations and inscriptions that are constructed or adopted as conventions within a community to support disciplinary practices via functions like those summarized by Latour (1990): communication, mobility, combination, selection, emphasis, and the like. Moreover, we prefer to restrict the term *model* to a class of representational forms that include an ensemble of particular properties. These properties help us understand representational and modeling practices as related yet distinguishable.

Certainly, using one system to represent another is an important part of what constitutes modeling. However, in addition, modeling includes the explicit and conscious separation of a model and its referent. That is, the reasoner is aware that the model is not just a copy; instead, some aspects of the referent are not captured in the model. When students construct, critique, and revise models, these representational choices are highlighted, and as children become more sophisticated connoisseurs of the effectiveness of representational choices for particular purposes, their reasoning edges closer to becoming model-like.

A particularly significant benchmark is the comparison of two or more models for their relative fit to the world. Comparing and evaluating fit requires first of all, the understanding that alternative models are possible (in itself, a substantive achievement for elementary school children, who often assume that their model is the model), and moreover, that the activity of modeling can be engaged in for the explicit purpose of deciding among rival models. Young children do not tend to spontaneously generate rival models, but with thoughtful instruction, they can (Grosslight, Unger, Jay, & Smith, 1991; Penner, Giles, Lehrer, & Schauble, 1997). This emphasis on residual as a necessary component of modeling provides the epistemic grounding to consider, for example, whether the results of an experiment indicate model effects or simply error.

Finally, we emphasize the role of *mathematization* in modeling. We are not arguing that all models are or should be mathematical, but as experience becomes progressively mathematized, it becomes both more mobile and more extensible—more model-like. By mobility, we mean that modeling processes can easily be transferred from one place or one process to another, and extensible means that the implications can be worked out in the model-world and then tested against the real world. Mathematization not only helps students see things that they otherwise might not; it also makes disconfirmation a real possibility that must be reckoned with when a solution is proposed. When measurements and representations are mathematized, conjectures and beliefs can be held to account. In classrooms where mathematization of the world is standard operating procedure, children can construct critical standards of evidence that constitute a basis for choosing among many plausible explanations or causes that may be advanced for a phenomenon of interest.

As we pointed out, children spontaneously use objects, drawings, and scale models to stand for other objects, but they do not necessarily select representations for the purpose of highlighting objects and relations that are theoretically important. They need experience in *conventionalizing* those representations, that is, using notations and inscriptions that are adopted as conventions for supporting the reasoning of a practicing community. Modeling needs to be practiced systematically, so that the forms and uses of a variety of models are explored and evaluated. Richard Lesh has suggested that being a good modeler is in large part a matter of having a number of fruitful models in your "hip pocket," and in addition, we add, having the propensity to bring them out on the appropriate occasion to solve novel problems. Acquiring such a collection and understanding its uses requires sustained work in a context where modeling has a continuing purpose and payoff.

Of course, we are not implying that modeling is an either/or thing. Rather, some representational activities are less model-like (e.g., all language includes reference, but we do not consider words to be models), and others are more model-like. Our research focuses on how development takes place as children acquire the practices and conventions that define important benchmarks along the continuum. In particular, we investigate how instruction can capitalize on resources at the less model-like end of the continuum by recruiting these pre-existing resources for purposes of meaning making. Often, the functions served at that end of the continuum are selection and elaboration of aspects of the situation that are relevant to the problem at hand. For example, most young school children can draw and can make physical copies of objects, often varying the scale. When children in one of our first-grade classrooms drew a series of pictures to capture changes in tomatoes that were left out in the sun, they began to attend carefully to features that they otherwise might have overlooked. To keep interpretable records, they had to decide together on a variety of representational conventions (Lehrer, Carpenter, Schauble, & Putz, 2000). How, for example, could they distinguish between color changes due to ripening and changes in texture ("squishiness")? As children began to ask questions that were increasingly interesting (like how one would know when ripening shifts to rotting), changes in their representational resources were observed.

We wish to learn how instruction can nurture these early resources over time, so that eventually, the class of students generates "cascades of inscription" (Latour, 1990) that come progressively closer to the kinds of knowledge, practices, and activities that are clearly model-like. Such practices at the model-like end of the continuum embody the critical qualities we have described: separation of model and referent, assessment of error and fit, conventionalization of external representations, incorporation into ongoing forms of disciplinary practice, evaluation of competing alternatives, and mathematization. It is essential to base instruction on what we know about children's thinking, but it is equally important to adopt the longer term view that is needed to stretch these early competencies into forms of thinking that are complex, multifaceted, and subject to development over years, rather than weeks or months. Consequently, we orient our work toward invention and revision of models with an eye toward helping children develop repertoires of powerful models that support understanding of mathematics and science.

CRAFTING DEVELOPMENTAL LANDSCAPES

As we consider how to craft developmental landscapes for modeling, we have found it useful to recall that at its most basic level, a model is a kind of analogy. In an analogy, the objects and relations in the model map on to corresponding objects and relations in the world. For example, the often-described Rutherford model of the atom is an analogy that suggests mappings between the objects and relations that make up the solar system and those that constitute the parts of the atom. Objects in this analogy include the sun and the nucleus, as well as the planets and electrons, respectively. Relations include the relative size of the body at the center (the sun and the nucleus are more massive than the planets and the electrons) and the relation of revolving around. Of course, not all features get preserved in a model; an important goal of modeling is to eliminate features that are not theoretically important. For instance, in the solar system atom model, "hot" is not included in the model, even though the sun is hot. Nor is there an expectation that the time for one revolution will be constant from one system to the other. For many models, we can define a set of transformations and explore their consequences by carrying them out in the model and then using the results as the basis for making a prediction about corresponding actions or transformations in the world.

Because models are analogies, and because little research had been completed on young children's model-based reasoning, we initially turned to the research on analogy for clues about effective ways of engaging young students in the modeling game. Like analogies (Gentner & Toupin, 1986), models are particularly likely to cue children's memory for systems or situations that share perceptual resemblance with features in the model, whether or not the model really is a good fit for the phenomenon. For example, when we asked first graders to use springs, dowels, Styrofoam™, and other materials to build a model that "works like your elbow," almost all students at first included balls and Popsicle sticks to represent hands and fingers (Penner, et al., 1997). Most children insisted on sticking a small ball onto their model to show "the bump where the elbow should be." Only after constructing models that "looked like" their elbows did the children turn their attention to function. Similarly, when asked to draw a net (a two-dimensional figure that represents a 3-D shape, unfolded) of a cereal box, one child omitted the bottom of the box and drew sides that were not rectangular, but carefully lettered "Kelloggs' Corn Flakes" on the front (Lehrer, Jacobson, Kemeny, & Strom, 1999).

Yet, similarity serves as an apt entry point for children's modeling activities, especially if the phenomenon under investigation is one in which similarity-based representations can give good initial purchase. The children who were concerned with representing hands and fingers eliminated these features in their second round of models, when the class's attention shifted to making a model that bent through the range (and only through the range) of the human elbow. (As one child explained, "Your real elbow gets stuck right there.") Similarly, attempts to fold the Kelloggs' Corn Flakes net into a cereal box provided feedback about the missing and distorted faces, and subsequent attempts successfully produced a rectangular prism.

Because they preserve similarity, physical models are fruitful places to begin the modeling game with children. We refer to these as *physical microcosms*, because they rely on physical resemblance for modeling the world. They include scale models of the playground, jar models of aquatic ecosystems, and Bottle Biology™ compost columns—all of these incorporate pieces of, or closely resemble, the systems they are intended to represent. The purpose of emphasizing physical microcosms with novices or young students is to firmly ground children's understanding in familiar objects and events. This familiarity supports mapping from the representing to the represented world by permitting children to rely on literal similarity to facilitate model-world relations. On these beginnings, however, teachers immediately begin to erect a common classroom history that expands students' representational repertoires and elaborates the forms and functions of models. (This does not mean that physical microcosms are abandoned, but rather that as students' repertoires grow, they are amplified and supplemented by other forms.)

Many *representational systems* begin their existence as physical microcosms, but then evolve, via inscription or notation until the properties of resemblance that initially sustained them fall away. For example, first graders recorded the heights of growing amaryllis bulbs by cutting green strips that matched the heights of the plants at successive days of measure (Lehrer & Schauble, in press-a). At first, the children insisted that all the strips be green (like stems), and that each be adorned with a large flower. Soon, however, the first-graders began to ask questions about the changes in height over time, an orientation that shifted their attention to successive differences in the heights of the strips at different days of measure. With this new focus, the need to distinguish among small differences in height made it important to be able to mount strips side by side so that they were contiguous, and when the flowers interfered with this goal, they were deemed no longer essential and were eliminated. When children decided they wanted to compare the growth of one plant to that of another, they found it important to be able to inspect several strips mounted on a timeline, which in turn, made it difficult to tell at a glance which strips represented which plants. Someone suggested that it might be easier to see a pattern in the clutter of strips if they used different-colored strips to represent different plants. Shortly thereafter, the records for the Red Lion bulb were represented by strips of red, not green. As questions like, "Which grew bigger?" evolved into questions like, "How much faster did Red Lion grow than Hercules?", quantification and precision became central concerns. Increasingly, the strips themselves, rather than the plants, became the focus of investigation. With each shift in these directions, the strips progressively lost the properties of

3. MODEL-BASED REASONING IN MATH AND SCIENCE

similarity that originally supported their generation and relied increasingly on conventions, like height = quantity.

Representational systems are usually grounded in resemblance between the model and the world, but the word "grounded" is important. Like the plant strips, these representations typically undergo fundamental transformations via inscription. Consider, for example, maps, in which the resemblance of the model to the world relies heavily on convention, rather than perception. Reading a typical street map depends on knowledge about conventions like the compass rose, the meaning of different line weights, the presence of a key, and often, other conventions, as well. These conventions influence our perception and often alter them. For example, Mercator projections are often employed by cartographers to preserve direction, thus facilitating navigation, but they fiercely distort area at higher latitudes, thereby altering (or perhaps reinforcing) geopolitical perception.

As third graders worked to map their school's playground, their first attempts looked more like pictures of playground equipment than maps, thus grounding perception in their experiences of this space. However, as the students attempted to use each other's maps to navigate in the space, they quickly appreciated the need for revisions so that the maps would reliably communicate needed information not available in the initial drawings. Over subsequent revisions, the maps increasingly incorporated conventions about scale, relative position, and polar coordinate systems. These conventions, which were negotiated within the classroom, effected the evolution of the drawings into maps—models of space, rather than artistic expressions (Lehrer & Pritchard, in press). The system of map-making was mobile and literally transportable, which made mappings of other spaces for other purposes feasible, and even expected.

Often, representations that are initially based on similarity differentiate over time as children's attention to or understanding of the phenomenon under investigation becomes more refined. Consider the earlier example in which first graders made drawings to serve as records of changes in tomatoes. Originally, students used color to capture the changes that they observed. The choice of this first representational device, color, was made on the basis of resemblance; that is, they used color to stand for color. However, after a couple of weeks, students noticed that color was not the only attribute that was changing—the tomatoes were also becoming squishy in some regions, and spots of mold eventually began to grow. To record the location and extent of these squishy spots, children decided to use shading and other kinds of surface patterns. This second representational convention, the surface pattern, preserved some aspects of the appearance of the tomato (i.e., location on the tomato and approximate proportion of area covered), but not others (e.g., the pictorial record did not literally resemble the tactile quality of squishiness).

Even as drawings and other pictorial systems preserve perceptual qualities of the objects they represent, students nonetheless select and foreground certain aspects and eliminate others. Moreover, they often participate in systems of inscription that extend the reach of any particular form. In a third-grade class, children's semester-long study of plants was anchored to a display in which photocopies of pressed plants were mounted along a timeline representing days of measure (Lehrer, Schauble, Carpenter, & Penner, 2000). The photocopied plants provided what Latour (1992) would call a "fix" of growth, so that the state of the plants in days gone by was always at hand throughout the investigation. Like all inscriptions, this one also selected for certain properties over others: it provided a record of attributes that children were investigating, including height, surface area and number of leaves, buds, and width of the plant, but did not preserve changes in color, weight, or volume. Moreover, the displays, mounted around the room, provided children with a continuing reminder of the referent of their other notations and inscriptions, which eventually expanded to include coordinate graphs, frequency displays, and three-dimensional models. Students' investigations of the plants started with careful observation of the real thing, then shifted to a representational system (the photocopies of pressed plants), which was, in turn, transformed into inscriptions and notations that highlighted other properties, especially mathematical properties. Although the transformation of forms can be defined as a progression, their interpretation is often supported by interplay among representational forms, even if these are indicated by gesture and not by word. Thus, representational models are almost never univocal; they speak within particular forms of practice (e.g., modeling change) and are supported by mutual embedding of reference (e.g., a coordinate on a graph, a pressed plant photocopy, a live plant).

Because representational systems have this quality—their feet in the ground and their heads in the air, so to speak—many, perhaps most, of the cases of the representational tools in our elementary classrooms fall into this category. We think of this category as quite extensive, ranging from the green strips and drawings of the first graders to the acquisition of some rather sophisticated conventions by older elementary students. The common property of representational tools at this level is that they are initially anchored in resemblance and physical models of the world, which are then stretched to mathematical description. Once stretched, the mathematical descriptions take on a life of their own. They become embodiments for other systems, fulfilling their promise as mobile and extensible modeling tools. Mathematization enables a transition toward modeling relations in systems without necessary couplings between models and the world based in literal resemblance. We refer to this more purely analogical emphasis on relational structure as "syntactic" modeling.

Syntactic models bear little or no physical similarity to the phenomenon that they represent. Instead, they summarize the essential functioning of a system with objects and relations that are not literally present in the phenomenon of interest. They cannot be literally pointed to, so these models emphasize structural relations, only. For example, coin flips or turns of a spinner can be used to model a bird's choice among different seeds at a feeder. Spinners do not look like birds, and the most obvious commonly shared attribute, motion, is not relevant to the relation of chance device. These kinds of models are important because they are so general, especially for describing and explaining behavior,

3. MODEL-BASED REASONING IN MATH AND SCIENCE

but are challenging for students and teachers alike. Because instruction does not begin by capitalizing on resemblance, the relations between the model and the phenomenon being modeled may be particularly difficult to understand. Moreover, mismapping, or importing of inappropriate objects and relations, is always a danger in analogical thinking, but may be a particularly characteristic pitfall of this form of modeling. Nevertheless, we can make some significant headway with children in the elementary grades.

For example, we have encouraged children in the fourth and fifth grades to consider distributions as signatures of biological processes. When plants first start to grow, the distribution of plant heights is apt to be positively skewed. This asymmetry arises because the initial value of the measure of height (0) produces a wall in the distribution. Other, more interesting processes of growth, like growth limits or variable timing, often produce skewed distributions (see Appendix). Our goal is to help students learn to use distributions as clues to growth processes and as constraints on causal models of growth. Explanations of growth should be consistent with the distributions observed. Distributions also provide students with ready means of representing natural variation. Such models of natural variation provide a necessary backdrop for experimentation. For example, if students believe that fertilizer will affect the heights of Fast Plants™, any differences observed between plants treated with different amounts of fertilizer must be considered in light of naturally occurring differences (plant-to-plant variation).

Of course, employing distribution as a signature for growth requires work at developing its mathematical foundations. We find it useful to consider these foundations in light of ongoing inquiry and data collection, so that the mathematics of distribution is developed in parallel with ongoing investigations of the natural world. This backdrop of investigation renders much of the mathematics more sensible. For instance, in one fifth grade class, students asked questions about how typical plants might change and how variable the change might be. They directed their efforts to developing representations that would communicate this dual aspect of "typical" and "how spread out" (questions grounded in their familiarity with the plants). Ideas about shape of distribution emerged from their efforts to develop these representations. For example, one sense of shape is foregrounded if students order the heights of plants at a particular day of growth and then represent each plant height as a vertical line on an ordered axis (see Appendix). Quite another sense of shape emerges if students first define an interval (a "bin") and then represent relative frequency of plant heights within that interval vertically, again on an ordered axis (see Appendix). By comparing and discussing what these representations, and others like it, communicated about typical and spread, students embarked on an exploration of the mathematics of distribution. Armed with this kind of conceptual tool, students could come to expect relationships between growth mechanisms and resulting distributions of organisms.

Finally, *hypothetical-deductive models* incorporate mechanisms that can produce previously unseen and often unpredicted behaviors, and are often implicated, albeit indirectly, as explanations in other systems of modeling. For example, the kinetic theory of gases describes properties like pressure and temperature as outcomes of random collisions among atoms and molecules. We have directed most of our attention to hypothetical-deductive models that rely on descriptions of agency from which the objects of modeling emerge (Resnick, 1994). Understanding these models is difficult because explanations at this level often violate intuitive sense. Many forms of insect behavior, for example, appear complex and well orchestrated, but can be modeled as interactions among agents following comparatively simple rules. It may seem obvious that termite colonies, which build complex mounds, must be operating according to a central plan, but students in the third grade were surprised that they could make something like a mound by following a few simple rules about picking up and dropping pieces of paper. How could a complex design be created with no guiding blueprint—not even a "boss" termite? Because so-called emergent properties like these cannot be easily predicted from the initial conditions and rules, children—and adults—often find these kinds of models difficult to understand (Resnick, 1994). We have conducted very little empirical research with models of this kind, but we think there may be potential to make progress with computational modeling tools like StarLogo™ (a parallel-processing version of Logo™). StarLogo™ permits users to construct objects, define the simple rules that the objects obey, and observe the results when the collection of objects is turned loose to interact. In emergent systems, the collection generates behaviors that seem more complex than the individual rules could predict.

In sum, we have sketched out a continuum of kinds of models, ranging from physical microcosms on one end, to hypothetical-deductive models, on the other. As one moves across this continuum, models rely less and less on literally including, copying, or resembling aspects of the world that they are designed to represent. At the hypothetical-deductive end of the continuum, the model generates phenomena that are not even part of the ontology of the original modeling language. For example, the traffic jam that results when individual drivers follow simple rules about speeding up and slowing down is not a phenomenon that is describable in the rule set.

This category system helps orient us to our concerns about development—not to stand as a claim about the way that development occurs, but to serve as a design framework for organizing our ideas about how best to design instruction over the long haul of students' education. Our experience suggests that grounding in similarity is frequently a wise instructional move. It provides connection and meaning to the more symbolic or syntactic systems that are built upon this base, and it helps students understand, when they acquire the conventions of a discipline, what kinds of problems those representational conventions were invented to solve. Of course, there are dangers to category systems. In this case, we have taken phenomena that we consider complex and continuous and put them into categories that seem qualitatively discrete and mutually exclusive. In that sense, the categorical structure may be a bit misleading. As we have described, what is important in instruction is that teachers assist children in moving along the continuum. It is that motion, or edging out, that interests us most, a focus that makes it necessary to attend to

3. MODEL-BASED REASONING IN MATH AND SCIENCE

long-term histories of learning, not any particular state or moment. By attending to learning in the long run, we hope to engender model invention and revision as an everyday practice of schooling.

We are far from accomplishing such an ambitious agenda, but in our collaboration with teachers, we have reached several milestones. As we look back, we can apprehend some progress toward a "pocketful of models" that Richard Lesh once suggested as benchmarks of progress for this agenda. In the realm of physical models, students have developed models of a variety of systems, ranging from elbows to insects, plants, and ecosystems. They have also used physical models to understand processes like decomposition and growth. As we noted, these models often began with physical resemblance but with artful teaching, students came to appreciate function as well as form. These physical models are nearly always accompanied by invention and progression of representational models, so that, for example, processes of change observed in compost columns are represented symbolically.

As we suggested, students have invented or appropriated a wide range of representational models. Perhaps these are best illustrated by considering how students have modeled change in the context of growth. Young children typically draw the changes that they see as organisms grow (or decay), but then enlarge and elaborate their drawings to include new features that come to their attention, like the squish of rotted fruit or the emergence of true leaves in cotyledonous plants. These drawings anchor selection of particular attributes which are then further amplified and isolated. Recall, for instance, the first graders who first used green paper strips to represent the heights of plant stems. By considering successive strips of paper, and by teacher focus and recruitment of attention to the successive differences in heights, change is re-represented not just as difference, but as successive difference. So what the eye is drawn to by looking at strips ordered by height comes to signify successive difference rather than simple succession. Consideration of successive differences opens the way for entertaining new questions of variable difference: Are the differences constant? If not, why not? This recognition of difference plays into children's representational competence (e.g., children solve arithmetic problems involving the separation of sets), but also stretches it to go beyond models of simple difference to begin to consider accumulating differences.

With older children (i.e., third grade and up), we raise the conceptual ante by introducing ratios of successive distances, so that students begin to consider rates as, for instance, height: time ratios. Rates are typically framed within Cartesian coordinates, so children have access to both numeric and graphical representations of change. Again, these representations often participate in a cascade of inscriptions that include the photocopied pressed plants mentioned earlier, drawings, and other related means of making sense of change, like paper cut-outs of rectangular prisms or cylinders to show changes in the volume of Fast Plants™ over the life cycle. Teachers often choose to introduce children to the idea of interval, so that students can consider how a particular rate changes when an interval of time expands or shrinks.

From these foundations, students usually go on to consider variable rates and search for explanations of why plant growth is not simply linear. As they develop these explanations, students make piece-wise linear approximations to functions like the logistic, which they refer to as the "S-shaped" curve. At this point, children can compare the growth curves of different attributes of the same organism, like roots and shoots. They find that although rates of growth at particular intervals may differ for roots and shoots, nevertheless there is a pattern; both follow the S-shape. Hence, mathematics serves to reorient student perceptions of the world, and the entailments are explored quite apart from any clear focus on resemblance (no plant looks like an S).

Students in the upper elementary grades take the rates, heights, and other measures of previous modeling experiences as input for further exploration of change (building on from these previous efforts). Students reexamine ideas about variation in light of distribution, and with this increased mathematical power, new questions arise. Why, for instance, might plant growth or animal growth vary more in some conditions than others? Does variability change over time? Is changing variability a characteristic of all attributes of an organism, of just some?

Distributional thinking is a tool for coupling the growth of organisms with the growth of populations. This coupling sets the stage for investigating other kinds of functional relationships. For example, fifth-grade students modeled the growth of populations under different assumptions of survivorship and reproduction (this work was conducted recently by Susan Carpenter in collaboration with fifth-grade teacher, Larry Gundlach). In turn, models like these set the stage for introducing other forms of hypothetical-deductive modeling. In collaboration with Mitch Resnick and Uri Wilensky, we intend soon to couple the study of distribution with agent-centered explanations of the shape of these distributions, so that distributions can be seen as emergent, and not simply as descriptive.

As these reflections illustrate, each time we learn something new about effective forums for modeling, new questions have arisen that needed to be addressed in collaboration with our teacher-partners. This cyclical nature of research, of course, is precisely the quality evident in the investigations being conducted by our participating students. Questions give rise to investigations, and the resulting evidence, expressed as representations, inscriptions, and models, pushes the questioning forward to focus on new issues that were not previously evident. We look forward to continuing these investigations with our teacher collaborators to determine whether and, if so, how this approach can help inform our mutual efforts to create effective instruction in mathematics and science.

Chapter 4

Piagetian Conceptual Systems and Models for Mathematizing Everyday Experiences

Richard Lesh
Purdue University

Guadalupe Carmona
Purdue University

According to the models and modeling perspectives that are emphasized throughout this book, conceptual models that students develop (to construct, describe, or explain their mathematical experiences) can be thought of as having both internal and external components. The internal components often are referred to as *constructs* or *conceptual systems*—depending on whether attention is focused on the system-as-a-whole or the elements within the system. The external components often are referred to as either *artifacts* or *representations*—depending on whether the purpose is to produce a system that is important in its own right, or to describe some other system.

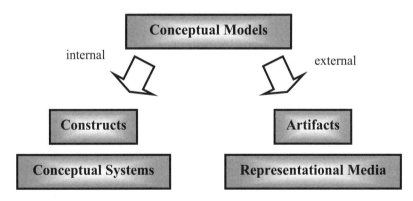

FIG. 4.1. Mathematical models have internal and external components.

71

This chapter focuses mainly on the internal components of conceptual models. That is, we'll emphasize *constructs* and *conceptual systems* rather than the *artifacts* and *representation systems* in which they are expressed. Similarly, we'll emphasize children's developing abilities to think *with* these constructs or conceptual systems (by focusing on students' abilities to make sense of mathematical experiences) rather than being preoccupied with students' abilities to think *about* them (by making the systems themselves objects of reflection, and by associating names and symbols to their parts, processes, relationships, and properties).

SIMILARITIES AND DIFFERENCES WITH PIAGET

Piaget and his followers inspired a great deal of the best research that mathematics educators have conducted on the nature of students' developing mathematical knowledge. For example, just as Piaget describes the sequences of stages that children go through to develop general cognitive structures for making sense of everyday experiences, mathematics educators have described learning trajectories that students go through in topic areas ranging from whole number arithmetic (Simon, 1995; Carpenter & Fennema, 1992; Fuson, 1992; Steffe, 1994), to rational numbers and proportional reasoning (Lesh, Post, & Behr, 1989), to early algebra and calculus (Kaput, in press).

Similarities between these two types of research are obvious. But, differences often are equally significant. So, the following table briefly describes a few of the most significant differences between Piaget's research and comparable research in mathematics education.

	PIAGET	**MATHEMATICS EDUCATORS**
Types of Constructs Emphasized	Piaget focused on general conceptual systems that apply to a wide range of specific contexts - and that develop through ordinary everyday experiences.	Math educators focus on highly specialized conceptual systems that often do not develop beyond primitive levels unless artificially rich mathematically experiences and tools are provided.
Ages or Ages Ranges Emphasized	Piaget focused on periods when major conceptual reorganizations tend to occur—such as at 7 or 12 years of age—when transitions occur from concrete operational to formal operational thinking.	Math educators focus on transition periods before, during, between, and after 7 and 12 years of age—and on developments that occur after periods of major conceptual reorganization.
Media for Expressing Constructs	Piaget focused on underlying conceptual mathematical systems rather than the language, diagrams, or other media in which they are expressed.	Math educators give a great deal of attention to the development of powerful and generative language, symbol systems, diagrams, and other culturally developed conceptual tools.
Develop- mental Stages	Piaget described ladder-like stages that span several years of cognitive development for a given construct—and he often speaks as if knowledge was organized around mathematical abstractions.	Models & Modeling Perspectives focus on sequences of modeling cycles that occur during specific problem-solving activities—and on conceptual tools that are designed for specific purposes and situations.

MANY MATHEMATICAL CONSTRUCTS ONLY BECOME MEANINGFUL AFTER RELEVANT CONCEPTUAL SYSTEMS FUNCTION AS WELL-ORGANIZED STRUCTURES

One reason why Piaget's work has been especially interesting to mathematics educators is because he emphasizes the fact that mathematics is the study of structure. It is about *seeing* at least as much as it is about *doing*. Or, alternatively, one could say that *doing* mathematics involves more than simply manipulating mathematical symbols. It involves interpreting situations mathematically – in the ways described in chapters 1 and 2 of this book.

Contrary to conclusions that many people have formed based on school experiences, mathematics is NOT simply a checklist of machine-like condition-action rules (definitions, facts, or skills) that need to be programmed in students' heads and executed flawlessly. In particular, when we ask what mathematical ideas and abilities are most important for students to learn as preparation for success beyond school, we should go beyond asking *What kind of procedures can they do?* to also ask *What kinds of situations can they describe or explain mathematically?*

When Piaget investigated the nature of children's developing mathematical knowledge, and when he investigated their abilities to make mathematical judgments and to describe or explain mathematical experiences, he showed that many of the most powerful underlying constructs only become meaningful after relevant conceptual systems begin to function as well organized systems-as-a-whole.

To investigate the nature of the conceptual systems that underlie children's mathematical reasoning, Piaget often used "conservation tasks" that require students to reason about properties that are (or are not) invariant under relevant systems of transformations. Thus, if a child is unable to reason correctly about a given invariance property, then this inability often provides strong evidence about the child's level of mastery of the underlying conceptual system.

One of the things that's most difficult to understand about models & modeling perspectives of mathematics learning, and about Piaget's description of knowledge development, is this notion that relevant conceptual systems must function as structured systems-as-a-wholes before the parts and properties within these system can take on their appropriate meanings. Therefore, because this notion is so foundational to most of the perspectives described throughout this book, this chapter focuses on Piaget-like conservation tasks.

The remainder of this section presents a series of conservation tasks for adults—so that readers can assess the validity of Piaget's claims by reflecting on their own reasoning. This approach is possible because Piaget's claims about knowledge development are not just statements about children's mathematical reasoning, they are statements about mathematical reasoning for adults as well.

CONSERVATION TASKS

There is common sense epistemology, based on folk wisdom about how elementary mathematical ideas evolve in the minds of children, which assumes that: *In the beginning there are units, and the first units are naked numbers that tell how much but not of what. Then, gradually, elementary quantities such as length (e.g., 1 foot) come to be understood; and, still later, these elementary quantities are followed by more complex quantities such as area or volume, or speed, which involve combinations of units (e.g., square feet, or cubic feet) or quotients of units (feet-per-second).*

Similarly, it is assumed that simple counting numbers are gradually extended to include negative numbers (e.g., -1), fractional numbers (e.g., 3/4), irrational numbers (e.g., π, √2), and imaginary numbers (e.g., √ -1).

But, common sense that is based on immediate perceptions also suggests that the sun revolves around the earth and that the natural state of objects is to be at rest. So, in mathematics just as in other sciences, a great deal of wisdom involves going beyond the use of common sense (that is based on isolated immediate perceptions) to use uncommon sense (that is based on patterns and regularities beneath or beyond the surface of things). So, it is productive to questions the validity of the preceding common sense assumptions.

Piaget's research showed that the notion of a unit is not the starting point for the conceptual development of students' understandings of whole numbers, fractions, lengths, areas, volumes, or other quantities. Rather, the unit and the relevant systems evolve simultaneously. ... To help make these ideas more meaningful, it is useful to consider the adult conservation tasks that are given in this section.

Example #1: Conservation of Number

Fig. 4.2a shows one of Piaget's "number conservation" tasks that is so famous that we will simply assume readers are familiar with it—and that they also are familiar with a variety of possible explanations that researchers have suggested for helping children understand why the number of counters remains invariant when the length of one of the rows is changed so the one-to-one correspondence is no longer as obvious. Our first adult-level task begins by asking you, the reader, to write down several explanations that you believe might be helpful to children on such a task. Then, we will ask that you test the adequacy of these "explanations" by applying them to a series of tasks that follow.

FIG. 4.2a

Does the red (top) row have the same number of counters as the blue (bottom) row?

4. PIAGET SYSTEMS AND MODELS

Some common suggestions about how to help children solve the preceding task include:
1. Ask the child to count the counters in the two rows. Then, spread the counters (or to rearrange them in other ways), and ask the child to count them again. This is intended to demonstrate that no matter how the counters are rearranged, the number of counters remains constant.
2. Emphasize the one-to-one correspondences between the two collections of counters by drawing dotted lines to match counters in two rows. This is intended to demonstrate that the one-to-one correspondence does not depend on the arrangement of the counters.
3. Point out to the child that: *When you spread the counters apart, you can reverse this process. That is, you can push them back so they match one-to-one.*

The adult-level conservation tasks that follow are helpful to examine the validity of each of these explanations.

Example #2: Checking Your Response to Example #1

To test the adequacy of your explanations in example #1, try applying them to the following situation which begins by using two identical 32 inch loops of string to make two identical squares on a pegboard (or on a grid like the one shown in the figure below). Then, to pose the task, gradually modify the right loop from a square shape to a rectangular shape as shown. The question is: "Are the number of "spaces" (small squares) inside the left loop the same as the number of spaces inside the right loop?" Before answering this question, readers might want to notice that "When you spread the loop apart, you can reverse this process. That is, you can push the loops back so they match exactly." But, does the number of "spaces" remain the same? [Note: You can check your answer by counting the spaces in each figure.]

FIG. 4.2b

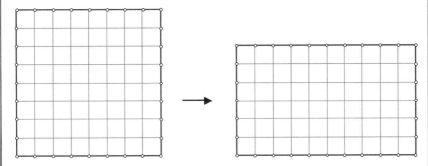

Is the number of small squares the same in the two figures?

Test your other "explanations" that seemed possible for the task shown in example #1. Does your explanation lead to correct answers for the task shown in example #2? Why or why not? If they lead to incorrect answers in one case, why should anybody believe that they'll lead to correct answers in other case?

In Example #2, the perimeter is the same for the two figures. But, there is a change in the number of spaces (or squares) that are enclosed by the loops. That is, even though the process can be reversed, the area of the second rectangle changes. ... Could the opposite situation also occur? That is, could a transformation have been performed so that the number of squares remains the same but the perimeter changes? Consider the following task.

Example #3: Checking Your Response to Example #2

Put an 8 x 8 array of squares inside the original square loop of string from example #2. Then, rearrange the squares to form a 4 x 16 array. Will the same loop of string fit around the border of both arrays? In this case, what properties do and do not remain invariant under the transformation?

FIG. 4.2c

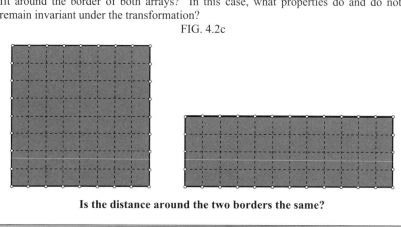

Is the distance around the two borders the same?

The central issue that these examples raise is: When you count something, or establish one-to-one correspondences, how do you know that the things being counted or matched are units of number, or area, or length? How do you know that the transformation has not changed these units in some fundamental way – so that they are no longer equivalent? ... Consider the following examples.

Example #4: Checking Your Response to Example #3

To present another task that's similar to the ones shown in examples 1, 2, and 3, begin with two 8 x 8 cardboard game boards like the ones shown in the following figure. Then, cut one of the game boards into three pieces as shown, and reassemble these three pieces (by moving them like puzzle pieces) to form a rectangular shape like the one shown on the right-hand side of the figure. Do the two boards have the same number of spaces? Count them and check. Then, notice that it is again true that "When you spread them apart, you can reverse this process and push them back so they match one-to-one." ... In this case, describe the transformation that was performed. What properties do and do not remain invariant under this transformation?

4. PIAGET SYSTEMS AND MODELS

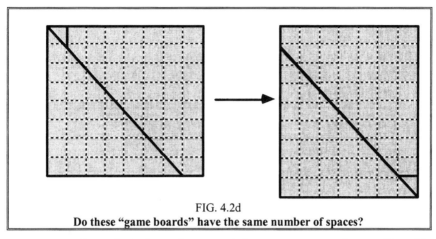

FIG. 4.2d
Do these "game boards" have the same number of spaces?

In example #4, both the perimeter and the area appear to change. But, this is not really true. The area does not really change. It is only the number of small "squares" that changes; and, this is because some of these "squares" are no longer the same size that they used to be – and they're not really squares. The original squares along the diagonal were cut and re-assembled to form near-squares. But, this was done cleverly. Just enough was shaved off of each of these squares so that when the pieces were re-assembled, the number of apparent-squares changed. ... If you look closely at part-to-part comparisons in the second figure, the small near-squares are no longer the same size.

Example #5: Checking Your Response to Example #4
On grid paper, cut out a triangle like the one shown on the left below. Then, cut this triangle into six pieces, as shown, and reassemble these six pieces to form the shape that is shown on the right. Does the area of the two shapes remain the same? Precisely what transformation was performed in this case? Which properties varied and which remained invariant?

Fig. 4.2e

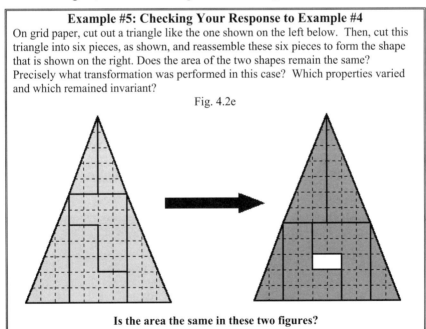

Is the area the same in these two figures?

In this case, the parts remain the same. But, the wholes change. That is, the near-triangle on the right is no longer a real triangle. The sides are no longer exactly straight. You can verify this because the slopes of the two triangular pieces are not identical. … Stated differently, if you make part-to-whole comparisons in the two figures, the wholes are not the same.

Summarizing:
- In Example #2, perimeter remains invariant but area changes.
- In Example #3, area remains invariant but perimeter changes.
- In Example #4, area remains the same while the number of spaces changes.
- In Example #5, the shaded parts of the figures have the same area, but the shape changes from being a triangle (before the transformation) to being a pentagon (that is a near-triangle) after the transformation.

One straightforward observation about the preceding sequence of tasks is, in fact, a tautology: *Whenever you transform something, something gets transformed.* Or: *Whenever you change something, something gets changed.* But, it's also true that, something also does NOT get transformed - or changed. If this weren't true, then we would not speak of a thing that is getting transformed. We'd simply refer to two completely different things. … Consequently, in the case of each of the tasks, 1-5, the relevant question to ask is: "Which properties remain invariant, and why?" … As the examples we'll give next continue to show, the answers to these questions often cannot be as simple minded as our "explanations" to children often suggest.

Another straightforward observation about Examples 1-5 is that, when a figure has a variety of properties, transformations can be created in which any one of these properties remain invariant while the others change. Therefore, for any given transformation, the relevant question to ask is: *Does the invariant property involve number, area, perimeter, shape, size, or some other quantity?* … The simple fact that a transformation can be undone says nothing whatsoever about which properties it preserves or changes. Similarly, neither counting nor one-to-one correspondences are legitimate unless the units being matched are actually equivalent.

To see more reasons why counting and one-to-one correspondences sometimes fail to be persuasive to children in the context of Piaget's number conservation tasks, consider the following standard procedure that college mathematics professors often use to "prove" to their students that:

(a) All of the rational numbers can be counted.
(b) There are just as many counting numbers as there are rational numbers.
(c) Counting numbers and rational numbers can be matched one-to-one.

4. PIAGET SYSTEMS AND MODELS

Example #6: One-to-one Correspondence:

Rational Numbers & Counting Numbers

First, notice that it is possible to organize the rational numbers so that every possible number of the form n/m corresponds to a cell in a table like the one shown below. Then, notice that, to count all of the cells in this table, one procedure that works is to follow in the order A(i), B(i), A(ii), A(iii), B(ii), C(i), D(i), C(ii), B(iii), A(iv), A(v), B(iv) … and to continue the same pattern.

FIG. 4.2f

	A	B	C	D	E	F	G	H	etc.
(i)	1/1	1/2	1/3	1/4	1/5	1/6	1/7	1/8	etc.
(ii)	2/1	2/2	2/3	2/4	2/5	2/6	2/7	2/8	etc
(iii)	3/1	3/2	3/3	3/4	3/5	3/6	3/7	3/8	etc.
(iv)	4/1	4/2	4/3	4/4	4/6	4/6	4/7	4/8	etc.
(v)	5/1	5/2	5/3	5/4	5/5	5/6	5/7	5/8	etc.
(vi)	6/1	6/2	6/3	6/4	6/5	6/6	6/7	6/8	etc.
etc.	etc.	etc.	etc.	etc.	etc.	etc.	etc.	etc.	

A two-dimensional table can be used to organize all of the rational numbers.

After using the procedure that's described above to establish a one-to-one correspondence between the counting numbers and the set of all numbers of the form n/m, notice that there are an infinite number of rational numbers between any two counting numbers on a number line like the one shown below. Are you persuaded that there are the same number of counting numbers as rational numbers?

FIG. 4.2g

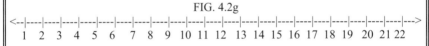

A number line can be used to organize all of the counting numbers.

In example #6, college students often give responses that are strikingly similar to those that kindergarteners give to Piaget's number conservation tasks. For example, college students often say things like: "*I still think there are more rationals because there are lots of rationals between any two counting numbers. Besides, you can just see that there are lots more rationals.*"

Such similarities should not be surprising. In many ways, example #6 is simply an infinite version of the "number conservation task" in example #1. In example #6, the sets are infinitely large, and the transformations are also infinite; but, the basic issues that are involved are nearly identical in example #1 and example #6. What students at both levels lack are sufficient experiences manipulating, combining, separating, organizing, and comparing the relevant kinds of sets—infinite sets in the case of college students, and finite sets in the case of kindergarteners. Consequently, when students focus on one part of the system, they often forget about others; when they focus on characteristics of collections (such as arrays), they often think about the parts in ways that are fuzzy and confused; when they treat two objects as being equivalent (e.g., by treating them as units of some quantity), they often forget about differences (such as those that are involved in ordering relationships); and, when they focus on one characteristic of objects (e.g, based on similarities), they often forget about others (e.g., based on differences). ... In other words, the conceptual systems that students are applying to the situations are not functioning as systems-as-a-whole.

The examples that follow illustrate another related point. That is, in many elementary situations, there is more than a single correct conception of quantities such as length, area, or volume; and, choices among these conceptions depend on both context and purpose. For example, if we consider a straight line to be the shortest path between two points, then, on a sphere (like the earth), it sometimes is sensible to consider a straight line to be part of a great circle on the surface. Or, if we consider Einstein's Theory of Relativity, then, in space, where the path of light determines the shortest path between two points, a straight line might be warped in ways that are not at all straight in the usual sense of the term. ... Other examples follow.

Example #7: What is the Area of a Mobius Strip?

Begin by coloring each of the spaces in a 1 X 10 array of squares. What is the area of this 1 X 10 array? Next, use the 1 X 10 array to make a Mobius Strip by twisting and taping the array as shown below (so the "A" corners are together and the "B" corners are together); and, color the squares on the Mobius Strip. What do you think should be considered to be the area of the Mobius Strip? Why?

FIG. 4.2h

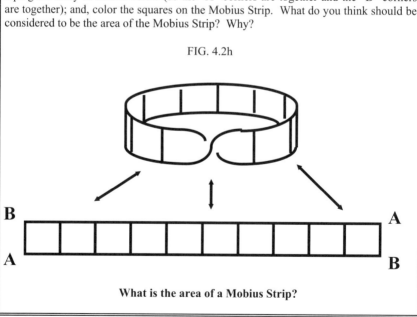

What is the area of a Mobius Strip?

4. PIAGET SYSTEMS AND MODELS 81

In this example, when we are asked about what should be considered the area of a Mobius strip, a relevant response might be: For what purpose do we want to find the area of this Mobius strip? If the purpose is to paint the Mobius strip, then the area we are interested in getting is 20. But, if we want to know how much paper will be needed to make a Mobius strip, then the area we need is *10*. So the correct answer (and the way we chose to think about the concept of area) may vary depending on the purpose.

Example #8: Lengths of Lines on a Computer Screen

Use a standard computer drawing program to draw and move a pair of line segments as shown. In the final figure, what do you think it should mean to say that the red (top) line is the same length as the blue (bottom) line? Notice that, especially if the resolution of your computer screen is low, the "line" will not even be straight at intermediate stages of rotation. Are there different possible answers to this question under different circumstances and for different purposes?

FIG. 4.2i

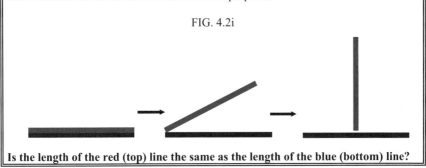

Is the length of the red (top) line the same as the length of the blue (bottom) line?

In this example, consider the fact that both lines are formed by an arranged collection of pixels. The first picture above suggests that the two lines are the same length. But, if we use the computer's "turn" command to rotate the top line, then the arrangement of pixels changes significantly for the top line. This is shown in the second and third pictures. Thus, in some ways, it is reasonable to consider the red (top) and the blue (bottom) lines to be the same length for each of the three pictures. But, in other ways, they are not even straight lines—much less lines that are same length. For example, if we look carefully at the pixels in the second picture, then the "line" looks more like stairsteps.

CONCLUSIONS BASED ON CONSERVATION TASKS

Conservation tasks involve making judgments about invariance properties that are associated with a variety of different systems of operations, relations, and/or transformations. They emphasize the fact that properties such as number, length, area, and volume (or other constructs that Piagetians have investigated) cannot be derived FROM a situation unless relevant systems of transformations, operations, and relations are applied TO these situations. That is, these constructs cannot have their appropriate mathematical meanings until the relevant operational-relational-transformational systems come to function as systemic wholes.

The system-dependent nature of mathematical constructs is even more obvious in the case of ideas such as *inertia* (in science) or *center of gravity* (in mathematics).

- *Center of Gravity*: Within a two- or three-dimensional figure, one point is given special status, and is called the center of gravity, because of systems of transformations and relationships that we have found it productive to use to analyze such figures. But, in fact, this point is just like every other point. It attains special status only because of systems that we apply to the objects and the space that surrounds them.
- *Inertia*: The "desire" of an object to remain in constant motion is a property that is attributed to objects to preserve the sensibility of systems of forces that we have found it productive to use in order to make sense of motions of objects.

Both of the preceding constructs are properties that are associated with the systems that we use to make sense of experience. They are not simply properties of the objects themselves.

Now, the surprising fact that Piaget and his followers have emphasized is that virtually all mathematical ideas are similar to inertia and centers of gravity in the sense that judgments about them depend on thinking in terms of systems of relations, operations, and transformations. Of course, this fact should come as no great surprise to mathematicians who long ago abandoned the notion that mathematics is about truth and reality, or properties of real objects. However, a somewhat surprising consequence of this fact is that the concept of a unit, for numbers and elementary quantities, often evolves surprisingly late in the thinking of students. This is because the notion of a unit of a given quantity presuppose reasoning that is based on systems of relationships and operations that must be coordinated into systemic wholes before properties such as equivalence, transitivity, and invariance have much meaning.

Piaget's conservation tasks were intended to show that quantitative thinking does not begin with the concept of a unit; and, the relevant conceptual systems begin their conceptual existences as poorly coordinated and undifferentiated actions that are only gradually coordinated, sorted out, refined, and organized to form relevant organizational/relational systems-as-a-whole. They do not appear fully formed during children's preschool or elementary school years.

SOME RELATIONSHIPS BETWEEN CHILDREN'S CONCEPTUAL SYSTEMS AND MATHEMATICIANS' FORMAL LOGICAL SYSTEMS

Piaget's tasks were intended to help him identify precisely what operational-relational-transformational system is associated with each concept that he chose to investigate. To pursue this research agenda, Piaget often based his tasks on

4. PIAGET SYSTEMS AND MODELS

properties of relevant formal mathematical systems (Piaget & Beth, 1966). For example, in the case of whole number concepts, consider Peano's Postulates for defining the Natural Numbers (1, 2, 3, ..., n, ...). Or, in the case of Piaget's famous number conservation tasks, consider the formal algebraic definition of an *integral domain* (which corresponds to the integers: ... , -3,-2,-1, 0, 1, 2, 3, ...).

Peano's Postulates Defining the Natural Numbers

N is a nonempty set $\rightarrow$:
1. $\exists$ i$\in$N
2. $\forall$ n$\in$N $\exists$ a unique n'$\in$N
3. $\forall$ n$\in$N n'$\neq$i (n' is called the "successor" of n)
4. $\forall$ n,m$\in$N n$\neq$m $\Rightarrow$ n'$\neq$m'
5. If A$\subseteq$N and i$\in$A, and if p$\in$A $\Rightarrow$ p'$\in$A, then A=N.

The Formal Definition of an Integral Domain

A set N is an integral domain if in N there are defined two binary operations, denoted by $\otimes$ and $\oplus$, and an equivalence relation, denoted by $\approx$, such that for any elements n, m, p in N the following properties are satisfied:

1. the closure property: m$\oplus$n is in N   and   m$\otimes$n is in N
2. the commutative property: m$\oplus$n$\approx$n$\oplus$m   and   m$\otimes$n$\approx$n$\otimes$m
3. the existence of identity elements: There exist two elements o and i in N such that:
                                     n$\oplus$o$\approx$n   and   m$\otimes$i $\approx$m
4. the existence of inverse elements: For every n in N, there exists an element n' in N such that: n$\oplus$n'$\approx$o
5. the associative property: (m$\oplus$n)$\oplus$p$\approx$m$\oplus$(n$\oplus$p)   and   (m$\otimes$n) $\otimes$p$\approx$m$\otimes$ (n$\otimes$p)
6. the distributive property: (m$\oplus$n)$\otimes$p$\approx$(m$\otimes$p)$\oplus$(n$\otimes$p)

The Definition of an Equivalence Relation

$\approx$ is an equivalence relation for a set N if, for every n, m, p in N, the following properties are satisfied:
1. the reflexive property: n$\approx$n
2. the symmetric property: If n$\approx$m then m$\approx$n
3. the transitive property: If n$\approx$m and m$\approx$p then n$\approx$p

---

These formal definitions illustrate several important points about the nature of mathematical systems. First, they can be expected to involve four components:
(a) elements,
(b) relations among elements,
(c) operations on elements, and
(d) principles governing the patterns and regularities that apply to the operations and relationships.

Second, in formally defined mathematical systems, the only meanings that are assumed to be associated with the elements, relations, or operations are those that are derived from properties of the system-as-a-whole. This is because pure mathematics is assumed to be about systems that are devoid of any content. For example, just as geometers may think of a pure line as having no thickness and infinite length (and as existing and functioning apart for the marks that are scratched on papers to represent them), it is also possible to think of all mathematical structures as being pure systems that exist and function apart from any of the spoken, written, concrete, graphic, or digital notation systems in which they may be expressed.

Unfortunately, when attention shifts from the construction of formal systems by mathematicians to the development of system-based intuitions by children, the children's conceptual systems do not have the luxury of beginning with undefined terms—whose formal meanings are derived from the systems in which they are embedded, and whose informal meanings are assumed to be understood without being defined. Nonetheless, to investigate the nature of children's pre-systemic thinking, it is useful to use concrete versions of the properties of the formal systems. For example, the table below suggests several Piaget-like tasks that could focus on commutativity, associativity, and transitivity; and, similar tasks are apparent for other properties.

	Number Concepts	Length Concepts
Commutative Property $m \oplus n \approx n \oplus m$		
Associative Property $(m \oplus n) \oplus p \approx m \oplus (n \oplus p)$		
Transitive Property If $m \approx n$ & $n \approx p$ then $m \approx p$		

MODELING CYCLES DURING A SPECIFIC MODEL-ELICITING ACTIVITY

Earlier on this chapter we mentioned that, whereas Piaget described stages that spanned several years of cognitive development for a given construct, models and modeling perspectives focus on sequences of modeling cycles that occur during the development of products in model-eliciting activities.

This section describes an example solution that a group of average ability seventh grade students developed for the *Quilt Problem* that is given in Appendix A for this chapter. The complete transcript is given in Appendix B. Another model-eliciting activity, called *Gravity Rules*, together with a complete transcript is given in Appendix C. ... From the point of view of this chapter, one point that is most interesting about both transcripts is that the students were developing a concept of a unit of measure that would be useful for the conceptual tool that they were attempting to develop. Another point that is interesting is that, in both cases, the students were not simply building-up a conceptual system, they were also sorting out different ways of thinking – and different conceptions of an appropriate unit of measure. Also, the final conceptual system that the problem solvers develop draw upon (and integrates) ideas from a variety of topic areas. In particular, the concept of unit that they develop does not fit into a single isolated topic area.

During the class period before the *Quilt Problem* was given, the whole class discussed a newspaper article about a local quilting club. The article showed photographs of several different kinds of quilts; and, it also described how club members made the quilts using patterns and templates like the one shown here for a diamond shaped piece.

FIG. 4.3a

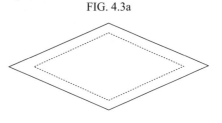

The problem statement described how quilt club members sometimes had difficulties when they tried to use photographs to make templates that were exactly the right size and shape to make quilts that club members found in books, newspapers, and magazines. So, the job for the students was to write a letter that did two things for the members of the quilting club. (a) First, the letter should describe procedures for making template pieces that were exactly the right size and shape for any quilt whose photograph they might find. (b) Second, the letter should include examples of how to follow their procedures by making templates for each of the pieces of the quilt that's shown on the following page.

Note: The quilt that's shown was for a double bed. So, the finished size needed to be approximately 78" by 93".

FIG. 4.3b An Example Photograph of a Quilt

4. PIAGET SYSTEMS AND MODELS

The three students (A, B, and C) whose transcript is given for this activity are seventh grade girls in a remedial mathematics class in a rural school in the Midwest. In the past, they had worked together on more than half a dozen other model-eliciting activities. In particular, they had worked on the *Big Foot Problem* that is described in chapter 1 of this book. Both, the *Big Foot Problem* and this *Quilt Problem* involve scaling-up. But, beyond the fact that they both involve some type of proportional reasoning (A/B = C/D), they focus the students' attention on quite different kinds of quantitative relationships.

The description that is given of the students' work is focused on the local conceptual development of these students on three types of quantitative relationships: whole–whole, part–whole, and part–part. Students progressively became aware of more relationships among and within the objects they focused on; and, their idea of unit evolved significantly as they gave different interpretations to the problem.

Interpretation #1a: Focusing on the Quantitative Relationship between the Two Shapes-as-a-Whole

The first task that students decided to focus on was to assign measurements to the different parts. At the beginning, Students B and C read the sizes from the printed information that was given in the statement of the problem, whereas A was the only one looking at the given picture of the quilt. Student A used a 6-inch ruler to measure the bottom edge of the quilt. The ruler marks included fourths, eighths, and sixteenths of an inch. However, because she did not measure very carefully, Student A incorrectly computed a length of 4.5 inch for the $4^5/_8$ inch that she read on the ruler.

Before these students had worked on the Big Foot Problem several weeks earlier, it was natural for them to begin the Quilt Problem by trying to use a similar scaling-up process. Thus, at first, they tried to determine the value of the scaling-up factor. The first relation that was perceived was a whole-whole, from the picture of the quilt to the quilt that had to be made (78 inches by 94 inches), without viewing the quilt as being made up of many smaller pieces that are related to one another. By comparing similar sides, 78 to 4.5 (without mentioning units), and using a calculator, they determined that the scaling-up factor was 17.33333..., even though this number caused them some conflict.

A: It's four-point-five now and it needs to be ... let's see ... 78. ... Where's the calculator. {She writes 4.5 and 78 on a piece of paper.
B: {B picks up a calculator and punches in 4.5} So, we need ...78....I'll try 12. ... Nope, that's just 54. I'll try 20. ... Nope, that's too big. 90. ... It's between 12 and 20.
A: Try 16.
B: 72. That's pretty good. 72.
A: Try 17.
B: It's 76.5. We need just a little more.
A: Try 17.5.

B: That's 78.75. too big.... I'll try 17.3
...{Several minutes pass while A and B continue to use trial and error to guess what number they should multiply by to get 78.}...
B: So it's 17.333333 ... Now what?
C: So, we multiply all the numbers that much.
B: That doesn't seem right. How can you multiply by 17.333333 ... {Are} You sure it's not just 17. ...
C: No, 17 isn't big enough.
B: What about 18?
C: Nope. That's too big. ... {long pause}... I guess that's it. 17.333333

Interpretation #1b: Scaling-up Individual Pieces within a Whole

Once the students found a scaling-up factor, they decided to apply it to the parts and the whole of the given shape. Earlier, there was no evidence that the whole was viewed as being aggregated from its parts—nor that the same scaling factor should apply uniformly to the whole and the parts. One of the parts they found needed to be scaled-up was of length ¾ (no units mentioned). When the equivalent (0.75) was multiplied by the scaling factor, it gave them 13, which made them feel more comfortable to have obtained an integer. The second measure that they needed to scale-up was ¼. When they multiplied the equivalent (0.25) by the scaling factor, they got 4.3333..., which again caused them trouble.

The students were beginning to recognize that scaling-up individual pieces is not enough to simply calculate a single scaling factor for the shape as-a-whole. They also needed to be able to use this scaling factor to scale-up individual pieces; the same factor must work for all pieces, and the pieces must fit together. These two issues can be considered as self-assessment for the students throughout the whole session. Thus, the students started to measure more carefully, to the nearest sixteenth of an inch, and began counting the intervals in sixteenths of an inch. But they still did not speak of their measurements by referring to a unit (inches, sixteenth of an inch, etc.)

A: So, let's make a piece. What piece we gonna make?
B: Let's try one of these squares. {She's looking at the large squares that have the stars inside.} ... How big is it? ... Let's see. {She is measuring the square with a ruler.} ... Great. It's about three-fourths. ... So, multiply ¾ times that number {17.333333}.
A: Le'me do it. {She takes the calculator.} ... What is it?
B: It's three-fourths.
A: How do I do three-fourths?
C: Punch in .75 dummy.
A: Oh yeah. Right.... .75 Times what?
B: Times 17.333333.... What do you get?
A: Ah, 13. that's good. 13.... Now what?

B: Try one of these little squares. {She takes the ruler and starts trying to measure one of the little squares in the picture.} ... It's hard {to measure}. It's about a fourth.
A: What's a fourth? ...
C: Point two five.
A: Oh yeah. .25 ... So .25 times 17.333333 .... Oh, god!
B: What?
A: It's four-point-three-three-three-three-three-three. ... Some thin's wrong.
C: Do it again. Le'me do it. Here. ... {She takes the calculator away from A.} ... What is it again?

A: .25 times that. ... 17.333333
C: I got 4.333333
A: That's what I got.
C: So? ....
B: So, how big we gonna make it?
A: I do'no. ... This is awful.
C: What'd we do wrong.
A: I do'no. ... We can't just make it close. The pieces wouldn't fit if we do that.
... {Long pause. The students are thinking. They seem to believe that the source of their problem is that they've calculated wrong (or measured wrong). That is, they assume that the number is wrong just because it's a "really ugly number" that they don't know how to use to scale-up other pieces.} ...
C: Let's measure it again. ... Gimme the ruler. ... It's. Let's see. It's four-and-three-fourths.
A: OK. Four-and-three-fourths. ... {She writes 4 ¾ on a piece of paper.} ... What's the other side?
C: It's. Let's see. ... Whoa! ... It's not quite five-and-three-fourths. ... What are these? {she's looking at the marks for sixteenths} ... Le'me see. ... Hmmm!
B: How many of those little lines are there? ... Let's see. ... {She takes the ruler and looks at it closely.} ... Where'd you mark?
C: Right there! {She points to the place on the rule where she had marked the length of the long side of the picture of the quilt.}
B: Let's see. ... There are .... 1, 2, 3, 4, 5, 6, 7, 8, 9 10, 11, 12, 13, 14, 15, 16. So there are 16 of the here. {She's pointing to the first inch.} ... Then there are 16 more here. {She's pointing to the second inch.} And 16, and 16, and 16. How many is that?
A: Do it again.
C: 16. 16. 16. 16. 16. ... What's that.
B: ...{B has been writing 16, 16, 16, 16, and 16 on a piece of paper} ... I'll add 'em up. ... {long pause} ... It's 80.
C: And, there are ... let's see ...1, 2, 3, 4, 5, 6, 7, 8, 9. 10. 11. There are 11 more here.

Interpretation #2: Mapping from one Unit to Another within Two Similar Pictures

When Student B started to make sets of sixteenths of an inch, she added them up, and determined there were 91. When she looked back at the statement, she saw that she needed 93, and determined that the two quantities were very close. She performed a similar procedure for the width of the quilt, and found that there were a total of 76, whereas they needed 78. Then, they found that the difference for both cases was 2. This lucky correspondence allowed them to adopt a procedure to map one sixteenth of an inch (from the picture) to one inch (of the original quilt). This new way of thinking focused on the relationships between the units that were used to measure the picture and the quilt. Still, Student A found a way to justify the 2 unit difference between the correspondences, by a quite sophisticated reasoning, referring to the original statement ...the finished size of the quilt should be about 78 inches by 93 inches, thinking explicitly about errors that can arise due to estimation. Even though student C referred at some point to inches, she went back to using naked numbers.

C: Hey, wait a minute. ... We got 91 and it was 93, and we got 76 and it was 78. We're just off by 2. ... We must've made some little mistake. ... Check again.
...{Long pause. For several minutes, the three students repeated the preceding procedure and got the same result.} ...
A: So, what're we gonna do. ...{long pause} ... Look, it just says ABOUT 73 and ABOUT 93. We can just, like, put a ribbon {border} around the edge ... that's one inch.
B: Ok.
C: What color?
A: Same as this {pointing to the outside edge of the picture of the quilt.}...
B: Ok. So, how big we gonna make these little guys? {Here, A is pointing to the little square in the picture of the quilt}.
C: {She takes the ruler and starts measuring the pieces in the picture of the quilt}... OK, this guy is 9 {She's been measuring the bottom border of the quilt.}
B: Ok. 9. {She writes 9 near the piece that was measured on the picture of the quilt.}
C: This is 2. {referring to the black stripe around the larger squares within the quilt}.
B: OK. 2. {She records this one on the picture of the quilt}.
C: And, this is three-fourths of an inch. ... Let's see. It's 12. {referring to the larger squares in the quilt picture}.
...{For the next few minutes, B and C continue measuring and labeling the picture of the quilt.} ...
A: So, how much is that altogether?
B: What do you mean?

4. PIAGET SYSTEMS AND MODELS 91

Intepretation #3a: Beginning to Focus on Part-Part Relationships

As the students began to determine the pieces of the quilt that were relevant to focus on, and measure and record the sizes of the pieces, they began to explore quantitative relationships involving part-part comparisons.

A: {She's pointing to the measurements that have been written on B's picture of the quilt.} 9 plus 2 plus 12 plus ...
B: Oh, I get it. Let's see. ... 9 and 2 is 11. 11and 12 is .. let's see, what is it. {She writes 11+12 on a piece of paper.} It's 23.
A: Ok. 23 plus 2 is 25. ... Then, 25 plus 12 is.. Hey, guys, let's use the calculator. ... {She picks up the calculator.} OK, C, read 'em off.

Interpretation #3b: Explicitly comparing Wholes and Sums-of-parts–and Comparing Measurements in Two Dimensions (height and width)

In this part, for the first time, the students explicitly expressed their recognition that when the whole and the parts are scaled-up, the sum of the scaled-parts must equal the scaled-whole. They also investigated, for the first time, whether the border of the quilt should be the same width for both, the top and the bottom of the quilt, and the left and right sides of the quilt. In particular, they were trying to make all of the borders be 9 sixteenths—even though, earlier, they had measured them to be slightly different in size.

C: OK. 9. 2. 12. 2. 12. 2. 12. 2. 12. 2. 12. 2. 9. ... What is it?
A: OK. It's 90.
B: 90! @#%!! We wanted 91. That's what we got before. 91.
{For the next ten minutes, A, B, and C, measure and re-measure the quilt picture, and the add and re-add the numbers. But, they can't get the measurements to agree. }

Interpretation #3c: Explicitly Comparing Parts to Parts

The students shifted toward emphasizing part-to-part comparisons. They were willing to adjust the size of the whole quilt as long as the parts fit together properly.

C: Look guys, let's just do what we were gonna do. Just add a little more ribbon around the outside.
B: Or, just make it a little short this way.
A: I hate that. I hate it when my feet stick out 'cause the blanket's too short.
B: Then, make the ribbon two. ... Then, it'll just be a little too big.
A: Yeah. That'll work. I like 'em a little big. That's better 'n too little.
C: Sounds good to me.

B: OK. That's OK. ... {Long pause} ... What about these other pieces? {She's pointing to the smaller squares and parallelograms within the larger squares.}

Interpretation #4: Part-part Comparisons Based on Shared Components (Sides)

The students started focusing on smaller pieces within the large squares by making part-to-part comparisons. Once they determined the size of the smaller square within the larger square (3 by 3), they recognized that the sides of the small squares should be the same as the sides of the kites and the triangles.

The students thought that they had completed the activity. When they announced that they had finished, the teacher reminded them of the second task to construct a template for each of the three shapes in the design.

B: OK. That's OK. ... {Long pause} ... What about these other pieces? {She's pointing to the smaller squares and parallelograms within the larger squares.}
C: {She picks up the ruler and starts measuring one of the smaller squares.} ... This is three. ... {long pause while she tries to measure the parallelogram} ... I can't tell quite what this is. ... Oh, I guess it's three too, 'cause it's the same as the side of the square. {She's noticing that the small squares share one side with the parallelograms.}
A: OK, we're done. That's all of 'em.
B: Hey, Mrs. {TEACHER}, we're done. We're done.

Teacher   Let's see. Have you made your templates yet?
B: What's a template?
Teacher   Look at the newspaper article again.

...{Several minutes pass while the students reread the newspaper article and the statement of the problem}...

Interpretation #5a: Part-Part Comparisons Based on Shared Shapes

The students gave a significant shift on the approach they began using for the templates. Instead of focusing on the calculations that they had previously done, they began finding more accurate geometric patterns and similarities that facilitated their task. Also, as they compared some parts with others, they began to focus on details involving angles. The relationships they described are quite elaborate: (a) the triangles are one half of the small squares (congruence of parts), (b) the strips are one half as wide as the small squares, and (c) the sum of the angles of the square, the triangle, and the kite make a straight line.

4. PIAGET SYSTEMS AND MODELS 93

C: Oh yeah. ... Ok, let's make 'em. This won't take long. ... {long pause} ... We don't need 'em for the stripes. And, we don't need 'em for the border. We just need 'em for the little square, and for these kite-shaped pieces.
B: What about the big squares?
C: OK, them too. So, how big is this little square.
B: It's three. ... Make it three inches.
C: That's easy. I'll get the scissors.
B: What should we make 'em with?
A: We can use this paper.

{For the next few minutes, A, B, and C worked together to measure and cut a small 3"x3" square out of the corner of a standard 8.5"x12" piece of paper. Then, B compares one of their 3"x3" squares with the template that was shown in the newspaper.}

B: Oh oh! We forgot the border. We gotta do it again.
C: Oh yeah. But, that'll be easy. Here, just trace it here. ... {She traces the square that she already had cut out. Then, she draws a border around it} .... How big does it need to be? {referring to the width of the borders for the templates} ... {long pause while all three students look at the picture in the newspaper article}. ... It doesn't say. ... That, looks about right. {pointing to what she had drawn.}
A: I think it just needs to be big enough to sew it. ... I think that's it.

Interpretation #5b: Noticing Sizes of Angles (as well as Size and General Shape)

At this point, the students became very precise at finding relationships among the pieces, and developing the templates. But at the same time, they were also increasingly willing to "tinker" with what appears to be true to make things fit together as they should (based on the increasingly complex system of quantitative relationships that the are attributing to the quilt and the picture).

B: Ok, so how we gonna make this piece. {referring to the parallelogram}? How do you know how to do this corner? {referring to the more pointed side of the parallelogram.}
A: Just guess. See how it works.

{For several minutes, A, B, and C all three made several small parallelogram templates (with borders) that "look about right". But, among the five different ones that they made, some were more thin and pointed whereas others were close to squares.}

A: Which one should we take?

{For several minutes, A, B, and C had a discussion about which of five possibilities "looked best". Finally, they selected one that they thought would work.}

A: Look. Let's try it. The pieces need to fit together to make a square. ... We can try ours and change it a little if we need to.

{For several more minutes, A, B, and C made 8 copies of their "best kite-shape" template and 4 templates for the small squares}
C: Oh oh! We forgot these {pointing to the four half-squares in each of the big squares}. ... And, don't we need one for the big square too ... to sew them on?
{For several more minutes, A, B, and C made 4 new kite-shaped templates, and one for the big squares. Then, they put all of the small pieces together to make one big square.}
Intepretation #6: Coordinating all of the previous quantitative relationships {whole-whole, part-whole, and part-part}

Once the students selected two comparable units for the small quilt and the real quilt, all other quantitative relationships focused on part-whole and part-part comparisons; including relationships between shapes and their angles and shared sides and lengths. They had a clear idea that the sum of the parts should be equal to the whole (big square), so when they put together the templates of the smaller pieces, compared it with a template of a big square, and found that there was a discrepancy, they decided to revise (redo) the templates of the small pieces.

B: That looks pretty good.
C: Yep, but it's just a little too big. Look. {C put the big square onto the square shape that had been made by putting the small pieces together—and, it was clear that the square made by the small pieces was somewhat larger than the template for the large square.}
A: I guess we gotta do it one more time ... but make the pieces a little smaller.
...{Long pause. A, B, and C begin to work on remaking the smaller pieces.}

Final Interpretation: A Two-stage Process

The three students ended up using a two-stage process that integrated a complex mathematical system.

C: I got it. I got it. ... We can get 'em exactly right if we fold this big square. ... Look.
...{For approximately the next ten minutes, C demonstrated how to fold the large paper square to make small pieces that fit together exactly. She did it using the steps shown below. Then, A, B, and C worked together to create templates for each of the pieces. }...

Stage 1 consisted of determining the size of big squares, borders, and thin strips; separating the squares and the borders.
Stage 2 consisted of using paper folding to determine the size of the small shapes within the large squares.

4. PIAGET SYSTEMS AND MODELS

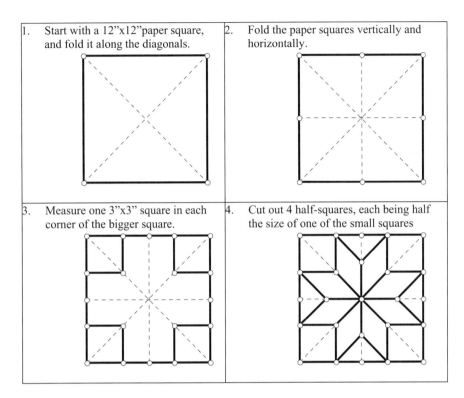

Finally, Student C made some last adjustments by modifying trial solutions. These adjustments were quite small and did not require analytic techniques.

C: It's almost perfect. ... But, not quite. I gotta make these little squares just a teeny-weeny bit smaller. Then, it'll be perfect.

From these interpretations, it is clear that the idea of unit not only evolved, but changed significantly in the students' thinking throughout the problem-solving session. The students were progressively encountering new relationships among parts and wholes, and their reference wholes also changed depending on the focus of the smaller task that they encountered during the session. This simultaneous awareness of the students on different systems of relationships caused their idea of unit to evolve. Their final product resulted in a very complex mathematical model that can be considered as a hybrid that integrates geometry, measurement, and estimation.

CONCLUSIONS

The main purpose of this chapter was to focus on the internal conceptual systems that students develop as part of their solutions to model-eliciting activities. Just like the cognitive structures that Piaget investigated, the constructs and conceptual systems that students develop during model-eliciting activities are holistic systems whose most important properties often are based on characteristics of the system-as-a-whole. Also, just like the cognitive structures that Piaget investigated, the systems that underlie students' conceptual models consist of: (i) elements, (ii) relations, (iii) operations, and (iv) patterns and regularities. Therefore, when we seek evidence that children have shifted from one model to another during the solution of model-eliciting activities, what we need to identify is evidence that they are thinking in terms of: (i) different elements, (ii) different relations, (iii) different operations, or (iv) different patterns and regularities.

One consequence of the holistic nature of students' conceptual system is that relevant units of measure often cannot have their appropriate meanings until students are able to think in terms of these systemic holes. Consequently, understandings about the units often evolve simultaneously with developments related to the relevant systems.

Piaget-like conservation tasks often are useful to help researchers and teachers identify the conceptual systems that students need to develop to understand various units of measure; and, to develop such tasks, it is useful to be familiar with the formal systems that mathematicians use to define the relevant constructs. On the other hand, when students develop conceptual models in the context of specific model-eliciting activities, they typically draw on ideas associated with a variety of formal mathematical systems. Furthermore, development often involves sorting out and refining processes at least as it involves construction or assembly.

Whereas Piaget-inspired researchers often speak of mathematical constructs (such as those related to various units of measure) as if they were specific manifestations of general/all-purpose cognitive structures, the *local conceptual developments* that we observe during *model-eliciting activities* suggest that the evolution of mathematical constructs tends to be more accurately characterized as gradual increases in local competence. That is, relevant conceptual systems tend to function first as situated models that apply to particular problem-solving situations. Then, these models are gradually extended to larger classes of problems as they become more sharable, more transportable, and more reusable. In other words, it is only at relatively mature levels of thinking that knowledge is organized around abstractions rather than around experiences.

Chapter 5

A Semiotic Look at Modeling Behavior

Paul E. Kehle
Illinois Wesleyan University

Frank K. Lester, Jr.
Indiana University

Consider the following problem:

The Suez Canal is 160 km long. The ships that use it are so large and the canal is so narrow that ships cannot pass each other. Traffic flow must be one-way in any one segment of the canal at any one time; however, the canal runs through two lakes and is wider in these two places, between 51 and 60 km and between 101 and 116 km as measured from the north end of the canal. In each of these lakes, 17 and 36 ships, respectively, may dock and pull out of the way of ships heading in the other direction.

For safety reasons, ships must travel at 14 km/hr and must maintain a 10-minute spacing between ships. Demand for travel through the canal is equally high in both directions; however, the northbound ships are more often fully loaded with cargo, making docking difficult. Design the most efficient schedule possible. Justify the optimality of your schedule.

This problem is complex and quite realistic and calls for a mathematical-modeling approach in its solution. Its complex and problematic nature becomes clear as you begin to try to solve it. It is realistic in that it is the very problem solved by Griffiths and Hassan (1978) in the mid-1970s for the Suez Canal Authority.

Although some numbers are present, the problem has no formal mathematical context. Yet, to arrive at a solution, it seems reasonable to construct some kind of formal mathematical representation to help the problem solver think about it. High school students making use of a Cartesian coordinate system and linear programming have solved this problem.

Mathematical modeling of the sort involved in solving the Suez Canal Problem is an involved, multi-phase process that begins when someone working in a complex realistic situation poses a specific problem. One conception of this process is shown in Fig. 5.1. To start solving the problem, the individual simplifies the complex setting by identifying the key concepts that seem to bear process is shown in Fig. 5.1. To start solving the problem, the individual simplifies the complex setting by identifying the key concepts that seem to bear most directly on the problem. This simplification phase involves making decisions about what can be ignored, developing a sense about how the essential concepts are connected, and results in a realistic model of the original situation. A realistic model is a model precisely because it is easier to examine, manipulate, and understand than the original situation.

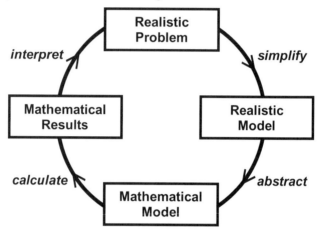

FIG. 5.1. The mathematical-modeling cycle based on Maki and Thompson (1973); note similarity with the diagram in chapter 1 of this volume.

Next comes the phase that introduces formal mathematical concepts and notations. This abstraction phase involves the selection of mathematical concepts to represent the essential features of the realistic model. Often the abstraction phase is guided by a sense of what a given representation makes possible in the subsequent calculation phase. The explicit representation of the original setting and problem in mathematical symbolism constitutes a mathematical model of the setting and problem.

Once a problem solver generates a mathematical representation of the original situation and a specific mathematical problem related to the representation, the mathematical problem acquires a meaning all its own, becoming an isolated, well-defined mathematical problem.

The third phase of the process involves manipulating or reasoning with the mathematical representation to deduce some mathematical conclusions. During this phase, a person's store of mathematical facts, skills, mathematical reasoning abilities, and so forth, come into play. For example, the problem might call for a solution of a system of equations and solving this system of equations does not depend on the original context of the initial problem. To date, mathematical problem-solving research has focused largely on the activity people engage in during this phase of the mathematical modeling cycle—that is, once the

5. A SEMIOTIC LOOK AT MODELING BEHAVIOR

mathematical representation has been developed—and this research has focused on problems already cast in a mathematical context.

Finally, the results of this mathematical reasoning are interpreted and applied in the original context of the problem. Often, the results are lacking in a way not anticipated as the model development phase began. What naturally results is a reiteration of the modeling cycle. Figure 5.1 is useful for thinking about mathematical modeling behavior, but, as discussed further below, it conveys a false sense of linearity to the modeling process. Both novice and experienced modelers often jump from one phase to another, back up a step, or iterate the entire cycle many times. And, as noted in Lesh and Doerr, chap. 1, this volume, it is pedagogically necessary for students to have multiple sets of iterated modeling experiences on a variety of related model-eliciting tasks for them to develop their understanding of both mathematical concepts and the modeling process.

Figure 5.2 depicts a simplified version of the modeling process. It highlights the focus of this chapter by drawing attention to two different kinds of representational contexts: the (often realistic) context of the original problem and the mathematical context containing the mathematical model.

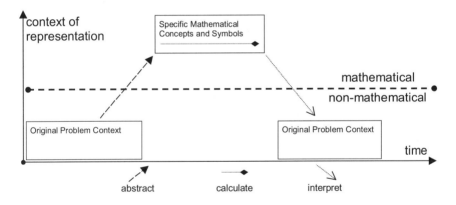

FIG. 5.2. The transition from non mathematical context(s) to mathematical context(s).

We have been involved in studying how students make the transition from the original problem context to a mathematical representation. In particular, we have been examining the inferences students make and how their interpretations of various contexts help or hinder their overall progress toward constructing a useful mathematical representation (and its attendant understanding).

Our goal is to provide an empirically grounded example of how mathematics educators can better attend to, describe, and reflect on students' attempts to generate useful mathematical representations while solving mathematical-modeling problems. Using a semiotic framework that draws attention to the ideas, conceptual frameworks, and modes of inference-making that students use or do not use, we analyze students' mathematical-modeling behavior in terms of crucial episodes of sense making.

The semiotic framework described below complements other frameworks that have been brought to bear on mathematical problem solving and mathematical thinking, for example, Polya (1945), Kilpatrick (1967), Lucas (1972), Lester (1978), Schoenfeld (1985), Sierpinska (1994), Pimm (1995), and others as discussed in Kehle (1998). In different ways, all of these previous frameworks left us wanting more detail. What were students doing when they were analyzing? What prompted their explorations? What aspects of a problem stimulated their planning? Imagining how we might direct a hypothetical student to become a more successful problem solver motivated these questions. Especially in the context of modeling problems, we wanted to do more than increase their metacognitive awareness—of what exactly should they be more aware? The Peircean semiotic framework we settled on is one we believe to be ultimately of use to teachers as they and their students engage in mathematical-modeling activity.

A PEIRCEAN SEMIOTIC FRAMEWORK

First and foremost, semiotics is about meaning making and the communication of meaning. Semiosis is the cognitive process of structuring our experiences and making sense of the world. As Lemke (1997) puts it:

> Semiosis is meaning-making [sic]; it is taking one thing as a sign for another, construing a thing, event, process, or phenomenon in relation to one or more others. Semiosis is selective contextualization; it is making something meaningful by seeing it as a part of some wholes rather than others, as being an alternative to some options rather than others, as being in some particular relation to some things rather than others. (pp. 39–40)

Unlike Sierpinska's (1994, cf. p. 69) emphasis on the network of mental representations of mathematical concepts that facilitates and results from meaning-making activities, Peirce's emphasis is on the fluid flow of one instance of meaning making to another.[1] Merrell (1997) stresses the point that meaning lies not with signs or mathematical symbols or concepts, but in the process of their formation and use:

> Peirce's semiotics by its very nature includes a theory of meaning. But semiotics is not about meaning in the ordinary way of taking it. It is about meaning engendered when signs are in their act of becoming signs, a becoming that includes sign interpreters as participating agents in the very semiosic process of becoming. I cannot overemphasize my contention that meaning is not in the signs, the things, or the head; it is in the processual rush of semiosis; it is always already on the go toward somewhere and somewhen. (p. xi)

---

[1] Charles S. Peirce, one of the most influential writers on semiotics, wrote in a style often as obscure as his ideas are stimulating. In this chapter we have relied on Peirce's work as found in Buchler (1955) and on Merrell's (1997) interpretations of Peirce's work.

5. A SEMIOTIC LOOK AT MODELING BEHAVIOR

This attitude and framework is precisely what can help mathematics educators take seriously the notion that mathematics is not a collection of facts and skills, but rather a way of thinking.

According to Peirce, all meaning making is mediated by signs, and semiotics is the study of how signs facilitate this meaning-making practice. For Peirce, a sign is anything that "stands to someone for something in some respect or capacity" (in Buchler, 1955, p. 99). Quite literally, anything can conceivably serve as a sign (e.g., a chair, words, mathematical symbols, a drawing, a dance, an ant, or a bit of lint—the list is truly all-encompassing!). Likewise, anything at all can be the object signified by a sign, also called the referent of a sign. Semiosis is the irreducible triadic interaction of these two elements together with a third, called the interpretant. The interpretant both consists of an individual's reaction to a sign and an object, and simultaneously defines the sign and object in question. In Peirce's view each of these three notions, sign, object, and interpretant exist only in dynamic relation with the other two. Discussing any effect produced by one, or two, without the others is impossible (for introductions to semiotics, see Houser, 1987; Deely, 1990; and Danesi, 1994). Figure 5.3 depicts the irreducible triadic interaction that constitutes semiosis.

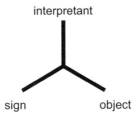

FIG. 5.3. The triadic nature of semiosis.

A concrete specific example helps to initially define these concepts for our purposes. Consider, as a simplistic example, the following signs: 42 – 22

The numerals and the horizontal line segment might be signs representing quantities and the operation of subtraction, which might be in turn the objects or referents of the signs. The interpretant that is present when most people confront these signs is an impulse to compute a difference or the blending of these individual signs into a larger sign signifying a subtraction problem. Indeed, very often the interpretant is another sign that serves as a link to yet another interpretant in what is perhaps a long chain of meanings or episodes of sense making. Notice that the signs and objects are determined by, at the same time they determine, the interpretant. With a different person, say a very young child, these three elements would be very differently entwined. Note that neither the symbol "20" nor the quantity 20 are necessarily the interpretants in this example. To reach either of these concepts ordinarily requires an inference. In this case the inference is the computation of a difference, or the manipulation of the originally presented signs. However, because many signs are connected with their objects and interpretants arbitrarily, purely by an agreed upon convention, it is possible that a person could come to immediately generate 20 as an interpretant to the signs $42 - 22$.[2]

To qualify as an interpretant in Peirce's sense, no intermediate inferences or manipulations are allowed. As we can see even from this simple example, the tracing of each interpretant in the flow of semiosis can be challenging, if not impossible. At other times, the identification of all three elements of a crucial episode of sense making is easy and meaningful. Perhaps most importantly for teachers and students concerned with learning, though, is the relevance of what we call stalled semiosis. These moments occur when we are confronted by signs, or a lack of signs, and no meaningful interpretants are generated.

The interpretant is the hardest element to explicate, largely because it is the least objective. The interpretant not only depends on the specific person confronting a sign but also on the specific circumstances surrounding each confrontation. A speed limit sign is a sign signifying certain well-defined driving behavior. However, depending on whether you are on your way to the store to pick up a loaf of bread or on your way to the emergency room with a poisoned child, your behavior in response to the sign might be very different. Part of what dictates your response are the other signs and interpretants that form a local constellation of semiotic moments or sign system that leads to different interpretants than another constellation might.

For example, taken by itself in isolation, $42 - 22$ would have no persistence as a sign. It persists and is a part of meaningful semiosis only because it can be seen as part of a larger sign system, perhaps that of arithmetic. The local meaning it might help create further depends on other sign systems, such as the one concerning employees available for various tasks, employees already assigned, and new tasks needing attention. The linking of two such sign systems is an act of mathematical abstraction.

Very broadly, the interpretant is the effect that a sign, its object, and the previous interpretant engaging a person have upon the person. We prefer to think of it as a potential for subsequent action—physical, cognitive, or emotional—that is sensed in a particular situation as a result of encountering the sign(s) in question. For example, suppose you see a drawing of a triangle (a sign) with the measures of two angles marked (more signs), the object the signs stand for might be the abstract idea of a triangle with certain dimensions. One possible response to such a figure sign is that you realize the possibility of computing the measure of the third angle. It is this potential you have for action that is the meaning underlying your encounter with the signs and what they represent. This meaning is built up from previous encounters with triangles and your gradual composing of sequences of thoughts (interpretants) linked to the relevant signs that yield the measure of the third angle.

Although the collection of what can be an interpretant is not as varied as is the collection of things that can be signs, many different things do emerge as interpretants: physical actions, emotions, and most important to this study, thoughts or ideas that become new signs in the "processual rush of semiosis."

---

[2] As further evidence of the variety of possible interpretants, we point out the case of the person whose immediate response to this example was to view it as part of a combination for a lock.

5. A SEMIOTIC LOOK AT MODELING BEHAVIOR 103

METACOGNITION AND SEMIOSIS

When viewed from a semiotic perspective, metacognition occurs when an instance of past semiosis becomes the object of new semiosis; see Fig. 5.4.

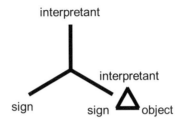

FIG. 5.4. A past instance of semiosis can become the object of new semiosis.

Consider this episode from Evan and Lisa's (pseudonyms for college seniors who were participants in our modeling research project) work on a modeling problem (Linear Irrigation, stated later in this chapter).[3]

Evan: Know what though?
Lisa: What?
Evan: We're thinking of just the circles as if they are not moving. And not the lines in which they move in.
Lisa: Okay.
Evan: You know what I mean?
Lisa: Yea, okay. Okay.
Evan: Like we are making this way too much harder than what...[trails off into silence]

This episode is interesting for several reasons, but it best exemplifies a metacognitive episode in Evan's semiosis and his attempt to share it with Lisa. (Although they did not pursue the implications of his metacognitive realization at this point in time, they returned to it later and it is crucial to their arriving at a good solution.) The signs that led him to his realization were the drawings he had been making of overlapping circles and the thoughts he and Lisa had been having about the areas of different parts of the drawings. His ability to use his thinking about the problem as new signs or objects supporting new instances of semiosis aided him in making sense of this problem and in solving it. Semiotician Donald Cunningham views this reflexivity, the ability of semiosis to double-back upon itself as a major strength of a semiotic framework (personal communication, February, 1997). It has important implications for teachers concerned with developing students as self-sufficient learners who need to be better able to think about how they think.

---

[3] All statements of the problems used in this research can be found online in Appendix A.

Reflecting on how an idea or series of ideas structure our thinking, as opposed to simply having the ideas, is an example of metacognition of past cognitive activity shaping future cognitive activity. This type of metacognition is different from the general metacognitive awareness we might have of whether we are making progress with a problem.

Isolated instances of semiosis are difficult and often impossible to interpret. Signs, the objects they stand for, and the interpretants that bind these together depend for their meaning on the larger sign system of which they are a part. On a very basic level the notion of distinct sign systems helps distinguish between a formal mathematical representation of a problem and a nonmathematical representation. By paying attention to the symbols, diagrams, words, and other signs used by the participants, we can offer evidence to support claims about what context they are working in at any given moment.

Sign systems are seldom mutually exclusive. In fact, the overlapping of sign systems and the intentional juxtapositioning of different signs and instances of semiosis lie at the heart of semiotic perspectives on learning and research (Shank, 1988). The way in which people consciously, or not, shift their cognitive activity among different contexts or sign systems lies at the center of our study. Minimally, there is the shift from the original problem context to a mathematical one, but intermediate contexts might also be utilized.

SEMIOSIS AND INFERENCE

Cognition within a sign system (or across sign systems) consists of drawing inferences based on the signs and interpretants formed during interaction with the given system(s). Peircean semiotics distinguishes three forms of inference: deduction, induction, and abduction (Cunningham, 1992; Deely, 1986). Drawing deductive inferences is the most commonly recognized of the three forms. The rules for deduction are presented in classical logic texts and constitute a syntax for inference making that takes signs as both beginning and ending points. If one follows the rules of logic, one's deductive conclusions are just as trustworthy as are their premises.

Semioticians use "induction" to denote not the mathematical process of inductive proof (actually an example of deduction), but rather the process of exploring to what degree our prior understanding of some general phenomenon can inform our understanding of what might be a particular instance of this phenomenon. It is the application of familiar episodes of semiosis to a specific situation not previously involved in the familiar episodes. Induction is closely related to the hypothesis-testing phase of the scientific method and represents the application, or mapping, of signs to some phenomenon. The result of an inductive act can be confirmation that some phenomenon is or is not an example of some abstract construct, or it can be the realization of a deficiency in the abstract construct. The result of an induction is the confirmation that a general way of thinking is useful and appropriate in a specific situation.

5. A SEMIOTIC LOOK AT MODELING BEHAVIOR

Abduction is what we do when we are confronted with a novel experience, one that does not make immediate sense. In such a situation, we guess, hypothesize, play hunches, gather clues, make diagnoses, and in general try to understand that which is new in terms of something familiar (Shank & Cunningham, 1996). Abductive acts take experience as their beginning point and result in the creation or adoption of signs to interpret the experience. The following is a textbook example of an abductive inference made by Evan as he and Lisa were working on the Linear Irrigation Problem.

Evan: How long would it take to provide a two inch watering to one strip of... One strip, okay.
Lisa: Well okay, so we're 2,200, a strip is 2,200 feet long, two inch watering. So we are two inches wide.
Evan: Inches deep.
Lisa: Two inches deep.
Evan: Okay so we have a vat.
Lisa: A vat! [chuckles]
Evan: We have to disillusion ourselves, basically sort to speak from the fact that we have this [pointing to drawing of a strip of land], it pivots beyond this point, because we'd just have to save it. It does that, deal with it as a whole.
Lisa: Right.
Evan: Because we could sit there for ever and try to figure out how that little area over there gets the two inches that it [the closest nozzle] never reaches.
Lisa: Right. So we've got to think that it covers everything all the time.

Evan's decision to think about the problem of watering a long strip of land with many moving nozzles as if it were a problem about filling a vat with one nozzle is an abductive inference. In particular he is drawing an analogy between the irrigation problem and a more familiar phenomenon, that of filling a rectangular tank. At the close of the modeling session, during a debriefing interview, Lisa reflected on this stage of their activity:

Paul: ...Did you ever think of the problem in terms of anything else? At any point in time was there anything else in your head that you were thinking this is...
Lisa: What I was thinking, but I didn't know how to verbalize it, I didn't, I was apprehensive too. Is that when we talked, when the problem stated 2 inches of watering, I was thinking two standing inches of watering. Are we talking how fast is the ground sucking up the water? Or do we have to take that into consideration, the absorption rate? You know, I was like, okay if we do, we don't have enough information. Probably don't know how to find that information. I mean I was thinking of those things, so I thought it was just so much easier just to think of it as a big vat or a big pool, you know. You just

have a 2 inch pool and you want to fill it up. You know, how do you, you know that's a lot easier than thinking okay, the ground is a bit dry for X many days and you've got to take in account of how fast the absorption is and what 2 inches of watering is it? Two inches standing or how do you measure, you know. I was thinking that, but I am like, let's stay clear of those ideas and just stick with the idea of just 2 inches, just poured over the land. I thought that was a good idea, the whole 2 inches. The vat idea!

Evan and Lisa's ability to think about the problem in terms of something that was more familiar to them helped them solve this problem. In fact, it was their inability to think about the problem in its original terms, an unfamiliar sign system concerning soil, absorption, moving nozzles, and overlapping spray patterns, that led them to seek out another way, or another sign system with which to make sense of the problem. The new sign system consisted of rectangular vats, a single source of water, volumes, and flow rates. This sign system closely overlapped with the more formal sign system concerning algebraic relationships between areas, volumes, and rates. To get to this sign system, required an abductive act by Evan and Lisa to think of the problem as something else.

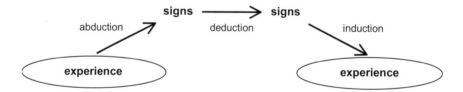

FIG. 5.5. Three modes of inference making employed in sense-making activities.

Figure 5.5 summarizes these three modes of inference making (adapted from Cunningham, 1992). Abduction begins primarily with experience and ends with the taking of something to stand for something else; it is the generation or appropriation of signs to explain novelty. Deduction is the manipulation of signs according to prescribed syntactical rules. Deductive inferences begin and end with signs and introduce no new ideas than those already present in the sign system. Induction begins primarily with a well-articulated sign system and applies it to a specific experience believed to be meaningfully understood in terms of the sign system.

In typical everyday experiences, we shift continuously from one experience to another, meshing abductions, deductions, and inductions as needed to make sense of, or just cope with, what we encounter in the wide world. For the most part, we carry out our inference making unselfconsciously. The seamless automatic aspect of semiosis lies at the core of human intelligence. At the core, it is joined dialectically by the more deliberate inventive capacity of semiosis that comes into play when we confront novelty—when our automatic responses meet with unexpected consequences.

CONTEXT PROTOCOLS

We organized our study of modeling activity of several pairs of college students (in this chapter we drawn upon two pair's work: Evan & Lisa, and Robin & Ellen) around a context protocol (or context parsing) produced from field notes and video and audio recordings. These context protocols identify the locally coherent sign systems the students were working in and among, and they summarize the types of inferences made by the participants. Closer examination of each context reveals the detailed nature of the signs, inferences, and semiotic activity present. The contexts are not mutually exclusive, nor as discrete as they appear, nor is their occurrence as linear in time as it appears. One such protocol is given in Fig. 5.6.

A drawback to this representation's strong temporal dimension is that it conveys a false sense of linearity to a pair's cognitive activity. Just because a person has apparently moved into a new cognitive context and appears to have left the ideas of a previous one behind does not mean that her or his current cognizing is unaffected by what has preceded it, nor does it mean that ideas left behind might not be recalled at any moment in the future. Such recollections could have mutually modifying effects, on the current context and on the previous idea. Despite this drawback, these protocols do chart useful middle ground between the extremes of microanalyses and global heuristics, each less suited for use on the fly in the classroom. Because the chart is just a point of departure for organizing a discussion of the modeling activity, it does not do everything. Transcripts must make up for a chart's shortcomings. A context protocol's main purpose is to provide points of reference and a macroscopic summary of modeling activity as stimuli to further discussion and analysis.

On the chart are arrows indicating various types of inference moves. These are best viewed as indicative of the dominant mode of inference making taking place within and between contexts. It is impossible to identify every inference made by the participants. What is useful for our purposes is to look for overarching styles of inference making, the presence of crucial inferences, and the lack of what would seem to be valuable and possible inferential moves. These are the aspects of a student's thinking that teachers should attend to most if they are to understand better and perhaps help improve the student's thinking.

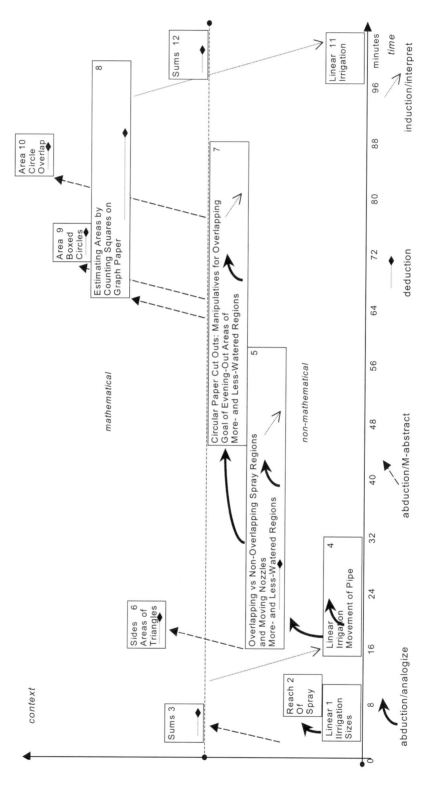

FIG. 5.6. A context protocol for Robin and Ellen's work on the Linear Irrigation problem.

5. A SEMIOTIC LOOK AT MODELING BEHAVIOR

Despite this drawback, the context protocols do chart a useful middle ground between the extremes of microanalyses and global heuristics, each less suited for use on the fly in the classroom. Because the chart is just a point of departure for organizing a discussion of the modeling activity, it does not do everything. Transcripts must make up for a chart's shortcomings. A context protocol's main purpose is to provide points of reference and a macroscopic summary of modeling activity as stimulus to further discussion and analysis.

Also on the chart are arrows indicating various types of inference moves. These are best viewed as indicative of the dominant mode of inference making taking place within and between contexts. It is impossible to identify every inference made by the participants. What is useful for our purposes was to look for over-arching styles of inference making, the presence of crucial inferences, and the lack of what would seem to be valuable and possible inferential modes, or specific inferences. These are the aspects of a student's thinking that teachers should attend to most if they are to understand better and perhaps help improve the student's thinking.

INTRODUCTION TO THE LINEAR IRRIGATION PROBLEM

One problem we used in our research was the Linear Irrigation Problem derived from one version of it found in *Mathematical Modeling in the Secondary School Curriculum* (NCTM, 1991).

> A farmer is designing a linear irrigation system. The fields are divided into strips that are each 2,200 ft long and 500 ft wide. A linear irrigation system consists of a pipe 500 ft long with nozzles spaced at regular intervals along the pipe. The pipe is mounted on wheels and is connected to a long flexible hose. With the water turned on, the pipe rolls slowly down a strip of field watering the crops as it moves. When it reaches the end of one strip, the pipe assembly is pivoted and moved down the adjacent strip of land.

Each of the nozzles sprays water uniformly over a disk-shaped region that is 20 feet in diameter. The farmer has 45 nozzles available with which to build the system. The goal is to water each strip of land as uniformly as possible.

Part 1: How many nozzles should the farmer use and how should they be placed on the pipe to provide the most uniform possible watering of a strip of land?

Part 2: If each nozzle delivers 14 gallons per minute, how long will it take to provide a 2 inch watering to one strip of land?

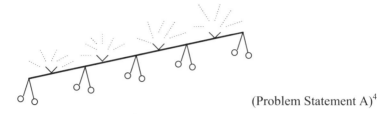

(Problem Statement A)[4]

At the heart of this problem lies the need to make sense of uniformity. Whatever sense is made of uniformity, one must develop a representation that does justice to the constraints of the irrigation context, while at the same time affording a means of assessing the uniformity of the watering provided by different arrangements of nozzles. What makes the problem difficult for most people, is the dynamic aspect of the irrigation system. A static fixation, if left unchecked, usually leads to a comparison of areas that receive a double watering with those that receive none. Often, this fixation is reinforced by a metaphorical link, an abductive one, to painting and the idea of one area receiving two coats of paint to another area's one. Unless the consequences of such ideas are pursued carefully, through inductive inferences that return attention to the fields, pipes, and water, it is hard to realize the trouble with this way of thinking about the problem.

EPISODES FROM ROBIN AND ELLEN'S MODELING ACTIVITY

Of all the pairs of college students whom we asked to work on the Linear Irrigation Problem, Robin and Ellen's work on the Linear Irrigation problem was the least successful in terms of their confidence in it, and in terms of the usefulness of their model when viewed from a richer, more comprehensive perspective. The case of Robin and Ellen's modeling activity on this problem is quite revealing, however, and is representative of the situations where a teacher's pedagogical intervention is needed. Indeed, the nature of their work is consistent with the majority of students we have taught who are new to this kind of mathematical activity and need a lot of initial guidance in becoming more self-sufficient. Though we succeeded, it was difficult to refrain from interacting with them in a more constructive manner.

Each of the following subsections primarily highlights one aspect of their work or one episode of sense making. A careful study of the transcripts reveals many other aspects and episodes that could be discussed. Our goal is not one of exhaustion, but rather to emphasize the potential of a semiotic framework. See appendix B for the complete transcription of their work.

---

[4] We note that the kinds of mathematical modeling problems used as prompts in the research discussed in this chapter are essentially equivalent to the "model-eliciting activities" described elsewhere in this book. (See Appendix A for statements of all problems).

5. A SEMIOTIC LOOK AT MODELING BEHAVIOR

Iconicity in Problem Choice: Signaling a Pattern

Robin and Ellen worked on the Linear Irrigation Problem during their second modeling session. At their first session, they had successfully solved the Alaskan Pipeline Problem and the Committee Scheduling Problem (see appendix A). After their first session, they confided that they were relieved to learn that the problems they encountered were so "doable" by them. They had feared that they would not be able to get anywhere with them. They eagerly agreed to return for the second session, and though session two was difficult for them, they chose to return for the final session.

At the beginning of their second modeling session, Robin and Ellen somewhat reluctantly agreed to work on the Linear Irrigation Problem instead of the Suez Canal Problem. Both students were much more tentative in their selection of problems than they had been at the first session when they confidently rejected the Museum Guard Problem as clearly involving geometry—something they wished to avoid. In the post modeling debriefing session, they explained their choice in terms of being familiar with irrigation systems (having seen them on farms) and not expecting the problem to involve geometry. Ellen did not see a big difference between the problems and thought both would require more common sense than mathematics. She was not particularly interested in either problem. Robin said she did not pay attention to the picture of the pipe with the nozzles, or else she might have vetoed it. She was drawn to the numbers (2200, 500, 20, 45, 14, 2) and expected their work to be more numerically oriented. During the session, both remarked that had they known the problem called for geometry they would have made a different selection.

Their choices provide examples of people responding strongly to the iconic nature of a sign. Fundamental to Peirce's conception of signs, are three characteristics (iconic, indexical, and symbolic) that determine a sign's nature: (1) the extent to which a sign resembles its object is a measure of its role as an icon (a child's doll), (2) the extent to which a sign is causally linked to its object is a measure of its role as an index (deer tracks in mud), and (3) the extent to which a sign refers to its object by agreed-upon convention is a measure of its status as a symbol (a wedding ring; Buchler, 1955, pp. 104–115). Note that a sign can partake of more than one of these characteristics, as in the case of the shadow of a cat (both iconic and indexical) or the call of a loon (both indexical and symbolic).

In Robin and Ellen's attempt to avoid geometry problems, they responded to the iconicity of a museum's floor plan insofar as they inferred from it a need to work with edges, angles, and shapes—the objects of geometry. Although numbers are usually viewed primarily as symbols, Robin and Ellen also responded to their presence in the Linear Irrigation Problem as icons resembling aspects of a sheet of arithmetical computations. This strong iconic mode of sign usage during problem selection is characteristic of Robin and Ellen's work as a whole and stands in contrast to the stronger symbolic mode of sign usage in other students' work.

The Modeling Cycle: A Cameo Appearance

During the first 8 minutes of their session, Robin and Ellen quickly moved through all of the major phases of the modeling process. Their avoidance of geometry and their initial naïve interpretation of uniformity led them in part, perhaps tacitly, to use a representation that allowed them to greatly simplify the problem. Their initial representation of the problem was a line, representing the pipe and the width of a strip of land, they used to help them determine how many nozzles are needed to cover the width of the field with water. This approach allowed them to make use of a sign system with which they were comfortable and confident: arithmetic and repeated sums in particular.

Their inferential trajectory through the first part of the context protocol begins with the abductive move to view the problem as one of coverage wherein the goal is to get all of the ground wet. Notice in the transcript the repetition of variations of the thought "spraying out 10 feet." Their primary concern is the reach of each nozzle and how much ground it can cover—most noticeably the part of the ground iconically cast into relief by the pipe. Their sense of reach and coverage is primarily linear at this point and uniformity to them means everything getting watered.

Next, they determine through a combination of deduction and induction that placing the nozzles to produce a tangentially linked row of circular spray patterns along the pipe will cover the field. The interpretant at work here is their self-known capacity to add together repeatedly a fixed quantity to make up a larger total. More deduction yields the number of nozzles that will cover the 500 ft width of the field. Finally, they return to the context of the farmer's supply of nozzles and the pipe to confirm that they have solved the problem. The following excerpt is of the first 8 minutes of their work. It begins with their work on the problem. (Excerpts are from Robin and Ellen, 11/20/97. See Kehle, 1998, for a fuller discussion of Robin and Ellen's work as well as the work of other pairs of students on other problems.)

13 R: Alright, first off we have the fields. There are 22, I'm just, understanding the problem.
14 E: I know.
15 R: Okay.
16 E: Hmm.
17 R: Okay. Alright, I need a picture. Alright, this is what I am thinking. And we have this thing here and we've got these little nozzles spaced on to here, right? And it's going this way because since it's 500 feet wide and the pipe itself is 500. Taking it over the field this way? [Draws a rectangle with a line across its width representing the pipe.]
18 P: Correct.
19 R: Okay.
20 E: Okay.
21 R: Are you following me?

5. A SEMIOTIC LOOK AT MODELING BEHAVIOR 113

22	E:	Yep. I'm just. I have to draw my picture too (chuckle). I can't see it. Okay. [Reproduces Robin's drawing.]
23	R:	Now then each one of these is going to.
24	E:	40, 20 feet, right.
25	R:	Well wait, the diameter is 20 feet. The radius will only be 10 feet then. Right?
26	E:	Um hum.
27	R:	I mean you know what I am saying? It's like, saying they will spray a circle and the whole distance across is 20. So actually it's only spraying out 10 feet.
28	E:	Right.
...		
37	R:	Alright, to get here the first one has to be placed at 10 feet, so all of these, these 10 feet are all watered right?
38	E:	Um hum.
39	R:	Well this one is gonna go 10 feet. But the one comes this ways is going to 10, so this one won't be placed until 30.
40	E:	Right
41	R:	And it's going to go that way, all the way down until we get to the other end. So we've got 10, 30, 50, and we need to get to 500, right?
42	M:	No, we need to get, right! (Laughter). Yeah, yeah, 500.
43	R:	Okay, so if we are going up by 20s but we are at odd numbers, we're going to end up at 490. Right? And that would work because if one is at 490 it's going to spray 10 feet this way. Okay, if we went 300.
44	E:	Right, right, right. I'm thinking of like, oh yeah! I was thinking about the disk thing going by 2 but it will still take care of it. Do you see what I am saying? We're going to be alright.
45	R:	Okay, and we are going to end up with one at 490 and it's still going to spray 10 feet this way, so that's going to cover the 500. So we just have to figure out if these are spaced, how many we need. Right?
...		
59	R:	End at 490. Yeah, okay. Yeah, that's right. Okay so it is 25 right? So he needs to use 25 of them.
60	E:	Spaced 20, except the first and the last one.
61	R:	The first one is going to go at 10.
62	E:	And then it's going to go every 20.
63	R:	Right. How many nozzles should the farmer use and how should they be placed on the pipe to provide the uniform possible watering of a strip of plant.
64	E:	Did we answer it? It's 25, the first at 10.

70 P: Okay. How uniform do you think the watering will be, that results from the placement?

Although rarely did any students generate a formal model with all its assumptions, properties, and limitations clearly articulated, the amount of time they had to work on a problem really didn't allow for this level of polished work. So, instead of referring to Robin and Ellen's line, adjacent spray regions, and sums as a formal model, we view it more as the abstract representation of the pipe, nozzles, fields, and watering patterns whose manipulation constituted their understanding of the problem. This view is consistent with a semiotic view of mathematical modeling because it emphasizes modeling as a way of thinking about a problem, rather than emphasizing some objective structure. (In formal algebraic terms the model might be: $s = d$ and $n = l/d$, where s, d, n, and l are respectively nozzle separation, diameter of spray, number of nozzles, and length of pipe.)

This view makes further sense if we recall that any set of signs that might constitute a formal model could signify, via various interpretants, very different things to different people, or the signs could fail to signify anything at all to someone. A model is really only a model if one knows how to think with it. It is in this manner that Robin and Ellen had a model that allowed them to make sense of the Linear Irrigation problem. They resolved the problem of how many nozzles to put where by focusing on the goal of simply getting the whole field wet. This focus stemmed from their interpreting the problem in a numerical sense rather than in a geometric sense. We believe this focus resulted from interpretant formation and inference making with a strong iconic mode to them, and a lack of familiarity with symbolic sign systems for working with the concept of uniformity in this irrigation context.

Making Sense of Unfamiliar Signs: Competing Sign Systems

Their abrupt solution of the problem, albeit of a much simplified interpretation, was surprising. Largely because they had ignored the issue of uniformity, they were asked how uniformly the crops would be watered with their placement of nozzles. This simple prompt stimulated almost two hours of additional work on part one of the problem.

The next parts of the protocol document their initial attempts to come to terms with the movement of the pipe and the behavior of overlapping sprays. During this period and for some time afterwards, they are searching for a way to represent and think about this behavior. In general, their work for the next hour (context boxes #4, #5, and #7 of Fig. 5.6) is largely abductive as they generate many hypotheses and are casting about for any kind of mental toehold on the problem. On the context protocol, this work ranges from the 8-minute mark to the 26-minute mark, beginning in context box #4 in Fig. 5. 6.

5. A SEMIOTIC LOOK AT MODELING BEHAVIOR 115

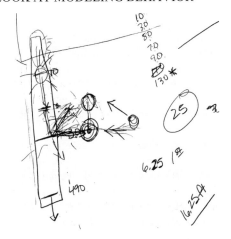

FIG. 5.7. A portion of Robin and Ellen's written work.

During this time, they were referring to the diagrams reproduced in Fig. 5.7 and 5.8. Their primary concern was about the amount of water that two different spots on the ground would receive as the pipe with nozzles on it, depicted at the left edge of Fig. 5.7, moved to the right on their paper. The "10" near the pipe represents the location of a nozzle (the circle to the left of the "10") and the "X" below it represents the edge of its spray. Below this "X" is the region of spray belonging to the second nozzle whose location is indicated by the dark horizontal line drawn across the pipe. To the right of this line are two sets of multiple lines, one set drawn to the right and the other set at an upward angle. These sets of lines indicate the spray from the second nozzle. The two spots on the ground, of concern to Robin and Ellen, are represented by the circles at the ends of these sets of lines. The third circle, further to the right, represents the second nozzle after it has moved past the spots on the ground.

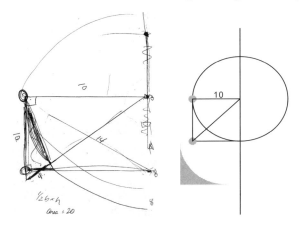

FIG. 5. 8. At left is a portion of Robin and Ellen's written work; at right a clarification of it made after they completed work on the problem.

They drew the larger diagram in Fig. 5.8 to help them think about the two spots on the ground, one in front of a nozzle, and one at the edge of the nozzle's spray. At right in Fig. 5.8 is a more accurate drawing rendering of the situation they were considering. Their demarcation of the area of interest (their triangle at the left of their figure) is inaccurate, yet their result of 20 is very close to the area of the region they meant to be finding (shaded region at the lower left of the drawing on the right), approximately 21.46.

The work associated with Fig. 5.8 is an example of signs signifying for the wrong reasons, because on a closer level of inspection the signs only stand for what was intended via mistakes that cancel one another. This misalignment could lead to subsequent confusion. The source of the misalignment is their drawing's lack of consistent scale (notice the two disparate lengths each marked 10). Later they use a ruler and compass to make accurate and useful diagrams.

In statement 106 (abbreviated ST 106), Ellen comes to understand Robin's concern about spots on the ground receiving different amounts of water. They reached this understanding by working with rough sketches and imagining movement of parts of the sketches relative to others. Following their return to the problem in ST 70 and through this point in their work, several important and complex ideas have appeared very quickly, and yet another appears in ST 120.

This phase of their work (Box #5, Fig. 5.6) is profuse with many signs that are competing for their attention. A partial list includes: a moving pipe with nozzles, different spots on the ground getting different amounts of water, simply getting the entire field wet (their earlier goal), the reach of the spray, the number of nozzles available and needed to span the 500-foot pipe, and the idea of overlapping spray patterns. Because these signs are not linked together in a familiar sign system, Robin and Ellen have to develop such a system. Most of these signs have quite complex objects and are more indexical and symbolic in nature than the earlier signs they used. This shift in the nature of the signs could be responsible in part for their difficulty in making sense of the situation—especially if they are more comfortable working in a more iconic mode. This preference manifests itself even more clearly toward the end of their session.

Furthermore, they are trying to develop a meaningful sign system with very little concrete realistic experience to draw upon (despite their claim of familiarity with irrigation systems). On another level, note that their familiarity with irrigation systems might itself be largely iconic. All of their comments to the effect of "I'm just trying to think," "I am thinking," "I think I'm getting it," and "I don't know" are telling of their struggle to find a way to think about a strange situation. They are very much in the process of forming a representation and associated way of thinking to help them make sense of the ideas involved. The most important development during this episode is their ability to distinguish different spots on the ground. By focusing more precisely on different spots on the ground, they are creating the potential to consider uniformity in terms of these locations, rather than in terms of a strip as a whole. Despite their progress in making sense of the problem, a clear representation and

5. A SEMIOTIC LOOK AT MODELING BEHAVIOR

a more stable, structured way of thinking about the problem do not emerge until the 54-minute mark, in ST 446.

Dominating Sign Systems: Runaway Semiosis or Planned Exploration?

The next segment of transcript displays a common phenomenon we observed in almost every session with every pair of students. It is related to the concept of competing sign systems and is an example of a familiar sign system quickly stimulating an episode of semiosis to the exclusion of other possible episodes. Often these episodes were initiated by noting familiar signs and were not part of an organized plan. Once again, and not only in the case of Robin and Ellen, these momentarily captivating episodes seem most often to follow a response (interpretant) to signs playing a largely iconic role. These episodes served two functions, exploring what a particular representation and its associated ways of thinking might afford, and providing at least some psychological comfort by providing the pair with familiar patterns of semiosis to complete. These functions were served to varying productive degrees in different episodes.

In the following episode, a triangle briefly took over Robin and Ellen's cognitive focus. In their case, noting a triangle and computing its area might have set the stage for an important later move on their part to compare various areas. It is also important because it returned them to an iconic mode of sign manipulation, a more comfortable mode for them that was to dominate the rest of their work.

120 R: Now I am trying to think. [Ignoring me, turning to Ellen.] Okay, we've almost got a triangle here.
121 E: I know, I see that too.
122 R: Okay. Well let's, wait. If this is 10 feet, the distance that I said that was going to get the least watered because of this, is going to be exactly in between the two nozzles, right? So that would mean this would be 10 feet.
123 E: Un huh.
124 R: Does that help any (chuckle)? I'm probably just making it more complicated. I'm sorry.
125 E: Okay (chuckle).
126 R: I know. (inaudible) A square + B square = C square. I'm sorry, I (inaudible).
127 E: Well this is a right [triangle]. If this [pointing to an angle] is right, because this one [pointing to the hypotenuse] is always longer than [pointing to the legs].
128 R: I know, but it shouldn't be that much longer.
129 E: I don't know (chuckle). You're right, it shouldn't be.
130 R: Okay.
131 E: No wait. Give me a calculator. Okay, it would be 100 + 100 is 200, squared of 200. [probably "square root"]
132 R: Oh yeah, I was thinking. (Inaudible - both speaking).

133 E: Right. Alright. Where's the squared? [again, probably "square root"]
134 R: Right here.
135 E: Where?
136 R: It's like you have to hit, I haven't used this thing since like, yeah I think. Yeah. I think, I don't know. Try it. Yeah, that looks good.
137 E: 14. (Laughter)
138 R: Okay, where does that get us though?
139 E: I don't know. I don't know. Completes our triangle. [excited satisfied inflection]
140 R: Okay.
141 E: Well there is a triangle here, that isn't. If we cut down, because even if we have. See, like there is almost a triangle here too with this.

Foreshadowing of a Well-Defined Goal: Transient Signs

In ST 148 (21-minute mark) we see foreshadowing of what becomes their final way of making sense of the problem and obtaining a solution. They continue to wrestle with the idea of comparing areas for a long time before finally developing a clear representation accompanied by ways of thinking about it and a specific goal in ST 446. Sometimes, as they worked, pairs adopted, adapted, or generated a useful sign system almost instantaneously for interpreting novel signs. Robin and Ellen's process of articulating a specific way of thinking about the Linear Irrigation Problem took a long time and never acquired the detail or comprehensiveness needed for a deeper understanding of the problem.

The good ideas they have had so far do not lead very far because they don't have a stable well-defined system relating and preserving them. As a result, many of the signs and inferences made in context box #5 (Figure 6) are transient ones. Over time and with repeated occurrences, some of these transient ideas persist and become enduring signs more readily available for future semiosis. We will see that their persistence depends on meaningful connections to other familiar concepts and on better representations.

Refining Representations and Ways of Thinking: Recurrence

The recurrence of the idea of comparing two areas can be traced through STs 148, 245, 318, 336, 394, and 432, and finally culminates, in ST 446, in the development of a good representation of the problem, more precise concepts related to the idea, and ways of working with the representation. ST 446 marks a pivotal point in their work, because from then on their work is much more constrained and consistently aimed at solving the problem as they decided to specify it. The path by which they arrived at this pivotal point in their work is long and circuitous. This segment of their work spans the period from the 21-minute mark to the 56-minute mark in the context protocol (Fig. 5.6). In general, it is best characterized by a refinement of the ideas and representations

5. A SEMIOTIC LOOK AT MODELING BEHAVIOR 119

that have surfaced so far in their work. Below we note only the highlights along their path.

After first raising the possibility, in ST 148, of comparing the area being watered to the area that is not being watered when the pipe is stationary, Robin and Ellen use a compass and ruler to draw a more careful representation of overlapping spray patterns. In working with this diagram, the comparison-of-areas idea recurs, and the clarity of the diagram allows them to confirm previous tentative ideas about how the water is distributed to different parts of the ground.

This more careful diagram allows them to reject both a nozzle separation of 20 feet and one of 11 feet as being too disparate in terms of water distribution as measured by areas receiving twice as much as others. The next highlight is a potentially revealing inductive inference made in ST 320. It is preceded by another recurrence of the comparison-of-areas idea.

In ST 320, an explicit awareness of the motion of the spray pattern past a spot on the ground is juxtaposed with their attention to the difference between different areas' positions relative to the region of overlapping sprays. This juxtaposition sets up the abductive inference in ST 432 that leads to their breakthrough in ST 446.

However, they never thought about the idea of looking at chord lengths. We expected it to emerge following STs 318 and 320. Much later, even when they drew a chord down the middle of an area of overlap, it failed to signify anything meaningful in terms of the amount of water different points on the ground would receive. This failure is due to their viewing the line segment only in terms of a context of a triangle and an area computation. Here we see a narrowness of focus constrained by a familiar sign system and the lack of a metacognitive move to consider what else that line segment might represent. We can only conjecture that their past experiences with chords are too faint to be meaningfully shaping current interpretants. Although, it is not only through the formal geometric concept of a chord that a less formal idea of a line segment connecting two points on a circle could be had; thinking about the rows of crops under a nozzle's spray is a common way that, in the first author's experience, many students spontaneously generate it.

In ST 336 Robin and Ellen simultaneously make clear analogical inferences to view the areas of concern in terms of a recently studied problem from their mathematics methods course. This analogical link served to more firmly establish what was to become their final way of making sense of the problem. Although they still encountered difficulties in solving the problem, they viewed weaknesses in their approach as being acceptable compromises, not fatal flaws.

From this point in the session forward, they spend most of their time trying different ways to calculate the areas of the spaces between the circles. Finally they settle for using graph paper and estimating the area by counting squares. The final revision of their thinking begins with Robin's request for scissors and her cutting out two blue paper circles. Robin and Ellen spend the next 5 minutes revisiting most of the ideas they have had thus far with the aid of the circular manipulatives. Some of their insights during this time are valuable and others are not. The next excerpt includes another crucial abduction in ST 432. It also includes Robin and Ellen's statement of the conditions, in terms of their representation of overlapping and non-covered areas, which resolves the problem for them. After ST 446, the vast bulk of their work is mathematical problem solving all done in terms of the representation they have settled on by this point in time. The challenge for them is calculating various areas. This work is guided largely by a guess-and-check heuristic that entails comparing areas for various nozzle separations before settling on a solution of 16.25 feet.

Their final goal becomes one of choosing a nozzle separation that makes the striped area equal to the sum of the areas of the shaded regions in the diagram at left (Fig. 5.9). Later, they improve this goal, in the context of their way of thinking about the problem, to be equalizing the shaded and striped areas in the diagram at right (Fig. 5.9).

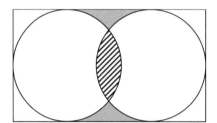 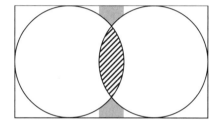

FIG. 5. 9. Robin and Lisa's goal: Equalizing areas.

Manipulatives: Facilitating an Indexical Mode of Thinking

In ST 442 is a clear statement of Robin's satisfaction in being able to manipulate the paper cut-outs representing two circular spray regions. Robin and Ellen were very attentive to the disks as they moved them across their paper, moving them closer together and farther apart. Watching them caused the first author to wonder how their thinking about the problem might be changing. It also stimulated an interest on our part in how to account semiotically for their use of the paper disks.

Although their decision to use an area of overlap to compensate for an area initially receiving no water solidified during their use of the paper disks, we can not be sure they would not have reached this conclusion in the absence of the disks. Other people have reached the same conclusion without the use of manipulative disks. In Robin and Ellen's case, their thinking with the disks took on a strong indexical mode. They were very intent upon how the movement of the disks caused different areas to be watered. By observing the moving disks, they refined their thinking about the problem and used inferences made via the moving disks to structure their subsequent work.

The physical movement of the paper disks itself was intimately connected to their thinking and represented a coupling of their thought to the physical signs they were trying to interpret. This activity exemplifies what Cunningham (personal communication, February 1997) has described as a distributed sense of self. In this view, Robin and Ellen's thinking is not restricted to their brains—its domain has been enlarged to include the interaction with the paper. This episode raises the more global question of where does a person's thinking end? At their brainstem? At their fingertips? At the edge of the paper circles? What we are seeing is how cognition, or semiosis, is strongly coupled to the environment the person is working in via sign systems.

The physical movement of the paper immediately constrained what interpretants could emerge in Robin and Ellen. At the same time, how they chose to move the paper constrained what could be cast into relief. This is a physical instantiation of the mental manipulation of ideas (signs of the nonphysical variety) that we call thinking. In Robin and Ellen's case, this manipulation, physical and mental, was too often not varied enough to stimulate other representations of the problem. Specifically, the absence of any signs of lines along which the circles moved or above which the circles passed (rows of crops) resulted in an impoverished sign system and final representation or model for this problem.

Final Representation: A Locally Coherent Sign System

Robin and Ellen's final representation for and way of thinking about the problem depend upon the significance they have attached to the idea of "making up for something" as it applies to areas that are going to be either watered singly or doubly. There is still unresolved tension in their approach, but it is a tension they have not been able to release and is something with which they have decided to live. The tension involves the lingering notion that everything depends on how long a spot on the ground will be under a nozzle's spray. Because they have been unable to relate this idea to their emerging sign system, the idea is transient and able to be ignored.

They arrive at their solution to the problem via a representation that has meaning for them in terms of whether the area of an overlapping region of spray is sufficient to compensate for a region that initially receives no water at the time other regions are receiving water from one nozzle. Robin and Ellen have approached the problem of uniformity on a largely iconic level composed of two different areas that are brought into congruence by an appropriate nozzle separation. This resolution makes only superficial use of the temporal aspect of the problem and is lacking comprehensiveness in terms of how water accumulates at different points on the ground. They did not pursue inductive confirmation of the consequences of this final approach to the problem largely because they could not think, during the modeling session, of any alternatives.

CONCLUSION

We have demonstrated how a semiotic framework can be used to attend to, describe, and reflect on the mathematical-modeling activity of one pair of undergraduates as they worked on one problem. If we have succeeded, you have noticed both common conceptual constructs of mathematical problem solving recast in semiotic terms (e.g., metacognition), and new constructs previously absent from studies of mathematical thinking (e.g., interplay of sign and inference types). In the space of this chapter, and within the confines of the case of Robin and Ellen, we can only hint at the potential of this semiotic perspective on mathematical modeling. In concluding, we draw upon data and insights gathered from across multiple cases of pairs of college students working on multiple mathematical modeling problems. Our primary purpose is to inspire others to consider researching, and articulating the pedagogical value of this semiotic perspective.

We have come to describe what seems to be desirable in modeling activity as a reflexive quality of attention. By way of clarification, we are not talking about attention from a deficit perspective—either it is, or is not present—as this perspective is problematic from a semiotic view for reasons we do not go into here. The reflexive quality of attention exhibited by the more successful modelers is distinguished, in part, by several characteristics that teachers might profitably cast into relief as their students engage in modeling activity:

- tolerance for, and interest in ambiguity and work across multiple sign systems, and among multiple levels of significance
- deliberate transition from iconic to symbolic modes of inference making
- intentional juxtapositioning of familiar and unfamiliar signs
- fluid merging of metacognitive shaping with primary cognition
- a playful disposition toward work with representations

In our continuing research, what is proving most valuable is the extension and refinement, afforded by the semiotic framework presented here, of teachers' abilities to name salient aspects of their own, and their students' cognizing during modeling (and other mathematical) tasks. In a satisfying circular way, given Peirce's own definition of a sign quoted earlier, naming a thing is the first step to perceiving, and analyzing it. It strikes us that much of what is required for expertise in a given domain often goes unnamed. Thus, when a preservice student of teaching in a class of the first author's exclaimed, "What is it I'm doing when I'm not deducing?," he saw it as a step toward an enriched ability to attend to her own cognition. And as this book undoubtedly makes clear mathematical modeling activity can not even get started without modes of inference making besides deduction.

ACKNOWLEDGMENTS

The research and perspectives discussed in this chapter are based in large part on the first author's doctoral dissertation completed under the direction of the second author in December 1998 (Kehle, 1998).

II
A MODELS AND MODELING PERSPECTIVE ON TEACHER DEVELOPMENT

Chapter 6

A Modeling Perspective on Teacher Development

Helen M. Doerr
Syracuse University

Richard Lesh
Purdue University

In this chapter, we discuss how the principles and theory that are the foundation for understanding the nature and development of student knowledge can provide a foundation for understanding the nature and development of teacher knowledge. Much of the early research on teachers' knowledge focused on what it is that teachers did and how particular teachers' actions affected student learning. Hidden in this emphasis are two flawed assumptions that have undergirded research on the development of teachers' knowledge: (a) that learning to teach is a matter of learning how to do what effective teachers do, and (b) that there is some generally unspecified image of a good teacher that novices should strive to emulate in action. This conceptualization of teachers' knowledge development as learning how and what it is that teachers do is problematic at several levels. Among other things, it focuses preservice teacher education and continued professional development on the pragmatics of skills, procedures, guidelines and rules of thumb that can be used in the classroom. As Lampert and Ball (1998) and others have noted, preservice teachers come to their teacher education programs with years of experience in observing what teachers do and this experience is, in fact, an obstacle for novice teachers to overcome in developing their own effective practice. The crux of the difficulty is not whether or not these preservice teachers have observed poor teaching. The difficulty is that the years of observing what it is that teachers do seldom provide preservice teachers with insight into how it is that teachers think or even what aspects of the teaching and learning process that teachers are thinking about.

Just as children, a central point for teachers' knowledge is that teaching mathematics is much more about seeing and interpreting the tasks of teaching than it is about doing them. Indeed a distinguishing characteristic of excellent teaching is reflected in the richness of the ways in which the teachers sees and interprets her practice not just in the actions that she takes. It is precisely a teacher's interpretations of a situation that influence when and why as well as what it is that the teacher does. The essence of the development of teachers' knowledge, therefore, is in the creation and continued refinement of sophisticated models or ways of interpreting the situations of teaching, learning and problem solving. These situations include such tasks as knowing how students learn important mathematical ideas about numbers, ratios, and functions, identifying what students already know, selecting appropriate activities and curricular materials that support the further development of students' ideas, and assessing the effectiveness of particular lessons. A models and modeling perspective on teachers' development brings to the foreground a focus on the ways teachers interpret their practice.

This then defines our work in teachers' professional development in terms of supporting teachers in their development of increasingly useful interpretations of situations. Our efforts focus on designing effective and sophisticated ways of helping teachers see and interpret children's thinking, the development of that thinking, and the numerous other aspects of practice.

> For example, the use of video cases in teacher education often focuses attention on what it is that the teacher does in a particular situation, but not on how the teacher thought about the context, what alternatives she considered, what purposes she had in mind or what elements of the situation she attended to and the meaning of those elements.

It is precisely the ways of seeing and interpreting situations for particular purposes that are the models that teachers create, extend, refine and apply in the context of classrooms. An understanding of teachers' knowledge needs to account for the varying contexts and purposes of teaching, the strengths and weaknesses of teachers' interpretations of those contexts and purposes, and the continued refinement of the ways of thinking that teachers bring to the classroom.

We begin by making some observations on the notion of expertise in teaching and its development, followed by our assumptions about the complexities of teaching. We then discuss the nature of teacher knowledge and our arguments about its development, drawing on Dienes' principles for instruction and the model development principles for student learning described in the opening chapters. Analogs to these principles are put forward to describe the nature and the development of teacher knowledge. Finally, we elaborate on the events and tasks which tend to foster the development of models and suggest some of the practical implications that this has for professional development of teachers.

The chapters that follow elaborate several designs for teacher professional development that draw on these principles for model development. In particular, several authors address the challenges in the design of professional development activities, the ways in which teachers express their thinking about student learning, the role of the teacher in facilitating the development of student thinking, and the role that assessment tasks can play in eliciting change in teachers' instructional activities. In the concluding section of this chapter, we identify some of the limitations of our current work as well as several areas that are in need of further research and development.

EXPERTISE IN TEACHING

As we have argued, knowing how to teach means knowing how to see and interpret a complex and ill-structured domain. Teacher knowledge is clearly not reducible to a list of skills and procedures that emphasize what it is that a good teacher does in a particular situation. The nature of teacher knowledge is much more about how teachers interpret the complexity and the situated variability of the practical problems of the classroom, how those interpretations evolve over time and across settings, and how and when those interpretations influence decisions and actions in the classroom. As we have stated earlier, it is not enough to see what a teacher does, we need to understand how and why the teacher was thinking in a given situation, that is, interpreting the salient features of the event, integrating them with past experiences, and anticipating actions, consequences, and subsequent interpretations.

These characteristics of expertise in a complex and ill-structured knowledge domain such as teaching suggest that there is not a single image of a good teacher nor is there a single trajectory in becoming a good teacher. Rather, expertise in teaching varies considerably across individuals and across settings as well as within a given individual and within a particular context. Despite that variability (and, as we discuss later, because of that variability), the knowledge base and expertise of teachers can (and does) develop in ways that can be seen as continually better, without necessarily having some particular, fixed endpoint in mind.

Because the image of good teaching and of teachers' knowledge is not a single, uniform, global image, the professional development of teachers should not be construed as the convergence of teachers and teaching toward some predetermined standard of excellence. This is not to suggest that we are rejecting notions of accountability or assessment of good teaching. On the contrary, it is suggesting that teachers need to be viewed as evolving experts. We need to ask how they are improving along the multiple dimensions of expertise in varying settings and contexts. We can examine the nature and the growth of teachers' knowledge by looking at ways in which it is continually improving in light of particular purposes and contexts.

THE NATURE OF TEACHERS' KNOWLEDGE

Over the past 20 years, substantial progress has been made in children's learning in those areas where we have had extensive, sustained, and successful research on children's knowledge in such areas as number sense, rational numbers, and functions. Numerous researchers have investigated both what it is that children know in each of these domains and how it is that their ideas develop. This research led to the development of such important documents as the National Council of Teachers of Mathematics principles and curriculum standards (1989, 2000), the creation and subsequent implementation of substantially new curricular materials (e.g., Investigations in Number, Data, and Space, ARISE, the Connected Mathematics Project, and others now commercially available), and the development of innovative teacher development programs to support teachers in implementing curricula and pedagogies which focused on the student's conceptual understandings (e.g. Carpenter, Fennema, Peterson, Chiang, & Loef, 1989, Schifter & Fosnot, 1993; Simon, 1995; and others). However, these programs for teachers' professional development have not been grounded in a similarly extensive research base on the nature of teachers' knowledge and its development; this research base (by comparison to the research base on children's knowledge) is still in its infancy. For this reason, we are arguing that for significant progress to be made in teaching, a theoretical understanding of the nature of teachers' knowledge is needed as the foundation for practical work that supports the development of that knowledge.

Recent developments in cognitive science and situated learning suggest that knowledge is situated and grounded in the contexts and constraints of practice (e.g., Borko, Mayfield, Marion, Flexer, & Cumbo, 1997; Lave & Wenger, 1991; Leinhardt, 1990, and others). Other researchers have emphasized that teaching is an ill-structured domain and that expertise in such a domain requires the flexible use of cognitive structures to accommodate partial information, changing or unclear goals, multiple perspectives, and uncertain consequences (Feltovich, Spiro, & Coulson, 1997; Spiro, Coulson, Feltovich, & Anderson, 1988). Together, these two perspectives bring to the foreground that experts' knowledge is pluralistic, multidimensional, variable, contextualized and continual. Teacher education research has tended to capture this situated variability in an emphasis on teachers' beliefs or belief systems which serve as filters for teachers' knowledge and account for the variability in teachers' practices and for the resistance to changes in practice, but not for the often found disconnect between professed beliefs and actual practice. The nature of teachers' knowledge is not a uniform, consistent or fixed set of constructs, but rather it is characterized by the following kinds of complexity and variability:

- *Knowledge is pluralistic.* Several correct models may be possible for interpreting any given situation. For example, in the O'Hare Airport problem, which is described in chapter 19, several models were possible to interpret, explain, or make predictions in that problem situation. Some of these models (across student groups) were contradictory whereas others had significant overlap; earlier models (within student groups) differed substantially from the later models. This multiple or competing models view of knowledge assumes that knowledge exists in pieces (diSessa, 1988) and that a number of reasoning patterns may be potentially available to fit a given situation. The particular choice of reasoning pattern is activated by salient details, global features, perceived structures, regularities or both. Multiple and overlapping lines of reasoning may be needed to adequately interpret or explain a given situation. A given individual may have multiple ways of thinking

6. MODELING PERSPECTIVE ON TEACHER DEVELOPMENT 129

about a given situation and multiple individuals may have multiple ways of thinking about the situation.

- *Knowledge is multidimensional.* Foundational models tend to be organized around situations and experiences as well as around abstract, conceptual categories. Real world experience often informs the principles on which important models are based, providing similar cases that define the scope of validity of a given model. Meanings, descriptions and explanations evolve along a number of dimensions such as concrete-to-abstract, simple-to-complex, external-to-internal, sequential-to-simultaneous, discrete-to-continuous, particular-to-general, and static-to-dynamic. Models are the systems that teachers use to make sense of multiple aspects of experiences along multiple dimensions.
- *Knowledge is variable.* The knowledge that an individual brings to bear in a given situation can vary considerably along the dimensions outlined. Any single individual may be thinking at various places along any of these dimensions. Depending on the particular purposes and goals at hand, a teacher may think in terms of the particulars or in terms of generalized strategies. With only partial information available, initial interpretations of a problem situation may be simple, static, and discrete. Two important implications of this variability are: (a) the recognition that in a complex domain (such as teaching) no single individual performs in all contexts at some uniformly high level of performance along all dimensions, and (b) that the variations within and among individuals can be an important source of dissonance that move the models or systems of interpretations in directions which are continually more refined and more generalized.
- *Knowledge is contextual.* Contrary to the assumption that tasks which are characterized by the same underlying structure should be comparable in difficulty, mathematics and science educators have found that a student's ability to use a given conceptual model often differs a great deal from one situation to another, depending on a variety of contextual factors and student characteristics (Carraher, Carraher, & Schliemann, 1985; Lave, 1988). In fact, in areas such as proportional reasoning, the level of reasoning often is not even consistent within a given task, if for example the number relationships or perceptual distractors are changed slightly (Lesh, Post, & Behr, 1989). Ideas such as proportional reasoning are first mastered in a small and restricted class of problems, and then competence gradually extends to larger classes of problem situations. Situations and real world experience inform the development of teachers' mathematical knowledge (about proportional reasoning and other mathematically significant conceptual systems such as functions or proof) as well as their knowledge about how children learn these mathematical ideas and about ways to teach these ideas.
- *Knowledge is continual.* Conceptual development is characterized more by gradually increasing competence and less by the acquisition of some

general, all-purpose cognitive structure. Researchers such as Sfard (1998) have drawn distinctions between those metaphors for knowledge, which emphasize the verb knowing and participation within a community and those metaphors for knowledge, which emphasize the noun knowledge and acquisition of structure and content. As Sfard suggests each perspective suggests important aspects of knowledge. For our purposes, these metaphors suggest that knowing can be seen as continually evolving as one participates in a community and that knowledge (except in the most trivial of cases) can be seen as gradually acquired over time.

One aspect of teachers' knowledge that has been the subject of much research is teachers' content (or subject matter) knowledge. Many of these studies have tended to have a deficiency point of view of teachers' knowledge. That is, teachers are found lacking in understanding of the number system or proportional reasoning, or are found to have impoverished understandings of complex mathematical notions such as functions. As Cooney (1999) has pointed out, it not entirely surprising that teachers' understanding of the content of school mathematics is limited, for often they have had no mathematical experiences or learning with that content since they themselves were school-aged children. As a result of these findings from research, teachers' professional development programs often focus on filling the gaps in teachers' content knowledge. From a modeling perspective, we want to shift the emphasis from repairing the deficiencies in teachers' mathematical knowledge to using what teachers already know to express, test, revise and refine their content knowledge and to extend that knowledge to more powerful forms for classroom teaching. This means, among other things, attending to the pedagogical dimension of their mathematical knowledge, as well as other equally important dimensions such as the psychological, the historical, the affective and the practical.

In Shulman's (1986) paper on the knowledge needed by those who teach, he put forward the notion of pedagogical content knowledge to fill the gap between knowledge of content and knowledge of pedagogy. According to Shulman, there are aspects of pedagogy which need to account for the specifics and particulars of a given discipline (such as mathematics). In other words, there is a pedagogical side to teachers' mathematical knowledge. Although this notion has certainly gained widespread acceptance in the research literature, the specific meaning of this concept has yet to be fully elaborated. One example of this meaning can be found in Yackel and Cobb's (1996) notion of socio-mathematical norms. These researchers have argued that mathematical argumentation and justification are not developed simply as social norms within a classroom, but rather they have a specifically mathematical characterization. What counts as a form of argument and the evidence used to support the argument is not the same in mathematics as it is in history, law, art, or science. Hence, a teacher needs to have a pedagogical understanding of how to develop and support the norms that govern participation in generating arguments that are mathematical in nature. The teacher's pedagogical knowledge about strategies for supporting classroom discussion is informed by the mathematical nature of the discussion. Despite the richness of this particular example, the nature and scope of teachers' pedagogical content knowledge has not yet been fully elaborated. Perhaps more importantly, how pedagogical content knowledge is developed needs to be documented and explained.

The focus on pedagogical content knowledge generally seems to emphasize how the knowledge of pedagogy needs to take into account aspects of the content domain. We wish to emphasize that to fully understand teachers' knowledge, we need to also focus on how the knowledge of content needs to take into account aspects of the pedagogical domain. That is, to understand the nature and development of teachers' knowledge, we need to examine not only the pedagogical side of mathematics, but also the mathematical side of pedagogy. One aspect of this emphasis can be found in the work of the Cognitively Guided Instruction (Fennema, Carpenter, Franke, Levi, Jacobs, & Empson, 1996) project and among others (e.g., Ball, 1993; Schifter & Fosnot, 1993; Simon & Schifter, 1991) who have focused both research and professional development efforts on helping teachers understand how children think about mathematical ideas and on how those ideas develop in children. However, simply understanding the ways in which a mathematical idea or a system of ideas (such as proportional reasoning or functions) may develop in children does not in and of itself translate into knowledge about how to teach in ways that support that development.

Teachers need to understand the landscape of children's conceptual development, which is not the same as understanding one way of thinking or a way of developing or a learning trajectory. Rather, teachers need to recognize that within a given classroom children can be engaged in multiple ways of interpreting a problem situation and have multiple paths for refining and revising their ideas. The task for the teacher, then, includes seeing the multiple ways that children might interpret a situation, understanding that their ideas might be revised along various dimensions (although not being tested or refined along other dimensions), and acting in ways that will support the children's development towards more refined, more generalized, more flexible, and more integrated ways of thinking. The knowledge that teachers need consists of at least the mathematical understanding of the idea, an understanding of how children's thinking might develop, and a knowledge of pedagogical strategies in relationship to both the mathematical development and the psychological development.

TEACHERS' MODELS

Teachers' models are systems of interpretation. These systems are the interpretative and explanatory frameworks that teachers use to see children's thinking, to respond in ways that further that thinking along multiple dimensions, to differentiate the nuances of particular contexts and situations, to see principles and more generalized understandings that cut across contexts and situations, and to support the continual revision of their own interpretations in light of evidence from experiences. Such systems are significantly more complex and broader in scope than the mathematical models which children create. Children's models are usually contained more narrowly within the domain of mathematics and with the setting of the school classroom. Teachers' models, on the other hand, include not only the complexity of multiple children's models, but also need to account for the nature of those models, for ways of fostering the development of those models, and for ways of communicating children's thinking to colleagues, parents, principals, and other stakeholders. Teachers need to know how mathematical ideas are developed in curricula, the strengths and weaknesses of various representational media in supporting the expression and development of mathematical structures and concepts, and aspects of the practical origins and current uses of mathematics. Furthermore, teachers' models are generally developed in situations where the available information is often partial or incomplete and where the consequences of actions are not always immediate and local but sometimes remote and global. This suggests that a key perspective for a view of any particular model or system of interpretation needs to hold in the foreground that the model is likely to change as additional information and new evidence leads to revisions.

Teachers' models are also more than systems of interpretation. Having an interpretation of a given situation is necessary but not sufficient for the interpretation to be considered a model. Like children's models, essential characteristics of teachers' models are that they are shareable and re-useable and that they serve some purpose. This allows us to distinguish, therefore, between an interpretation or way of thinking about a particular situation that is only a particular instance of the teacher's reflection on his or her work and a system of interpretation that can be shared with other teachers, used in other contexts and measured in its usefulness for some defined purpose. In subsequent chapters, other authors discuss the role of shared and reused tools created by teachers for purposes of observing and assessing student work. The tools and artifacts created by teachers reveal their interpretative systems and, at the same time, provide a starting point, which allows the usefulness of their tools to be evaluated by both teachers and researchers in ways which potentially support the development of more refined and integrated models of teaching. Teachers' models are shared and evolving systems of interpretation that are reused for some defined purpose.

A modeling perspective on teachers' knowledge allows us to go beyond the limits of constructs such as pedagogical content knowledge or knowledge of the development of children's ideas. Teachers' knowledge needs to include at least the knowledge of psychological development, logical (or mathematical) development, curricular development, representational (or media) development and practical development. In chapters 22, and 19 (this volume) others discuss in more detail some of the noncognitive factors related to the development of children's models, which in turn the teacher needs to know. It is in these ways that teachers' knowledge is significantly broader in scope and complexity than the knowledge of children's ideas and their development. In particular, teachers' knowledge needs to include an understanding of the multiplicity of children's models as they develop along many dimensions. At the same time, teachers' models are not single models that conform to some predetermined standard of excellence, but rather they too are models that develop along many dimensions. The difficulty posed in the introduction to this chapter addressed the limitations of focusing on what it is that teachers do and on assuming that there is a single good image of teaching. A modeling perspective focuses on how teachers' interpret the practice of teaching and on how teachers can become better, without having a single endpoint in mind for all teachers.

MODEL-ELICITING ACTIVITIES FOR TEACHERS

Similar to the principles put forth for the development of model-eliciting activities for children, the following principles are intended to guide the development of those activities for teachers that both elicit and foster the development of their models. Teachers come to teaching and learning with models (or systems of interpretation or ways of thinking) already in place. Many of these models are implicit and often not shared with colleagues or re-used in other situations. Other models would appear to be particularly resistive to change or untested against new situations. The overall goals of model-eliciting activities for teachers are: (a) to have teachers reveal their current ways of thinking; (b) to test, revise, and refine those ways of thinking for some particular purpose; (c) to share with colleagues for replication; and (d) to reuse their ways of thinking in multiple contexts. These principles are intended to guide the design of activities that would achieve those goals:

1. *Reality Principle.* The situations which we ask teachers to interpret must be in the context of their actual practice. Asking teachers to interpret student work from their own classroom, or to analyze student's thinking over time on a concept that they teach, or to develop an assessment task that would reveal how students are thinking would engage teachers in activities embedded in their own practice. Pulling teachers out of their classrooms to develop local curriculum standards would violate this principle.
2. *Multilevel Principle.* Tasks for teachers should address the multiple levels or aspects of the teaching and learning environment. Teachers most often need to simultaneously address mathematical content, pedagogical strategies, and psychological aspects of a teaching and learning situation. Simply addressing student thinking is not enough.
3. *Multiple Contexts Principle.* The variability in the settings, the students, and the mathematical contexts of teaching needs to be accounted for. Teachers' knowledge varies across contexts and the multiple dimensions of those contexts, so a model eliciting activity for teachers should include variations in context that require interpretation and analysis. This leads to thinking in ways that are increasingly generalizable.
4. *Sharing Principle.* Ideas about teaching and learning need to be shared among multiple teachers and reused by those other teachers. Particularly powerful tasks for teachers are those that come from other teachers and can be used by other teachers. This in turn leads to revision and refinement.
5. *Self-evaluation Principle.* Are the purposes against which success can be evaluated sufficiently clear? Fuzzy statements of educational goals can

preclude effective judgements about teachers' actions or interpretations. More importantly, teachers need to able to judge for themselves whether their interpretations and consequent actions (such as teaching plans or assessment strategies) are moving towards desired ends in particular contexts.

These principles are intended to guide the design of activities for teachers that become embedded in the context of their practice (or emerging practice, as is the case for preservice teachers) and that enable the continued development of systems of interpretation in ways that are continually better. Better, in this case, is determined by the "fit" with goals, purposes and contexts that are explicit and shared with colleagues and the larger community of schooling.

MODEL DEVELOPMENT FOR TEACHERS

The principles and theories that explain the development of student learning can be used to explain the development of teacher learning. Analogs of Dienes' multiple embodiment principles for instruction and the principles of model development for student learning can take into account the complex nature of teaching and suggest ways in which teachers can continually learn through their activity. These instructional design principles (discussed in an earlier chapter) emphasized how embodiments of concrete materials can provide the foundation for the development of powerful mathematical reasoning in children. Our central claim is that those same instructional principles that we want teachers to use when teaching their students can guide our own work in teaching teachers.

In elaborating these principles for teachers, we need to take into account two important facts: (a) the models (or systems of interpretation) that teachers need to develop are far more complex than students' mathematical models, and (b) teachers' models tend to be much more difficult to name or point to. That is, teachers' models are both inclusive of and more extensive than, say, students' models of "proportional reasoning." Teachers' models include an understanding of students' models, but also an understanding of the development of students' model, notions about the curricular development of the concept of proportional reasoning, and pedagogical strategies for teaching the concept in various settings to students with varying backgrounds. Hence, teachers' models are less likely to have simple names that encompass and convey the meaning of a significant portion of a teachers' knowledge about proportional reasoning. Nonetheless, Dienes's design principles for how concrete materials and multiple embodiments should be used in instruction for children would appear to have important analogs in the instruction of teachers.

The Construction Principle. Many recent curriculum development efforts have included conceptually rich concrete activities to teach children. But often the related materials intended to teach teachers rely extensively on language-heavy explanations. Unlike materials for children which emphasize deeper and higher order understandings, teacher materials often involve little more than lower-level skills or "how to's." Just as materials for children focus on developing ideas from concrete-to-abstract, from intuitions-to-formalizations, and from specifics-to-generalizations, materials for teachers should address these multiple dimensions of the teachers' conceptual growth. Teachers' professional development activities should provide teachers with concrete opportunities to explore mathematical meanings, interpret students' thinking, plan for instructional activities, select materials, structure activities and assess student performance. Teachers should be engaged in activities, based on their own classroom experiences with their own students, that support them in seeking patterns and regularities within their instructional frameworks and that focus on the relationships among the tasks of teaching and learning.

The Multiple Embodiment and Perceptual Variability Principles. Teachers' knowledge is variable and uneven along the multiple dimensions of knowledge discussed earlier in this chapter. Among any group of teachers, their knowledge covers overlapping aspects of mathematical, psychological, pedagogical, and curricular knowledge and experience. This provides an opportunity for different teachers to bring different perspectives to any teaching and learning situation. By juxtaposing multiple perspectives of particularly rich situations, teachers can notice more about the details of the situation, can use considerably different organizational schemes to think about the important patterns and regularities, and can act in the situation in different ways. As teachers encounter similarities and differences from one perspective or embodiment to another, they improve in their abilities to make useful and insightful interpretations of students' concrete learning experiences (Post, Behr, Lesh & Harel, 1988; Cramer & Lesh, 1988). The juxtaposition of multiple perspectives can shift the teacher's focus from the details to the big picture, from isolated elements in a situation to interacting relationships, or from particular events to generalized relationships.

The Dynamic Principle. This principle suggests that the knowledge that teachers need should move from understanding relationships that are static to those which are dynamic. Some of the most important mathematical ideas that children need to develop involve mappings and transformations of data sets (rather than isolated points of data) and interpreting change and variation. Similarly, teachers' knowledge needs to attend to the continuous change and development of children's thinking not simply to the state of that thinking at a discrete point in time. Teachers need to understand how individual children's thinking is supported and constrained by the activities of the classroom in which the children interact with tools, tasks, peers, and the teacher.

Taken together, these principles and the model eliciting principles (discussed earlier in this chapter) suggest the characteristics of the events and the tasks which can foster teachers' model development in directions that are continually better without the necessity of some uniform notion or universal description of best teaching.

EVENTS AND TASKS THAT FOSTER MODEL DEVELOPMENT

In describing a models and modeling perspective on the development of childrens' thinking, we have argued that learning is fostered by sequences of modeling activities that provide mathematically rich contexts that elicit significant models in the first place. These models can be explored for their own sake, applied to new contexts, and extended in ways that are increasingly generalized. Teachers' models are developed in similar ways. The kinds of events which drive the development of teachers' models (or ways of thinking about teaching) include: (a) perturbing existing models or eliciting new ways of thinking, (b) resolving mismatches within the model or with reality, (c) sharing in ways that allow others to replicate, and (d) sustaining the model across a broad set of contexts. The development which occurs results in models that are continuously better "fits" for the dynamics of practice among groups of practitioners in a diversity settings. Our efforts in devising effective ways of working with teachers have focused on the design and development of tasks, which are likely to create environments in which the aforementioned kinds of events can occur.

In almost all cases, we have found it necessary to create contexts in which teachers can express their current ways of thinking. Often, this involves putting teachers in practice-based situations where they need to interpret or assess students' thinking, to plan appropriate sequences of instructional activities in light of students' current understandings, or to select, modify or extend tasks based on the results of student model eliciting activities. All three of these situations (assessing, planning, and selecting) are deeply embedded in teachers' everyday practice. The more closely aligned these situations are with teachers' actual students (as opposed to hypothetical student learning), the more effective they are in revealing teachers' current ways of thinking. Establishing communities where teachers' interpretations can be seen from multiple perspectives (such as on the details or on the big picture) or addressed at multiple levels (such as the mathematical or the pedagogical) or for multiple purposes (such as for introducing a concept or extending it) provides the potential for disturbing current ways of thinking and for resolving mismatches between the interpretation and the experienced realities of other teachers. We have found the following kinds of teacher-level tasks useful:

- Develop a library of exemplary and illuminating results of students' responses to modeling activities.
- Design reliable procedures for assessing the strengths and weaknesses of students' work.
- Devise ways or create tools for helping students assess the quality of their own work.
- Refine strategies for making useful observations about students' works-in-progress.
- Create a "quick notes" sheet for other teachers to use with students' modeling activities.
- Modify or extend an instructional task in light of students' responses to a performance assessment activity.

Tasks such as these have the potential to elicit expressions from teachers of their current ways of thinking about the tasks of teaching and learning. The difficulty

6. MODELING PERSPECTIVE ON TEACHER DEVELOPMENT

of achieving this expression of current ways of thinking is substantial. As Stigler and Hiebert (1999) have observed, teaching is a cultural activity and, as such, it is difficult to gain the perspective necessary to express what is taken as implicit and obvious in the daily rhythm of teachers' activities. However, when tasks are embedded in current practice, it is possible for teachers to engage in testing, revising, and refining their ideas as they are used in their own classrooms. For example, in developing a library of exemplary and illuminating responses by students, teachers reveal how they are interpreting the mathematical content, the context, and the value of the results that students produce. As the teachers select, organize and compare student work, they reveal how they are seeing the students' mathematical ideas. This may lead to mismatches between their expectations of some students based on notions and perceptions of students' abilities. It may lead to seeing students give mathematical interpretations of problem situations that the teacher had not seen. It is the resolution of such mismatches that provided the impetus for the development of teachers' knowledge.

The notions of sharing and replicating systems of interpretation among colleagues and sustaining the meanings across multiple contexts can be illustrated by an example of tool design and development. One such tool might be procedures for assessing the relative strengths and weaknesses of students' work; another might be strategies for making observations about students' works-in-progress that support the students in making refinements and revisions that are improvements on their current interpretations. Tools, strategies and procedures have the characteristic that they can be shared with others and used in multiple settings. The fields of teaching and teacher education have been plagued by fads of reform which come and go without contributing to the body of teachers' knowledge in ways which allow for that knowledge to become continually better. By engaging teachers (and teacher educators) in the shared use of tools, strategies and procedures by multiple teachers in multiple contexts, the tools, strategies, and procedures can be revised and refined in ways which can become understood as generalized across contexts as well as tuned in specific ways to particular contexts. These generalized understandings are models (or systems of interpretation) that carry the kinds of variability and multidimensionality that we described earlier as essential characteristics of the nature of teachers' knowledge.

We have found, for example, that multimedia case studies of teaching are particularly effective in eliciting shared understandings of effective practice among preservice secondary teachers (Doerr, Masingila, & Teich, 2000). The multimedia cases of practice were intended to capture the records of practice of a middle school classroom taught by an experienced mathematics teacher. Unlike video cases, which focus on what it is that the teacher does in a particular situation, the multimedia cases explicitly included the lesson plans of the teacher and interviews with her before and after each lesson. The plans and the interviews revealed the teacher's intentions for the lessons and her reasons for taking particular actions and developing mathematical ideas in particular ways. The teacher discussed classroom issues that concerned her following each lesson, and in that way revealed how and what it is that the she was attending to in the classroom. In this way, the preservice teachers who are interpreting the teaching and learning that is taking place can move beyond a focus on what it is the teacher does to a focus on how and why the teacher is interpreting the teaching and learning situation. In this way, a case study carries the story of practice that holds images of both the actions of the teacher and the ways in which the teacher thought about her actions.

SOME IMPLICATIONS FOR TEACHERS' PROFESSIONAL DEVELOPMENT

A modeling perspective on the nature and development of teachers' knowledge has several significant implications for us in our continuing work as teacher educators. In designing rich environments for teachers' professional development, our emphasis is on providing teachers with more powerful ways of seeing and interpreting the elements of practice. Rather than telling teachers what to do or recommending courses of action, we have shifted our work towards ways that support teachers in their efforts to see and interpret the complexities of teaching and learning. This emphasis on interpretation includes not only interpreting the ways in which students might learn, but also interpreting the mathematics itself, and its pedagogical and logical development, the relevant curricular materials, and possible ways of proceeding with a sequence of learning activities.

The complexity of teachers' knowledge, and the variability in the contexts in which that knowledge is used, suggests that teachers need multiple models or a collection of cases, which carry the images of possible interpretations of situations for teaching and learning. Such stories of practice enable a practitioner to simultaneously see the multiple dimensions and variability in a particular context and to reason about those elements in ways that can be selected and tested in the teacher's own experiences. This provides a means for the teacher to observe and measure whether their interpretations are useful in particular contexts for particular purposes. Over time and with experience, a teachers' collection of cases should provide that teacher with multiple, flexible ways of reasoning about practice. The cases that teachers collect and the tools that they develop are means by which teachers themselves can move their own practices forward, when placed in particular contexts for specified purposes.

The modeling perspective we have taken allows us to separate statements about teachers from statements about teachers' knowledge. This shift is reflected in our ways of working with teachers themselves, where the emphasis is on the ways that teachers interpret situations and the meanings that they ascribe to the patterns of interactions, rather than on the actions taken in a situation. This shift provides a new potential for a language for professional development and professional dialogue that moves away from evaluative statements about teachers towards descriptive and measurable statements about teachers' thinking.

Finally, an important implication of a theoretical grounding in a modeling perspective is that efforts for teachers' professional development must attend to the variability in teachers' knowledge and in its growth. We have seen that a

6. MODELING PERSPECTIVE ON TEACHER DEVELOPMENT

modeling perspective on children's thinking and its development forefronts the variability in the abilities, perceptions, and interpretations that children bring to mathematically rich contexts. Similarly, teachers bring variability in their abilities, perceptions and interpretations of rich contexts for teaching and learning. Not all teachers begin at the same "starting point" nor do they follow some predetermined set of paths for learning to interpret and analyze the complex and ill-structured domain of practice. Rather, teachers vary in their skills, knowledge and growth in ways that are contextualized, multidimensional and multilevel. Hence, programs for teachers' development need to provide opportunities for teachers' learning that address this variability and its growth.

CONCLUSIONS

We began this chapter by arguing that the theoretical perspectives that are the grounding for how it is that children learn mathematics can provide the same grounding for understanding the nature of teachers' knowledge and its development. Just as for children, where the critical feature of their learning is how they interpret or mathematize experienced phenomena, the critical feature of teachers' knowledge is how they see and interpret the teaching and learning of children. Teachers' models, however, are characterized by a complexity that exceeds that which we usually associate with children's mathematical models. We have developed principles and guidelines for the tasks and events which can foster the development of teachers' models. In so doing, we have found it necessary to go beyond traditional constructivist philosophy to focus on a modeling theory which has focused greater attention on the nature of the conceptual models that we want teachers to acquire and has made analogies as clear as possible between model eliciting activities for children and model eliciting activities for teachers. Much practical work remains to be done that meets the substantial challenges of finding effective ways of revealing teachers thinking so that it can be revised, refined, shared and sustained in ways that will contribute to a growing body of knowledge about what it is that teachers need to know.

Chapter 7

A Modeling Approach for Providing Teacher Development

Roberta Y. Schorr
Rutgers University–Newark

Richard Lesh
Purdue University

Over the last several years, many professional development projects have been implemented intending to help teachers build new ways of thinking about their mathematics teaching. The goal of these projects, generally speaking, is to help teachers move away from more traditional instructional approaches which emphasize memorization and the execution of rules and procedures, and move toward instructional practices, which provide students with the opportunity to build concepts and ideas as they are engaged in meaningful mathematical activities (NCTM, 1989, 2000). Moving away from teaching techniques that simply enable students to repeat, often without much understanding, various specific algorithms, rules, or procedures toward helping them develop deeper and more powerful understandings and abilities, is not an easy task. It "would require vast changes in what most teachers know and believe" (Cohen & Barnes, 1993, p. 246). "Teachers who take this path must have unusual knowledge and skills. They must be able to comprehend students' thinking, their interpretations of problems, their mistakes they must have the capacity to probe thoughtfully and tactfully. These and other capacities would not be needed if teachers relied on texts and worksheets" (Cohen, 1988, p. 75).

To help teachers make the transition toward teaching for deeper and more powerful understandings, many teacher development projects have emphasized specific strategies, processes, or beliefs that teachers should have or use. These can include the behavioral objectives, process objectives, and affective objectives of instruction. However, projects that emphasize these objectives often fall short of the intended goal—increased student learning because these objectives, taken alone or in combination, fail to include the cognitive objectives of instruction, where the emphasis is on generalizations and higher order understandings—both for teachers and for students. Some projects emphasize, for instance, the use of concrete manipulatives, different representational systems, and student collaboration. However, in many cases, even when teachers do make specific changes in teaching practices to include the above aforementioned ideas their actual teaching abilities and knowledge do not change. For example, Spillane and Zeuli (1999) note that many of the teachers involved in a study exploring patterns of practice in the context of national and state mathematics reform:

> used a variety of concrete materials to teach mathematics. Pictorial representations were also evident. Further, these teachers also used a combination of whole-class, small-group, and individual instruction. The conception of mathematical knowledge that dominated the tasks and discourse in these teachers classrooms however, did not] suggest to students that knowing and doing mathematics involved anything more than memorizing procedures and using them to compute right answers (p. 16).

In another large scale study involving fourth grade teachers, many of whom are involved in ongoing long term teacher development projects, Schorr, Firestone, and Monfils (in progress) have preliminary analyses which suggests that teachers are adopting surface features of standards-based reform without substantially increasing the intellectual demands of their teaching. Many of the teachers involved in this study are quite aware of the pedagogical practices emphasized in state and national standards, and have indeed developed more productive attitudes about mathematics teaching and learning, however, actual classroom practice appears to remain quite traditional. Worse yet, many new or standards-based approaches may be misapplied or misinterpreted. Spillane and Zeuli (1999) cite that teachers in their study often used phrases such as "hands-on activities," "conceptual understanding," "multiple representations," "teaching problem solving," "real-life problems," "student understanding," "why questions," and 'mathematical connections' to describe their mathematics teaching, but indeed actual classroom practice revealed that although the teacher's words appeared to be reform oriented, actual implementation was not. The following excerpt taken from the Spillane and Zeuli study will illustrate this point:

> Ms. Townsend stated that her students "do a lot of discussion" because it gives them the ability to question and explain [their thinking]." What was striking, however, was when Ms. Townsend referred to how discussion was instantiated in her teaching, she gave an example that was not connected to principled mathematical ideas. Her example concerned the rules for a board game students were designing: "I was trying to think of an example. They had a [board game] rule that if a student did something, then you lose three turns. And I said, "how does your partner move when they get those extra turns? they hadn't thought of that part yet." (p. 12)

7. MODELING FOR TEACHER DEVELOPMENT

There are two main points that can be inferred from this:

- Many teachers, who are very good at using the words of reform, often teach in ways that do not encourage conceptual understanding.
- It is not enough to simply observe a behavior, we also need to gain insight into why a teacher did what they did, and how they determined when to do it. This helps us to understand what teachers can do in specific situations as well as in a variety of situations.

The preceding two comments are meant to suggest that we must simultaneously focus on what people are actually doing at the same time as we focus on how they are thinking about what they are doing. This may appear to present a dilemma for teacher educators. How can we get out of this dilemma? One way to resolve this is to focus on thought-revealing activities for teachers in which they develop conceptual tools where they express their ways of thinking in the form of artifacts that are: purposeful (so that they can be tested and revised or refined in directions that are more effective); and sharable and reusable (so that they reflect general not just specific understandings). As an example of what these artifacts might be, consider the following (a more complete list appears later in this chapter): an instrument for helping teachers make more insightful observations of their students mathematical thinking as they work on mathematical problems; developing procedures for helping students assess the strengths and weaknesses of their own work; and, developing and refining a set of exemplary and illuminating results of students' responses to thoughtful mathematical problems. What lies at the heart of the approach that we are advocating is that it goes beyond helping teachers to reveal their own thinking, to extending, refining and sharing their thinking.

Our research (Lesh, Amit, & Schorr, 1997; Schorr, 2000; Schorr & Alston, 1999; Lesh, 1998) and the research of others (e.g., Cobb, Wood, Yackel, & McNeal, 1993; Cohen & Ball, 1990; Fennema, Carpenter, Franke, Levi, Jacobs, & Empson, 1996; Klein and Tirosh, 2000; Simon, Tzur, Heinz, Kinzel, and Smith, 2000) has shown that it is absolutely essential for teachers, like their students, to be provided with experiences that allow them to deepen, extend and share their own knowledge and understanding of the content of mathematics, the ways in which students build mathematical ideas, and the pedagogical implications of teaching mathematics in a manner which encourages the development of powerful mathematical models. Indeed, when considering teacher development, an assumption underlying this work is that the same principles that apply to student learning should also be applied to teacher learning. For example:

- If we believe that students' knowledge includes not only skills, processes, and attitudes, but also includes models for describing and explaining mathematically rich problem solving experiences, then it is also reasonable to entertain the notion that teachers' knowledge should include not only skills, practices, and dispositions, but also should

include models for describing and explaining mathematical teaching and learning
- If we believe that students should learn through concrete problem solving experiences, then it is also reasonable to entertain the notion that teachers should learn through personally meaningful problem solving experiences (note: it is important to note that what is problematic for teachers is not necessarily what is problematic for students in a given situation.)

We want to underscore our belief that effective teacher development should provide opportunities for teachers to develop insight into their students' thinking (Schorr & Ginsburg, 2000; Schorr & Lesh, 1998; Ginsburg, 1998;) as well as the ability to analyze, interpret, generalize, extend, and share their evolving models for teaching in a wide variety of settings and situations. It is important to emphasize the role that understanding student thinking plays within this paradigm. The teacher who has insight into student's thinking can appreciate the sense in students' interpretations and representations of mathematical ideas, and can deal with them constructively. By contrast, the teacher who lacks understanding of student's thinking is left in a kind of pedagogical delusional state: the teacher understands a concept in a certain way, thinks that concept is being taught to the student, but the student is either not learning it at all, or in fact learning an entirely different concept from the one the teacher has assumed. In either case, there is a wide gap between the mind of the teacher and the mind of the student. The teacher tends to deal with what is seen as the student's failure to learn by "shouting louder"—that is, by redoubling efforts to teach the concept (as interpreted by the teacher)—and remains unaware that the student is in fact attempting to learn something entirely different. In our experience, such gaps between teachers' minds and student' minds are widespread, and characterize teaching from preschool through university (Schorr & Ginsburg, 2000). But, we wish to emphasize that telling teachers about students' thinking, is no more effective than telling students about a complex mathematical idea and expecting them to understand it. Indeed, teachers may change specific behaviors, teaching strategies, or curricular materials, while still missing key ideas and understandings about their students' ways of thinking.

Central to our work in teacher development is the notion that the teaching performances that we care about involve complex and interrelated achievements and knowledge. This is not simply true for teachers, but rather for all individuals—whether they are students, teachers, administrators, teacher educators, or researchers, and whether they act alone, or in groups, classes, schools, school districts, or universities. In addition, our assumption is that teacher development projects should not establish, a priori, a preconceived notion regarding the single best type of teacher (or student). All teachers (all individuals) have complex profiles that include specific strengths and weakness. Furthermore, all teachers can be effective in some ways or places under some conditions, whereas being less effective in others. We have all seen teachers that are quite effective when they are teaching certain students under certain

7. MODELING FOR TEACHER DEVELOPMENT

conditions, while not being nearly as effective when the students, grade levels, content domains, or conditions change. A key point that must be emphasized is that all teachers, no matter the level, should continue to develop, because there can be no final or fixed state of expertise or excellence. In fact, as teachers develop, they often notice new things about themselves and the others that they interact with. For example, as teachers develop, they begin to notice new things about their students, which in turn causes them to revise their approaches to instruction, curriculum, and assessment. As this happens, their students begin to change, thus resulting in further changes in the teacher.

In this chapter, we describe our approach to teacher education—an approach that has proven to be especially effective in helping teachers build new models for the teaching and learning of mathematics. The central hypothesis of this work is that teachers must be provided with opportunities to both develop their own personal and instructionally relevant theories of how students interpret mathematics and then devise better ways of teaching it. This chapter provides documentation suggesting that thought-revealing activities for students can be used as a way to provide powerful and effective thought-revealing experiences for teachers, and that these thought-revealing experiences lead to the development of new instructional models for the teaching and learning of mathematics (see chap. 6, this volume,). We discuss ways in which teacher development opportunities can be constructed that use these activities to simultaneously help teachers increase their own content knowledge, develop a deeper understanding of how children learn mathematical concepts, and ultimately use that knowledge to make instructional decisions. More specifically, when used effectively, we have found that this type of teacher development can lead to changes in teachers' world views about the nature of students' developing mathematical knowledge (and about the nature of real life situations in which mathematics is useful, and about the many kinds of abilities that contribute to success in the preceding kinds of problem-solving situations), and therefore be particularly effective in helping teachers to change their classroom practices.

The approach to teacher development that we are advocating is dramatically different from many other approaches, which tend to emphasize activities that take teachers away from teaching. For example, some projects ask teachers to develop new problem tasks or instructional units for students built around specific themes or content areas. Although this may be interesting for teachers, and indeed they may develop increased understanding of the content that they are considering, the actual development of meaningful and effective curriculum units is a difficult task—a task that seldom builds upon the strengths of teachers' own experiences. Activities like writing curriculum units tends to take teachers away from what they are good at—teaching students. Besides that, writing curriculum units does not help teachers gain expertise in understanding their own student's thinking, or learn more about what students know about mathematics, or relate what they have learned about student's thinking to ongoing classroom activities and instruction.

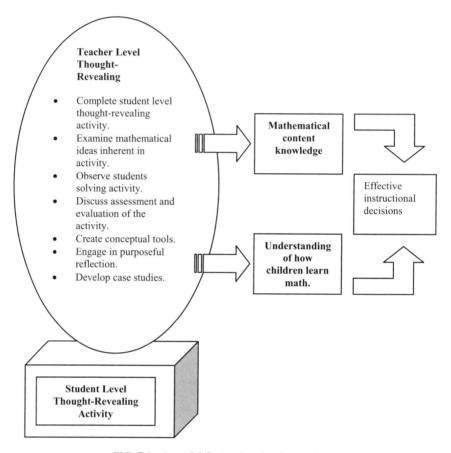

FIG. 7.1. A model for teacher development.

A MODELING APPROACH TO TEACHER DEVELOPMENT

Before continuing, it is important to clarify how this approach fits within the context of models and modeling. The models that teachers (and all individuals) have are organized around situations and experiences (at least as much as around abstractions) and they interpret new situations by mapping them into their own internal descriptive or explanatory systems (models). New models for teaching mathematics do not emerge instantaneously; rather they develop over time and in stages. Just as with students, when teachers look at something for the first time, or in a different way, the models that they use to make sense of the situations (e.g., to make predictions that guide actions) may be shallow, barren, distorted, or less stable versions of later models. Once the given situation has been mapped into an internal model, transformations within the model can take place that can produce a prediction within the modeled situation. This in turn can lead to further predictions, descriptions, or explanations for use back in the situation. Again, as with all learners, the models that teachers have develop in stages. Early conceptualizations or models can be fuzzy, or even distorted versions of later models, and several alternative models may be available to interpret a given situation. For example, when teachers first begin to analyze their students' mathematical thinking, they may miss, oversimplify, or misinterpret many of their students' ideas and conceptualizations. As they gain expertise, they begin to notice new and often deeper conceptions and misconceptions that their students may have about mathematical ideas. This in turn causes them to (re) interpret given situations and events relating to their own instructional practices and begin to use new and selected cues and information to reconsider many decisions regarding for example: (a) the appropriateness of mathematical activities for students; (b) the selection of curricular materials; (c) the best ways to implement these activities in their own classrooms (e.g. whether or not to use small groups); (d) the use of alternative methods for assessing their students' mathematical and scientific thinking and achievement; and (e) the use of appropriate technologies (including manipulative materials, calculators, computers, etc.). Because early conceptualizations or models for new classroom practices can be fuzzy or even distorted versions of later models, careful attention must be paid to the type of thought-revealing activities that are selected at any given time. It is worth underscoring the complexity and difficulty that most people have in reconceptualizing their ideas regarding the aforementioned. Many researchers have noted that teachers in transition and the researchers who work alongside them must constantly struggle to make the transformation from perspectives that can be difficult to move beyond (Simon, Tzur, Heinz, Kinzel, & Smith, 2000, p. 579)

THOUGHT-REVEALING ACTIVITIES FOR TEACHERS

Among the many types of thought-revealing activities for teachers that we have investigated, those that have proven to be most effective at promoting development do several things simultaneously, they positively influence what is taught and how it is taught; take into account the evolving nature of the learning that takes place—both for teachers and for students; and, help teachers to become more reflective and insightful about the effects of their efforts, as they continuously revise, refine, extend, and share their evolving models for teaching and learning. These activities allow teachers to actually use their own teaching experiences as the basis for professional growth. The key component, however, is that all activities help teachers to go beyond merely revealing their thinking, to extending, refining, and sharing their thinking.

Selecting thought-revealing activities for teachers must involve several interrelated steps. First, it is important to find a situation that elicits a relevant model. Next, involvement in the situation should result in an artifact or conceptual tool that can be used to provide evidence of the model. Finally, this should result in sharability and reusability by the teacher.

Following are some teacher development activities that have proven to be particularly effective in this regard. Note that many of the bulleted items do not result in a visual document or physical tool; still they can result in recorded information which can be referred to or shared. These activities involve teachers in:

- sharing observations and insights that they made of their own students as they were working to solve thought revealing activities;
- identifying the strengths and weaknesses of students' results;
- assessing the quality of students' work;
- developing strategies for leading discussions in which students assess the strengths and weaknesses of their own work;
- sharing one-to-one interviews with selected students:
- identifying follow-up instructional activities;
- developing and refining procedures for assessing the strengths and weaknesses of students' work, and for helping students assess the quality of their own work;
- developing and refining procedures for making insightful observations about students' work-in-progress; and,
- developing and refining effective procedures for individualizing instruction.

For example, consider the first item on the bulleted list, which refers to sharing observations of students. In this case, the teachers can develop a list of things to observe when students are working on challenging problems. For example, one group of teachers included the following items on their list:

- Ability to monitor and assess the product, refining it appropriately, continuing through solution cycles as necessary.
- Working within a group to analyze the problem, share insights, questioning both the process and the results.
- Roles assumed within the group
- Intellectual characteristics shown—reflection, confidence, curiosity, and so forth.

The teachers were then able to use the list in their own classrooms, share the results with others, and then use their own reflections and the feedback of others to revise and refine their list. The process can be repeated, each time focusing on new insights that are revealed as increasingly complex models evolve.

USING A MULTITIERED TEACHING STUDY TO INVESTIGATE TEACHER DEVELOPMENT

Using the approach to teacher development that was described in the preceding section, we now share the results of a multitiered teaching study in which this approach was used. Our overall research design involved a sequence of activities in which teachers interacted with students, other teachers, and researchers. These interactions focused on thought-revealing activities, which repeatedly challenged participants (both students and teachers) to reveal, test, refine, revise, and extend important aspects of their ways of thinking. The thought-revealing activities were intended to promote learning, and simultaneously produce trails of documentation that show important aspects about the nature of the constructs that develop for both students and their teachers. In this way, researchers were able to continuously adapt, modify and extend their own thinking about the effects of their project, and consequently make informed decisions about the thought-revealing activities that would be used for the teachers.

In this research design, one level of the project focused on teachers' interpretations the nature of students' developing interpretations of thought-revealing activities. The solutions that students produced involved the development of mathematically significant models that described or explained problem solving situations such as fractions, ratios, rates, proportions, or other important mathematical constructs. A second level focused on teachers developing conceptions about the teaching and learning process. Another level focused on researchers' developing conceptions about the nature of teachers' developing knowledge and abilities. At each tier, the students, teachers, and researchers were continually challenged to reveal, test, reject, revise, refine, and extend their ways of thinking. Thus, they were all able to develop in directions that they themselves could judge to be continually better— even in the absence of a predefined conception of "best." Consequently, in this teaching project, it was possible to create conditions that optimized the chances that development would occur without dictating its direction.

Three particularly useful activities that emerged involved:

- having teachers generate observation guidelines that they and their colleagues could use to make significant observations about students as they were working in groups;
- working together to identify the strengths and weaknesses of students' results that could be used to assess the quality of the products that students produced; and
- developing ways for students to assess, revise and refine their own work.

In our teacher education design, teachers were generally asked, while in a workshop setting with researchers present, to solve a series of thought-revealing activities that could later be used with their own students. Once they had solved the problem activity, they would share ideas with each other and the researchers about the main mathematical ideas, proposed implementation strategies, and anticipated student outcomes. Next, they were asked to implement these activities in their own classrooms (with researchers present for some activities). At follow-up workshops, teachers were asked to share their students' work on the activities, their thoughts about the mathematical ideas that were elicited in their students, implementation issues, and their assessments of student work. They were also asked to prepare observation forms that colleagues could use to record significant observations about students as they worked in groups. They then had a chance to test this out in the context of their own classrooms, and revise their thinking accordingly based upon their experiences.

Before sharing actual classroom findings, three characteristics should be mentioned that relate to investigations involving the development of teachers and students. First, their development tends to be highly interdependent. Although for some purposes it is possible to deal with the development of students and teachers as though they were isolated and functioning independently from one another, in reality, their development tends to be inter-dependent and unable to be disassociated. For example, when teachers first implement thought revealing activities in their own classrooms, the responses that students produce may not be as mathematically sophisticated as those produced later on (as can be seen in the following examples). As teachers develop new ways of implementing these and other activities, and new ways of building upon the mathematical thinking of their students, the responses that students produce become qualitatively better. As the responses become better, the teachers are able to continue to refine and revise their ways of thinking about their students' work, as well as the teaching and learning process. Second, development for students and teachers is not simply the result of passive acceptance of information. In both cases, the learners are developing, revising, refining and extending their models for thinking about mathematics, teaching, learning as a result of their own experiences or all of them. Third, researchers tend to be integral parts of the systems they are hoping to understand and explain. That is, researchers are also revising, extending, and refining their models for thinking about the ways in which students and teachers build ideas. Consequently, researchers should use this information to design new and more appropriate learning experiences. Researchers are not external to the process; rather they are inextricably intertwined in the process.

METHODS

This particular project involved a group of upper-elementary teachers involved in a 3-year teacher development intervention. As background, the teachers had been selected by their principals to form a task force representing four small regional schools. The purpose of the intervention was to help the teachers consider and then implement, more meaningful forms of mathematics instruction and assessment, first with their own students and then with their colleagues. The teachers met regularly with researchers in workshop sessions where teachers were presented with challenging problem tasks and asked to work together to produce solutions that represent important mathematical ideas. After sharing their own ideas and representations, they agreed to use these or similar tasks in their own classrooms. During classroom implementation, often with researchers present, teachers were encouraged to recognize and analyze their students' evolving ways of thinking about these mathematical ideas. They studied and assessed their students' work, and selected particularly interesting products to share in subsequent workshop sessions. Studying these samples of student work, as a group, provided the opportunity to both consider the development of these ideas in students, and to discuss the pedagogical implications of using this approach in their own classrooms.

The data collected included the following: (a) The teachers' responses to questions asked periodically about what they perceived to be the main ideas in particular problem activities, and important student behaviors to observe and document during classroom problem-solving sessions; (b) reflections of researchers based upon their observations, both of teachers and students, during classroom problem-solving sessions; (c) students' work from thought revealing activities that were analyzed during working group sessions; (d) students' reflections on their own and other students' work; (e) the teachers' criteria for analyzing and assessing the children's work for each task, and their continuing development of generalized criteria for assessing individual children's developing knowledge; and (f) researcher field notes taken while working with teachers in classrooms and workshop sessions.

RESULTS

To illustrate some of the points, we consider data taken from teachers' own classroom implementations of thought-revealing activities for students that occurred at the beginning and at the end of this project. Results indicated that when students' problem-solving activities lead to the development of complex products, and when teachers attempted to make sense of these products, their interpretations became more refined and sophisticated over time. In particular, the amount and type of information that they noticed about their students thinking and their own instructional practices was strongly influenced by their assumptions about the nature of mathematical constructs that might be useful for addressing a given task, and factors that influence assessments of success in real life situations outside of schools. Consequently, their interpretations of student work changed dramatically over time, and interestingly enough, they all tended to move in similar directions relating to the overall mathematical integrity and quality of their students' work. In one particular activity used with fifth and sixth graders at the beginning of the project, students were given a newspaper article with a picture of a school banner. The banner had a picture of a lion whose face was made up of various geometric figures. The students were told that the actual banner had been lost, and it was their job to help fellow students construct a pattern for a new full size banner that would look exactly like the one in the picture.

Initially, the teachers were asked to solve the problem in a workshop setting with researchers present. When they solved the problem activity, the teachers noted that they had "used proportions to help scale-up the newspaper drawing to make the full banner". There was a short discussion about how the problem would be implemented in the classroom. This discussion focused on allowing small groups of children to work together as they generated solutions to the problem (without the teacher describing or prescribing how she or a student she had selected, would solve the problem). The teachers decided that it would be best to allow the students to choose their own strategies, and develop their own solutions without teacher intervention. They agreed to share student products, and identify exemplary or illuminating examples of student work.

The teachers then implemented the activity in their own classrooms. Shortly after, they again met with researchers in a workshop setting to share classroom results. The following are examples of student products that the teachers identified as exemplary:

> *Product I:* We made a little pattern for the lioness out of construction paper. Then we traced a cymbal for the head. Then we traced a plastic lid for the ears. For the eyes we traced the bottom of a glue bottle. For the nose we used a ruler. For the mouth we used a protractor.
>
> *Product II:* Making the lioness is not hard at all. All we did was make five circles, one for the head, two for the eyes and two for the ears, and just put half of the ears behind the head to make it look like ears. Then we cut out one triangle for the nose, and one trapezoid for the rest of the nose.

One teacher (who reflected the consensus of the group) noted that these were "very good student products because the students used many different objects to construct the lion head." She continued, "I really liked the way that the students were engaged in the activity, they used lots of different objects to construct the circles and other shapes." Throughout the entire time that the students' work was being discussed, none of the teachers noted that the lions drawn by the students for the banner were not at all proportional to the original lion in the newspaper clipping—even though the teachers had identified that as one of the most important mathematical ideas in the problem activity.

The same teachers were then asked to design an observation tool that could be used for the purpose of observing and assessing the students as they work in groups to solve problems, they responded with the following types of items, which reflected the consensus of the group and was taken directly from the teachers own notes:

- [Make sure that students are] on task, understand the directions, [are] all involved, groups/pairs [are] getting along.

7. MODELING FOR TEACHER DEVELOPMENT 153

- [Did students] answer all questions thoroughly, remain on task, work cooperatively, help each other, and follow directions.

The teaching project continued on for approximately 2 years. During that time, teachers met approximately once a month with researchers. At each meeting they would solve thought-revealing activities that could later be used with their own students, and participate in teacher-level thought-revealing activities which included:

- Observing their own students as they were working to solve thought-revealing activities.
- Identifying the strengths and weaknesses of students' results.
- Assessing the quality of students' work.
- Identifying follow-up instructional activities.
- Developing and refining reliable procedures for making insightful observations about students' work-in-progress.

To highlight the differences in teachers between the beginning and the end of the project, consider the changes in the types of observations that they recorded as important to consider as their students worked in groups to solve problems (taken directly from teacher notes and recorded comments that reflected the consensus of the group):

- Identification of the problem and understanding of the task.
- Choosing and using tools appropriately.
- Creativity in selecting and using strategies for solving the problem.
- Organization of problem data.
- Quality of the product (how it looks visually, organization, completeness).
- Persuasiveness of the justification for all parts of the results.
- Mathematical perseverance—staying with the task.
- Ability to monitor and assess the product, refining it appropriately, continuing through solution cycles as necessary.
- Working within a group to analyze the problem, share insights, questioning both the process and the results.
- Roles assumed within the group
- Intellectual characteristics shown—reflection, confidence, curiosity, etc.
- Evidence of knowledge and skill with particular mathematical concepts.
- Differences in strategies chosen.
- Questioning that the solution "makes sense"—that it in fact appropriately answers the question posed and would be satisfactory to the client.

Notice that at the beginning of the project, the teachers were primarily concerned with making sure that their students were on task, working well with their peers, and following directions. As time progressed, they became more concerned with the mathematical ideas and quality of work that their students were doing. For instance, over time they began to notice the different ways in which their students approached the problem-solving situations and the ways in which the students were able to "monitor and assess the product, refining it appropriately, continuing through solution cycles as necessary". The teachers also became increasingly concerned with the degree to which students could defend and justify their solutions. They stressed the importance of being sure that student solutions did indeed make sense in the context of the problem. Although teachers were still concerned about their students working well together, the nature of what they noticed about student collaboration changed. At the beginning of the project, most teachers were concerned with being sure that students got along, and that the more able students were helping their peers. As time went on, they became concerned with how students interacted, and how students worked together to "analyze the problem, share insights, and question both the process and results." Another critical change that is not apparent when viewing the aforementioned lists is that over time, their conceptualizations about what constituted mathematical ability had changed. Many teachers noted that new students had emerged as having talents and abilities that had gone unnoticed before, and students that had traditionally done well in mathematics classes did not always outperform their peers.

In order to highlight the nature of the changes that took place in both the level and quality of student work, and nature of the changes in teachers, we now present a more in depth look at one particular teacher. We focus on how she helped students to reflect on, and assess their own work. This activity was an outgrowth of one of the teacher-level thought revealing activities that was continually used. This particular teacher designed an "Assessing Myself" sheet for her students to use as a means of reflection about their own problem solving. Each student filled out this sheet after all the groups had shared their solutions with the class. She included the following questions on the sheet, stating she felt that these questions were important for her students to consider in order to help them become more reflective about their own problem solving and mathematical skills and ideas:

- What were you able to contribute to the solution of this problem?
- What kind of outside information, math, tools, or materials helped your group solve this problem?
- What kind of math skills and ideas did you use working on this problem?
- If you could change your product, what would you do to make it better? Explain how these revisions might improve your products - feel free to use ideas from other groups, or ideas that you heard in your group that didn't get used.
- What did you learn while working on this project?

The problem activity, "Olympic Proportions," was a pilot version of one of the PACKETS Investigations for Upper Elementary Grades, developed at

7. MODELING FOR TEACHER DEVELOPMENT

Educational Testing Service with funding from the National Science Foundation.[1] The teachers had worked on the investigation themselves during one of the workshops and this teacher had chosen to implement it with her 5th grade students because it dealt with ideas about fractions and ratio in a context that required the analysis and interpretation of data. The problem materials provided to each group included photographs of five athletes and a drawing of a "Greek Ideal Athlete". The students were asked to compare various body parts of each to the Greek Ideal Athlete, using the athlete's head length as the unit of measure, and then decide which athletes were closest to the Greek Ideal. Children worked in groups to complete their solutions and write a report describing and justifying their results and explaining their method. Group 1's solution included the following:

We judged the athletes in the following way. First we made a ruler using the heads of the athletes. After we had done that for each athlete we divided each of the heads into five equal sections. Then we scored each athlete the following way. Each fifth of a head that the athlete is away from the ideal is a point. So, for example if the ideal was five heads and the athlete is four and three-fifths away from the ideal he would receive two points because he was two-fifths away from the ideal. After all the ideals and all the athletes had been scored we added up all the points. The athlete with the lowest score is *Classic Man*. This is a good way to decide who is *Classic Man* because it uses a fairly simple scoring system. The only problem we encountered was when the ideals were three and three-fourths and one and one-half because the denominators aren't five. For three and three-fourths we decided, since the difference is so small it wouldn't make much of a difference, to consider it three and four-fifths. For one and one-half we considered it one and two-fifths and half of a fifth which it is equal to. To use this method on other photos just follow the steps we followed while we were working on this problem.

Their solution was accompanied by two charts, one indicating the actual measurements, and the second indicating the actual scores.

The solution of a second group of students included the following:

1) We took one of the strips of paper, lined it up with the persons' head, and made a mark at the chin. 2) Then we moved the #1 mark to the top of the head, and made a second mark at the chin. 3) We did this until we had 8 marks. Last, we measured all the athletes, then compared their results to the Greek measurements....You can do this with other photos by doing steps 1–3, and then measuring the body parts as shown on the recording paper. Example: Lower leg. Then see how close each person came to the Greek expectations. You can figure this out by seeing how many of the persons numbers match the Greek numbers. If some don't

---

[1] Reprinted by permission of Eductional Testing Service the copyright owner.

match, see how far away it was, then add up the numbers of fractions to see who is closest without being to far away. Example: 1/2 + 1.

This product also included a bar graph that showed the total accumulated difference from the ideal for each of the athletes. The athlete with the smallest accumulated difference was considered to be the closest to the ideal.

In analyzing the reflections of the students, this teacher and the teacher educators felt that several were particularly interesting. Two are highlighted next.

In the first case, one member of the student group noticed that the measures of various body parts in relation to the head, in most cases, included fractions that were closer to fourths than to fifths. As she considered the fourth question in the Assessing Myself sheet, she wrote the following: "If I could change our product I would have used a 4 for our scoring system because all the fractional ideals were or could easily be converted to fourths. This would have made our product more accurate because we wouldn't have had to estimate." In responding to the third question, she had identified the math skills and ideas used in the following way "The math skills and ideas we used were fractions converting fractions, comparing fractions, and estamation [sic]".

This student originally divided her unit of measure (the head of the athlete's body) into fifths, which required her to "estimate" her measurements to the closest fifth. Her "estimates" appear to be carefully considered, however. She in fact appeared to invent a symbolic representation for one half using fifths. Upon reflection, she realized that dividing the head into fourths would have been a more efficient, or certainly easier, approach to the problem.

In the second case, one of the students from the group, in response to question 4 of the Assessing Myself sheet, stated that "I would see how much each person missed each cadigory [sic]. I would see by how they missed it and if it was by to much I would eliminate that person." His comments reflect the fact that his original solution had been based on accumulated differences. His group product did not account for major differences from the ideal for any one body part within the total accumulation. He is now suggesting that he would consider eliminating a person if one body part missed the ideal "by too much" regardless of the accumulated difference.

The two responses to the Assessing Myself sheet appears to provide evidence that these questions did indeed provide the students with an opportunity to reflect on their own strategies for solving the problem. It also gave them an opportunity to reflect on the strategies offered by other students, and consider how they might revise and refine their work accordingly. This, along with the written products and classroom observations, provided the teacher with useful information about their mathematical thinking. As a result of the first example, she gained important insights regarding the student's knowledge of fractions. The second provided her with a basis to build ideas about range, measures of central tendency, and variance.

Before concluding, a final comment must be made about the differences in student products taken at the beginning of the project, as evidenced by the responses to the Lion problem and those taken toward the end of the project as evidenced by those in response to the Olympic Proportions problem. Students at the same level produced both sets of results. As one examines the small sample presented here, which are typical, it becomes increasingly apparent that the nature of the students' responses changed dramatically over time—both in terms of mathematical sophistication and overall quality of response. This provides some evidence that as the teachers changed, so did the students, and vice versa. This documentation can be used to highlight the highly interdependent nature of student–teacher change.

CONCLUSIONS

In conclusion, the data above suggests that the continuing cycle of studying, implementing, and assessing the students' results of thought revealing activities was itself a thought revealing activity. One product of this activity for teachers included the Assessing Myself sheet. The value of this product was an outgrowth of this teachers thoughtfulness about her teaching, and her ability to create a classroom environment where meaningful mathematics could take place, and students could use tools like the Assessing Myself sheet in a meaningful manner.

Results of this project indicate that over the course of the project, many of the teachers: (a) changed their perceptions regarding the most important behaviors to observe when students are engaged in problem activities; (b) changed their views on what they considered to be strengths and weaknesses of student responses; (c) changed their views on how to help students reflect on, and assess their own work; and (d) reconsidered their notions regarding the use of the assessment information gathered from these activities. Of course each of the teachers evolved in a unique way—as one would reasonably expect. This list is merely intended to cite the ways in which most of the teachers appeared to have made transitions. Future research needs to be done to ascertain the level and depth of these changes over time.

Finally, because the activities that were used in this project take advantage of thought-revealing activities for students, where the student work and classroom activity becomes the basis for the continuous development, teachers were not taken away from their own teaching experience to build new ideas about teaching and learning mathematics. In fact, that is why we often refer to this type of teacher development as being "on-the-job" teacher development.

Chapter 8

A Modeling Approach to Describe Teacher Knowledge

Karen Koellner Clark
Georgia State University

Richard Lesh
Purdue University

This chapter describes how modeling principles and theory can be used to understand the nature and development of teacher knowledge and what it means for a teacher to develop mathematics content, pedagogy, and an understanding of how students develop mathematical ideas. Pedagogical content knowledge has been described as what it takes for a teacher to understand a concept and the strategies implemented to help students acquire particular conceptual ideas (Schulman, 1986). Most researchers agree that pedagogical content knowledge goes beyond what children need to know and do in order to understand a particular concept. We describe an approach to professional development that aids in understanding what it means for a teacher to understand particular mathematics concepts. In particular we focus on middle school mathematics content and pedagogy and the nature of teachers' developing knowledge.

When we look at teachers' knowledge, we investigate beyond and do not merely look at skills, global processes, attitudes and beliefs. Instead we emphasize the conceptual systems that teachers use to make sense of complex teaching and learning situations. That is, we emphasize the models that teachers use to interpret teaching and learning experiences. In this chapter, we explore effective modeling activities for teachers and show how a modeling approach for teacher development can be implemented as well as how it could be used to document the nature of teacher-level understanding of particular mathematics topics using multitier professional development.

It is reasonable to assume that understanding the nature of teacher-level understanding involves going beyond logical connections among mathematical ideas to also include at least three other dimensions; psychological connections, instructional connections, and historical connections. Psychological connections focus on how ideas may develop in the minds of children and adults. Instructional connections focus on how the ideas can be developed using available curriculum materials (textbooks, software, multimedia experiences, concrete problem-solving activities, etc.). Historical connections describe the circumstances by which an idea was derived and how a conceptual idea is placed in a historical schema. This connection details the person's historical development of the concept and their understandings that contributed to their learning about particular mathematical concepts. A main goal of multitier professional development, which is described in detail in the next section, is to provide a context for students to develop models to make sense of mathematical problem-solving situations. At the same time they provide a context for teachers to develop models as they make sense of students' modeling behavior. In other words, a key to our approach is that students develop models to make sense of mathematical problem-solving situations and teachers develop models to make sense of students' modeling behavior. The models that teachers construct establish the foundation for teacher development including distinctions regarding their teacher knowledge, pedagogy, and knowledge of student thinking. It is these models and the instructional process used that we present in this chapter.

MULTITIER PROGRAM DEVELOPMENT

Multitier teaching experiments can also be thought of multitier program development and/or implementation designs that focus on the interacting development of students, teachers and researchers and/or teacher educators (see Fig. 8.1). In other words as students are engaged in problem-solving situations that repeatedly challenge them to reveal, test, refine, and revise important aspects of mathematical constructs, teachers are focused on their own thought revealing problems that focus their attention on their students' modeling behavior. Further as the teacher revises and refines his or her model, this in turn affects the students models and vice versa. At the same time researchers and, or teacher educators are focused on the nature of teachers' and students' developing knowledge and abilities which in turn are constantly affecting each other. For example, a teacher model-eliciting problem might require teachers to construct a tool for classroom decision making such as a "ways of thinking sheet" to document student strategies, conceptions and misconceptions (as described by Schorr & Lesh, chap. 7, this volume). Another model-eliciting activity for teachers could be to design observation sheets or concept maps with a purpose.

In a series of workshops, we had in-service teachers participate in this type of professional development for a 1-year period. They agreed to take part with the understanding that they would use student model-eliciting problems in their classrooms as the basis to better understand their own thinking regarding the skills and concepts of their curriculum and their students' mathematical thinking. They agreed to identify underlying concepts and skills elicited from each student model-eliciting problem as well as state and district standards that were aligned with these concepts.

8. MODELING TO DESCRIBE TEACHER KNOWLEDGE

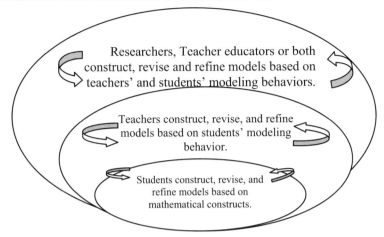

FIG. 8.1. Multitier program development.

Thus, the modeling eliciting problem for teachers was to create a concept map for each student model-eliciting problem they implemented in their classroom which would be sharable and reusable by other teachers as a curricular guide and also as a tool for instructional and assessment decisions.

Constructing a concept map as an assessment tool is not a new medium for researchers. Since the seminal work of Novak in the early 1970s, concept mapping has been widely used in science (Horton, McConney, Gallo, Woods, Senn, & Hamelin, 1993; Novak & Gowin, 1984). However, in mathematics education research there has been relatively few studies that have used concept mapping as a research tool. Williams (1998) used concept mapping to assess conceptual knowledge of functions. Whereas Doerr & Bowers (1999) used concept maps to reveal the thinking of preservice teachers about the concept of function. Using concept maps as a model-eliciting activity for teachers not only allowed teachers a tool to make sense of their own thinking, but it was purposeful in that teachers were designing the concept maps to serve as curricular guides for student model-eliciting problems. Giving concept maps a purpose makes them distinctly different from how they were used in previous studies and is foundational in their functionality as a model-eliciting activity for teachers.

A basic premise to multitier professional development is that teacher development should be analogous to student development. Thus, model-eliciting problems for teachers are based on the same instructional design principles as those that we advocate for children. In fact we have found that for almost any principle we apply to children there is an analogous situation for teachers. For example, we advocate small collaborative groups for student learning and similarly we advocate small collaborative groups for teacher development. Small group arrangements provide teachers an outlet to share, revise, and refine their understandings of particular mathematical concepts as well as pedagogical decisions used in their classrooms. Their task of creating a concept map was complex in its own right and teachers took advantage of each other's knowledge in order to make sense of their own thinking about mathematical content as well as instructional decisions.

Each of our workshops was set up to first introduce a new student model-eliciting problem as the teachers concentrated on their model-eliciting activity of constructing a purposeful concept map. The teachers began by solving the problem in groups of three or four. Then they would discuss the skills and concepts elicited from their collaborative solutions to the problem in order to begin to construct a concept map for that problem. During the following week they would implement the model-eliciting problem with their class where they were to try and better understand and document not only their students' thinking, but also deepen their understanding of the mathematics identified in student generated solutions. At the following workshop, they would present their ideas to the other teachers to further revise and refine their models. There were several ways that teachers tested their current ways of thinking. In fact, there were three modes of external criteria including feedback from teacher educators, feedback from their peers, and trial by fire in that they implemented these problems in their own classrooms with their own students and were able to self reflect and critique their own thinking. Teachers and teacher educators both identified changes in teacher behavior in their classroom. For example, changes were evident in their written feedback to students as well as their verbal feedback in their classrooms.

At one of the initial workshops the teachers worked on a thought-revealing activity called *CD Deals* (see Digital Appendix A), a PACKETS investigation developed by Educational Testing Services with support from the National Science Foundation.[1] The teachers constructed concept maps to illustrate the concepts and skills found within student generated solutions. After analyzing student generated solutions and reflecting on her curriculum and standards, one teacher identified three main concepts elicited by this problem. Her concept map (Fig. 8.2 see digital Appendixes for larger image) illustrates that problem solving, communication, and averaging were the underlying concepts she found within this thought-revealing problem after solving the problem with the other teachers and examining her students' solutions. The concept map illustrated was the final version that went through six iterations where the teacher revised and refined her map.

Teachers went through iterative models that increased in stability and sophistication over time. In other words, the conceptual maps were embodiments of teachers' ways of thinking in much the same way that students' responses to model-eliciting activities should embody significant parts of students' ways of thinking. Moreover, throughout the year the teachers' concept maps of various problems became more and more sophisticated. Thus, not only were we able to document iterative modeling cycles among one concept map, but we identified modeling cycles throughout the year of the professional development.

---

[1] Reprinted by permission of Educational Testing Service, the copyright owner.

8. MODELING TO DESCRIBE TEACHER KNOWLEDGE

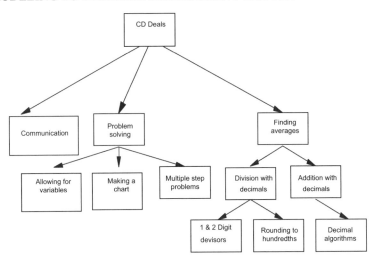

FIG. 8.2. Cathy's concept map for cd deals.

For example, in the latter part of the workshops, the teachers solved a thought-revealing activity called *Penny Pitch* developed by Educational Testing Services with support from the National Science Foundation and implemented it in their classrooms (see Digital Appendix B). The concept maps for this problem typically were more sophisticated as illustrated in Fig. 8.3 (see digital Appendixes for larger image).

Using the previous concept maps that one teacher created (see Fig. 8.2 and 8.3), the following indicators of change were identified. These indicators were typical of the conceptual changes documented in all of the teachers:

- More skills and concepts included on concepts maps
- Names for categories more detailed
- Ideas, goals and standards included
- More interrelationships identified by arrows
- More ideas bundled together
- Labels and verb included to show how items were related

For more concept maps that indicate change, please see Digital Appendix D. Further evidence of change in teacher behavior was noted in their classroom practices and in their feedback given to students.

Multitier professional development can incorporate several layers of modeling behaviors. The model-eliciting activity for teachers, constructing a purposeful concept map, had several layers of modeling. First teachers focused on constructing a concept map that served as a model to illustrate the interrelatedness of rational number concepts embedded within the responses that students generate to model-eliciting problem for students.

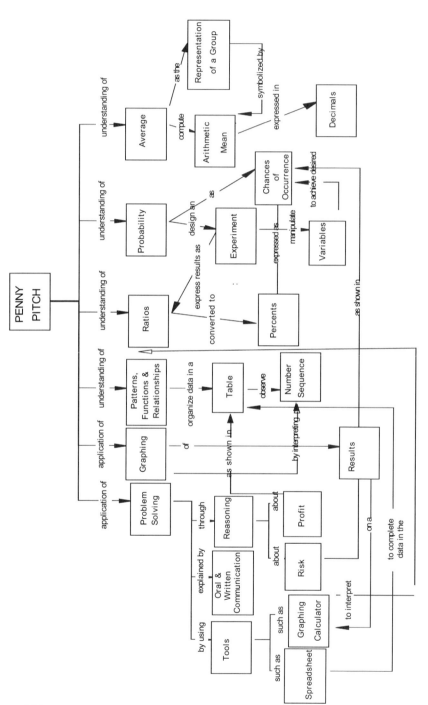

FIG. 8.3. Cathy's concept map for penny pitch.

They also identified state standards aligned with these skills and concepts and included these on their concept maps. Second, as they implemented the student model-eliciting problems within their own classrooms they constructed models of student thinking that included both misconceptions and student generated strategies for solving model-eliciting problems. These were in the form of note taking journals or lists of strategies and misconceptions. Third, they developed sets of smaller word problems and symbolically represented problems that served as forms of assessment or homework sequences to reinforce the skills and concepts found within particular model-eliciting problems. Thus, these layers provided several alternate ways to test the power and usefulness of different conceptual maps.

The modeling approach provided thought-revealing activities for teachers that were based on model-eliciting problems for their students, and provided effective "on-the-job classroom based teacher development activities" based on the same instructional design principles as those that we advocate for children. The thought-revealing activities produced models that were continually revised and refined as teachers worked through student model-eliciting problems and implemented these with their own students. They concentrated on clarifying and or elaborating their own ideas including mathematical content, pedagogy, and knowledge of student thinking while at the same time making connections among and between their previous models and elaborating on each others thoughts and ideas. Consequently, as with the student models described previously (see chap. 2, this volume) the models that the teachers constructed left a trail of auditable documentation that begins to describe the nature of teacher knowledge. This approach provided an alternative means to study teacher knowledge compared to naïve "expert-novice" studies as well as "funneling" approaches to teacher development and assessment.

TEACHER MODEL-ELICITING ACTIVITIES: DESIGNING PURPOSEFUL CONCEPT MAPS AND ANALYZING TYPICAL MODELING CYCLES

We believe that expertise in teaching goes beyond a checklist of condition-action rules that describe what "good teachers do" in a variety of situations. Rather we believe that seeing is as important as doing. One of the hallmarks of good teaching is that the teacher usually sees a lot more than others, so developing model-eliciting activities for teachers that elicit these schemes for interpreting experiences (i.e., models) is an important goal of teacher education.

The activity of constructing purposeful concept maps with in-service teachers served as a foundation for their personal inquiry into their curriculum, pedagogy, and their students thinking. Purposeful concept maps became the vehicle for deep and thoughtful discussion, teacher development, and an effective means for interpreting experiences.

Using the modeling approach to identify skills and concepts as well as the state standards embedded within student-generated solutions to student model-eliciting problems proved to be a nontrivial problem and a thought-revealing activity that was quite challenging for teachers. Initially, they identified particular strategies that they would use to solve the problems. Further, when they explored the concepts and skills in more depth they realized that collaboratively they could identify many different ways to look at particular problems. In fact, just as students typically move through multiple modeling cycles when solving model-eliciting problems, the teachers also moved through multiple modeling cycles. Through these cycles the teachers' descriptions, explanations, and predictions were gradually refined, revised, or rejected based on collaborative feedback from group discourse or reflections after implementation. Further, these cycles happened not only within a workshop session but also throughout the study. It is typical for teachers, just as their students, to vacillate back and forth through the cycles from more naïve notions to more sophisticated ideas when given a novel situation. To provide a general example, not content specific, the following modeling cycles were salient across each of the model-eliciting problems presented to teachers. These modeling cycles show how teachers moved through patterns of thought. They are illustrated by snippets of abbreviated transcripts from small group discussions to illustrate some initial main points but full transcripts, content specific, can be analyzed in greater detail in Digital Appendix C. The teacher educator, Karen Koellner Clark, is indicated by the letter K in these abbreviated transcripts. These excerpts are selections from the transcript of a 90-minute portion from the second workshop.

Interpretation 1: Teachers literally identified problem-solving components from the standards rather than deeper mathematics found within the student generated solutions to model-eliciting problems.

When first identifying the skills and concepts embedded within a student model-eliciting problem, teachers first identified skills and concepts from a problem-solving perspective. The following excerpt from a small group discussion in a workshop illustrates the first most naïve model.

K: Which standards does this problem address?
J: I think it is M.6.1[number of the state standard] which is "solves problems, reasons, and estimates throughout mathematics." It is definitely problem solving. I can see my students using the four step problem solving method or logical thinking to solve this problem.
B: I agree I think it is "solves nonroutine problems for which the answer is not obvious and relates concepts and skills to practical applications."
J: Yes, I could use this problem in the beginning of the year when we cover problem solving it is aligned with the first four chapters of our text.

8. MODELING TO DESCRIBE TEACHER KNOWLEDGE 167

Interpretation 2: Teachers identified computation only as opposed to bigger mathematical ideas.

In the second model of the cycle, the teachers identified the mathematics in the problem literally as the computation used to solve the problem.

- B: It is also M.6.9 uses addition, subtraction, multiplication, and division in computation and problem solving with whole number, decimals, fractions, and mixed numbers with like and unlike denominators.
- J: Yes, multiplication. My students would use multiplication to solve this type of problem.
- B: Or doubling or even repeated addition.

Interpretation 3: Teachers grappled with understanding the difference between skills, concepts, and strategies.

The third model illustrates the teachers differentiating between the levels of mathematical tasks called for by a problem as well as the language for discussing them.

- M: Yes, I can see my students using repeated addition or multiplication to solve this problem.
- K: Yes, but is that a skill or a strategy?
- B: I think it is both.
- K: I think that may be the point. They are both in a way but what about the four step approach isn't that a strategy? I think that M.6.9 are skills and actually ratio is the bigger idea or concept that the students are developing.
- M: Yes, I agree.
- J: I am not sure I see it that way because some of my students would use repeated addition and never know what a ratio is—I am confused about what the difference is between a skill, concept, and a strategy. I can identify them but I think the same thing would fit in more than one category.

Interpretation 4: Through discussion and collaboration big ideas about each mathematical topic were identified.

In the fourth model of the cycle the teachers make sense of the big ideas or concepts that can be identified within the problem.

- B: Yes, but they are basically using a ratio table to solve the problem so the big idea is ratio and proportion.
- K: Definitely. Again if they used a ratio table then that might be their strategy or tool that helped them come to grips with the problem. What other standards were included?

M: Patterns and relationships, number theory, data collection, fractions, and proportions. Okay patterns and relationships because they would find patterns ....
K: Okay so here some of my students might use a ratio table and find a pattern that would be patterns and relationships so those are kind of linked and when you say fractions do you mean ratio or how do you see fractions?
B: Well, I guess they might write a ratio using fractional notation but you are right they are not using fractional reasoning they are using ratio reasoning. In fact, I think most of my children would use ratio reasoning as opposed to proportional reasoning.
K: How would you know what kind of reasoning they are using?
B: I'm not sure I will have to listen to them and listen to their strategies.

Interpretation 5: The differentiation between concepts and skills is identified and the Georgia QCCS were aligned with the skills.

In the fifth model, teachers identify the big ideas, concepts, skills, and their alignment with particular state standards.

J: If ratio and proportion is a big idea then what skills go underneath that?
B: Well, I think ratio notation.
M: Well, it is much broader than that I think. I mean, it first depends on the problem and for this problem they might have used repeated addition as a skill or multiplication of ratios as a skill or setting up a proportion as a skill.
B: Oh, I see what you mean—yeah, those are the skills. I guess they do fit underneath ratio and proportion. It's interesting because I can think of a lot of things that include ratio and proportion that would be different from this problem like a unit on measurement.
J: Well, its easy to connect the QCCS to these skills because they match perfectly.
M: Yes, I think the QCCS are aligned with skills which is another reason why it seems like we are encouraged to teach proceduralized it is like the QCCS are proceduralized.

Interpretation 6: Deeper mathematical concepts were identified and connections between additive and multiplicative reasoning addressed.

In the sixth model student strategies are predicted and discussed. The additive and multiplicative nature of their thinking was identified.

K: My students would use the ratio 2 to 3 to make sense of this solution.
M: Yes, so the big idea is proportionality and there would be different children at different levels that might look at it differently.

8. MODELING TO DESCRIBE TEACHER KNOWLEDGE

B: Yeah, my children would use repeated addition. However, they might get mixed up because the number is so big—in fact if some saw the pattern they would use the pattern to find the solution. But this would be considered additive reasoning I guess, in that, they would use the constant ratio and repeatedly add until they get to the end.

M: Still, others could use the ratio and then doubling or multiplication to find the solution. Or if they realized the proportionality of the situation they could use a situation to find the solution. I think all of this would be considered multiplicative.

K: Well, perhaps you would have to listen to how the children reasoned about the problem to decide whether their reasoning is additive or multiplicative. It seems that there is a mid point where they might use multiplication but they are still thinking additively but multiplication is faster. Then on the other hand if they set the problem up into a proportion then it would be considered multiplicative.

These abbreviated excerpts were typical of the modeling cycles that teachers went through after they solved a student model-eliciting problem and were collaborating with the skills and concepts that the problems addressed. Moreover, the excerpts, although general in nature, illustrate the cyclic process that occurs when teachers collaboratively focus on a model-eliciting activity. It appears that the key factors in their growth involve the rich, focused task, as well as the collaborative and reflective nature of the activity.

After this initial workshop the teachers implemented these problems in their classrooms. It was at this stage when they focused on student thinking, which in turn significantly impacted their previous models. As the teachers grappled with implementing the model-eliciting problems in their classrooms, one of the most important aspects that added to their deeper understanding of the problem was their focus on student thinking. As the teachers attended to the strategies their students used to solve problems and their explanations they used to relay their ideas to other students in their groups much about the mathematics their students knew and applied. When teachers focused on student thinking they in turn learned much about the skills and concepts identified within student generated solutions as well as student misconceptions. As teachers documented student thinking they learned much about the mathematics that their students' were reasoning about. The more they were able to learn from their in-class documentation, the more they were able to refine and revise their concept maps. One of the most profound findings the teachers made, when constructing their concept maps, were the holes in their student's mathematical understandings and how isolated and compartmentalized their students' understanding of particular mathematics topics was. And in many cases, the teachers' found this in their own understanding of mathematical topics. Thus another dimension of teacher knowledge was documented. As we analyzed the factors that impacted teacher knowledge and development it became clear that the patterns that emerged came from distinct dimensions. These dimensions are described and discussed in the following section.

DEEPENING TEACHER KNOWLEDGE IN THREE DIMENSIONS: LOGICAL, PSYCHOLOGICAL, HISTORICAL, AND INSTRUCTIONAL

It is reasonable to suggest that what teachers need to know about particular mathematics topics goes beyond what is required for student-level understanding. However, there is no blueprint that describes the necessary components that a teacher needs to know in order to successfully teach particular middle school mathematics concepts and skills. Through our multitier program development we have identified multiple dimensions that we believe illustrate teacher-level understanding. In other words the models that teachers construct were multidimensional in that they included more than logical mathematical understanding or how mathematical ideas are developed logically beginning with basic axioms, definitions, and assumptions and deriving other constructs or theorems. We identified psychological dimensions, instructional dimensions, as well as historical dimensions (see Fig. 8.4). Thus, we believe that it is these dimensions and possibly others that are critical in documenting descriptions of teachers' developing ways of thinking.

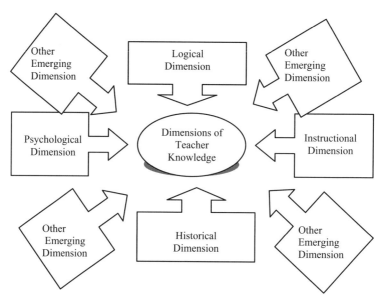

FIG. 8.4. Dimensions of teacher knowledge identified through multitier program development.

When the teachers constructed purposeful concept maps they identified and described how students thought about particular concepts including student strategies, common misconceptions, as well as teaching strategies to optimize student learning. Figure 8.5 (see digital Appendixes for larger image) shows one purposeful concept map that captured not only the major concepts involved

8. MODELING TO DESCRIBE TEACHER KNOWLEDGE 171

within the problem, but also the most efficient means of solving the problem and the technology to aid in the solution process. Focusing teacher attention on the psychological dimension of learning mathematics topics is a powerful means of teacher development that is aligned with our belief about how students learn mathematics best.

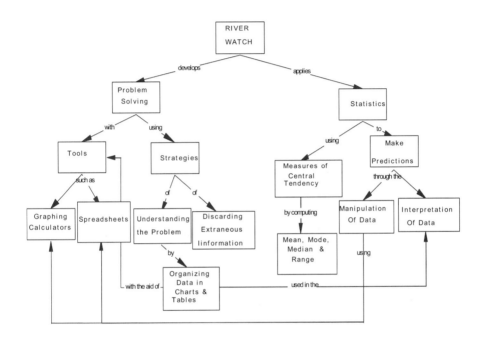

FIG. 8.5. River watch concept map.

HISTORICAL DIMENSION

The historical dimension focuses on how teachers' or students' solutions or ideas were derived historically. The context or situation that they identify with particular concepts or ideas is essential for complete understanding of their conception(s). Understanding students thinking from a historical perspective can help teachers make important connections for their students. When constructing the purposeful concept maps occasionally teachers denoted historical connections and when they did they were better able to access children's prior knowledge and recognize key interrelationships to help children make connections to prior knowledge and previous situations involving the same ideas or concepts. It was typical for teachers to make references to students' previous experiences that were highlighted on their concept maps. For example, the teacher who created the River Watch concept map (Fig. 8.5) helped her students make connections to their previous experiences with finding the measures of central tendency as well as when they had previously used spreadsheets and graphing calculators to find mean, median, and mode (documented in field notes, March, 2000).

INSTRUCTIONAL DIMENSION

The instructional dimension focuses on how the ideas or concepts can be developed using available curricular materials including books, software, multimedia experiences, concrete problem-solving activities, and manipulatives. Teachers showed growth in this area by choosing optimal texts, activities, software, or problems that reached most of their students. Further, they were able to identify key interrelationships among concepts and skills and help their students construct these ideas as well. This was the dimension where we documented the most growth from teachers.

Using our teacher model-eliciting activity of constructing purposeful concept maps we identified varying levels of the instructional dimension. For example, some teachers were able to identify curricular themes that were foundational throughout their curriculum. They were able to construct multiple problem sets pulling from a variety of sources to reinforce particular model-eliciting problems because they were able to see the big picture and make connections among and between many of the domains of rational number.

MULTITIER PROFESSIONAL DEVELOPMENT: DESCRIBING THE NATURE OF TEACHER KNOWLEDGE

Multitier professional development is designed to focus teachers' attention on students' modeling behavior. The model-eliciting task for teachers of designing a purposeful concept map was actually comprised of multiple dimensions of interpreting student modeling. For example, not only did teachers design the concept maps but they also documented student thinking and student generated strategies using a journal. Further they constructed sets of smaller word problems that served as assessment tools or homework sequences that reinforced the concepts and skills presented in the model-eliciting activities. It was through these multiple experiences that the multiple dimensions of teacher knowledge (psychological, historical, and instructional) were identified. Teacher modeling behavior like student modeling behavior is complex, continuous, and cyclic in nature. Therefore, different teachers bring their own personal experiences to model-eliciting tasks and growth and development are individualistic. However, it is the context that provides the vehicle for professional development.

We believe that on-the-job classroom development is the means to effective teacher development. This approach is quite different from traditional approaches that use a pull out model or those that identify expert standards that everyone is suppose to strive towards emulating. Multitier professional

8. MODELING TO DESCRIBE TEACHER KNOWLEDGE

development allows all teachers to grow without the constraints of the previous models. It takes into account that teacher knowledge is pluralistic, multidimensional, variable, and contextual. Model-eliciting tasks for teachers provide a context in which teachers are able to discuss, extend and refine not only their pedagogy but their content knowledge as well. Teachers are able to analyze their own ideas and then refine, revise, or reject these ideas based on implementation with their students and communication with other educators.

Constructing the purposeful concept maps allowed teachers to think deeply about their curriculum and the standards they were required to teach. They were able to conjecture how their students would solve particular problems and anticipate misconceptions that in turn promoted confidence and effective execution and facilitation of the problems.

As teachers grapple with model-eliciting activities auditable trails of documentation are left behind that illustrate logical, historical, psychological, and instructional dimensions of knowledge. It is our belief that using multitier professional development is key to documenting the nature of teacher knowledge of particular middle school topics such as fractions, ratios, and proportional reasoning.

THEORETICAL IMPLICATIONS

Multitier professional development provides teachers and their students with modeling experiences in which constructions, explanations, or descriptions serve as the products in these types of problems. Further their products reveal (both the teachers and students) their ways of thinking. The model-eliciting problem for students serves as baseline in that each teacher implements the problem in his or her classroom and at the same time they are solving their own teacher model-eliciting problem in which the central focus of the teacher problem is student thinking. When the teachers collaborate in small groups they are able to discuss their understanding of mathematical content as well as pedagogical decisions they deem important. Multitier professional development provides multiple occasions for reflection both individually, as well as collaboratively. It is these occasions that iterations of teacher modeling cycles can be documented by the teacher educator and/or researcher.

In the case of teachers constructing purposeful concept maps, teacher thinking was documented in multiple cycles among each concept map they constructed, as well as over the course of the year. Teacher knowledge was identified in multiple dimensions including, but not limited to the mathematical dimension, instructional dimension, historical dimension, as well as the psychological dimension. We identified growth in teacher knowledge not only from the continued development and more elaborately detailed concept maps, but also through their instructional decisions and actions in the classroom.

Chapter 9

Task-Analysis Cycles as Tools for Supporting Students' Mathematical Development

Kay McClain
Vanderbilt University

My purpose in this chapter is to describe an iterative process of task analysis and its role in supporting students' mathematical development. In doing so, I will present an episode taken from an eighth-grade classroom in which my colleagues and I conducted a 12-week teaching experiment during the fall semester of 1998.[1] During the course of the teaching experiment, the research team took responsibility for all aspects of the class including the teaching. (I assumed primary responsibility for teaching. I am therefore the teacher quoted in the episode.) The goal of the teaching experiment was to support students' development of ways to reason about bivariate data as they developed statistical understandings related to exploratory data analysis. This teaching experiment was a follow-up to an earlier classroom teaching experiment conducted with some of the same students during the fall semester of the previous year. Over the course of the two teaching experiments our goal was to investigate ways to proactively support middle-school students' development of statistical reasoning. In particular, the teaching experiment conducted with the students as seventh graders focused on univariate data sets and had as its goal supporting students' understanding of the notion of distribution (for a detailed analysis see Cobb, 1999; McClain, Cobb, & Gravemeijer, 2000). Our goal for the eighth-grade classroom was to build on this earlier work and extend it to bivariate data sets (for a detailed analysis see Cobb, McClain, & Gravemeijer, 2000).

It is important to clarify that the approach my colleagues and I take to conducting classroom teaching experiments involves explicit attention to (a) the proactive role of the teacher in supporting students' mathematical development,

---

[1] The research team was composed of Paul Cobb, Kay McClain, Koeno Gravemeijer, Cliff Konold, Jose Cortina, Lynn Hodge, Maggie McGatha, Carrie Tzou, and Nora Shuart.

(b) students' contributions in a classroom where productive mathematical argumentation is valued, and (c) the use of innovative instructional materials including appropriate tools and models. Within this paradigm each of these aspects is viewed as a critical part of the classroom microculture. As a result, attention is given to each in both the planning and conducting of the classroom teaching experiment attention is given to each in both the planning and conducting of the classroom teaching experiment. It is therefore important to situate the analysis of the classroom episode reported in this chapter within the larger context of the classroom teaching experiment. The particular focus of my analysis is on the role of whole-class discussions in helping to initiate shifts in students' ways of reasoning toward more efficient and sophisticated arguments, models, and inscriptions (cf. Lehrer, Schauble, Carpenter, & Penner, 2000). These discussions are part of the third of three phases of the task-analysis process. The first phase can be described as *posing the task*. This part of the process entails a lengthy discussion during which the teacher and students discuss what information would be needed and how the data should be generated (for a detailed analysis of the data creation process, see Tzou, 2000). *Student analysis of data and development of arguments* follow these data creation discussions. As part of this phase, the students are typically asked to write a report or argument and develop a model or inscription to substantiate their argument. The last phase of this process involves a whole-class *discussion and critique of the arguments*. The task-analysis process is described below in Fig. 9.1.

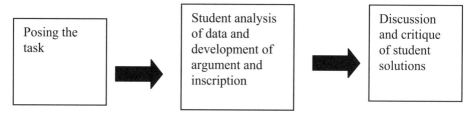

FIG. 9.1: The task-analysis process.

Within each of these three phases, there is an implied relationship between the negotiation of classroom social and sociomathematical norms and the students' mathematical development. In particular, the first phase is characterized by students actively involved in the data creation process during which they clarify for themselves (a) the question to be answered or dilemma to be resolved, and (b) the ramifications of the data collection procedures on their analysis. In the second phase, the students work in small groups on their analyses while the teacher carefully monitors their activity in order to make informed decisions about appropriate ways to orchestrate the subsequent whole-class discussion. The final phase is characterized by students explaining and justifying the results of their analysis in a whole-class discussion that is planfully orchestrated by the teacher. The teacher's role in this process is both to continually support the

9. TASK-ANALYSIS CYCLES AS TOOLS

negotiation of norms for productive mathematical argumentation and to ensure that the discussions have mathematical validity with respect to the overarching goals for the sequence of instructional tasks. This is not to imply that discussions are funneled or scripted. The image that results is that of the teacher constantly judging the nature and quality of the discussion against the mathematical agenda in order to ensure that the mathematical issues under discussion offer means of supporting the students' development. For this reason, students are selected to share their arguments so that significant mathematical issues can emerge in the course of discussion. This implies a very proactive role for the teacher that includes careful monitoring of small-group work so that informed decisions can be made. In particular, the teacher's goal during the second phase of the task analysis process involves working to understand the varied and diverse solutions that students are developing to solve the task. The teacher is not trying to funnel students to a certain process or solution or ensure that all students understand in a similar manner. The information gained from understanding students' solutions is then used as the basis of the teacher's decision-making process in the subsequent whole-class discussion. This understanding of the students' solutions provides the teacher with the ability to orchestrate a whole-class discussion in a manner that simultaneously builds from students' contributions and supports the mathematical agenda. This view of whole-class discussion stands in stark contrast to an open-ended session where all students are allowed to share their solutions without concern for potential mathematical contributions. These elaborations to the task-analysis process are described in Fig. 9.2.

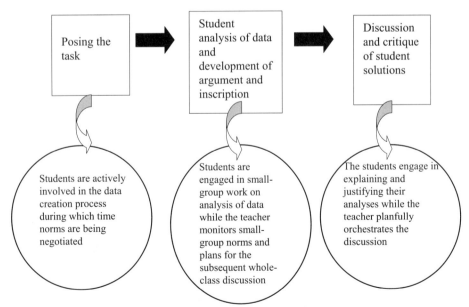

FIG. 9.2. The elaborated task-analysis process.

Although Fig. 9.2 accurately depicts the three phases of the task-analysis process, it does not acknowledge the critical role of the iterative process of task analysis in supporting students' development. In particular, it is as the students engage in a sequence of tasks that they improve and refine their ways of analyzing data that results in cycles of the task analysis process. For this reason, a better representation of the task-analysis process would account for the student learning that occurred in the first iteration or cycle of this process as a second task was posed. This cyclic nature of the task-analysis process is described in Fig. 9.3.

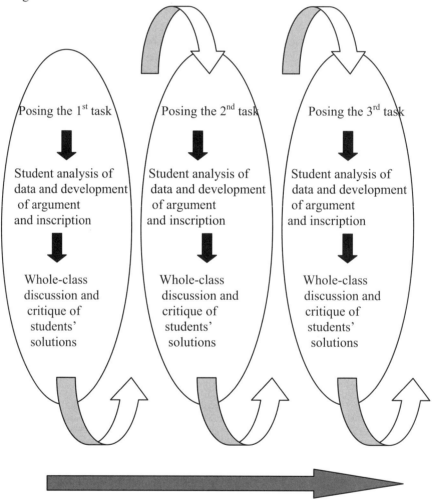

Progression toward a mathematical endpoint as defined by conjectured learning trajectory.

FIG. 9.3. Cyclic process of task analysis.

9. TASK-ANALYSIS CYCLES AS TOOLS

The model of classroom activity outlined by the iterative task-analysis cycles is grounded not only in the premise that the sequence of instructional tasks is developed to support the emergence of a conjectured learning trajectory, but also that the teacher plays a proactive role in guiding and orchestrating the classroom. It is on this latter issue that I focus my analysis. In the following sections of this chapter, I first outline the intent of the instructional sequence used in the eighth-grade classroom and briefly outline the conjectured learning trajectory. Against this background I then describe the classroom and present an episode intended to highlight the importance of the whole-class discussion in the task-analysis process in supporting students' mathematical development.

INSTRUCTIONAL SEQUENCE

The particular ways of working in the classroom that my colleagues and I employ falls under the general heading of *design research*. Design research involves cycles of research and development where ongoing classroom-based research feeds back to inform instructional design decisions in a cyclic manner. As part of this process, prior to entering the classroom we develop a conjectured learning trajectory that serves to guide the development of the instructional sequence. This trajectory is a conjecture about the path of the mathematical development of the classroom community. It focuses on pivotal mathematical issues that the research team hopes will emerge into classroom mathematical practices (for a detailed description of the evolution of classroom mathematical practices, see Cobb, Stephan, McClain, & Gravemeijer, 2001). The sequence as it is sketched out prior to entering the classroom is provisional at best and is constantly being revised in light of informal assessments of students' current ways of reasoning. In this way, ongoing, informal analyses serve to ground daily instructional decisions as we work toward a mathematical endpoint.

In planning for the eighth-grade teaching experiment, it was important for the research team to take the mathematical endpoints of the seventh-grade teaching experiment as starting points. I should therefore clarify that by the end of the seventh-grade teaching experiment, almost all of the students could reason multiplicatively about distributions of data (cf. Cobb, 1999; McClain, Cobb, & Gravemeijer, 2000). For example, interviews conducted shortly after the seventh-grade teaching experiment was completed document that most students could readily interpret graphs of unequal data sets organized into equal interval widths, (analogue of histograms) and into four equal groups (analogue of box-and-whiskers plots) in terms of trends and patterns in the distribution of the data points.

With regard to the endpoints of the eighth-grade sequence, we intended that the students would interpret standard graphs of bivariate data sets such as scatter plots as texts about the situations from which the data were generated (e.g., the number of years of education and salary of a group of males). If this occurred, the students would conjecture that measures of two attributes of a situation that were judged relevant with respect to a question or issue covaried in some way, and that the nature of their relationship could be determined by analyzing the scatter plot. We therefore hoped that the students would come to interpret the cloud of dots on a scatter plot as a bivariate distribution.

Our initial goal for the eighth-grade sequence was that the students would come to inscribe two measures of a single case as a dot in two-dimensional space (e.g., years of education and salary). The initial instructional activities therefore involved the students developing their own inscriptions of two sets of measures in order to resolve a question or understand a phenomenon. Although the students developed a number of different types of inscriptions, this first phase of the experiment was relatively unproblematic.

Our instructional intent for the second phase of the teaching experiment was that the students would come to interpret scatter plots in terms of trends or patterns in how bivariate data were distributed. We introduced a computer tool at this point in the teaching experiment in order to enable the students to structure and organize bivariate data in a variety ways. In our design efforts, we viewed the use of computer tools as an integral aspect of statistical reasoning. This was based on our belief that students would need ways to structure and organize large sets of data in order to facilitate their exploration. Therefore, the computer tool we designed was intended to support students' emerging mathematical notions while simultaneously providing them with tools for data analysis. (The computer tool developed for use in the instructional sequence provides a variety of options that can be explored via on web at http://peabody.vanderbilt.edu/depts/tandl/mted/minitools/minitools.html.)

Our initial classroom observations indicated that as the students used the computer tool, they quickly began to trace lines on the data display to indicate global trends in the data. Further, they could readily describe these trends in terms of the rate of change of the dependent variable with respect to the independent variable. For instance, as the number of years increased, the CO_2 levels increased. (This data set can be found by downloading the "years and CO_2 levels" data on the third minitool). At first glance, the students' reasoning might therefore appear to be relatively sophisticated. However, on closer examination, it appeared that the trends they described involved merely collapsing scatter plots into a single line. In other words, the data did not seem to constitute a bivariate distribution for the students. Because our goal was that students might come to view data in terns of bivariate distributions, we worked to support the notion in terms of viewing vertical "slices" of the graph instead of global "line fitting." To this end, tasks that were posed to the students during the majority of the class sessions entailed what we came to call "stacked data." For instance, the graphs generated by plotting the years of education and salary of a group of males would result in stacks of data points on the corresponding years as shown in Fig. 9.4.

9. TASK-ANALYSIS CYCLES AS TOOLS

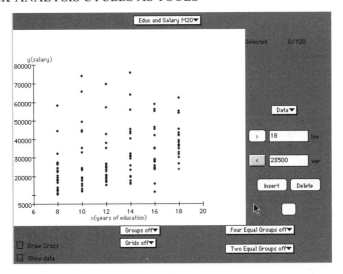

FIG. 9.4. Data display of stacked data showing years of education and salary.

(This data set can be viewed by downloading "years of education and salary" in the third minitool.) Our goal was that students would come to think of these sequences of stacks or slices in terms of a sequence of univariate distributions. In particular, they would use the grids, two equal groups, and four equal groups options on the computer minitool to structure these slices and then reason about the trends or patterns that emerged across the slices. (The reader is encouraged to structure data sets using the two equal groups and four equal group options on the third minitool before proceeding.) The final phase of the sequence as guided by the conjectured learning trajectory would then involve a transition from stacked data to cloud-like scatter plots.

Analysis of episodes taken from the eighth-grade classroom indicates that the students were in fact able to reason in sophisticated ways about bivariate data presented in stacks (cf. Cobb, McClain, & Gravemeijer, 2000). In particular, they were able to use the tool to structure the data so that they could describe the patterns, even with the data hidden. However, as the end of the teaching experiment approached, we found ourselves with less than adequate time to fully investigate students' transitions from stacked data to more traditional scatter plots. As a result, we decided to continue working with stacked data and then to use the final class sessions to pose a task to the students that involved bivariate data in scatter plots. Our intent was to investigate the students' ways of reasoning about scatter plots as bivariate distributions against the background of their prior problem-solving activity with stacked data.

CLASSROOM ANALYSIS

The final task posed to the students during the eighth-grade classroom teaching experiment involved their analyzing data from two speed-reading programs in order to make a judgment about which program was more effective. In posing the task, the students and I (as the teacher) engaged in a lengthy discussion about how such a decision could be made. The students delineated features of each program they would want to measure in order to make a comparison and discussed how these measures could be generated. In the process, the students created the design specification for conducting a study that would generate the data they deemed necessary to make an informed decision. This included discussion of issues such as sample size, testing procedures, accounting for variability within and across samples, and accuracy of timing. Against this background, the students were given pre- and post-course reading scores on 138 people from one program (program G1) and 156 people from another (program G2). (This data can be accessed by downloading the data via the third minitool.) The task was then to analyze the data in order to determine which program was more effective. In doing so, students were asked to develop a written report and create an inscription that would substantiate their argument. The students then spent the remainder of the class session working in pairs at the computers to analyze data and develop arguments. As they did so, I circulated around the room to monitor their activity as I planned for the subsequent whole-class discussion.

As was stated earlier, whole-class discussions comprise an integral aspect of the task-analysis process. As such, part of the teacher's role[2] includes making decisions about which solutions, ways of reasoning or both should be highlighted. This requires that the teacher carefully monitor students' small-group activity with an eye toward planning for the subsequent discussion. As a result, in this phase of the task-analysis process I was focused on understanding the students' diverse strategies so I could build from them as I planned for the whole-class discussion.

In planning this task, the research team had anticipated that students would structure the data using either the four equal groups option or the grids. In doing so, we conjectured that the students would reason about the slices and then make decisions about the effectiveness of the program across the groups of participants. For instance, they might argue that the majority of the people who read at between 300 and 400 words per minute prior to taking program G1 were, afterwards, able to read more than 300 words a minute. We anticipated that they would make arguments about groups of people clustered by their pre-test scores (i.e. reasoning about vertical slices of the data across the two data sets).

---

[2] The way that the research team functions while conducting a classroom teaching experiment involves one team member taking primary responsibility for the teaching. However, on numerous occasions a second member participates in asking questions during whole-class discussions. In addition, additional members of the research team may circulate around the room to monitor students' activity as they work in small groups.

9. TASK-ANALYSIS CYCLES AS TOOLS

Our focus would then be on how they reasoned about the data within the slices since the data would not be linear as in earlier tasks.

What members of the research team found, however, as we circulated among the groups was that all but one of the groups had structured the data using the cross (see Fig. 9.5 and Fig. 9.6). We found this surprising in that few of the students had used the cross option in prior analyses. In addition, it had rarely been the focus in whole-class discussions. Nonetheless, they were using the cross to reason about the four quadrants, making comparisons across the two data sets. For us, this was a less persuasive way to reason about the data as it left many questions unanswered. Further, we speculated that the students were not viewing the data sets as bivariate distributions, but simply reasoning about qualitative differences in the data in each quadrant. As a result of assessing their activity, I felt it was important to begin with the students' ways of organizing the data, but then attempt to initiate shifts in their ways of reasoning by posing questions that could not be answered using displays of the data generated by structuring the data with the cross.

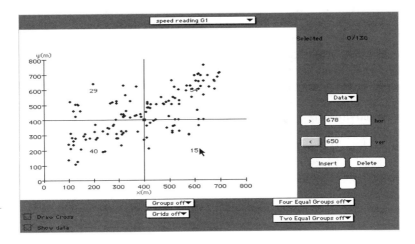

FIG. 9.5. Speed reading data on program G1 with cross option.

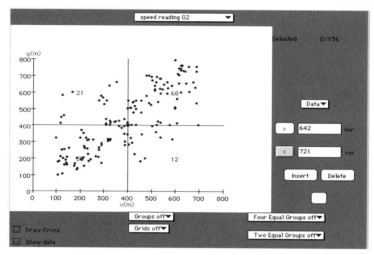

FIG. 9.6. Speed reading data on program G2 with cross option.

Throughout both teaching experiments, the students explained and justified the results of their task analysis in whole-class discussions. This was facilitated by the use of a computer projection system that allowed students to display the data sets to the entire class, showing the ways that they had structured the data. In addition, the students produced written arguments along with models or inscriptions that they developed in the course of their analysis. In doing so they had to develop ways to show trends and patterns that supported their argument without reproducing the data sets in their entirety. Their ability to engage in such activities had developed through cycles of task analysis during which whole-class discussions were focused on critiques of their analyses. In this way, each subsequent task was viewed as building on the students' prior activity as described in Fig. 9.3.

It is also important to clarify that the norms for argumentation were relatively stable by this time as they had been negotiated over the course of two teaching experiments (cf. McClain, McGatha, & Hodge, 2000). Students understood their obligation to explain so that other members of the class understood their justification for their way of structuring the data. Likewise, listening students understood their obligation to ask questions when they did not understand another's way of reasoning. Even though, in most cases, the students agreed on the outcome or decision resulting from their analysis of the data, they nonetheless typically engaged in lengthy discussions that focused on ways of structuring the data and models or inscriptions that could "best" be used to validate the argument.

The following day, the first pair of students chosen to share their analysis in whole-class discussion and critique of student solutions had partitioned the data using the cross and then reasoned about the four quadrants in terms of whether it was a "good" quadrant or a "bad" quadrant (see Fig. 9.7). Dave and Susan

9. TASK-ANALYSIS CYCLES AS TOOLS

were chosen as the first in the series of solutions to be highlighted because I judged that their argument was based on qualitative distinctions in the quadrants and did not appear to be viewing the quadrants as creating two slices of the data. Further, making direct additive comparisons about the number of participants in each quadrant was problematic due to the fact that the two data sets contained unequal numbers of data points. From their inscription, the students did not appear to have created a way to reason about the data in multiplicative terms. My goal was therefore to raise issues and questions that could not be reconciled with this type of model in an attempt to shift the discussion towards ways of reasoning about the data in terms of bivariate distributions which were structured multiplicatively.

good	stayed the same		good	stayed the same
29	54		21	68
40	15		55	12
about the same	worse off		about the same	worse off

FIG. 9.7. Data structured with cross and quadrants defined.

As Dave and Susan explained their inscription and argued that program G1 was better because it had more participants in the good quadrant, one of the other students in the class challenged their argument. Kyra noted that the two programs did not have the same number of participants so direct additive comparisons would not be a valid argument. Dave responded by noting that Program G2 had more participants, but Program G1 had more participants whose scores were in what he described as the "good" quadrant.

Kyra: One has more than the other.
Teacher(McClain): One group has more than the other so . . .
Kyra: So that's not a fair way of reasoning.
Dave: Well, this one, G2 has more and this one [points to G1] still has more in this [points to the good] quadrant.

Ryan then posed the following, "But how do you know that whatever ones are left out of G1 wouldn't all be down in the lower quadrant?" Dave and Susan responded by continuing to make comparisons of corresponding quadrants across the data sets. However, as they continued, their argument became less clear and very cumbersome. In making their arguments, they came to realize that very little information could be gleaned from the inscription they had created. Even though they found ways to make comparisons across each corresponding quadrant, they struggled to make a decision about the relative overall effectiveness of the two programs. Further, when I asked what they knew about people who started out reading over 400 words per minute, they were only able to generalize that they would either stay the same or get worse with either program.

At this point, I asked Ryan to share his way, which involved using the cross option in a similar manner but then finding the percentage of participants in each program as shown in Fig. 9.8. After the percentages were recorded on the board, the students agreed that Dave and Susan's initial analysis had been confirmed by calculating the percentages since program G1 had 21% of its participants in the good quadrant whereas program G2 had only 14% of its participants there. However, when I questioned them, the students were unable to use this model of the data to offer any additional information about what results people who read over 400 words per minute could expect from either program.

	good	stayed the same		good	stayed the same
	29	54		21	68
	21%	39%		14%	44%
	29%	11%		35%	8%
	40	15		55	12
	about same	worse off		about same	worse off

FIG. 9.8. Data with percentages noted.

I then posed a question intended to push the students to see the limitations of their current ways of structuring the data.

Teacher (McClain): So if you are running one of these programs and somebody comes in and they are reading 700 words a minute, what are you going to guarantee them?
Ryan: Outta this [points to models on the white board]? Nothing.
Teacher (McClain): Yeah!
Ryan: Shoot!

After some discussion, the students agreed that all that could be said was that a person already reading 700 words per minute would stay the same or get worse—the same information that could be given to anyone reading over 400 words per minute. I then asked what they could guarantee a person who was reading 300 words per minute. The students argued that all they could say was that the person would either stay the same or get better. However, in making this argument, Ryan said that some of the people who originally read 300 words per minute actually went down. He then declared, "So what *can* you say?"

In response, I asked Brad and Mike, the two students who had structured the data using the four equal groups option, to share their way of thinking about the data. They used the computer projection system to show the data structured into four equal groups as shown in Fig. 9.9 and 9.10.

9. TASK-ANALYSIS CYCLES AS TOOLS

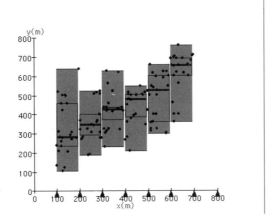

FIG. 9.9. Speed reading data from program G1 partitioned into four equal groups.

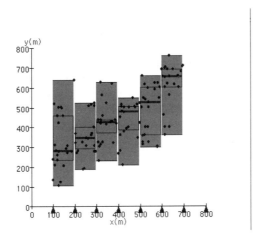

FIG. 9.10. Speed reading data from program G2 partitioned into four equal groups.

I continued by posing the following question:

> Teacher(McClain): I asked you a question about people who were reading 300 words a minute. Is there a way that I could use what Brad and Mike did to look more specifically at the people who were reading about 300 words a minute?

The students began by identifying what slice of the graph represented those people who entered each program reading 300 words per minute. After some discussion, they agreed that by using the four equal groups option you could say that in program G1, 75% of the people who were reading 300 words per minute improved. However, they noted that with program G2, only 50% of the people reading 300 words per minute improved. They then argued that if you knew you were reading at a rate of 300 words per minute, you should enroll in program G1. Next I asked the students which way of structuring the data would be most useful if you were personally trying to make a decision about which program to select. The students unanimously agreed that the four equal groups option allowed you to answer questions about specific groups of people. They noted that if they were going to enroll in a program, they would want to see the data as structured by the four equal groups in order to be able to answer the question about their particular situation. One student commented that "with the four equal groups you can see where a percentage of the people are." Here he was noting that the four equal groups showed what percentage of the people with any given pretest improved as a result of either program.

DISCUSSION

As stated earlier, in planning the instructional sequence to be used in the eighth-grade classroom teaching experiment, the research team had developed a hypothetical learning trajectory that guided the development of instructional tasks. However, this trajectory was continually subject to modifications based on ongoing analysis of the students' activity on a daily basis. As a result, revisions were made against the background of the hypothetical trajectory but with the information provided by informal analyses of students' task-analysis activity. Similarly, my decision-making process in the classroom was informed by the mathematical agenda but constantly being revised and modified in action based on the students' contributions. This is in keeping with earlier analyses that describe mathematical teaching as pursuing a potentially-revisable agenda as informed by inferences about students' beliefs and understandings (cf. Ball, 1993; Carpenter & Fennema, 1991; Cobb, Yackel, & Wood, 1991; Maher, 1987; Simon & Schifter, 1991; Thompson, 1992). To this end, I worked to capitalize on the students' contributions to advance the mathematical agenda. In doing so, I first took their current understandings as starting points and then worked to support their ways of reasoning about the problem context in more sophisticated ways. In this way, I was able to start with the students' models and inscriptions and build toward the envisioned mathematical endpoint, that of reasoning about data in terms of bivariate distributions.[3] The students' data

---

[3] Although the whole-class discussion focused on the distributions within slices, the students did not engage in conversations where they reasoned about the data across the slices. It could be argued that the nature of the problem did not lend itself to students reading trends and patterns across a single data set. I was, in fact, unable to pose questions that would require that the students reason in such a way. For this reason, we would question the appropriateness of the task in assessing students' ability

models and ways of structuring data were critical in supporting this effort. In planning for the classroom teaching experiment, the research team maintained that statistics would serve as an ideal area of mathematical activity in which to explore the role that symbolizing and modeling play in enabling students to make profound conceptual reorganizations. In this episode we see evidence to support this conjecture in that the students' models of their analyses served as thinking devices (Wertsch & Toma, 1995) as they reasoned about the adequacy of both their arguments and their ways of structuring the data.

It is also important to highlight the critical role of prior task-analysis cycles in supporting the mathematical agenda. During this episode, I was able to build on the students' prior task-analysis activity of investigating data sets and developing models and inscriptions in order to initiate a shift in their ways of reasoning about the speed reading data. Because the students had developed an understanding of data structured using the four equal groups option during prior task-analysis activities, questions which could not be answered with more simple inscriptions were made accessible by building from students' earlier analyses. This highlights the importance of the cyclic nature of the task-analysis process located within the learning trajectory in supporting students' emerging understandings (see Fig. 9.3).

The task-analysis cycle as described in this chapter focuses on students' activity situated within the mathematical content domain of statistical data analysis. However, I would argue that the same process embedded in iterative cycles would be productive in any content area. The key point is that students need to solve significant mathematical tasks that are part of a coherent instructional plan, sequence, or agenda. The students' problem-solving activity then needs to be the focus of a deliberately facilitated whole-class discussion during which the teacher takes a proactive role in guiding the emergence of significant mathematical issues. The process should be iterative in that students' prior problem-solving activity should be a basis for subsequent activity. This contributes to the negotiation of productive classroom norms and supports the teacher's efforts in guiding the students' mathematical development.

ACKNOWLEDGMENTS

The research reported in this paper was supported by the National Science Foundation under grant no. REC-9814898 and by the Office of Educational Research and Improvement through the National Center under grant no. R305A60007. The opinions do not necessarily reflect the views of either the Foundation or the Center.

---

to reason about traditional scatter plots as bivariate distributions. The most that could be argued is that the students did appear to view slices of the scatter plot in terms of distributions.

Chapter 10

Explanations Why? The Role of Explanations in Answers to (Assessment) Problems

Martin van Reeuwijk
Utrecht University

Monica Wijers
Utrecht University

Reformed mathematics curricula demand a shift from students giving only product oriented answers to students providing insight into the processes that lead to the answers as well.

In order to assess students' understanding and be able to trace the growth of their understanding over time, classroom assessment is an important instrument. As Romberg and Shafer (1999) indicate, the assessment program should capitalize on the information that teachers gather through informal assessment methods during instruction rather than assessing only written responses on assessments quizzes or tests.

Extensive research shows that improved classroom assessment can lead to raised standards (Black and William, 1998). In order to actually raise the standards, teachers must provide frequent assessment feedback and advise the individual student on what he or she can do to improve (id). Of course feedback can only be given if students' answers are such that feedback on them is possible. This implies that answers consist of more than just a number or a right-wrong statement. Only if an answer shows some kind of reasoning, thinking, or a strategy, feedback can be given that enhances the learning.

One of the frequently used strategies to get this type of student work is asking students to explain their answer.[1] However, (assessment) problems do not get any better by just asking for justifications or explanations. The written problems themselves, or the way they are orally presented by the teacher, should

---

[1] Analysis of several of the NSF sponsored mathematics curricula, especially those for the middle school (MathThematics, MathScape, Connected Math Project, Mathematics in Context), showed that the prompt "Expain your answer" is used very often (more on the web, see Appendix A).

be such that it is clear to the students what the nature of the answer must be and for what purpose, and what sort of an explanation (if any) needs to be included. Otherwise, students do not have a way to know when they are finished and have no way to validate their answer (Lesh, 2001; Lesh in Wijers & van Reeuwijk, personal correspondence, October 1997).

The shift towards more process oriented instruction and assessment also implies a change in teachers' and students' dispositions towards what is valued as important in mathematics class. In other words, it requires different mathematical norms (Carpenter & Lehrer, 1999, Cobb & Yackel 1996) in which expressing and communicating the thinking process are important forms of mathematical activity. Kemme also sees communicating as an important aspect of understanding mathematics. He follows van Hiele when he states: 'One step further in the learning-teaching process, when the student has understood what has been explained (taught) and has taken ownership, he should be able to explain the knowledge himself to another'. Furthermore: 'When a student has gained understanding or insight, he should be able to use this insight to solve non-routine problems, and to explain why and how the insight is used' (Kemme, 1990).

Whether good and useful explanations are given, and whether students actually develop understanding, depends heavily, according to Carpenter and Lehrer (1999), on the normative practices in the classrooms. These help determine whether the appropriate tasks and tools are used for the purpose of understanding. Carpenter and Lehrer argue that almost any task can be used to promote understanding, and that the important issue is for students to engage in the task for the purpose of fostering understanding and not just for completing the task. Ideally, their claim may be true, but we think good tasks help facilitate learning for understanding, and to establish appropriate classroom norms. In this chapter, we describe some examples of assessment problems that were used in classrooms that promote understanding (see Appendix B). In these classrooms, students were expected to provide explanations in order to reveal their thinking and understanding. Our goal was to focus on the question of whether the prompt "explain your answer" helps to get explanations that reveal thinking and understanding? What is the influence of classroom norms on how students react to this prompt? How does "give an explanation" differ from "show your work?" What do teachers actually expect when they ask students for a (written) explanation? And, what do students actually write down? These issues are discussed in this chapter, where our attention focuses on four examples of assessment problems. Our hypothesis is that one should be very careful with the use of prompts.[2] Ideally, students should always show their work; this should be part of the classroom culture and social norms. Also, ideally, the need for explanations to be given should be implicitly embedded in a well-defined task.

---

[2] The exact wording of the two prompts "Show your work" and "Explain your answer" can differ. The wording is often problem specific.

10. EXPLANATIONS AND ASSESSMENT PROBLEMS

BACKGROUND OF THE STUDY

The examples in this chapter come from a study that was carried out with a small number of middle school mathematics teachers working in an urban inner city school in the New England area (see Appendix C).[3] The two teachers in the study were experienced middle school teachers who had worked together on other projects as well.

For these teachers it was the first year that they were working with the reformed Mathematics in Context (1998) curriculum (see Appendix D). In reformed mathematics curricula learning and teaching for understanding is an important theme. Furthermore, one of the overall goals of the curriculum is to develop classroom norms that support learning for understanding.

For teachers implementing a new curriculum, it is often difficult to oversee all the aspects of the underlying philosophy. It requires time to take ownership of a new curriculum. In this process, teachers often hold on to beliefs and dogmas that they believe are characteristic for the new approach—and that they believe contribute to learning for understanding. Examples of such beliefs that we found in our study included the following:

- Quizzes are forbidden
- Lecturing in front of the classroom is not allowed.
- All strategies must be valued equally (even if they are wrong).
- Students should always explain their answers.

This last belief often is supported by the fact that many problems in the instructional materials and in the assessments with reformed curricula end with the phrase "show your work" or "explain your answer." One of the reasons that these prompts are so frequently used in reformed curricula is that they should contribute to establishing new classroom norms that support and facilitate learning for understanding. The prompts in the instructional materials, together with the attention a teacher pays in class to the importance of explaining and showing work, will change teachers' and students' dispositions towards mathematics.

However, the frequent use of these prompts may lead to the belief that adding such a prompt in itself is sufficient to promote learning for understanding—and that it leads to student answers that give insight into their thinking processes. However, as we argued earlier, and as is shown in examples throughout this chapter, a simple modification of a problem by adding a prompt is inadequate. Classroom practices and norms (and implicitly the teacher's professional knowledge) should change accordingly.

Our experience shows that, when new classroom norms have been established, it is not needed to explicitly use these prompts so frequently, because it should become a natural habit for students to show their work and to

---

[3] Besides the teachers in the New England area, some of the research was carried out with two teachers at a mid-Western suburban middle school. In the last example we use some student work from these teachers.

explain an answer. The meaning of the prompt "explain your answer" evolves. From a way to have students develop an attitude of automatically showing their work, it becomes a question that asks for an explicit explanation.

ASSESSMENT IN PRACTICE

Observations and interviews (see Appendix E) in the first few weeks of the school year showed that teachers had to find ways to deal with the beliefs described in the preceding section. Both teachers tried to establish norms to create classrooms that promote understanding. The teachers were constantly stressing the importance of students showing their work, writing in correct English, and giving explanations. Strategies that seemed to work were shared; students were asked to show their work in front of the class using the overhead; and, student work was discussed in whole class or group discussions. A mixture of class work and homework was used for assessment purposes.

The two teachers highly valued students explaining their thinking and reasoning as is illustrated by the following excerpts from interviews:

> T1: I think they learn a lot of each other. Sometimes when I explain something and another kid explains the same thing that I'm explaining, they get it better from the students than they get it from me.
> T2: [...] because they have other kids in their groups to help them explain the information if they need to in their own language.
> T2: Another personal goal is that the kids learn to communicate in mathematical language.

The teachers were very open to changing their classroom culture and norms, and that contributed to a fruitful research atmosphere.

PROMPTS OR NO PROMPTS?

In the next part of this chapter, we focus on examples of assessment problems used by these two teachers. The examples are taken from an instructional unit on geometry. This MiC unit—called Reallotment (see Appendix F)—was the first unit from the new curriculum the teachers were using. We use problems from the end-of-unit assessment, and from the unit itself, to explore the use of prompts. The problems are relatively small, and all of them were designed to get insight into students' thinking and reasoning. We use the examples to demonstrate how the explicit use of the prompts "Explain your answer" and "Show your work" are used; with what intentions, how it may help establishing new social norms, and what other effects these prompts may have.

10. EXPLANATIONS AND ASSESSMENT PROBLEMS

Example 1: Urba and Cursa

The problem presented in Fig. 10.1 was the first problem in the end-of-unit test with the unit Reallotment (see Appendix G). The explicit prompt "Explain your answer" is needed because giving an explanation in this case is not an obvious part of the answer.

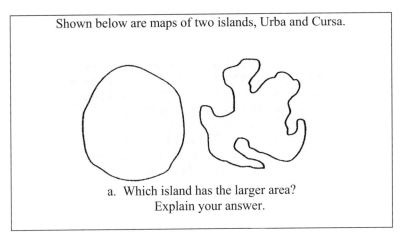

FIG. 10.1. Urba and Cursa problem.

Almost all students did indeed give a written explanation (see Appendix H) as well as an answer. Most often the structure of their responses was "Urba is larger, because…" The expectation was that without the prompt, many students would only give the name of the larger island. For this problem there is no obvious work to show, because most students reason, and the answer should reflect this reasoning.

As an alternative to the preceding approach, it is possible to rewrite the Urba and Cursa problem in such a way that an explanation indeed is an integrated part of the answer. The new text could look like the following.

> Carlos owns two small islands Urba and Cursa. Carlos wants to grow crop on the island with the larger area. That will give him the most produce. Write a letter to Carlos in which you give him a clear advise which island to grow crop on.

Some differences of such an alternative version with the version in Fig. 10.1 are are:

- Students need more time to engage in the context of the problem.
- The wording of the problem is more elaborate and complex so students need more reading skills.
- For the answer in the form of a letter, students need more writing skills.
- Writing a letter is more time consuming.

On the other hand, more mathematical thinking is involved in the alternative problem:

- Students need to make sense of the context, this involves mathematizing.
- Students need reading skills and reading mathematics is different than reading poetry, or other types of material.
- Students writing a letter involves mathematical communication which is different than writing in other genre.

Because in the alternative version more processes and stages are involved—many of which involve mathematical thinking (planning, monitoring, assessing progress, etc.)—more time is needed by the students to do this problem. It can be questioned whether—in the context of the end of unit test—this problem is intended or should be intended to cover all these skills and processes. This certainly was not the aim of the original problem, especially not because it was the first problem in the end-of-unit test: a simple, not too difficult and complex problem, to have students get a comfortable start of the assessment. What we want to show with this example is that an explicit prompt can be omitted by rewriting the problem. But one should always consider if the rewritten problem still covers the same goals.

Example 2: The area of a rectangle.

The following problem (see Fig. 10.2) comes from the same end of unit assessment as the Urba and Cursa problem in Fig. 10.1.

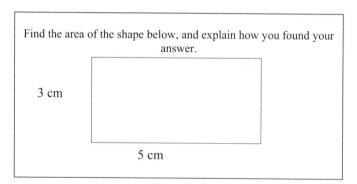

FIG. 10.2. An area problem.

In this problem, students' skills to compute area are assessed. In class, while working through the unit Reallotment, students worked with different models for area. The main characteristic of the unit Reallotment is the broad development of the concept of area. The concept is not limited to simple shapes, but right from the start it is connected to irregular shapes and three-dimensional forms. Shapes are qualitatively compared and ordered according to their area

10. EXPLANATIONS AND ASSESSMENT PROBLEMS

without actually calculating it. The Urba and Cursa problem is an example of how this can be done.

In Reallotment, area is represented by and expressed in various quantities like cost, weight, the amount of cookie-dough, the amount of paint needed to cover it, and so forth. These quantities are often more concrete to students. By using these intermediate quantities students develop different ways of thinking about quantitative relationships and develop a broader view on area.

In the unit students also developed and used different strategies (see Appendix I) for calculating areas: a strategy in which the shape is covered with little squares, a cut-and-paste strategy (realloting), leading to a more formal and general strategy of using formulas for the area of known shapes (square, triangle, rectangle, and parallelogram).

In class much attention was paid to students sharing their strategy. Because it was a geometry unit, showing the strategy often was done by using a drawing. Different strategies were discussed in class, and attention was paid to the quality of students' written work that they kept in binders. The teacher mainly checked student work for completeness and neatness, and some of it was graded. Occasionally, once in a while students explicitly practiced writing down their strategies. In their answer to the problem in figure 10.2, students were expected to write "3 x 5 = 15" or "3 x 5= 15 cm^2 as their full answer.

The explicit prompt "explain how you found your answer" served no other purpose than having students show their work. The students, however, interpreted this prompt as a question for an explicit explanation, and as a challenge to do more than simply show their work. Obviously, the students had already developed an attitude to automatically show their work. Therefore, this explicit prompt caused confusion. Students seemed to feel insecure about what to explain, and to whom, and for what purpose:

Is "3 x 5 = 15" a good enough explanation?
Do we need to explain *why* 3 x 5 = 15?
Do we have to show that we know the table of multiplication for 3 and 5?

Many students found a way out of the confusion, for instance by using more than one strategy to find the area or by adding an explanation (see Appendix J) in words that they "multiplied 3 and 5 and this equaled 15" or by drawing in the small squares and showing that there are indeed 15 of them. Clearly, the teachers' expectations about students' answers and the corresponding answers were not in line. That was caused by the different interpretations of the prompt.

In the study classrooms we expected—based on our observations and on other written student work—that without the prompt, most students would have shown their work by writing "3 x 5 = 15."

We tentatively conclude that adding the prompt was done with a certain carelessness. The wording was not carefully chosen. The purpose of the prompt was more to support or establish a set of classroom norms than to reveal students' thinking. For the students, however, these norms had already been established; they automatically showed their work. They thought that because of the explicit prompt, they had to show or explain more.

Example 3: Constructing shapes

The problem in Fig. 10.3 comes like the first two problems from the same end-of-unit test. It is a so-called "own production" or "own construction" problem. This type of problem asks students to construct or produce something. In this example they have to construct two shapes (a parallelogram and a triangle) with a given area. This implies that students need think about area in another way. Some people tend to believe that such production-construction problems are too hard to do for most students because of this turning around process in their thinking. However, this is not the case: since the construction space is big, almost all students can actually do such problems although they do so on different levels of formalization, sophistication, and abstraction. Students can show what they know and what they can do. Students' solutions can reveal much about their thinking, strategies, and the level they work on.

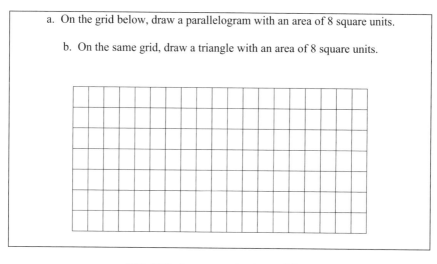

FIG. 10.3. An own production problem.

There is no explicit prompt in this problem. The reason is that a student's solution in itself shows if the student has a good understanding of the concept of area and of the strategies for finding area. If, for instance, a student always calculates areas by counting and rearranging small pieces, the student discovers that he or she cannot use this strategy to do this problem and therefore needs to turn to other interpretations of area.

Students' solutions (see Appendix K) often show how they constructed their shapes and, how they checked the area to be 8 square units or both. It is clear that the need to show some of the work is an integrated part of the solution. Giving an extra explanation is hard to do—try it yourself—and does not add much to the solution. Adding such a prompt probably weakens this problem.

10. EXPLANATIONS AND ASSESSMENT PROBLEMS

Example 4: Meeting teachers' personal goals

With this last example we want to show how a teacher can bend the purpose of a problem to fit her own goals. Important aspects of learning with understanding are communication and reflection. These need to be practiced. Of course practice takes place in class discussions as part of establishing the classroom norms, but practice in writing also is necessary.

A personal goal of the teachers in this study was to have students explicitly practice writing explanations in the form of whole sentences.

> T1: One of my goals personally is to get the kids write more. […] They weren't expressing their thoughts in writing. They were using too many diagrams, so my goal was to get them to write more of their ideas out. They're starting to do that more and more than last year but they're still wanting to write as little as possible so their descriptions are kind of unclear at times. So my goal is to have them write enough to make it a clear explanation.
>
> T2: I believe it is better for the kids to have math in this way, because they are forced to write down their thoughts even though [their writing] may be unclear now, by the end of the two year stretch their writing will be getting better. They won't be just throwing out an answer, they'll be able to write it and think about their thinking.

The problem used by the teachers was taken from the unit *Reallotment* to be used as a quiz. This quiz was given to the students after working on area problems in which they used the less formal strategies such as counting, halving, moving pieces, cut-and-paste. The problem (see Fig. 10.4) consists of 16 shapes of which the area has to be found.

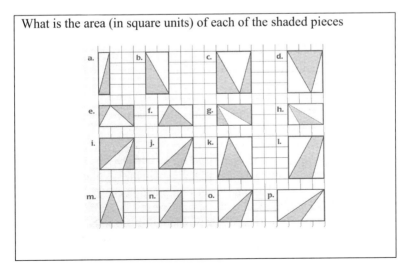

FIG. 10.4. Problem from *Reallotment*.

In this problem, no prompt for providing explanations or showing work is added. The Teacher Guide (see Appendix L) includes the following information about the purpose of this problem.

> If students simply count the squares in the triangles, encourage them to use a strategy such as halving or subtracting unshaded portions [...]. Some students may see that the area of figure **c** can be used to find the area of figure **d** [...]. Students may see and use other relationships between the shaded and unshaded areas of different figures.

It is clear from these comments that students are expected to show their work, or at least must be able and prepared to do so. The teacher can only give feedback to individual students and adjust her teaching if students show their work. This implies that the classroom norms need to be such that students always show their work. However, the teachers in the study who used this problem for assessment had another purpose.

The first teacher orally stated that she wanted her students to find the area for all 16 shapes, and to write an explanation on how they did this for each shape. She added that it should be a written explanation in which they had to use whole sentences. A drawing or a diagram would not be valued as a written explanation. She also stated that she wanted to use this quiz mainly to assess how well they could write about mathematics, so the written explanations would determine half the mark and the area itself the other half.

Student work shows that many students found the area, but wrote an explanation—as a whole sentence—that did not relate to their strategy and often did not even relate to the specific problem. For example, these written explanations most often mentioned general strategies like "cut and paste," "moving the pieces and counting all the small squares," and "using a formula."

Also, in some students' work, it was clear that explanations written for one shape were copied for all other shapes. Due to this, explanations were very vague and unspecific (see Appendix M).

With the teacher, we can conclude that the writing goal is reasonably met. The goal of communicating mathematics is only partially met, but the implicit original goal of getting insight in students' thinking and reasoning about area had been pushed to the background. Furthermore, scoring and grading all 16 explanations for each student turned out to be a demanding task, whereas the explanations were not really informative about students' skills and insight. Because not all the goals of the problem were obvious (not all criteria that the teacher would use to judge the quality of the work were clear to the students), the real problem for the students was to both solve the problem and also satisfy the (unspecified) criteria. The impact of this practice can be very negative for students who grow tired of playing the teacher's game of "guess what I want.".

The second teacher who used this problem decided to have students pick 5 out of the 16 problems, one from each of the first three rows and 2 from the last row. Students had to find the area and write explanations only for these 5

10. EXPLANATIONS AND ASSESSMENT PROBLEMS

shapes. She gave the same oral instruction about the form of the written explanations.

This worked better; but still, the written explanations were often quite vague about what students actually did. There were however more explanations fitting the specific problem, and there was less copying. Also, student work was much quicker and easier to score, although it still provided the same information about students' skills and insight.

Another teacher participating in the same study, but at a different school, used a more refined version of this problem to meet the same goals. The added written instructions were: "Choose any 10" and "Choose one from any of the problems above and describe (not just name) your strategy".

In this version of the problem, the goals were much better met. The prompt "describe your strategy," which is a problem-specific variation on "explain (how you found) your answer" caused students to actually do so. Because they have to do so for only one shape, the explanation was specific for the chosen problem. Students' work (see Appendix N) indeed shows that explanations were less vague. Furthermore, the choice of the shape gives some extra information about the students' level of mathematical understanding and abstraction: a student may pick an easy shape, like a, b or n, to feel sure or to do it in an easy way, another student may pick a hard one, like l, o or p, as a challenge or to show that he can do it. In general it prompts students to show what they know.

The problem in this version is also much easier to score and the information that is gathered with this version of the problem is at least as much as in versions 1 and 2.

REFLECTION

It is important for the teacher to be aware of the prompts used in the curriculum and how they can help or hinder establishing classroom norms. Expectations of the teacher must be made explicit. By making the expectations explicit, for example by letting students know what they are expected to show when they solve problems, a teacher can get much more information about students' mathematical understanding—and can be more effective at raising standards of performance.

The examples of student work that have been given in this chapter reveal that students (and teachers) sometimes seem to interpret the prompt "Explain (how you found) your answer" as "show your work or method". This is often caused by the fact that, in many situations, "explain your answer" is thoughtlessly added to problems where no explanation can be given beyond simply showing the work. This prompt then seems to interfere with the established classroom norms. It is likely that students show their work even if no prompt is added. The "explain how you found your answer" confuses students as to what is expected of them. It is unclear to students what is meant by giving an explanation.

So, when establishing norms, it is important to be careful with adding specific prompts that may interfere with the more implicit norms.

For existing problems it may be wise to check ones own goals and expectations against the wording of the problem, and to focus especially on the prompts (a teacher may want to add, change or delete some). The fourth example shows the teacher adding specific prompts to have students communicate their reasoning. That is, the teacher used the problem to have the students practice writing abilities. In this case, students knew exactly what the teacher meant by giving an explanation. But, it was evidently not clear to them why this explanation had to be given other than to practice writing. The focus shifted more towards these writing abilities than to students thinking and reasoning about area.

We can conclude that written explanations and the showing of the work can help establish classroom norms for classrooms that promote learning for understanding. However, these prompts do not guarantee the teacher insight of the thinking processes. Even if the problems are fit for the intended purposes and the students know exactly what is expected of them, other methods of gathering information (interviews, class discussion, observations, and presentations) often should be used together with written work to get an overall picture.

With respect to assessment, the same arguments hold. A balanced set of assessment is necessary to fully assess students' understanding. How a student interprets explicit or implicit prompts to show the work or the process that leads to an answer differs from classroom to classroom. Therefore, classroom assessment is crucial because the teacher can take into account the classroom norms.

We recommend that teachers (and curriculum developers) be careful in using prompts and be clear about their expectations. The dilemma of what to make explicit and what to leave implicit is hard to resolve. There is no simple guideline that says when to make everything that should be included in the answer completely explicit and when to leave expectations implicit. However, it is important to state the problem in such a way that students know what is expected of them. Ideally the prompt "show your work" can be omitted because showing the work has become part of the classroom norms and culture, and the prompt "explain your answer" can be omitted because the explanation is an integrated part of the answer to the problem.

III
MODELS AND MODELING AS VIEWED BY HEAVY USERS OF MATHEMATICS

Chapter 11

What Mathematical Abilities Are Needed for Success Beyond School in a Technology-Based Age of Information?

Richard Lesh
Purdue University

Judith S. Zawojewski
Purdue University

Guadalupe Carmona
Purdue University

WHAT IS THE TCCT CENTER?

Purdue's *Center for 21st Century Conceptual Tools* (TCCT) was created in 1999, based on combined support from the AT&T Foundation, the Lucent Technologies Foundation, and a continuing allocation in Purdue's budget from the State of Indiana. TCCT's overall research mission is to enlist collaborating teams of evolving experts (that include teachers, parents, policy makers, business leaders, and university-based scientists in fields preparing leaders for future-oriented professions) to investigate:

- What is the nature of the most important elementary-but-powerful understandings and abilities that are likely to be needed as foundations for success beyond school in a technology-based age of information?
- What is the nature of typical problem-solving situations in which students must learn to function effectively when mathematics and science constructs are used in the everyday lives of students, their families, or friends?
- How can we enlist the input, understanding, and support of parents, teachers, school administrators, community leaders, and policy makers—during the process of generating answers to the preceding questions?

In particular, TCCT focuses on ways that the traditional 3Rs (Reading, wRiting, and aRithmetic) need to be reconceptualized to meet the demands of the new millennium—as well as considering ways that these 3Rs need to be extended to include four more Rs: Representational Fluency, (Scientific) Reasoning, Reflection, and Responsibility.

BY-PRODUCTS OF TCCT

By-products of TCCT investigations include theory-based and experience-tested prototypes for materials, programs, and procedures to emphasize the following five goals:

- Provide early democratic access to powerful conceptual tools (constructs or capabilities) that enable all students to achieve extraordinarily productive results—in simulations of real life problem solving situations that are typical of those in which mathematical and scientific thinking is needed (beyond school) in a technology-based age of information.
- Assess deeper and higher order understandings that are likely to provide the foundations for success beyond schools in the 21st century.
- Recognize and reward a broader range of mathematical and scientific abilities or both that are needed for success in problem-solving/decision-making in real life situations.
- Identify and encourage a broader range of students whose abilities and achievements often have not been apparent in settings involving traditional tests, textbooks, and teaching.
- Provide a variety of ways for students (or adults outside of school settings) to document their achievements and abilities in ways that open doors for opportunities such as entry into desirable schools, professions, and jobs.

FOCUS OF TCCT

TCCT focuses on mathematics, science, reading, and writing because these are the academic areas that tend to have the greatest impact on issues of equity and opportunity. Also, we focus on middle school students and teachers principally because, according to tradition, the middle school curriculum represents society's implicit definition of what every citizen should know and be able to do for informed citizenship—before entering into specialized training for specific careers or advanced schooling. However, our investigations tend to be multidisciplinary and cross-disciplinary because TCCT gives special attention to real life problem solving situations in which relevant ways of thinking seldom fall into neat disciplinary categories. Similarly, even though TCCT focuses on middle school students and teachers, this emphasis imposes few restrictions on the content topics that we investigate. This is partly because the middle school curriculum is a watershed that provides both a capstone for topics addressed in the elementary school curriculum as well as a cornerstone of topics addressed in the secondary school curriculum and beyond. But, it's also because, when we adopt the spirit of Jerome Bruner's (1960) famous claim that "*Any child can be taught the foundations of any topic at any time in some appropriate form!*" we have seen little evidence that middle school children do not have the potential to develop powerful constructs related to virtually any topic in the traditional K through 14 curriculum (Kaput, 1994; Lesh, 2002).

In general, the ways that TCCT is able to get extraordinary achievements from ordinary children involve: (a) building on relevant conceptual schemes that already are available in concrete or intuitive forms; (b) challenging children to express their ways of thinking in forms that they themselves can test and revise, refine, or adapt in directions that are more useful and powerful; and (c) making accessible powerful "conceptual amplifiers" in the form of language, diagrams, or other symbolic description tools that enable children to express their current ways of thinking in form that is as powerful as possible (Lesh, 1998). Consequently, when these approaches are adopted, and when a broader range of understanding and abilities that contribute to success are recognized, many students who are especially productive represent populations of students that are highly disadvantaged and underrepresented in fields that are heavy users of mathematics, science and technology.

Rather than investigating the extent to which children are able to follow teachers' preferred ways of thinking (expressed in the form of carefully guided sequences of questions, examples, and explanations that funnel students' thinking along preplanned learning trajectories), we investigate the way students express, test, and revise their own ways of thinking. For this purpose, we have designed a collection of *construct-eliciting activities* (Lesh, Hoover, Hole, Kelly, & Post, 2000) that are designed to put students in situations where they clearly recognize the need for the conceptual tool that we want them to design, and where they themselves are able to judge the usefulness, and alternative ways of thinking.

IN WHAT WAYS DO WE BELIEVE THAT TRADITIONAL CONCEPTIONS OF THE 3RS NEED TO CHANGE?

To some extent, this question was intended to have been answered by curriculum standards documents that have been produced in recent years by professional organizations such as the National Council of Teacher of Mathematics (2000) or the American Association for the Advancement of Sciences (1993), or by governmental organizations such as Indiana's State Department of Education (2000). However, these documents have been formulated mainly by people representing schools, teachers, or university-based experts in relevant disciplines; and, people whose views generally have been ignored include those whose jobs and lives do not center around schools— such as well as scientists or professionals in fields such as engineering or business or agriculture that are heavy users of mathematics and technology. Consequently, it is not surprising that the central concerns of "school people" have focused almost entirely on finding ways to make incremental changes in the traditional curriculum—rather than wiping the slate clean by asking "what is needed for success beyond school?" Also, because few attempts have been made to enlist the input, understanding, and support of parents, policy makers, business leaders, and others whose lives do not center around schools, it is not surprising that these latter people often have led backlash movements proclaiming "back to basics" as a theme to oppose many of the changes recommended by academic content specialists and school people.

Unlike investigations that are aimed at making incremental improvements in the traditional school curriculum, TCCT's investigations have focused special attention on the question: "What is needed for success beyond school?" Also, TCCT research is especially attentive to the views of parents, business leaders, and policy makers—and, in particular, to people in future-oriented professions that are increasingly recognized to be heavy users of mathematics and technology. Finally, unlike most research projects that have investigated the nature of basic understandings and abilities, TCCT research has tried to avoid becoming so preoccupied with minimum competencies associated with low-level or entry-level jobs (i.e., street vendors, gas station attendants) the appropriate attention is not given to the kind of powerful understandings and abilities that are needed for long-term success in desirable professions and lives. We've also tried to avoid becoming so preoccupied with avoiding failure in school (by focusing on low-level facts and skills) that we fail to give appropriate attention to preparing for success beyond school (by focusing on powerful conceptual tools).

WHAT IS DIFFERENT ABOUT "BACK TO BASICS" AND "FOUNDATIONS FOR THE FUTURE?"

Results from recent international tests consistently show that American students score significantly below their counterparts in most other countries with prosperous modern economies (Glen, 2000). As a consequence, the United States is (again) preoccupied with trying to catch up with respect to basics from an industrial age. Yet, countries with wiser strategies are shifting their attention toward preparation for success in the 21st century—where the goals of instruction clearly must include more than checklists of low-level facts and skills of the type emphasized on machine-scored tests. Implications of the United States' approach are clear. If America ever catches up, it once again will be behind. One reason this is true is because it is widely known that teaching to tests tends to have strong negative influences on what is taught and how it is taught (Lamon & Lesh, 1994). This is because short-answer tests generally do not measure the most important deeper and higher order understandings that teachers are attempting to teach and that students are attempting to learn, whereas they emphasize low-level facts and skills that are only important if they lead to more significant understanding.

Having made the preceding claims, it is important to emphasize the fact that TCCT does not advocate abandoning fundamentals, nor does it advocate giving less emphasis to issues of accountability. In fact:

- *Concerning fundamentals:* TCCT assumes that abandoning fundamentals would be as foolish in mathematics, science, reading, writing, or communicating as it would be in basketball, cooking, carpentry, or other fields that involve understandings that are characterized by complex systems of thinking. But, it is not necessary to master the names and skills associated with every item at Sears before students can begin to cook or to build things; and, Indiana did not get to be the heart of basketball country by never allowing its children to scrimmage until they complete 12 years consisting of nothing but drills on skills. No doubt, if basketball were taught the way mathematics is taught, there might be a brief period in which students' basic basketball skills would appear to improve. But, over a longer period of time, the effects surely would be negative. What is needed is a sensible mix of complexity and fundamentals. Both must evolve in parallel and interact; and, one does not come before (or without) the other. Also, it is important to consider the possibility that basics from the past may not provide adequate foundations for the future—and that isolated skills out of context are not the same as integrated systems in context.

- *Concerning accountability:* Greater accountability is one of the things that TCCT develops prototypes to support. But, it is important to recognize that the United States spends far more money on testing than any other country. Yet, our students' performance on tests is among the least impressive in the world. So, whereas it is wise to hold schools and teachers accountable for student achievement, teaching to tests tends to be the disease for which it purports to be the cure. In fields like engineering or business management, it is popular to say that "You get what you measure." This is why, in such fields, great care tends to be taken to measure outcomes that are considered to be most important; and, it also is why, when the performance of complex systems is assessed, indirect or artificial indicators seldom are favored over direct measures of desired outcomes. As an analogy, consider the fact that thermometers and clocks are used to measure temperature and time; yet, no sensible person believes that simply causing the mercury to rise, or causing the hands of the clock to move, will do anything significant to change the weather or the duration of events. On the other hand, when people think about schools and education, it is common to assume that indicators of achievement (that are embodied in checklists of condition-action rules) provide the only means available to define goals of instruction.

A common excuse that is given for teaching-to-tests is the (false) notion that high scores on standardized tests are the main factors taken into account for getting admitted into desirable colleges, professions, or jobs. Yet, when TCCT researchers interviewed the Director of Purdue's Office of Admissions, he told us the same things we have heard at Northwestern University, Princeton University, the University of Wisconsin, and other leading institutions where we have had first-hand experience about admissions policies. That is:

> "At Purdue, scores on standardized tests rank seventh among seven criteria that are considered in decisions about admissions. But, even then, we use them mainly to make positive decisions in cases where little other information is available. Factors that we consider first are based on courses students have taken, grades in courses, rank in class, recommendations from teachers, and other information that reflects complex achievements. If TCCT can help students develop convincing credentials focused on complex projects like those that are stressed throughout campus here at Purdue, you can be sure we'll give this information a high priority in admissions decisions—especially if you can help us identify students with high potential among under-represented populations in the sciences and engineering."
> Dr. Douglas Christiansen, Purdue's Director of Admissions

Because of the preceding kinds of opportunities, the TCCT Center's concerns about equity go beyond traditional views that focus on closing gaps (between the performances of underprivileged versus privileged groups of students) as they are defined by traditional textbooks and standardized tests. Instead, we emphasize opening doors to success in a technology-based age of information. We emphasize this approach because, as long as traditional views of equity are preoccupied with catching up (based on obsolete notions about goals and prerequisites), we miss the opportunity to go beyond and lead (based on future-oriented conceptions of mathematics, science, learning, communication, and problem solving). Conversely, when strategies for helping disadvantaged students emphasize teaching to tests that focus on mainly low-level facts and skills, or when strategies for dealing with gifted children focus on racing ahead along narrow and shallow conceptual paths, the results risk hurting the very students they are intended to help. Problems occur when tests go beyond being used as indicators of success to being used as definitions of success. Teaching to tests is not a practice that most instructors would use in their own courses. Why would we expect it to work better in a school or district as a whole?

In fields ranging from art, to athletics, to cooking, to carpentry, when people ask: "Should instruction emphasize basic skills; or, should students be encouraged to get involved in complex projects?" The answer is obvious: "Both are important!" Furthermore, it is usually clear that the payoff from complex experiences tends to be greatest when the conceptual or procedural systems that students are encouraged to test, revise, and refine are those that are elementary-but-deep—by having the greatest power and range of applicability. For this reason, if the goal is to optimize students' chances of success in desirable colleges, professions, and jobs, TCCT asks "Why focus on the least significant criteria that schools like Purdue use to make admissions decisions? Why not focus on factors that are most significant?"

WHAT RESEARCH METHODOLOGIES DO WE USE TO IDENTIFY "FOUNDATIONS FOR THE FUTURE?"

To investigate the nature of the most important mathematical understandings and abilities that are likely to be needed for success beyond school in the 21st century, TCCT researchers want to avoid simply trusting our own instincts and abilities to make ethnographic observations. When we have tried to use such approaches in the past (Lesh, Landau, & Hamilton, 1983), the following questions arose, and our answers seemed to depend too heavily on our own preconceived notions about what it means to "think mathematically."

- Where should we look? In grocery stores? In carpentry shops? In car dealerships? In gymnasiums? In jobs that require the use of advanced technologies?
- Whom should we observe? Street vendors? Cooks? Architects? Engineers? Farmers? People playing computer games? People reading newspapers? People reading maps? People at entry-level positions, or people who are adapting to new positions? Or, veterans for whom well established routines have been developed to address nearly all difficulties that arise?
- When should we observe these people? When they are calculating? When they are estimating sizes, distances, or time intervals? When they are working with numbers and written symbols? When they are working with graphics, shapes, paths, locations, trends, or patterns? When they are deciding what information to collect about decision-making issues that involve some kind of quantities? When they are using calculators or computers? When they are describing, explaining, or predicting the behaviors of complex systems? When they are planning, monitoring, assessing, justifying steps during the construction of some complex system? When they are reporting intermediate or final results of the preceding construction process?
- What should we count as mathematical activities? For example, in fields where mathematics is widely considered to be useful, a large part of expertise consists of developing routines that reduce large classes of tasks to situations that are no longer problematic. As a result, what was once a problem becomes only an exercise. Therefore, the following kinds of questions arise. "Is an exercise with numeric symbols necessarily more mathematical than a structurally equivalent exercise with patterns of musical notes, or Cuisenaire rods, or ingredients in cooking?"

Answers to the previous kinds of questions often expose prejudices about what it means to think mathematically; and, they also tend to expose (previously unexamined) assumptions about the nature of real life situations in which mathematics is useful. Therefore, because it is precisely these assumptions that TCCT is intended to question and investigate, it does not seem justifiable to begin by assuming that the correct answers are known by someone who is dubbed to be an expert based on our own preconceived notions about the nature of answers to the questions that are being investigated. Instead, TCCT enlists input from a variety of *evolving* experts who include not only teachers and curriculum specialists but also parents, policy makers, professors, and others who may have important views that should not be ignored about: *what is needed for success in the 21st century?* (Lesh, 2002). Then, we enlist these evolving experts in sequences of situations in which their views must be expressed in forms that are tested and revised iteratively.

In this book, authors such as Koellner, McClain, and Schorr describe how evolving expert studies often use thought-revealing activities for students to provide the basis for equally thought-revealing activities for teachers and others (Lesh, R., Hoover, M., Hole, B., Kelly, A., & Post, T., 2000; Lesh & Kelly, 2000). For example, in research is based on models & modeling perspectives of the type emphasize throughout this book:

- Thought-revealing activities for students often involve simulations of real life problem solving situations in which the goal is to produce powerful conceptual tools (or conceptual systems) for constructing, describing, explaining, manipulating, predicting, or controlling complex systems.
- Thought-revealing activities for teachers (parents, policy makers) often involve developing assessment activities that participants believe to be simulations of real life situations in which mathematics are used in everyday situations in the 21st century. Or, they may involve developing sharable and reusable tools that teachers can use to make sense of students' work in the preceding simulations of real life problem-solving situations. These tools may include:
    o *Observation forms* to gather information about the roles and processes that contribute to students' success in the preceding activities.
    o *Ways of thinking sheets* to identify strengths and weaknesses of products that students produce—and to give appropriate feedback and directions for improvement.
    o *Quality assessment guides* for assessing the relative quality of alternative products that students produce.
    o *Guidelines for conducting mock job interviews* based on students' portfolios of work produced during *thought-revealing activities*—and focusing abilities valued by employers in future-oriented professions.

When the preceding kinds of tools are tested and revised, three distinct forms of feedback often are available: (a) feedback from researchers. For example, when participants hear about results that others have produced in past projects, they might say, "I never thought of that!" (b) feedback from peers. For example, when participants see results that other participants produce, they might say, "That's a good idea, I should have done that!" (c) feedback about how the tool actually worked. For example, when the tool is used with students, participants might say "what I thought would happen, didn't happen."

11. MATHEMATICAL ABILITIES IN TECHNOLOGY-BASED AGE

When TCCT investigates a variety of peoples' views about abilities that are likely to be needed for success in the 21st century, we do not begin with the assumption that it is enough simply to send out questionnaires—or to conduct interviews that presuppose that the opinions expressed are well informed and based on thoughtful reflection. Instead, we enlist diverse teams of 3 to 5 "evolving experts" to work together during semester-long sequences of workshops and experiences that are aimed at designing, testing, and revising learning-and-assessment tools that express participants' current beliefs about the nature of:

- *problem-solving experiences* that (they believe) are typical of those in which mathematics is useful outside of school in a technology-based age of information.
- *a small number of powerful constructs or conceptual tools* that (they believe) are accessible to virtually all students—but that enable ordinary students to achieve extraordinary results in the preceding simulations of real life problem solving situations.
- *deeper and higher order understandings of the preceding constructs and conceptual tools* that are sharable, reusable, and easily modifiable for a variety purposes.
- *a broader range of mathematical/scientific abilities* that (they believe) contribute to success in (simulations of) real life problem-solving-decision-making situations.
- *a broader range of students* whose outstanding abilities and achievements have not been apparent in settings involving traditional tests, textbooks, and teaching.

Note: If trial tools that are produced are not useful for the preceding purposes, then this is an indicator that the tools themselves need to be redesigned.

TCCT's evolving expert methodologies recognize that: (a) different experts often hold significantly different views about the nature of mathematics, learning, and problem solving; (b) none of the preceding people have exclusive insights about truth regarding the preceding beliefs; and (c) all of the preceding people have ways of thinking that tend to evolve significantly if they are engaged in activities that repeatedly require them to express their views in forms that go through sequences of testing-and-revision cycles in which formative feedback and consensus building influence final conclusions that are reached. To investigate such issues, TCCT assumes that what is needed is a process that encourages development at the same time that evolving views are taken seriously. What is needed are procedures in which the views of diverse participants are expressed in forms that must be expressed, tested, and revised or refined repeatedly and iteratively (Lesh & Clarke, 2000; Lesh & Kelly, 2000).

In evolving expert studies, the aim is not simply to label one type of teacher (or stakeholder) as an "expert"—nor to characterize novices in terms of deficiencies compared with some fixed and final ideal. Instead, evolving expert procedures provide contexts in which:

- Participants continually articulate, examine, compare, and test their current conceptions about the nature of mathematics, teaching, learning, and problem solving—by expressing their ways of thinking in the form of conceptual tools for accomplishing specific tasks.
- To test, revise, and refine the preceding tools, formative feedback and consensus building processes provide ways so that participants can decide for themselves which directions they needed to develop in order to improve their tools—and their ways of thinking.
- As participants develop tools (and ways of thinking) that they themselves are able to judge to be more effective for addressing specified goals, they simultaneously produce auditable trails of documentation that reveal the nature of what is being learned.

Using the preceding approach, teachers and others who participate in TCCT's evolving expert activities are truly collaborators in the development of a more refined and more sophisticated conception of what it means to develop understandings and abilities of the type that can be expected to be most useful in real life problem-solving situations.

WHY IS PURDUE AN IDEAL PLACE TO INVESTIGATE WHAT'S NEEDED FOR SUCCESS IN THE 21ST CENTURY?

Purdue has a distinctive identity as one of the United States' leading research universities focusing on applied sciences, engineering, and technology—in future-oriented fields that range from aeronautical engineering, to business management, to the agricultural sciences. Purdue stands for content quality; it stands for solutions that work; and it has pioneered sophisticated working relationships among scientists and those who are heavy users of mathematics, science and technology in their businesses and lives. Furthermore, ever since Amelia Earhart worked at Purdue to recruit women and minorities into the sciences and engineering, Purdue has provided national leadership in issues related to diversity and equity in the sciences. But, most importantly for the purposes of the TCCT Center, Purdue is a university that is filled with content specialists whose job is to prepare students for future-oriented jobs—and who know what it means to say that the most important goals of instruction (and assessment) typically consist of helping their students develop powerful models and conceptual tools for making (and making sense of) complex systems. For example, see the chapters in this book in which Oakes & Rud, Aliprantis & Carmona, or Kardos are the lead authors. These leaders emphasize that some of the most effective ways to help students develop the preceding competencies and conceptual systems is through the use of case studies—or simulations of real life problem solving situations in which students develop, test, and refine sharable conceptual tools for dealing with classes of structurally similar problem solving situations. They also emphasize that, when their students are interviewed for jobs, the abilities that are emphasized focus on: (a) the ability to communicate and work effectively within teams of diverse specialists; (b) the ability to adopt and adapt rapidly evolving conceptual tools; (c) the ability to construct, describe, and explain complex systems; and (d) the ability to cope with problems related to complex systems.

Comments in the preceding paragraphs do not imply that, if we simply walk into the offices of random professors in leading research universities, then their views about teaching, learning, or problem solving should be expected to be thoughtful or enlightened. In fact, it is well known that some university professors have been leading opponents of standards-based curriculum reforms. But, one reason why this is true is because school curriculum reform initiatives typically make few efforts to enlist the understanding and support of parents, policymakers, and others who are not professional educators. Therefore, these nonschool people often end up opposing proposed curriculum reforms. But, research in Purdue's TCCT Center is showing that this need not be the case. For example, when professors (or parents, or policymakers, or business leaders, or other taxpayers) have participated as evolving experts whose views about teaching, learning, and goals of instruction must go through several test-and-revision cycles, then the views that they finally express often become quite sophisticated and supportive for productive curriculum reform efforts (e.g., Lesh, Hoover, & Kelly, 1993).

The history of mathematics education in the United States has been characterized by a series of pendulum shifts in which, each decade or so, schools alternate between focusing on *behavioral objectives* (BOs: low-level facts and skills) and *process objectives* (POs: content-independent problem solving processes)— with occasional attention also being given to *affective objectives* (AOs: motivation, feelings, and values). But, throughout this history, the kinds of objectives that continually are neglected are *cognitive objectives* (COs: models and conceptual systems for constructing, describing, and explaining complex systems). Yet, leading scientists who are participating in TCCT research suggest that it is precisely these COs that are most likely to be needed for success in the next millennium.

WHAT ARE SOME EXAMPLES OF ABILITIES NEEDED FOR SUCCESS IN A TECHNOLOGY-BASED AGE OF INFORMATION?

When people speak about appropriate roles for calculators, computers, and other technology-based tools in instruction, their comments often seem to be based on the implicit assumption that the world outside of schools has remained unchanged—at least since the industrial revolution—and as though the main thing that new technologies do is provide "crutches" that allow students to avoid work (such as mental computation or pencil-&-paper computations) that they should be able to do without these artificial supports. But, technology-based tools do a great deal more than provide new ways to do old tasks. For example, they also create new kinds of problems solving situations in which mathematics is useful; and, they radically expand the kinds of mathematical understandings and abilities that contribute to success in these situations.

In fact, one of the most essential characteristics of a technology-based *age of information* is that the constructs (and conceptual tools) that humans develop to make sense of their experiences also mold and shape the world in which these experiences occur. Consequently, many of the most important mathematical "objects" that impact the everyday lives of ordinary people are complex, dynamic, interacting *systems* that are products of human constructions—and that range in size from large-scale communication and economic systems, to small-scale systems for scheduling, organizing, and accounting in everyday activities. Therefore, people who are able to create (and make sense of) these complex systems tend to enjoy many opportunities; whereas, those who don't risk being victimized by credit card plans or other systems created by humans. The result is that a revolution is occurring in what is "basic" in the core curriculum areas-reading, writing, and mathematics—as technology-based tools are now used on a daily basis:

- in fields ranging from the sciences to the arts and the humanities,
- in professions ranging from agriculture to business and engineering, and
- in employment positions ranging from entry-level to the highest levels of leadership.

To see evidence of the kind of changes that are being introduced into our lives by advanced technologies, we only need to look at a daily newspaper such as *USA Today*. In topic areas ranging from editorials, to sports, to business, to entertainment, to advertisements, to weather, the articles in these newspapers often look more like computer displays than like traditional pages of printed prose. They are filled with tables, graphs, formulas, and charts that are intended to describe, explain, or predict patterns or regularities associated with complex and dynamically changing systems; and, the kinds of quantities that they refer to go far beyond simple counts and measures to also involve sophisticated uses of mathematical "objects" ranging from rates, to ratios, to percentages, to proportions, to continuously changing quantities, to accumulating quantities, to vector valued quantities, to lists, to sequences, to arrays, or to coordinates. Furthermore, the graphic and dynamic displays of iteratively interacting functional relationships often cannot be described adequately using simple algebraic, statistical, or logical formulas. For example, consider the section of modern newspapers that gives advertisements for automobiles, and think back about how these advertisements looked 20 years ago. They have changed dramatically. Today, it is often difficult to determine the actual price of cars that

11. MATHEMATICAL ABILITIES IN TECHNOLOGY-BASED AGE

are shown! Instead of giving prices, what is given are mind boggling varieties of loans, leases, and buy-back plans that may include many options about down payments, monthly payments, and billing periods. Why have these changes occurred? One simple answer is—because of the kind of graphing spreadsheets that were described in Fig. 1.7 in chapter 1. That is, by using the graphing and computational capabilities of modern spreadsheets (or calculators), it is often easy for car dealers to work out sophisticated plans for customers to take possession of the car of their dreams. Furthermore, even thought such tasks often require no more than a few basic ideas from elementary mathematics, the levels and types of understandings and abilities that are needed are quite different than those that are based on quaint notions that elementary mathematics is only about computing with (pairs of) whole numbers. For example, instead of operating on *pieces* of information, operations often are carried out on whole *lists* of data. Instead of simple one-directional "input-output" rules, the kind of functions that are involved often are iterative and recursive (sometimes involving sophisticated feedback loops); and, the results that are produced often involve multi-media displays that include a variety of written, spoken, constructed, or drawn media. Therefore, representational fluency is at the heart of what it means to "understand" many of the most important underlying mathematical constructs; and, some of the most important mathematical abilities that are needed emphasize: (i) mathematizing (quantifying, dimensionalizing, coordinatizing, organizing) information in forms so that "canned" routines and tools can be used, (ii) interpreting results that are produced by "canned" tools, and (iii) analyzing the assumptions that alternative tools presuppose—so that wise decisions will be made about which tools to use in different circumstances.

Some of the most important things that technology has done, both in education and in the world beyond schools, have been to radically increase the sophistication, levels, and types of systems that humans create. Furthermore, when new systems are created in the world, new conceptual systems often are needed to make sense of their behaviors—so that behaviors can be predicted, manipulated, or controlled. Also, as new conceptual systems evolve, they often are used to create new cycles of systems development in the world.

Another important thing that technology has done is create an explosion of representational media that can be used to describe, explain, and construct complex systems. Furthermore, at the same time that technology has decreased the computational demands on humans, it has radically increased the interpretation and communication demands. For example, beyond schools, when people work in teams using technology-based tools, and when their goals involve making (and making sense of) complex systems: (i) new types of mathematical quantities, relationships, and representation systems often become important (such as those dealing with continuously changing quantities, accumulating quantities, and iterative and recursive functions), (ii) new levels and types of understandings tend to be emphasized (such as those that emphasize communication and representation), and (iii) different stages of problem solving may be emphasized (such as those that involve partitioning complex problems into modular pieces, and planning, communicating, monitoring, and assessing intermediate results).

Transcripts that accompany chapters throughout this book provide more details about the kind of understandings and abilities that are needed for success beyond school in a technology-based age of information. For the purposes of this chapter, consider *The Summer Reading Program Problem* that follows.

The Summer Reading Program

The St. John Public Library and Morgantown Middle School are sponsoring a summer reading program. Students in grades 6-9 will read books to collect points and win prizes. The winner in each class will be the student with the most reading points. A collection of approved books already has been selected and put on reserve. The chart below is a sample of the books in the collection.

Title	Author	Reading Level (By Grade)	Pages	Student's Scores on Written Reports	A Brief Description of the Book
Sarah, Plain and Tall	*Patricia MacLachlan*	*4*	*58*		Note: On a "fold out" page, two or three sentences were given to describe each book. -- Was it a history book, a sports book, an adventure book, etc.
Awesome Athletes (Sports Illustrated for Kids)	*Multiple Authors*	*5*	*288*		
A Tale of Two Cities	*Charles Dickens*	*9*	*384*		
Much Ado About Nothing	*William Shakespeare*	*10*	*75*		
Get Real (Sweet Valley Jr. High, No. 1)	*Jamie Suzanne & Francine Pascal*	*6*	*144*		

Students who enroll in the program often read between ten and twenty books over the summer. The contest committee is trying to figure out a fair way to assign points to each student. Margret Scott, the program director, said "Whatever procedure is used, we want to take into account: (1) the number of books, (2) the variety of the books, (3) the difficulty of the books, (4) the lengths of the books, and (5) the quality of the written reports."

Note: The students are given grades of A+, A, A-, B+, B, B-, C+, C, C-, D, or F for the quality of their written reports

---

YOUR TASK: Write a letter to Margaret Scott explaining how to assign points to each student for all of the books that the students reads and writes about during the summer reading program.

11. MATHEMATICAL ABILITIES IN TECHNOLOGY-BASED AGE

The *Summer Reading Program Problem* is an example of a middle school version of a "case study" that we first saw being used at Purdue University's Krannert Graduate School of Management. Notice that it's similar to many problems that occur when:

- business managers develop ways to quantify constructs like: the "productivity" of workers, or the "efficiency" of departments within a company, or the "cost-effectiveness" of a possible initiatives,
- teachers calculate grades for students by combining performance measures from quizzes, tests, projects, and laboratory assignments—or when they devise "scoring rubrics" to asses students' work on complex tasks, or
- publications such as *places rated almanacs* or *consumer guides* assessments (compare, rank) complex systems such as products, places, people, businesses, or sports teams.

The *Summer Reading Program Problem* is an example of a "case study for kids" that was based on a case study that we first observed being used in Northwestern University's Kellogg School of Management. The original graduate-level "case study" was designed to challenge graduate students to develop powerful ways of thinking about *cost–benefit trends*.

> YOUR TASK: Write a letter to Margaret Scott explaining how to assign points to each student for all of the books that the students reads and writes about during the summer reading program.

We sometimes refer to problems like the *Summer Reading Problem* as "construct development problems" because, to produce the product that's needed, the basic difficulty that problem solvers confront involves developing some sort of an "index of reading productivity" for each participant in the reading program. This index needs to combine qualitative and/or quantitative information about: (i) the number of books read, (ii) the variety of books read, (iii) the difficulty of books read, (iv) the lengths of books read, and (v) the quality of reports written—as well as (possibly) other factors such as: (vi) the "weights"(or "importance values") that could be assigned to each of the preceding factors, or (vii) a "diversity rating" that could be assigned to the collection of books that each participant reads.

note: In the *Summer Reading Problem,* it might make sense to <u>multiply</u> the number of books by the difficulty level of each book. But, it might make sense to *add* scores from reading and scores from written reports. In general, to combine other types of information, students must ask themselves "Does it make sense to add, to subtract, to multiply, to divide, or to use some other procedure such as vector addition?" In other words, one of the main things that's problematic involves deciding which operation to use.

One important point to emphasize about "construct development problems" is that, even though such problems almost never occur in textbooks or tests, it's fairly obvious they occur frequently in "real life" situations where mathematics is used; and, it's also obvious that they represent just a small portion of the class of problems in which the products students are challenged to produce go beyond being a simple numeric answers (e.g., 1,000 dollars) to involve a the development of conceptual tools that can be:

- *used* to generate answers to a whole class of questions,
- *modified* to be useful in a variety of situations, and
- *shared* with other people for other purposes.

Beyond providing routines for computations, the tool also may involve:

- *descriptions* (e.g., using texts, tables, or graphs to describe relationships among variables),
- *explanations* (e.g., about how, when, and why to do something),
- *justifications* (e.g., concerning decisions that must be made about trade-offs involving factors such as quality and quantity or diversity), and/or
- *constructions* (e.g., of a "construct" such as "reading productivity").

Therefore, when mathematics instruction focuses on problem solving situations in which the products that are needed include the preceding kinds of conceptual tools, straightforward ways emerge for dealing with many of the most important components of what it means to develop deeper and higher-order understandings of the constructs that the tools embody.

Appendix A gives an example of the kind of products that middle school students often produce in response to *The Summer Reading Program Problem*. Just as for other transcripts that accompany chapters throughout this book, the response that is shown is typical in the sense that children who have been classified as below average (based on performance in situations involving traditional tests, textbooks, and teaching) often invent (or significantly revise or extend) constructs and ways of thinking that are far more sophisticated than anybody ever dared to try to teach to them. ...Solutions to *The Summer Reading Program Problem* generally involve the following new ways of thinking about old ideas.

- Relevant quantities include: quantities assigned to qualitative information, weights assigning values to different kinds of quantities, and quantities involving both magnitude and direction. In addition, relevant arithmetic operations go beyond operating on pieces of information to be carried out on whole lists (or matrices) of data. Furthermore, the operations may act iteratively, or recursively, to drive decisions based on telescoping processes. Also, relevant results may go beyond numeric answers to include multimedia graphic displays focusing on patterns or trends.
- At the same time that relevant conceptual technologies (e.g., spreadsheets, graphing calculators) may reduce the burden of computing

results, they also tend to radically increase difficulties associated with: (a) describing (e.g., quantifying, dimensionalizing, coordinating, organizing, or in general mathematizing) situations in forms so that the conceptual tools can be used, (b) interpreting results that are produced by "canned" tools, or (c) analyzing the assumptions that alternative tools presuppose—so that wise decisions can be made about which tools to use in different circumstances.

- Because the conceptual tools that the statement of the problem asks students to produce must be sharable, manipulatable, modifiable and reusable in a variety of situations, it is important to go beyond thinking with these tools to also think about them (and the assumptions that they implicitly presuppose). Also, because the problem solver is a team, rather than being an isolated individual, communication capabilities are important—and so are abilities needed to break complex problems and procedures into components that can be addressed by different people in the group. Finally, because the conceptual tools that are needed generally must be refined and revised using a series of testing-and-revising cycles, relevant abilities involve planning, monitoring, and assessing intermediate results and procedures.

Students who prove to be especially productive in such problem-solving situations often are not those who have histories of high scores on traditional tests. This is because the understandings and abilities that contribute to success often are quite different than those that have been emphasized in traditional textbooks and tests. For example, even though many of the same basic mathematical ideas are important (such as those involving rational numbers, proportional reasoning, and measurement), attention often shifts beyond asking *What computations can students do?* toward asking *What kind of situations can students describe (in forms so that computational tools can be used)?*

If we ask—*What kind of mathematical understandings and abilities will be needed for success beyond school in a technology-based age of information?*—then the kind of examples that I've given so far should make it clear that the kind of mathematical conceptual tools that are needed often must be based on more than algebra from the time of Descartes, geometry from the time of Euclid, calculus from the time of Newton, and shopkeeper arithmetic from an industrial age. For example, mathematical topics that are both useful and accessible to students may include basic ideas from discrete mathematics, complexity theory, systems analysis, or the mathematics of motion—where the emphasis is on multi-media displays, representational fluency, iterative and recursive functions, and dynamic systems. Nonetheless, in general, to provide powerful foundations for success in a the new millennium, the kind understandings and abilities that appear to be most needed are not about the introduction of new topics as much as they are about broader, deeper, and higher-order treatments of traditional topics such as rational numbers, proportions, and elementary functions that have been part of the traditional elementary mathematics curriculum, but that have been treated in ways that are far too narrow and shallow for the purposes that concern us here.

In other words, using model-eliciting activities that are designed using the principles that were described in chapter 2 of this book, what TCCT researchers are observing coincides with what Purdue professors have observed when college students (or on-the-job adults) work on similar case studies (or simulations of real life problem-solving situations in which students develop, test, and refine sharable and reusable conceptual tools for dealing with classes of structurally similar problem solving situations. That is, thinking mathematically is about interpreting and communicating at least as much as it is about computing; and, in particular, it is about making (i.e., constructing) and making sense of complex systems. Furthermore, productive people need to be able to work on teams of diverse specialists, using sophisticated tools, on complex tasks in which planning, monitoring, and assessing activities are critical.

Years ago, before personal computers became household items that are as familiar to many people as telephones and television, many people predicted that computers would never become popular. A reason for this prediction was that the preceding people imagined that what people would do with computers would be to compute. But, the popularity of computers began to explode when people started using them for desktop publishing, for e-mail, and for internet communication. Clearly, in the minds of most people, computers not about computing; and, they're not even about individual-to-computer interactions that are exclusively mediated by written symbols. Instead, they're about people-to-people communication; they're about multi-media communication; and, they're about sharing increasingly complex constructed artifacts. Similar transitions need to take place in schools—where computers and calculators continue to involve mainly individual-to-gizmo interactions, static symbolic mono-media, and one-way transmissions. Consequently, through it's multi-media, interactive, distance education programs, the TCCT Center is exploring new ways to use technology-based computation and communication tools to promote new opportunities for development for students ranging from preschool through adults. By recognizing a broader range of abilities that contribute to success beyond school, TCCT research is helping to document the extraordinary achievements of a broader range of people with exceptional potential.

Chapter 12

The EPICS Model in Engineering Education: Perspectives on Problem-Solving Abilities Needed for Success Beyond Schools

William Oakes
Purdue University

Anthony G. Rud, Jr.
Purdue University

Engineers use mathematics to produce products and services that benefit society. Engineering educators create learning environments where students develop models to apply basic mathematical and scientific principles to design new products and services to meet society's demands. Design requires students to take models and apply them to very different contexts in order to synthesize new solutions to technical challenges. In order to keep pace with the increase in technology and knowledge students must have, there is a growing need for students in K through 12 programs to be able to take complex mathematical models and apply them to diverse problems.

This chapter discusses how engineering students learn to apply their technical knowledge to real problems with social significance through an innovative undergraduate curriculum. We begin by describing how engineering education has changed to meet new workplace demands. Technical skills in mathematics and science are still the bedrock for engineers, but the so-called "soft" skills of teamwork, communication, and collaboration are also required for success in today's global economy. We discuss a program at our university that meets this demand through a curriculum that utilizes modeling eliciting activities to teach the mathematically oriented design skills through a community based, experiential learning program called EPICS (Engineering Projects in Community Service).

THE CHANGING FACE OF ENGINEERING EDUCATION

Today's engineers are being asked to do more than just make calculations. They are expected to work on multidisciplinary teams, manage multiple projects, and complete in a global market place. The explosion of technology has created an environment where the knowledge student's gain in college may be considered obsolete within only a few years. The Internet and the advancement of computer tools have radically changed the way engineers work and the businesses in which they work.

To meet these demands, engineering education has been undergoing changes that have been led by the National Science Foundation, the Accreditation Board of Engineering and Technology, as well as corporations. The National Science Foundation has established initiatives that encourage universities to participate in collaborative efforts called coalitions (Appendix A) that have produced many innovations that have brought more active learning and problem solving into curricula as well as shown how integrated approaches to learning can enhance the educational experience of undergraduate students. Efforts have also been established to disseminate such reforms to institutions around the country (Appendix B).

Another impetus for change within engineering is the Accreditation Board for Engineering and Technology (ABET, Appendix C) the national accreditation body for engineering programs. ABET has recently established a new accreditation philosophy called Engineering Criteria (EC) 2000. The new philosophy is one where engineering schools define educational outcomes for their specific programs and then demonstrate how they meet these outcomes. Outcomes include hard technical skills such as the ability to create models from basic mathematical and scientific principles as well as using these models to synthesize new designs.

There is also an increase in the importance of so-called soft skills. These skills include the ability to work on multidisciplinary teams and to communicate effectively. This reflects the importance of working on teams to design the complex systems that require coordinated efforts of many people or organizations.

U.S. corporations have also been active in the reform efforts. Corporations are also under increasing pressure from advancing technology as well as global competition. They need college graduates who have a strong understanding of mathematical and scientific principles and who can solve complex problems, analyze situations and work on multidisciplinary and even global teams to remain competitive. One of the corporate leaders in the reform effort has been the Boeing Company, which has developed a desired list of attributes of an engineer (Appendix D): These characteristics illustrate the changing work environment for engineers, which requires strong technical skills as well as the soft skills that ABET is emphasizing. Specific reform efforts have been presented through the American Society for Engineering Education at the Annual Conference (Appendix E) and the Frontiers in Education Conference (Appendix F).

12. EPICS MODEL IN ENGINEERING EDUCATION

ENGINEERING PROJECTS IN COMMUNITY SERVICE (EPICS)

As engineering educators seek ways to provide classroom experiences that prepare students for the kinds of challenges they will face after graduation, many efforts have been made to include "real" problems in the classrooms. These are typically open-ended problems that have more than one solution. The source is often a company that hires students from that school. Although these problems may work well, there are difficulties that include:

1. They require significant investment on the part of the faculty and the industrial contact to identify problems that are suited to the expertise of the students and can be accomplished within a normal 15-week semester.
2. Typically, students are not given vital projects. The result is that the projects often seem tangential and not fully relevant to the business. If a design is really important, the company probably has engineers working on it and can not wait for a student solution.
3. Students do not see their designs used and products developed as a result of their designs.
4. Company representatives are not located close enough to campus nor have the time to be integrally involved in the design process.

Although engineering educators are seeking ways to bring real problems into the classroom, community service agencies face a future in which they must rely to a great extent upon technology for the delivery, coordination, accounting, and improvement of the services they provide. They often possess neither the expertise to use nor the budget to design and acquire a technological solution that is suited to their mission. They thus need the help of people with strong technical backgrounds.

These two challenges create the opportunity to meet both needs in a creative and effective way. The Engineering Projects in Community Service (EPICS) program provides a service-learning structure that enables community service agencies to work with engineering educators and thereby satisfy each other's needs. An example is a current project with ten agencies of the Tippecanoe County (Indiana) Homelessness Prevention Network. These agencies have a need to be able to share information between agencies in a secure manner to protect the confidentiality of those who are served by the agencies. However, by sharing information, services can be coordinated to produce more effective care for the community. Computer science and engineering students possess the technical ability to address this need.

The EPICS structure supports such long-term projects in which teams of undergraduates in engineering are matched with community service agencies such as the Homelessness Prevention Network (HPN) that request technical assistance. Under the guidance of faculty in engineering, student teams work closely over many years with their partner community service agencies to define, design, build, and deploy the systems the agencies need. The results are systems that have a significant, lasting impact on the community service agencies and the people they serve.

For the HPN project, a team of students was established in 1995 and have deployed a secure, distributed database for the local agencies and have been issuing new versions of their software (Appendix G). For the first time in Tippecanoe County, an accurate count of the homeless and families who are at risk has been possible.

Through this program, the EPICS students learn many valuable lessons in engineering, including the role of the partner, or customer, in defining an engineering project; the necessity of teamwork; the difficulty of managing and leading large projects; the need for skills and knowledge from many different disciplines; and the art of solving technical problems. The HPN team spent most of the first year of the project working with the agencies to define the database so that all the agencies could use it when it was finished. They also learn many valuable lessons in citizenship, including the role of community service in our society; the significant impact that their engineering skills can have on their community; and that assisting others leads to their own substantial growth as individuals, engineers, and citizens. Students who have worked with homelessness for a number of years report continuing community activities after graduation.

Purdue's establishment of a Twenty-first Century Conceptual Tools (TCCT) center is well suited as a partner with the EPICS project. The hallmark of TCCT is leadership for systemic reform of K through 12 education. TCCT works to achieve this through a focus on the skills needed by all to succeed in a technology based age of information. The EPICS program is focused on the enhancement and strengthening of technical engineering skills within the context of a community outreach program. The undergraduate students in the EPICS program develop an awareness of civic responsibility, not through lectures and even discussion on this topic in humanities and social sciences classes, but through the use of their technical engineering skills while solving real life problems within their own community. EPICS is a suitable partner for TCCT because it deals with problems in their entirety and complexity, and demands that students seek practical solutions using a variety of disciplinary approaches.

ORIGIN AND RATIONALE FOR EPICS

EPICS is but one example of the main function of an engineer, namely design. The teaching of design has been the subject of a great deal of effort and innovation. Design has been traditionally taught with the context of a single semester design course under the guidance of a faculty member and usually in a student's senior year. In the School of Electrical and Computer Engineering at

12. EPICS MODEL IN ENGINEERING EDUCATION

Purdue University, the need for a different kind of design experience was recognized by a group of faculty that was lead by Professors Edward Coyle and Leah Jamieson. They realized that the traditional design course did not fully simulate the kind of design experience that graduates would be exposed to nor did it provide the rich learning environment students needed.

One of the main barriers to creating a richer learning experience was the single semester course, which artificially limited the process and projects and required the faculty member to make decisions for the students in areas such as the initial problem definition. This is done because the students could spend most or even the entire semester understanding the problem and the context in which it must be solved. In engineering, it is critical to fully understand the context in which a design will be done and this requires a thorough problem definition

As an illustration, one of the EPICS teams is paired with the Imagination Station, a local children's museum to develop systems to aid in science, mathematics and technology education. The team was established in 1997 and has delivered several exhibits (Appendix H). Often, the first semester of a project is spent refining an idea and understanding the educational context in which the new exhibit is to be built. The design of the exhibits involves complex systems which must be integrated with each other as well as the educational aspects the exhibit is intended to convey. The engineering students must understand how the exhibit is to be used and what display materials must accompany the exhibit. Ideas are generated and refined as the team interacts with their project partners, the staff of the Imagination Station, to create a thorough problem definition for their design. The multi semester system provides the luxury of an iterative approach to developing a complete problem definition.

Another limitation of traditional design courses is that students are typically put into homogeneous groups (e.g., **all** seniors in electrical engineering). However, both industrial partners such as Boeing and ABET recognized the need to work on diverse and multidisciplinary teams. In such an environment, students would work in groups of freshmen, sophomores, juniors, and seniors from different disciplines.

Vertical integration provides an environment where the seniors act as the technical leaders and mentor the younger students. In any design process, there are tasks that require varying degrees of technical expertise. The students in such a vertically integrated team would need to identify which tasks need what level of technical expertise and divide the work accordingly, thereby using higher level thinking skills.

Similarly, on a multidisciplinary team tasks would need to be identified that are appropriate for each team member. For example, while developing an electromechanical display for the local children's museum, mechanical and electrical engineers might work on different aspects of the project but need to be coordinating their efforts. Students from education might work on the integration of learning objectives into the overall display while visual design students laid out the overall display. By working with students from different backgrounds and expertise, they learn to appreciate what each other brings to the project. They also understand that to design truly complex systems requires many different sets of skills and expertise. What we have found is that the students on teams such as the Imagination Station, understand the need for diverse skills and abilities on their teams and actively work to recruit students with such expertise.

The last step in creating a richer design experience required actual projects. The most common source for such engineering project has been industrial projects. As highlighted earlier, there are several difficulties associated with industrial projects. The result of these difficulties was that the alternative kind of design experience remained in the idea phase. That is until a program through the U.S. Department of Education was announced that encouraged community service projects. The community project was the missing piece they sought. Professors Coyle and Jamieson spearheaded a group of faculty who looked at using local, community organizations as the customers for design projects. At first, some questioned whether there were technical projects within the community service agencies. An initial meeting with the United Way resulted in over 20 local projects being identified. In 1995, the EPICS program was born. The result has been a plethora of projects from agencies such as the Homelessness Prevention Network and the Imagination Station. Both teams currently have enough future work identified for several years into the future.

EPICS DESIGN TEAMS

Initially, electrical and computer engineering students were placed onto five teams that all had community organizations as customers that are called project partners. There was a realization that to do real projects, they needed more than just electrical and computer engineers. The first teams included some mechanical engineering students that had somehow heard of the program and registered for the course. These students demonstrated the need for and the benefit of multidisciplinary teams. The program quickly moved to include other engineering disciplines as well as non-engineering students. As the program has continued to grow, the number of disciplines has continued to grow as well up to 20 academic departments in the 2000 academic year.

The teams that work with the Wabash Center Children's Services to develop electro-mechanical toys and play environments for children with physical disabilities have developed an understanding of how important it is to understand how the products are to be used and the needs of the therapists (Appendix I). The success of these teams has lead to the creation of an EPICS course number in the department of Child Development and Family Studies Department. The devices developed by these teams have greatly increased the capabilities of the Wabash Center to serve their clients.

We find that students who excel in the EPICS environment do not necessarily excel in the traditional classroom. We have numerous examples of our outstanding team leaders who manage projects who do not have high grade point averages. This was driven home when we were asked to recommend a student for an award. The criteria was that they be part of a service-learning program and had a grade point average of 3.3 or above. We made a quick list of a half-dozen outstanding students on our teams that had the required major. When we looked up their grade point averages to make sure they were eligible, only one had met the grade point average requirement. Three of the six had below a B average and two others were barely above a 3.0. All six were leaders on their respective teams but had not excelled by traditional standards in traditional classes. These students did very well in a team oriented and more open-ended structure that EPICS provides.

The traditional methods of evaluating students did not reflect their ability to be effective practicing engineers. The learning styles needed to be a practicing engineer are not restricted to those that coincide with traditional classrooms. There are numerous examples of alumni who did not do well as undergraduates, entering the workforce and becoming very successful similar to what we see in the EPICS program. In some ways, the EPICS program more closely simulates the environment that students will face upon graduation. More importantly, it provides a diverse experience to complement the traditional classroom experience and widens the opportunities for students with skills to be effective engineers to be successful students.

This realization of a need for a broader scope to address societal problems links EPICS to an important theme in current educational practice. There is a large movement within both K through 12 and higher education to incorporate what is called service learning into curricula. Service learning seeks to expose students to community service and the needs of the local community while enhancing academic study within a discipline.

EPICS has quickly become a recognized program within the service learning community even though it was not initially designed as a service-learning program to promote community service. EPICS was designed to provide a quality design experience for students. However, they must gain an understanding of the community needs that are being addressed by their community partner to understand the context of the problems they are solving. In engineering design terms, this is called understanding the customer. They need to understand the context in which their design is to be implemented. The result is that they become aware of community needs they are helping to address. A deeper dimension to the EPICS experience is that the participating students also become aware of ways in which their professional expertise can be used within the community agencies they serve.

Feedback from alumni and their employers is that the EPICS experience is an excellent simulation of what they will experience upon graduation and thereby a great preparation for their careers. It is important to consider how traditional classroom experiences artificially limits the potential pool of science and engineering students entering college. Tomorrow's economy will require an increased number of technically trained professionals. However, the number of high school graduates who are interested in these fields are not keeping pace. Of particular concern is the large numbers of women and minorities who lose interest in math and therefore do not have the mathematical background or interest in such careers. Programs such as EPICS and other service-learning programs can increase this interest as students experience learning environments that are more conducive to their individual learning styles as well as being able to see interesting applications of their work. Studies have shown that students in middle or high school have a hard time seeing the connections between social issues such as homelessness and a career in computers. Programs such as EPICS seek to provide these links.

Within the engineering community, there are still skeptics who question the service learning approach to design. At one meeting, an industrial engineering alumnus commented that it was great that we were helping them be better people and better engineers, but from his perspective as an employer, he wanted the focus to be on being better engineers. The alumnus did not see that both goals are linked. By putting the engineering problems within a community service context, the community service awareness and growth comes about as a byproduct of being a good design engineer without taking time out of the design process to study how to become a better person.

A key component to service learning is a significant reflection upon a student's activities. These activities are not common to engineering programs or in mathematics. They have been incorporated into the EPICS program as part of the grading and assessment activities and serve as a learning tool to get students to reflect upon their work. These provide students with opportunities to write about what they learned and experience.

CURRENT PROGRAM AT PURDUE UNIVERSITY

The EPICS program has grown significantly from its beginnings to include over 300 students working on 24 teams each semester in the 2001 academic year. Each team has one or more advisers, one or more teaching assistants, and 10 to 20 students. These teams are paired with one or more community organizations. Students register for either one or two credits per semester and are encouraged to register for multiple semesters. This creates continuity whereby projects that span multiple semesters are possible. A typical team works on three projects being worked on in a semester.

The students come from over 20 academic departments: electrical, computer, mechanical, aerospace, materials, civil, industrial, chemical, interdisciplinary engineering, computer science, biology, chemistry, sociology, visual design, audiology, psychology, management, elementary and secondary teacher education, nursing, child development and family studies, and animal science and natural resources. Students from all the disciplines participate in the design and assessment of the complex systems created by the EPICS teams. They leave the program with an increase in awareness of the importance of effective teamwork in such complex systems as well as communication between teammates and the customer.

12. EPICS MODEL IN ENGINEERING EDUCATION 231

Other examples of team projects that illustrate the scope of EPICS at Purdue follow. Further information on these and other projects as well as the EPICS program in general can be found at Appendix J website:

Project Title: Habitat For Humanity *Project Partner:* Tippecanoe County Habitat for Humanity. *Tasks:* Design energy management systems to minimize home operating costs. Develop new construction techniques and investigate new construction materials. *Impact:* More energy-efficient housing for Habitat families. *Facts:* Begun fall 1996, Disciplines: Electrical Engineering, Computer Engineering, Civil Engineering, and Mechanical Engineering. Appendix K.

Project Title: Speech-Language and Audiology Clinics. *Project Partner:* The M. D. Steer Audiology and Speech-Language Center. *Tasks:* Integrate a speech recognition system with computer graphics and games to encourage language development. Design and build a working model of the vocal system, for clinicians to use in educating laryngectomy patients. Automate calculation of speaking rate from clinical sessions. *Impact:* New services for the clinic's client; improved feedback to speech clients. *Facts:* Begun fall 1995, Disciplines: Electrical Engineering, Computer Engineering, Computer Science, mechanical Engineering, Industrial Engineering, and Audiology. Appendix L.

Project Title: Happy Hollow Elementary School. *Partner:* Happy Hollow Elementary School. *Tasks:* Develop systems to aid in science, mathematics, and technology education. *Impact:* Improved educational resources for the community. *Facts:* Begun fall 1997, Disciplines: Electrical Engineering, Computer Engineering, Mechanical Engineering, Mechanical Systems Engineering (MSE?), Industrial Engineering, Computer Science, Visual Design, Sociology, and Education. Appendix M.

THE PHASES OF AN EPICS PROJECT

Each EPICS project involves five phases: Finding Project Partners, Assembling a Project Team, Project Proposal, System Design and Development, and System Deployment and Support Appendix N. Each phase requires different kinds of problem solving and model building. As an illustration of how these phases are implemented, the Constructed Wetlands (CS) EPICS team will be used as an example. Appendix O

Phase 1—Finding Project Partners: Each EPICS project addresses the technology-based problems of one or more service organizations in the local community. Agencies with appropriate problems must therefore be found.

When planning for the EPICS Program started in the fall of 1994, we were able to contact many different service agencies at the same time by making a presentation about the program and its goals at the monthly meeting of the directors of all local United Way agencies. This single presentation led to many discussions with individual agencies and a long list of potential projects.

From this list of potential projects, those best suited for the EPICS Program were selected. Projects are selected based on:

- Significance—not all projects can be undertaken, so those that should provide the greatest benefit to the community are selected;
- Level of Technology—projects must be challenging to, but within the capabilities of, undergraduates in engineering and the other disciplines involved in EPICS;
- Expected Duration—although projects may have components that can be completed in a semester or less, each project must be long-term, requiring two or more years of effort from a team of approximately eight undergraduates. Because the first round of projects that grew out of the United Way presentation, the source of new projects has been varied. Faculty has initiated some projects, while students have suggested others. As the program has become known in the community, several projects have been proposed by local community service organizations.

Each year, the EPICS co-directors and faculty, using the significance, level of technology, and expected duration criteria, select new projects. Once a project has been selected for the EPICS Program, the service agency that will be directly involved is designated the Project Partner.

Additional projects are identified through discussions with other organizations. These have been initiated by EPICS and also by local organizations that have heard about EPICS. In 1997, a faculty member from the Department of Forestry and Natural Resources approached EPICS with a proposal to design and build a constructed wetland to mitigate runoff from the Purdue University agricultural farms. It was quickly determined that this was an excellent fit for the EPICS program and the Department of Forestry and Natural Resources and the Purdue University farms would serve as the project partners. An adviser in Civil Engineering was identified and the team was born in the fall of 1998.

Phase 2—Assembling a Project Team Once a project and Project Partner have been identified, a student team is organized. Advertising the project in undergraduate classes and on the World Wide Web leads to the team being organized. Eight to fifteen students are chosen for each Project Team. Depending on the needs of the Project Partner, teams may reflect a single engineering discipline or may be multidisciplinary, including students from two or more engineering fields.

The team must be vertically integrated with a mix of freshmen, sophomores, juniors and seniors. Each student is requested to participate in the project for as many semesters as possible. The combination of a vertically integrated team and long-term student participation ensures continuity in projects from semester to semester and year to year. Projects can thus last many years if new students, especially sophomores, are recruited for the project as team members graduate.

The CW team presented an additional challenge, as it required mostly Civil Engineering students who were not widely represented within EPICS. A recruitment campaign was launched and the team was begun. The initial team was comprised of freshmen through seniors from Electrical Engineering, Civil Engineering, Interdisciplinary Engineering, as well as Chemistry and Biology.

12. EPICS MODEL IN ENGINEERING EDUCATION 233

A Civil Engineering faculty member, as well as an environmental engineer from a local company advises the team.

In the first 5 years of operation, 700 students have participated in the EPICS Program. Retention has been excellent, with over 70% of the students who were available to return to the program in the following semester (i.e., were not graduating or off campus on a co-op assignment) did so.

Phase 3—The Project Proposal During the first semester of a project, the Project Team meets several times with its Project Partner and the EPICS faculty to define the project and determine its goals. During this phase the Project Team learns about the mission, needs, and priorities of the Project Partner. A key aspect of this phase is identifying projects that satisfy three criteria: they are needed by the Project Partner, they require engineering design, and they are a reasonable match to the team's capabilities. Also, to ensure that the students build confidence and the Project Partners see progress, the teams are encouraged to pursue a mix of long-term and short-term projects. Short-term projects generally require only one or two semesters to complete; long-term projects take 2 or more years. This process of project definition culminates in a written proposal and presentation in the fourth week of the semester. The proposal is critiqued during a lab session, with detailed feedback provided in the areas of organization, content, technical approach, and writing. The proposal must be approved by the EPICS faculty and then be accepted by the Project Partner.

The CW team's proposal had the complication of determining what the team could accomplish and what needed to be subcontracted out. The wetland would need significant excavating involving heavy equipment that was not available to the students. The wetland would be made up of two cells each approximately 40 feet wide and 800 feet in length. The proposal forced the students to break the large problem of the wetland into much smaller segments that could be accomplished by the team itself or contracted out.

Phase 4—System Design and Development Starting from week five of the first semester of a project, the Project Team's goal is to produce a prototype of the hardware–software systems discussed in the proposal. Interaction with the Project Partner continues in order to ensure that the systems being designed and developed are as desired. The formal portion of this interaction takes the form of a written progress report and an oral presentation delivered by the Project Team to the EPICS faculty and the Project Partner at the middle and end of each semester. The progress reports must meet the same standards as the proposals. The Project Team demonstrates the current state of their systems to a team of EPICS faculty every 5 weeks for the duration of the project. This phase of a project lasts as many semesters as necessary for the team to complete the project to the satisfaction of the Project Partner.

The CW team's wetland was designed and begun in the first academic year. Much of the work in the first year involved modeling the wetland, itself and the design criteria needed. Students were forced to learn about things not covered in their courses such as plant selection, media to fill in the wetland and other factors specific to the wetland. The advisers helped to guide the students as they probed for the correct criteria. The team built small models and an aquarium to test plants.

Phase 5—System Deployment and Support The ultimate goal of each Project Team is to deliver a system to the Project Partner. After fielding a prototype, the team must train representatives of the partner in the use of the system, collect feedback, and make any reasonable changes requested by the partner. One of the hallmarks of the EPICS Program is that the systems designed and built by the students are deployed in the field, where they provide real, needed service to the community. It has been our experience that after a team fields a project, other projects are identified and the teams continue their relationship with the agency.

The CW team deployed the wetland and is continuing to improve on the initial design. While most EPICS teams continue to work on projects such as software, where a new version can always be written, or with a community partner that has other project ideas, a large construction project such as this, may have a logical ending point. When the question of ending the project was posed to the team, they responded by developing monitoring mechanisms to study and wetland itself and adding observation platforms to create an educational center around the wetland. The student team understood the value in what they had created and are seeking to leverage their work to benefit the educational community.

GOALS OF THE EPICS PROGRAM

Goals of the EPICS Program include: (a) providing students with multi year, team-based design and development experience; (b) teaching students, by direct experience, how to interact with each other and with customers to define, design, build, and deploy systems that solve real problems; (c) and showing engineering students how their expertise can benefit their community.

Whereas many service-learning programs expose students to community service, the EPICS program connects students' future professional expertise with community service. An example of this is the team that works with Habitat for Humanity. Most of the students who register for that team have worked with Habitat for Humanity as a volunteer. They all realize that they can donate a weekend to help roof a house or hang drywall. They have not, however, had opportunities nor realize that as engineers they can provide services for Habitat (Appendix P). The engineering students have created computer models that analyze the energy utilization of the homes that are being built and can recommend modifications. These changes make the housing less expensive to operate and thereby increase the value that Habitat is providing.

A number of skills that students need to become successful, both as engineers and as members of their community are emphasized:

- *Design and Modeling:* Each project is chosen to have significant technical challenges. The students have an opportunity to develop and use lessons

12. EPICS MODEL IN ENGINEERING EDUCATION

learned in their classes and develop models to apply them on real problems in the context of designs for community service organizations.

- *Communications:* Each project requires written reports, oral proposal and progress presentations, oral communications with project partner and consultants, and intra-team communications.
- *Organizational ability:* Because the scope and size of a project is much larger than would be possible in traditional courses (where much time is spent covering course material), students have to apply what they have learned to less well-defined problems across a variety of disciplines. This experience should encourage the development of the students' analytical thinking and organizational skills.
- *Teamwork experience:* Because projects are large, teamwork is an essential component. Students learn that the outcome of their project depends on the efforts of every member of the team. They learn how to divide up a large problem, assign and schedule sub-tasks, and integrate the pieces into a working solution.
- *Resourcefulness:* The multi class projects should encourage students to pursue non-traditional educational resources. Projects involve upper-class students in the process of guiding the less experienced students. Students are encouraged to learn from each other, as well as from the project partner and academic consultants who have experience related to the projects.
- *Resource management:* Each team develops a proposal for the equipment and space requirements for the project, and must to take into account the resources of the project partner. As a result, students should understand that practical as well as theoretical issues must be addressed throughout the design and development process.
- *Sponsor awareness:* Projects are being drawn from local community service agencies. Sponsor satisfaction is an important criterion for judging project success. Hence, the projects increase the students' awareness of the importance of the customer in producing a high quality product.
- *Expanded awareness:* By dealing with people outside of the academic community, the students become more aware of the world outside of Purdue that they are preparing to enter.
- *Professional ethics:* Professional conduct, both in relation to the project partner and within the team itself, is an essential component of successful projects. Students have to maintain an awareness of ethical principles while meeting the demands of the project.

The Importance of Fundamental Designs and Models

Although a broad array of abilities is needed for engineers and there is an increase in interest in the "soft" skills of engineering, it is still critical that they have a firm understanding of the fundamental mathematical and scientific concepts. Because modern employers are demanding that engineers be more well rounded, a basic understanding of fundamentals and the ability to apply this to the design of complex systems is still essential.

To address this need for fundamental skills, the engineering education community has created curricula that are heavily oriented to mathematics and science. The students sometimes struggle with understanding these concepts and applying them to new situations. There is a need to provide contexts for the theories they study. Engineering is full of applications, but there is a huge difference between a textbook example and a real example the students can touch and feel. The old paradigm was to continue to drill students in the concepts to establish this grounding in fundamentals. What has been found, however, is that many of the students need to make concrete connections to applications to really understand the concepts.

The EPICS program has provided many opportunities for students make this sort of connection and for the lack of connections in other courses to surface. A senior Electrical Engineering student at Purdue University is a good example. He was nearly a perfect A student who later went on to graduate school. This student was asked to evaluate a circuit that had been built the previous semester and had broken over the summer. This task involved troubleshooting the circuit to see where it had broken. He was unable to do this at first. He had received A's in his courses that covered circuits, but when asked to evaluate an actual circuit with which he was familiar, he was unable to make the connection.

The idea of developing models to solve problems is exactly what engineering education is about and what EPICS is accomplishing. The EPICS environment provides an excellent environment as the community service applications are generally out of the traditional applications covered in classes. The issues faces by EPICS teams involve the same technological issues as traditional applications, but often with a new slant. Adapting an electromechanical device so that a physically challenged child can operate it effectively is not what most engineering students think of as the work of an engineer. This out-of-the-box thinking forces the students to evaluate the models they have for applying their knowledge to new problems.

ASSESSMENT OF EPICS PROJECTS

The EPICS has integrated several types of assessment into the program to provide information on the effectiveness of the program and the performance of individual students. Because all of the work in the EPICS program is done as a team, there is a significant challenge of breaking out individual effort for evaluation. One of the most effective means of evaluating individual efforts is

12. EPICS MODEL IN ENGINEERING EDUCATION

with a design notebook. Students are required to keep a design notebook and record all their work in that notebook. They record the calculations they perform, as well as the thought processed they went through to make the calculations. They also record all meetings and other information related to the project. The notebooks function as a journal during the semester. The notebooks are evaluate three times during the semester. At the end of the semester, the notebooks serve as a portfolio of the student's work and thought process during the project. Examples of the notebooks and the criteria given to students are found in Appendix Q.

Another very effective means to determine individual contributions is through peer evaluation. (Appendix R). Each EPICS team evaluates each other twice per semester. This has been a successful way to identify problems, as well as outstanding students.

Summative evaluations have been performed each semester to measure the program objectives. The results have been compiled and presented at numerous forums. (The papers can be accessed at Appendix S). The program has been shown to be an effective means of exposing students to the design process and to teaming and is well aligned with ABET criteria.

Free Response Evaluations (Appendix T) are also used in the summative evaluations. One of the most interesting results is how students respond to the different learning environment that EPICS provides. The responses of the students indicate that they do not see how the team advisers are really teaching. They associate teaching with lecturing and the experiential learning environment doesn't register that there is teaching going on. When probed through focus groups, the students to come to this realization, but not through traditional summative evaluation. The students are keenly aware of the soft skills that they are learning. In one semester, out of 202 responses, 199 listed teamwork as one of the three most important things that they had learned from the course. They tend not to list the improvement in their technical skills unless probed further.

Other evaluation mechanisms are incorporated into the grading process. Each team gives a mid semester and end of the semester presentation which are evaluated by the advisers and other students. Each student also completes semester goals as the beginning of the semester and a self-assessment at the end. (Appendix U).

THE NATIONAL SCOPE OF EPICS: BEYOND PURDUE UNIVERSITY

By the fall of 1998, EPICS Programs were also underway at the University of Notre Dame (Appendix V) and Iowa State University of Science and Technology (Appendix W). The existence of EPICS programs at several sites opened the possibility of addressing community and educational needs that extend beyond those of a university and its local community. These included Habitat for Humanity, the Salvation Army, the Red Cross, and the YWCA. EPICS project teams working with these agencies in different cities could address such national-scale problems as homelessness, substandard housing, and disaster relief.

One multi site EPICS project in progress, the Homelessness Prevention Network (HPN) project, was initiated in the spring of 1998. It involves two EPICS teams, one at Purdue (Appendix G) and one at Notre Dame (Appendix X). Both are working with a number of agencies in their home cities of Lafayette and South Bend, Indiana. The local goal for each team is to enable its partner agencies to share demographic and services-provided information about their clients. The agencies could then produce duplicate-free counts of homeless individuals and families, meaningful data on the use and effectiveness of services, and a record for each client that can be used for case-management across all agencies and all available services.

The common goal of these two HPN teams is the sharing of data on homelessness between Lafayette and South Bend. Success at this task will enable city-to-city comparisons, help track migration patterns, and determine which services are the most effective. The extension of this project to national scope could provide the first accurate characterization of homelessness throughout the United States and lead to better-informed public policy in the area of homelessness.

The potential benefits of many different national-scale EPICS projects—pursued by a national-scale coalition of EPICS sites—led to the creation in September 1999 of the National EPICS Program. The universities participating in this program for the 2001 academic year are Purdue, Notre Dame, Iowa State, the University of Wisconsin–Madison, Georgia Institute of Technology, Penn State and Butler University. Seed funding and other support for this effort has come from the National Science Foundation, Microsoft Corporation, and the Corporation for National Service.

Moving to national-scale projects requires students to produce new models for solving more complex projects without being colocated with all team members. This is exactly the environment they will enter upon graduation however. Technology has produced a work environment where global partners work together to produce new products and services often without actually meeting. These new models in problems solving continue to enhance the educational experience the students receive as well as leveraging university resources to address larger social issues.

CONCLUSION

The EPICS program has proven to be effective in exposing students to the engineering design process, in addition to being a sound pedagogy for engineering design. The program teaches by direct experience how a student can use their technical expertise to benefit the community. This is true whether the students are in engineering, liberal arts, or other majors.

As Lesh and Doerr point out in the first chapter, "model-eliciting activities can be designed to lead to the significant forms of learning." Models as

12. EPICS MODEL IN ENGINEERING EDUCATION 239

instructional strategies "focus on helping students develop powerful models and conceptual tools for making (and making sense of) complex systems." The topics for the EPICS program, such as providing technical assistance to a homelessness prevention network, allow students to see how their mathematically oriented technical abilities can be used to design models that solve a real-world problem. The EPICS program helps the students not only gain technical expertise, but also work in a collaborative team that solves complex problems. By setting these problems in the community, the students are presented with a human context for their technical solutions.

Chapter 13

The Case for Cases

Geza Kardos
Carleton University

How do I pass on the experiences of a lifetime to students? How do I teach students to analyze complex situations? How do I teach them to make realistic decisions? These are the questions that I asked myself when I first started teaching engineering 35 years ago, after 20 years of engineering practice. Any answers I came up with at the time were unsatisfactory. Textbook problems, whether as student assignments or class demonstrations, seemed dry and pedantic. Problems I constructed from my own experience added something, but were still unsatisfactory.

The solution to my dilemma came when I had an opportunity to learn about the *Case method* in engineering at a National Science Foundation sponsored summer program at Stanford University. I had already been exposed to the case method at a short business course while working as an engineer. Even then, I was struck by the beauty of the method and wondered why it was not used in engineering.

It was in the case method that I found the answers to my dilemma. With the use of cases I was able to guide students through their learning using my professional experience. Since then, I have used some form of the case method in my teaching for more than 25 years. The following examples that I sight are engineering examples, but this does not mean that case use is limited to engineering. Real engineering is about modeling, problem solving, and decision making. Education for any profession that requires problem solving and decision making can be enhanced with the use of cases. I would like to pass on to you what I have learned and discovered about cases and their uses over the last 25 years. I have taught courses entirely through cases and I have used cases to supplement traditional courses. I have learned about case use from colleges and my students. Much of what I have here comes from case workshops presented throughout the world.

Here I explore why and how cases contribute to a student's education in a unique way. I define what I and my colleagues mean by a case. Cases may be used in different ways. I examine a number of these. Then I present some student and faculty opinions as to what can be learned from cases. Finally I present what I believe constitutes a good case based upon my experience in using cases and writing cases.

WHY CASES?

In traditional teaching, attempts are made to introduce the fundamentals of models and model building using various combinations of conventional lectures, examples, and textbook problems. Using this approach, students, may not be required to build their own models in spite of the fact that model-building experience tends to be essential for the students to work in their profession. For example, to operate as a professional, problems must be recognized and identified, models must be selected and adapted, predictions must be generated and decisions must be made, and actions must be taken or recommended. All of these activities occur within a context that is unique and complex. Within any real context the data may be insufficient, obscured, or redundant. Decisions must be made regardless of the difficulties. These are the kinds of problems students will be faced with once they leave school. Yet, experiences that are restricted to solving textbook problems do not educate students to deal with these characteristics of real situations. Textbook problems, for the most part, are developed as drill in a specific technique, drill is necessary but it is not sufficient.

Fortunately, many students, through experience, eventually make the transition from academic problems to real problems. But unfortunately, transitional learning may demand a high price because, good judgment comes from experience, and the most valuable experience comes from poor judgment. Therefore, to spare our students from the cost of bad judgments, and to ease the transition from school to real life problems, it is often wise to try to give students realistic experiences where the costs are not so high. One way to do this is to have them deal with realistic problems during their school years. That is, we can put them in situations where they struggle with cases that reflect the real world.

Using stories, such as parables, legends, fables, and biographies, has been a long accepted approach to teaching and learning. Good stories may or may not involve the reader with places and situations that are new to them. Stories cannot only entertain but can carry a message; they can instruct; and, they can provide information and insights in a manner that cultivates retention. Good stories also sometimes require that the listeners make broader interpretations and associations from the story.

Written cases are in some ways similar to short stories. Most cases are stories centered on individuals who must make or have made choices with respect to problem situations presented in cases. Often, cases are open-ended and complex, and students are expected to identify with the participants and to develop their own solutions and recommendations about the problems while

13. THE CASE FOR CASES 243

providing analysis and rationale to support their actions. A strength of case-based education is that students look at problems within a context and not in isolation. They must consider all the factors and select the best procedure to deal with them. They become part of the problem resolution.

Cases have long been a part of professional education in fields like medicine and law. For example, in medicine, the case method predates the scientific approach. There of course, the teaching hospital provides the student with hands-on cases. Medical practitioners learn by doing. In law, the situation is similar. For example, in English-speaking countries, law practice is based upon case law. Consequently legal education must be case based. Legal cases are readily available because they are drawn from court records. Shortly after the turn of the century, the business school at Harvard started developing the case method in emulation of their law school. But, prewritten business cases were not available. Therefore, Harvard introduced case writing as part of the case program.

In Engineering, teachers had long used anecdotes from their professional practice for teaching. Consequently, a number of textbooks have appeared over the years with stories and examples from engineering practice to validate the technical content. In the early 1960s, engineering cases were produced in pamphlet form mainly at Stanford and Berkeley. They introduced the case method into engineering education, modeled after that of the Harvard business school. From these beginnings, the case method has defused through most fields of professional education.

What Is A Case?

Some case users have defined a case as any written material that can be used as a basis for discussion, problem assignment, or analysis. Although this definition has a ring of truth to it, it is far too inclusive and tends to obscure the differences between cases and any other form of teaching material. For my purposes, I prefer to define a case to be: An account of a real activity, situation, or problem containing some of the background and complexities actually encountered by the individuals who have to do the job.

A case is preferably about real problems taken from life. It presents the story of a real experience, from the viewpoint of one or more of the participants. The case context, in addition to the narrative, may present additional information in the form of sketches, drawings, photographs, calculations, test results, scheduling data, production processes, budget information, and other pertinent data. The focus in a case is to tell how people achieved their results rather than demonstrating the validity or elegance of a solution. A case shows individuals interacting in a meaningful way to accomplish a task. The documentation is often in segments with each portion terminating at a critical decision point.

A case can be about a feasibility study, a design appraisal, an operational performance assessment, a failure investigation, a redesign, and so forth; or, it can involve some type of scientific research. The case does not necessarily present new knowledge (although it may do so). Rather it should show the nature of the problem, possibly how it was solved or how decisions are made. The details of the modeling may or may not be present.

A case is not a technical paper nor a news report. Technical papers are written to persuade the reader, to demonstrate the clearest and simplest rational about the work done. They represent a summary of the author's ideas. A news report is the reporter's interpretation and analysis of what happened. News reports and technical papers can be part of a case. But, by themselves, they do not constitute a case.

Even though the terms case and case study are often used loosely and interchangeably, I prefer to differentiate between the two. For us, a case is simply the description of a situation, an event or an occurrence as it actually took place without the intrusion of the writer. A case study examines events and draws conclusions from them based upon the writer's interpretation. The case study is openly subjective, whereas the case tries to be objective. For example, in a learning situation, the objective case is presented to the students and they are expected to produce a subjective case study. Many case studies (Appendix 2) are available that use historical examples to present the author's insight, analysis and conclusions. Such case studies are valuable learning material, but their objective is to present the author's thesis, they leave little for the student to do.

Finally, there is the term case method this refers to using cases in a classroom setting where cases are the principal method of teaching.

HOW CASES ARE USED

There are a variety of ways that cases may be used to add spice to courses. How many cases to use, and how to use them, depends on the course objectives and teacher's skill. Cases can be used with a variety of objectives. These include, problem identification, problem definition, problem solving, providing insight into practice, recognizing application of basic principles in technology, modeling complex situations, and making decisions, I have used cases for all of these. I have used cases in most of ways described here, whether teaching a course completely by the "case method" or a regular lecture course. Regardless of where and how, cases always seemed to provide an interesting learning experience for the students.

Class Discussion

The heart of the "case method" is using cases as the basis for class discussions. Students are assigned a case, or part of a case, to study and to prepare for the class discussion. Questions may be assigned to direct their study. Or, students may be asked simply to read the case and be prepared to discuss what they find important. Students may be asked to develop a conceptual solution (or a relatively detailed one with calculations and sketches, etc.) and to be prepared to present and defend it. Then, during the next class, they are required to present their assessments, and discuss it. If the case is in more than one part, after the

13. THE CASE FOR CASES

first discussion, the next part of the case is given to the student for further study, or to show students how the issues in the case evolved.

The class discussion becomes a forum for group problem solving and decision making where each member of the class, including the teacher, has an opportunity to contribute to the solution as well as the responsibility to convince their classmates that it is useful. Most cases provide opportunities to raise questions and search for answers, and because cases are based on reality, there is no single "correct" question or single "correct" solution. Cases, like good problems, may admit many possible solutions; they may depend on many disciplines. Many cases include questions of human behavior and ethics in addition to the technical problems, these must be considered in the final resolution of the issues.

In case discussions, the role of the teacher changes from the roles assumed in traditional practices (Appendix 3). They no longer act as the source of wisdom and truth. Rather, their role is more like that of an interlocutor. Their job is to keep the discussion going and remaining fruitful. They are no longer, the sage on the stage but the guide on the side. For example, in a good case discussion, it should appear to the students (and any outside observers) that the students are doing all the thinking, most of the talking, and all of the relevant discoveries. The objective is for students to seem to originate the ideas, organize the discussion, establish priorities, and cover the material in the time available – without apparent intrusive interference from the instructor. Students must become actively involved so that the discussion is emotionally satisfying, intellectually productive, and sometimes brilliant enough to provide new insights. The instructor's role is to manage the learning so that the discussion becomes all of these things to students.

In class discussions, students soon identify with the people involved in the case. They find they are required to extend their knowledge and back ground; and, they are challenged to apply their knowledge and experience to the specific problem at hand. This produces a focus for class discussions that cannot be achieved without cases. By having students and their peers wrestle with issues in the case, they learn to think, to detect useful information, to recognize false arguments, to construct a coherent problem definition, to see more than one side of a situation, to decide which is best, and, to formulate and defend their chosen course of action. Opportunities to practice these skills are rare outside the case method.

Students' usual educational experience tends to condition them to regard the instruction period as a time for absorption or recitation. They expect the instructor to provide "answers" or show approval for the students "correct answer." Student concern is largely to provide the response that the student perceives the teacher wants, and thereby earn a good grade. In case discussion, this pattern does not apply. When the teacher acts as an interlocutor, neither giving answers nor approving answers, the students soon learn that they are on their own.

Students soon recognize that successful treatment of case situations often means dealing with an accumulation of facts and feelings that must be treated as a whole. They find they must apply their abilities and knowledge to facts and issues in the case. They must deal with both technical and managerial issues. The case will not present them with well-structured problems, nor can they be sure that the solution presented by the principles in the case is correct. The objective in the discussion is to develop in the student, independent thinking, and decision making skills through practice. In the discussion, the emphasis is on the practice of these skills.

It should be noted that, in a discussion mode, the use of only one case is rarely profitable since open ended discussion is a new experience for students. Generally, they need at least one case to gain experience and get some feeling for the dimensions of this kind of learning. Use of at least two, preferably three or more, cases is preferred.

The discussion mode is particularly useful in having students learn about problem definition, model development, professional practice, and the subtle way that science can aid in decision-making. These skills are very difficult to present in lectures where students either see them as a set of rules to follow or as a "philosophy" of limited practical value.

A teacher's preparation for case discussion is often more demanding than preparation for a lecture. For a lecture, teachers control the schedule of the presentation and can sidetrack any questions that may not seem immediately relevant. With case discussion, the teacher must be thoroughly familiar with the case and any auxiliary information; they must anticipate where discussion may lead; and, they must be prepared to deal with alternate scenarios. If the case is in several parts, the teacher must be familiar with where the case is going so that, as discussion develops, the class will at least touch on the direction in which the case actually develops. What is more, the teacher must be prepared to abandon all their preparations when the discussion opens unanticipated but useful and exciting learning directions. More effort? Perhaps, but the results are much more fun and enjoyable.

Cases As Projects

Cases provide a ready source of projects within a realistic context. The project can be to complete the part or whole of the project described in the case. It can be to undertake some ancillary project not detailed but relevant to the case. Students can be asked to update the analysis in the light of new technology or do an alternate analysis for the case. The project may be to provide an alternate solution to that in the case. Or, by judicious editing of the case details, the student can be asked to define and solve the problem presented by the case. In extracting projects from a case, the teacher has an opportunity to assign projects that have a realistic context. For example ECL 122 (Appendix 4) "Pumping Waste Acid Deep Underground" provides numerous projects that can be assigned.

This case is more a real problem statement than it is a history of an action. The disposal of waste pickle liquor, a mixture of hydrochloric acid and sulphuric acids, is a problem in steel-making. The mixture is highly corrosive, and it's also

13. THE CASE FOR CASES

inconvenient to move or to store. Terry Abbey is the engineer who was placed in charge of this waste acid disposal project about midway through the project.

A well was to be drilled into a sandstone layer deep underground. A layer permeated with salt water. Other similar wells had been drilled. but not without problems. With a change in pumping demand, a subsequent change in the fluid head within the down pipe resulted in the inward collapse of the down pipe on more than one occasion.

The case study begins and ends with an exposition of the problem. The successful or unsuccessful conclusion is not given. The utility of the case study is in its description of a real problem. It is a simple one, not too difficult to understand.

What questions should be asked? What are the pressures, flow rates, and acid densities? What is the worst design condition that can be imagined? The obvious projects that can be drawn are the ones stated in the case description. But, in addition, questions about the validity of such waste acid disposition can be raised with the subsequent project to assess what will happen to the acid once it is inserted underground.

Cases for New or Complementary Technology

With rapidly the increasing knowledge, students should develop patterns of self instruction and learning, since most of their useful knowledge will have to be acquired after graduation. As far as practicable, they can learn new or ancillary technologies from textbooks or library assignments. Students find such assignments, relatively dull and profit little from them unless there is a specific purpose for the effort. Cases can be used to provide motivation for the acquisition of additional knowledge, often without the students knowing that's the purpose of the exercise.

For example, in a course in Machine Design, when the topic of journal bearings was due, students were asked to consider the first part of ECL 68 (5), "Tape Recorder Capstan Shaft." Before they were given any instruction in bearing design.

This short case describes design of a capstan shaft, which must operate smoothly in a precision tape recorder, it stops at the point where bearings must be selected. It also explains how the engineer on the job designed the bearings and, in particular, how he solved problems of "flutter" due to lubrication. (Parts A, B).

Students were asked to analyze the bearing, which was low cost, reliable, and without "flutter" over a given speed range, to see where the problem lay. Since the case was a real problem, it contained explicit and implicit dimensional, functional, and financial limitations within which students had to apply textbook theory. They tackled the bearing design using their own interpretation of the text material.

Cases as Problems

Two difficulties with textbook problems are; they are primarily for providing drill in manipulating the equations in some specific sections of the text, and that students do not learn to identify and model the problems from reality. Problems can be abstracted from cases and used as textbook problems. Alternatively, sections of a case can be assigned with a specified objective, with the students being required to define and model the problems.

For example ECL 151 (Appendix 6), "Underwater Pipeline," is a source of several text book type problems as outlined by the reviewer. These problems have an advantage, although the students can treat them as textbook problems, they have an opportunity to discuss their solutions with the whole class. And the solution of one problem leads to the posing of additional problems before they are finished with the case.

This case should be useful in any course that emphasizes the application of analytical techniques to real problems as opposed to the usual textbook format of "given-find." Specifically, the case involves the determination of the elastic curve of a statically determinate beam. Thus, it is appropriate for students with this background. The beam problem alone is interesting, because the slope and deflection at either end are known, but the length of the beam is unknown! This is quite disconcerting to some students.

The case is divided into three parts. Part A describes the proposed concept and the constraints involved, which are the pipe size, the maximum allowable stress, and the minimum allowable radius of curvature. Part A can be given to the students with the assignment of determining whether or not the stress and curvature constraints are met. At the next class session, hopefully, at least one student will have an analysis to present. In any event, Part B can be distributed and the analytical model set up and discussed.

Part B concludes with the rejection of the proposed solution by the company management. In his report, the analyst stated that non-linear beam theory would have been more appropriate, but he felt that the linear theory he used was adequate. The management group, presumably being nontechnical people, rejected the analysis in the hope that the non-linear theory would prove the original concept adequate. Several discussion topics arise here: the difference between the linear and non-linear theory, how to go about solving the non-linear equation, the possibility of the analyst not mentioning the non-linear theory in his report. This session could end with the students being told to assume that they had available a hundred feet of pipe. They could be asked to devise a test to confirm the linear theory; how would they set it up, what would they measure, and what modifications to the original analysis should be made to apply it to their proposed test.

Part C describes a test the analyst made in his back yard, using pipe he had previously purchased for a lawn sprinkler system. The students can discuss their proposed tests and compare their ideas with what was actually done.

Cases and Lectures

Cases can be used with lectures to provide examples of the application of lecture topics, especially in advanced topics where students may not have experience.

13. THE CASE FOR CASES 249

For example, the case ECL 87 (Appendix 7), "Design of a Creme-Lite Mixer," was given as a written assignment to analyze of the design methodology to junior year students.

This case should be useful for a course in mechanical design. It may also be used for graphics or design methodology.

A mechanical engineer is asked to design a large mixing tank. However, the fluid changes its viscosity rapidly in the temperature range being used. Moreover, no precise data is available on its viscosity. As in many design situations, data are incomplete. The engineer takes a very practical approach to getting viscosity data and proceeds with the design. The design calculations for the tank, paddles, gearbox, and motor are included in the appendices. Thus, the case can be used to illustrate the application of previously taught elements of pressure vessel design, fluid dynamic drag, torsional stressing, and heat transfer.

The case has a use in the teaching of graphics also. In part B, the engineer's sketches and key dimensions are given. Students can be asked to draw views of the tank and paddle assembly in various degrees of detail. Thus, students in graphics can work on a really engineering problem, making many minor decisions on layout, form, and dimensions. Afterwards. Part C can be presented with detailed drawings of the final apparatus.

Although not intended as such, this case also is suitable for a study of methodology. It is recorded the way it happened. Teamwork was used. Criteria were established in various ways. Arbitrary and rational decisions were made. There is analysis and synthesis.

In the following year in their Heat Transfer course, a different teacher took advantage of students' familiarity with this case, and used the case's heat transfer problem to show how the topics of the course could be used for better analysis of the system behavior. As each topic was introduced, reference was made to its application to the problem in ECL 87 (Appendix 7).

Cases and the Laboratory

For laboratory work, one enterprising teacher gave copies of ECL 42 (Appendix 8) "Failures at Welded Joints in a Hopper Trailer" to the class. This case focuses on service failures in welded joints of hopper trailers which are pulled by large diesel trucks. He then took the students into the materials laboratory and showed them some steel plates in the corner (which he had obtained from the scrap yard). They were told that the steel was from the failed hoppers in the case. He gave them the task to determine what was the cause of failure. From that point they were turned loose in the laboratory to decide what tests should be made, to make them, to obtain the appropriate information to solve the problem, and to write a report on their findings and recommendations. This approach (a very successful one) provided far more interest and motivation than "standard" or "routine" laboratory exercises.

Cases as History

Many modern students lack a historical perspective of the nature of scientific and technical work. Cases can be used to provide background. A series of cases can be assigned to sensitize students to the what and how of our technical world. For they too stand on the shoulders of Giants. Typical cases for this purpose might include; ECL 172 (Appendix 9), "This Must Be Done," the story of Wittle's invention of the jet engine; ECL 167 (Appendix 10) "Methods of Lessening the Consumption of Steam and Consequently of Fuel in Fire Engines," the story of James Watt's invention of the steam condenser; and ECL 1-14 (Appendix 11), "The Wright Brothers Airplane" detailing the process that the Wright brothers went through to achieve flight. Reading a series of these cases the students soon appreciate the nature of creative work in science and technology. This is where case studies, can be useful.

Armchair Cases

The term "armchair case" has been coined to describe cases that are put together out of the teachers or an author's imagination. I personally do not care for these kinds of cases. They very rarely ring true, and students quickly recognize that they are artificial. To get the most from cases with students, it is important to be able to say that a case is true, regardless of how improbable it may seem.

Nevertheless, there are occasions where a case suitable for the course is not available. Therefore, it may be necessary to resort to an armchair case. Here, I recommend that, if possible, an attempt be made to adapt some existing reality-based case or a news item for the purpose. To maintain credibility, one must be honest with the students and let them know that in this instance you are using a fictitious example. The less the armchair Case diverges from a reality the truer the Case will seem.

Recap

Cases provide a learning resource that can be used in a variety of ways. How and where cases are used depends on the course subject matter and objectives, the nature of the class, and the instructor. It should be restated that there is no right or wrong way to use cases. Some ways may work better than others, each class is different and cases must be handled in the way that best suits that class. I have suggested a variety of ways of using cases to bring reality into the classroom and to enhance student learning through the experience of others. Using cases does not mean, however, that the teacher must commit to a completely new or different style in the classroom. Cases are an additional resource that can add a new dimension to what is already being done. Or, they can open a completely new avenue that can provide unforeseen rewards for both students and instructors. The use of cases is limited only by the extent of creativity and ingenuity of the teacher.

13. THE CASE FOR CASES   251

WHAT IS LEARNED FROM CASES

Over a number of years, while teaching a course completely with cases, at the end of the term I passed out a questionnaire to find out what the students thought they had learned in the course. The results consistently showed students felt that cases, in order of importance, helped them: (1) discriminate between fact and opinion; (2) search for more alternative solutions; (3) identify and define practical problems; (4) spot key facts amid less relevant data; (5) tolerate ideas and errors of others; (6) appreciate what engineers do and what technical facts they use; and (7) understand the criteria for evaluating design and for judging procedures.

More subjectively, engineering professors, who were surveyed believed that the use of cases as teaching tools: (1) motivates students toward industry rather than academic pursuits; (2) teaches problem formulation and setting priorities; (3) puts theory into the world of application and exposes students to important concepts which have not been incorporated into textbooks; (4) provides a bridge between the generalized condensed form of academic information and the specific application orientation requirement of industry; and (5) teaches learning since cases use a natural (need to know) mode of learning.

WHAT MAKES A GOOD CASE?

To get the most for your students with cases it is important to select or tailor the cases to match the learning objectives and to make the case interesting for them. If the case is not interesting and challenging it becomes just another exercise to satisfy the teacher. The first criteria in identifying good cases is that cases unlike case studies do not include the observations, opinions or conclusions of the case writer, unless the writer is telling their own story. Case studies present a message. Cases are narratives from which students are to draw their own conclusions.

Ceram (Appendix 12), referring to Paul de Kruif's "Microbe Hunters"(Appendix 13) (a case study), expressed the view that writing about science does not have to be pedantic when he said: "Even the most highly involved scientific problems can be quite simple and understandably presented if their working out is described as a dramatic process." That means, in effect, leading the reader by the hand along the same road that the scientists themselves have traveled from the moment truth was first glimpsed until the goal was gained. De Kruif found, "that an account of the detours, crossways, and blind alleys that had confused the scientists—because of their mortal fallibility, because human intelligence failed at times to measure up to the task, because they were victimized by disturbing accidents and obstructive outside influences -- could achieve a dynamic and dramatic quality capable of evoking an uncanny tension in the reader." This is also what makes a case interesting and challenging to the student. The following are some of the things that I have found make good cases for student use.

Cases may vary in length from a few too many pages. But, the best cases generally are long enough to include some nontechnical aspects (people, time constraints, etc.), yet short enough to be studied in a few hours and be discussed in one or two class sessions of one hour each. Longer cases are better for assignments for students to prepare the written case study, individually or in groups.

Real world problems are often interdisciplinary (different technical fields and/or nontechnical fields); good cases reflect this. A case should describe some of the historical and technical background of the problems and issues in the case. It may include some of the analysis made and conclusions drawn by the individuals in the case. The details should be sufficient to ensure that a reader can fully appreciate the background and readily understand the way toward a solution of issues.

To be believable to the reader, an ideal case would include: 1) Setting, 2) Personalities, 3) Sequence of events, 4) Problems, 5) Conflicts, and 6) Application. Most usable cases include some of these. That is, a good case does not just include the bare bones of the problem to be solved. The problem is presented within its relevant context.

True to life cases may reflect all sorts of complexity, failures and successes, old and new techniques, theoretical and empirical results. In a good case, facts are not changed to show "how it should have been done," that is the roll of the case study. Cases must have a ring of truth to them. Only by a presentation of harsh reality can the student learn from the practitioner's experience. What was done wrong and why, as well as what was done right, is important in a case.

Modeling and problem solving is done by people. Therefore, the more interesting cases are presented from the viewpoint of individuals. What did they see as the problem? What was their background? What facts were available to them? What events led up to the situation? What resources were available? What were the constraints on their actions? These are often expressed in quotations, as their opinions. Good cases unfold like an adventure or mystery story in which the reader becomes engrossed because they care and want to know what happened. But, this is tempered by the fact that cases are primarily about technical issues. The case's narrative stream carries, but does not dominate, the technical aspects.

Quotations by individuals in the case are used to: 1) Express their opinions; 2) State important issues; 3) Express their personal philosophy; 4) Establish character; 5) Express differences of opinion; and 6) Increasing believability.

To capture the interest of the student good cases are not before and after examples; they are not photographic slices of life. In the sense used here, a simple record of final results does not constitute a case. Real problems rarely occur in neatly packaged set pieces as textbook problems do. In practice, solution methods are chosen which best suit the needs and context of a project. Good cases show this.

Information in real situations is received in a time sequence, even information which is internalized. The sequence in which information is received, and insight achieved, usually influences how a problem is solved.

13. THE CASE FOR CASES 253

Good cases should attempt to preserve the sequence and time frame. In the same way, decisions are not delayed until all the facts are in; they are made little by little, as new data is acquired the decisions are modified. Therefore, the sequencing of decisions should be apparent in the case.

Where a great deal of data is important, the data should be in an Appendix to preserve the narrative stream. The case should present the data as it became available to the people in the case. General background data is used sparingly, and useful technology easily available in textbooks is not included.

One unique way cases differ from stories is that they seldom have a clear cutoff point. Real life problems do not end or finish cleanly and concisely; things happen, actions are taken, projects run into one another, loose ends abound. Solutions may be incomplete or only partially successful. Sometimes action is taken and the problem goes away. Cases often end this way. This provides opportunities for the students to speculate and propose actions in the light of their own analysis. This is one feature that differentiates cases from case studies. Case studies always have endings, conclusions etc.

SUMMARY

There you have it. Those are my views on cases and case use. I have tried to show you why I believe cases are a necessary and useful augmentation to any education. There are a variety of ways to use cases from class discussions to motivation for laboratory experiments. Case use need not be confined to a single mode. I have differentiated between an objective case, a subjective case study, and the case method. I presented my view that a good case should be based on reality with people solving real problems through time. These are the things that work for me. I suggest that if you are interested in bringing outside reality into the classroom you try cases of the kind that I have discussed. If you find what I have suggested doesn't work you modify it for your own purposes and grow your own. Cases are a medium for interaction between you and your students. Finally, let me suggest that you try a case with your own class, ECL 282 (Appendix 14) "A Matter of Gravity", it only requires a knowledge of elementary Physics and a little algebra. Enjoy.

Chapter 14

Introduction to an Economic Problem: A Models and Modeling Perspective

Charalambos D. Aliprantis
Purdue University

Guadalupe Carmona
Purdue University

In this chapter, we present an activity for middle school students that was adopted from an economics problem typically used in calculus courses at a university level. It introduced students to a real-life situation, and encouraged them to develop and explain a mathematical model that would help solve a real-life problem. The results demonstrate convincingly that in spite of the common belief, most students can learn useful mathematics in schools. When given the opportunity, and the adequate task and environment, students are able to even go beyond expectancies from their teachers or school standards.

One of the main reasons why a university is a place where knowledge is generated and disseminated is because it provides an enriching environment where ideas can be discussed from a variety of disciplines by subject experts. This chapter is the result of numerous conversations between a professor from the School of Management and School of Mathematics, and a graduate student from the School of Education, who met in an economics course at Purdue University. Both authors converged on the following questions:

1. How can we provide students with rich experiences so that they can develop powerful mathematical ideas?
2. What does it mean to develop powerful mathematical ideas?
3. What does it mean for students to be proficient in their mathematical knowledge?

Coming from different backgrounds, we, the authors, were able to delve into these questions, and enrich each others points of view, and, as a result, provide an introduction to an economic problem from a models and modeling perspective.

THE STARTING POINT

We began our discussions by agreeing upon two main premises taken as a starting point for our conversations. First, we believe it is a myth that only a few students can understand basic mathematic ideas. A quote from Bruner (1960, p. 33) states, "any idea can be represented honestly and usefully in the thought forms of children of school age, and [that] these representations can later be made more powerful and precise the more easily by virtue of this early learning."

The traditional educators "advocate curriculum standards that stress specific, clearly identified mathematical skills, as well as step-by-step procedures for solving problems," (Goldin, in press). These educators also pay careful attention to the answers that students attain and the level of correctness that they demonstrate. Drill and practice methods constitute a huge portion of the time in the classroom to ensure the correct methods in order to achieve the correct answers. Reform educators, on the other hand, advocate curriculum standards in which higher level mathematical reasoning processes are stressed. These include "students finding patterns, making connections, communicating mathematically, and engaging in real-life, contextualized, and open-ended problem solving," (Goldin, p. 5). It is through this open-minded interpretation of education that different ways of students' thinking are verified and encouraged and where a broader variety of students are acknowledged, especially those that are capable, but considered remedial by the traditional standards.

In this chapter, we claim that by being concerned solely with the product of a mathematical procedure, then we are limiting students' thinking to a very narrow view of their capabilities. Thus, we are concerned with the mathematical process of how students attain that product as much as we are interested in their final product. In fact, this can be done from a models and modeling perspective, because when students solve a model-eliciting activity, the process is the product (Doerr and Lesh, chap. 6, this volume).

The second premise is based on a quote attributed to Aristotle: "You've proved that you learned something when you're able to teach it to others." By this, we mean that it is as important for students to develop mathematical ideas, as it is to communicate them to others. This process will not only allow them to converse with others, but will also promote student's development and refinement of these and other ideas. The communication process encourages students to question their own understanding of the ideas and contrast them with others, which in turn, allows them to constantly understand, revise, and refine.

When students enter an undergraduate program where mathematic courses are required, they are usually asked to justify their responses. This is an extremely difficult task for students, who are mostly used to multiple-choice tests, and are used to situations where only correct answers are rewarded. The emphasis in their middle and high school mathematic courses is focused on preparing students to achieve high scores on standardized tests. These tests are usually multiple choice, where further justifications or explanations are not

14. ECONOMIC MODELS 257

required. Rarely are these types of tests concerned with what students are required to do in their university mathematic courses, and even less to what they have to do when they obtain a job. Thus, there is a large discrepancy between what students are required to do in their middle and high school mathematic courses and what they are expected to do in their university courses and at working settings.

In order to prepare them for future jobs, faculty in universities believe it is important for students to develop their own mathematical ideas from real-life situations, as well as to justify how they obtained their results. In this case, the answer that students give to a problem is as important as the procedure they used to arrive at the solution. This requires from students two important skills: (a) to be able to develop a mathematical model from a real-life situation, and (b) to be able to explain this model to someone else. Neither of these two tasks seem to be a part of most middle and high school mathematic programs. This shows an urgent need for schools to incorporate mathematical activities related to real-life situations that provide students with similar experiences to the ones they encounter in their university mathematic courses as well as in their future jobs.

In this chapter, we present an activity for middle school students that was used with three seventh grade groups. It introduced students to a real-life situation, and encouraged them to develop and explain a mathematical model that would help solve a real-life problem. The activity had to be solved in teams of three students. When students were given the opportunity to explain and justify their thinking to others, it provided teachers with a way to follow students' understanding of the situation. That is, students' mathematical thinking was revealed to the teachers through the explanation that they describe, which provides a powerful evidence of students' mathematical knowledge.

HISTORIC HOTELS: AN ECONOMIC MODEL

One of the basic mathematic ideas that students learn in their high school math courses, and that is recursively utilized in most mathematic courses at an undergraduate and graduate level is quadratic functions. We decided that this powerful mathematic idea could be a good start to our interpretation of students' development of mathematical knowledge.

Based on an economic problem (Aliprantis, 1999) typically used on undergraduate calculus courses that deals with economic concepts, like profit, cost, price, maximization, and equilibrium; and mathematical concepts like recognition of variables, relation between variables (linear and quadratic relations), product of linear relations, and maximization; the two authors designed a model-eliciting activity (Lesh, Hoover, Hole, Kelly, & Post, 2000) that would allow middle school students to develop a mathematical and economic model approaching the concepts mentioned. This activity, Historic Hotels (Appendix A), was intended to encourage students to develop the model from a real-life situation, and incorporated elements so that they could share their model with others.

The real-life context of this activity was given as a newspaper article that described a historic hotel in Indiana, and how it had changed owners through time. The article mentions how it has been difficult for all of them to maintain the historic architecture and ambience of the hotel, in addition to other responsibilities that a good hotel manager is required to do.

The problem statement describes how Mr. Frank Graham, from Elkhart District in Indiana, inherited a historic hotel. He would like to keep it, but is unwilling because of his lack of hotel management experience. The whole community of Elkhart is willing to help him out because this historic hotel represents a major attraction for visitors, and thus, sources of income for everyone in the town. As part of the community, Elkhart Middle School has been assigned to help determine how much should be charged for each of the 80 rooms in the hotel in order to maximize Mr. Graham's profit. From previous experience, they have been told that all rooms are occupied when the daily rate is $60 per room. Each occupied room has a $4 cost for service and maintenance per day. They also have been told that for every dollar increase in the daily $60 rate, there is a vacant room.

Students solving this model-eliciting activity are required to develop a tool for the students in Elkhart Middle School that can help Mr. Graham solve his problem, giving complete instructions on how to use. This tool should be useful even if hotel prices and costs rise, for example, 10 years from now. Students must describe their product through some type of representational media, in order to communicate their tool and its use to the other students and to Mr. Graham.

We implemented a model-eliciting activity in a mid-western public middle school; more precisely with three groups of typical seventh grade students who did not have previous instruction in algebra (one group was remedial). The two teachers had previous experience in implementing model-eliciting activities in their classrooms, but for the students involved in this study, this was their first time.

The implementation consisted of two parts. First, we handed out a copy of the newspaper article, with focusing questions included, the day before the activity took place in the classroom. Therefore, the students were able to read the article as a homework assignment and had a chance to think about the topic that they would be working on the next day in class. The following day, two blocks of time, approximately 50 minutes each, were used for the activity. The first block of time was used for solving the problem and the second for the students to present their solutions to the remainder of the students.

The students worked in teams of three to five to solve the problem, and there were a total of 12 teams. The idea was for the students to work with each other and develop a solution to the problem that has been posed. The teachers were asked to observe the students and not intervene in the students' work time. We did not want the teachers showing students ideal procedures that could assist them in solving the problem since this would defeat the whole purpose of the activity. This request did not pose a problem for the two teachers involved, because they had previous experience implementing these types of activities.

14. ECONOMIC MODELS

At the beginning of the activity, students were asked to present their solution on an overhead transparency and then show and explain their solution to the class. After the students worked on the problem for about 50 minutes they were requested to give their presentations. At the end of each presentation, the teachers and the remainder of the class had the opportunity to ask questions of the students who were presenting their solutions.

For the purposes of analysis, we collected all of the students' written work. This included all of the paper that the students used while solving the problem and especially their written solutions to the problem and presentations. In addition, one of the researchers wrote field notes as an outside observer and we included this as part of our data.

Once students' work and field notes were collected, we developed different types of coding students' work, according to the representational systems that students used when solving this activity. To do so, one of us looked at each of the team's work individually. After classifying the students' work, we reviewed the students' actual work. We looked for consistency in their reasoning and errors in their mathematics or method. After looking at all of their work, we concluded whether the students found the correct answer or not. Even though this is not the point of the exercise, it was important to note and record. The coding was then discussed, and we both agreed on how the data was going to finally be coded.

The coding was done both, inductively and deductively. We developed a first coding, from Goldin (in press, p. 25). Then, we realized that we did not have evidence for affective representational systems (Goldin, p. 23), because we had not collected videotape for our data. Thus, we went through all of students' work, and developed the following coding:

Type of Representational System

    A: Algebraic
    C: Chart
    G: Graph
    L: List
    P: Pre-algebraic
    T: Text (includes prose with numbers and signs like $).

Based on our data, we also found that students developed different procedures to solve the problem. The method we used to code their procedures is shown here:

Mathematical procedures

1. Multiply the price by the number of occupied rooms. Multiply the maintenance fee by the number of occupied rooms. Subtract the maintenance cost from the total cost of the occupied rooms.

2. Systematical comparison of different numbers of occupied rooms with the corresponding price
   2a: and lowering number of occupied rooms by 5.
   2b: and lowering number of occupied rooms by 1.
3. Attempt to formalize algebraically an iterative process.
4. Unfinished procedure.
5. Introduction of formal terminology (economic/mathematical).

Because in model-eliciting activities the product is the process, we were also interested in how consistent students were in their procedures. That is, if they actually solved the problem the way they described the procedure to Mr. Graham.

Consistency in Procedure
   Yes
   No

This model-eliciting activity allows for different solutions, depending on how the students interpret the cost for maintenance, so we were also interested in analyzing whether the students wrote an answer (this was not necessary, because the problem asked students to develop a tool and not necessarily to give a numerical answer).

Correct Answer was given
   Yes
   No

After students' work was coded, we summarized our results in a table format that is shown in the next section for results.

RESULTS

The results of the analysis are summarized in the following chart:

Team	1	2	3	4	5	6	7	8	9	10	11	12
Representation	L T	A L	L T	G P T	T	P T	T	L T	P	A C	T	P T
Mathematical Procedures	1 2a	1 2b 4	1 2b	1 2b 5	1	1 2b	1	1 2a 2b	1	1 2b 5	4	1
Consistency	Yes	Yes	No*	Yes	Yes	Yes	No	Yes	No	No	No	Yes
Answer	No	No	Yes	Yes	No	Yes	Yes	No	N/A	No	No	No
Contextual Influence					💰	✎			🖩	📖		

FIG. 14.1. Table of Summarized Results. Historic Hotels Problem.

14. ECONOMIC MODELS

Relevant Comments From Students' Work

Team 4. *This group was formed by three girls who were in the remedial seventh grade class. They developed their list of operations, and systematically compared the total profit, with different combinations of number of rooms with the corresponding price. They started calculating this list on a separate piece of paper, where they found that the total profit followed a pattern. The values increased, until they got to a certain point (the solution: 68 rooms at $72 per room), and then they started to decrease. When they found this pattern, they immediately asked for a transparency to prepare their presentation. As they started to copy their list to the transparency, they wanted to make sure that their calculations were correctly performed, so they recalculated line by line. As part of their notation, they used dollar signs ($) before the four dollar maintenance fee. As part of their systematization, it was clear that when two of the columns were decreased by one (the number of occupied rooms), the other column increased by one (the price of the room). Unfortunately, they were running out of time, and when they got to line 16, where they were combining 65 occupied rooms at $75, they got confused with their own notation, confusing $4 with 54. This caused them to continue decreasing the values for that column to 53, 52, 51, …Of course, their numerical result for the total profit was affected by this process, and as a result, their hypothesis for the response and patterns found for the maximization of the function was not validated. (Appendix B).

Team 6. ỗ Students on this team presented a result for the total profit and added the following note: "This is if they took his taxes out of his pay check." After giving the total profit. This response was very illustrating for us, because it gave us information on how important context is for students when they are solving these types of problems. From this comment we were able to appreciate students' concerns about real-life problems (like taxes), and how they are giving a solution where they are conscious that Mr. Graham will probably need to pay his taxes, and how he should be aware that this method of solving the problem will give him the daily profit, but without calculating the taxes he will have to pay (Appendix C and D).

Team 5. 📖 These students introduced formal terminology from the Economics Field. For example: net pay, gross pay, and so forth. (Appendix E).

Team 7. ✍ These students clearly and concisely mapped out the tool through prose. They had to develop a tool that would help someone external (Mr. Graham), and they developed a very thorough description of what he was supposed to do in order to solve his problem. (Appendix F).

Team 8. ▮ This group used decimal places for the dependent variable (cost of the room) in currency units. (Appendix G).

Representational Systems

In some cases, the representations that the students used were a mixture of one or more of the basic descriptions. We believe this to be an important finding. When students are working with different representations, it is typical that most textbooks or computer activities ask them to give their response in a specific representational system. For example, "give your answer in a table," "from this chart, develop the graph," and so forth. What we believe is relevant from students working on this activity is the fact that students were not told or suggested to present their response in a certain form. Nevertheless, they used what they considered most appropriate for the task that they had to develop, and it is interesting to see how: (a) They came with their own representations, that are considered very powerful in the field of mathematics, and (b) about 70% of the students used more than one representational system to express their model, which implies that students were also able to map from one representation to another, and vice versa.

General Solution Patterns

At the beginning of the activity students seem to spend quite some time analyzing what the problem is. At first it is not clear that Mr. Graham would make more profit if not all of the rooms are occupied. Until students are able to overcome this initial assumption, they start operating with different combinations of occupied rooms and its corresponding room rate, and they start to see certain patterns in their response. After students operate on two or three combinations, they find that as the number of occupied rooms decrease, the profit increases. One might even think that it would be easy for students to generalize this as a rule fairly quickly and start decreasing the number of rooms as much as possible. Nevertheless, it is quite clear that when students are working on their solution, they are working with number of occupied rooms, and not only with numbers; simultaneously, they are working with a corresponding room rate, and not just a number with dollar units. Thus, it seems that the context helps the students not to generalize the rule, and continue a slow decrease in the number of rooms. This process, and constant comparison of results for different combinations, allows students to observe interesting patterns in their data, like how the profit increases up to a certain point (maximum), and then starts to decrease. Even though for these students these ideas are not yet formalized in terms of, for example, properties of a quadratic equation, a parabola, or maximization, the fact that these students start realizing these patterns at an early stage in their scholar education should provide them with more powerful tools for when they do encounter these mathematical concepts in a more formalized manner.

Procedures

We noticed throughout the course of our analysis that two main procedures were used by the students to help solve the problem. This was an important distinction for the quest of understanding why students act and think in the ways that they do. The first major mathematical procedure that the students used is taking into account the $4 maintenance fee as part of the original $60 cost of the room. Therefore, when figuring out the maximum profit, they began with $56

14. ECONOMIC MODELS 263

and started exploring values from there. On the other hand, other students thought that the original price of the room, $60, did not include the $4 maintenance fee. Therefore, these students began with a price of $64 when figuring out the hotel manager's profit. These two procedures, therefore, elicited two different answers depending on the ways that the students interpreted the problem.

Even though not all of the groups were consistent in their response, or obtained the "correct" answer, the models that they produced were considerably powerful, from the mathematical standpoint; and all of the students excelled, by far, the expectations from the teachers, the standards (NCTM, 2000), and even our initial hypotheses (as researchers).

Bruner states that "any subject can be taught effectively in some intellectually honest form to any child at any stage of development," he also adds that, "no evidence exists to contradict it." The fact that among the three classes there was not a single student that did not understand what the problem was, or that none of the groups were unable to develop a useful response is an important finding, especially if we are considering a task that was based on a typical calculus maximization problem solved by a seventh grade mathematics class. We believe this is considerable evidence that supports Bruner's statement.

CONCLUSIONS

The students developed, in solving the problem, a broad variety of representational systems that helped them express their mathematical ideas through the refinement of their model. Nine of the twelve teams utilized text in their work and most of the students listed their data in one way or another. Roughly half of the teams were consistent in their mathematical procedures. In addition, roughly half of the teams arrived at the correct answer based on the method used, as discussed previously. Although not all of the groups got the correct answer, their work shows very powerful mathematical reasoning. Their ideas about representational systems such as notation, terminology, and so forth are not fully developed, yet show results that were by far beyond expectations from their teachers, or from educational standards (NCTM, 2000). Assessing students' mathematical capabilities by only looking at the correct answer is giving a very limited view of the mathematical ideas that students have developed and what they are capable of accomplishing.

IMPLICATIONS

The implementation of this model-eliciting task was very encouraging from the standpoint of view of researchers, teachers, students, and program evaluators. The role of communication in the modeling activity was an essential part of the task. Working in teams allowed students to develop and refine useful mathematical models and to provide documentation of their learning. The fact that students are not only able to develop their own mathematical and economic ideas, that go beyond expectations from the whole scholar community, but also give evidence of this learning is very promising, especially if we consider that most of these capable students are classified as "below average". This is consistent with our initial premises that all students are capable of learning and developing powerful mathematical concepts, and that communicating them to others promotes the development and refinement of their ideas.

As life changes due, among other things, to technology and globalization, the educational system must also change. Early preparation of students for tasks similar to the ones they will encounter in their future education or job is not only possible (and productive), but should be a requirement.

Chapter 15

A Models and Modeling Perspective on Technology-Based Representational Media

Tristan Johnson
Purdue University

Richard Lesh
Purdue University

Developing productive ways to use technology to support learning has been a key focus for research and instructional development for many content areas within and beyond K through 12 schools. This chapter describes an educational technology perspective that was developed mainly in adult education settings beyond K through 12 schools, but that also is proving to be especially relevant to models and modeling activities in mathematics education in schools. In particular, the two authors have collaborated in the development of projects that focused on simulations of "real life" mathematical problem solving activities for middle school students and teachers.

The main purpose of this chapter is to examine the use technology-based representational media (eMedia—electronic media) when they are used as expressive tools for learners and problem solvers. Thus, as Fig. 15.1 suggests, we highlight the use of eMedia as tools for communicating and modeling; we focus on the use of these media in problem solving situations that might reasonably occur in students' everyday lives beyond school; and, we focus on functions that emphasize looking through these media rather than looking at them. That is, technology-based media are treated as being transparent at least as much are they are opaque; and, rather than being preoccupied with person-to-gadget communication, we focus on people-to-people communication - or on situations where people communicate with themselves—by externalizing or expressing internal conceptual schemes so that they can be examined, tested, modified and adapted for a variety of situations and purposes.

As Fig. 15.1 suggests, for us, models are conceptual systems that are expressed using a variety of interacting representational media—which may involve traditional media such as written symbols or spoken language or diagrams, but which may also involve eMedia such as computer-based animations, simulations, or graphics. In any case, purposes of these models generally involve: (a) constructing, describing, or explaining other system(s), or (b) communicating with others—or with oneself in some other situation at some later point in time.

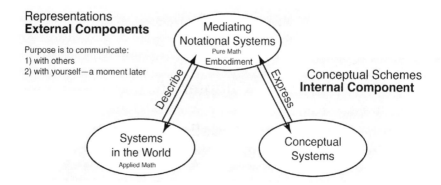

FIG. 15.1: Models: internal and external components.

This chapter has two main sections: (1) examples of communicating and modeling and (2) theoretical perspectives. In the first section we refer to several examples of students' work on model-eliciting activities to show spontaneous ways that students often use eMedia as communication and modeling tools in simulations of "real life" problem solving activities. In the second section, we discuss how models have both internal and external characteristics, how the meaning of underlying conceptual systems often are distributed among a variety of interacting representational media, and how relevant representational media interact in series or in parallel to lead to coherent and useful interpretations of complex learning and problem solving situations. As the final part of the second section, we describe how eMedia can be used as tools to represent and transform conceptual schemes, how eMedia can significantly expand the variety of representations that we have available, how eMedia interact with traditional representational media, and how eMedia often emphasize dynamic and manipulatable relationships in the situations that they are used to describe, explain, manipulate, or predict.

A Distinctive Educational Technology Perspective

For the past 10 years, Tristan Johnson, the first author of this chapter, has been working in the field of educational technology. His primary experiences emphasized instructional design for adult audiences using technology-based media for a variety of business topics and settings. As an educational technologist, he repeatedly asked the question: What instructional advantages are achieved with the use of the computer? But, one of the facts that he's observed is that technology often opens up completely new opportunities that simply were not possible using inert and non-interactive media.

Having seen both instructionally sound and unsound uses of technology, he has concluded that many of the most appropriate uses of technology focus on technology as a cognitive tool (Jonassen & Reeves, 1996). That is, instead of the traditional use of computers as information conveyors, knowledge

15. TECHNOLOGY-BASED REPRESENTATIONAL MEDIA

communicators, or tutors for students, he agrees with Jonassen and Reeves who emphasize the use of computer-based cognitive tools as intellectual partners to facilitate critical thinking and higher order learning. In particular, in this chapter, we focus on the use of technology as a medium to enable and facilitate: (a) communicating—where the media are used to express (to others or to oneself) one's current ways of thinking, and (b) modeling—where the media are used to interpret (or make sense of) "real life" situations.

As we adopt the preceding emphasis on communication and modeling, it is interesting to notice that, historically, as personal computers have become increasingly ubiquitous in our society, communication has been one of the foremost reasons for increased usage. For example, desktop publishing was one of the first reasons why many people used personal computers. Then, later, the advent of e-mail provided another major stimulus to growth; and, so did, communication through web browsers. In other words, for most people, communication has always been at the heart of their reasons for using computers. And, today, the desktop and laptop computers of the past are rapidly merging into the palmtops and cell phones of the present.

PART ONE: COMMUNICATING AND MODELING

If we look at the kind of traditional media that were emphasized in chapter 24 in this book, or the media that were referred to in Fig. 1.6 in chapter 1, then eMedia can be thought of as constituting a new layer sitting on top of traditional representational media (see Fig. 15.2). That is, for each of the types of traditional media described in Fig. 1.6 of chapter 1, there exists a corresponding type of eMedia. But, whereas traditional media tend to emphasize static objects, relationships, and events, their corresponding eMedia are easier to manipulate; they tend to be more dynamic (focusing on actions and transformations); they are interactive (by responding to actions on them), and they are linked (so that actions carried out in one medium are reflected automatically in other media) (Kaput, in press).

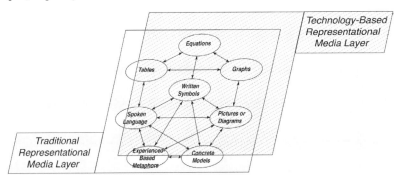

FIG. 15.2. Traditional and technology-based representational media layers.

In the Fall of 1999, the authors began working on various projects that emphasized models and modeling. In particular, we started designing and developing a digital library of model-eliciting activities and accompanying tools and resources (see chap. 2, this volume) which included *"how to" toolkits* - and other tools and resources that are described in chapter 2 of this book (also see the Web Appendix to this chapter) They also included a variety of resources to support distance-learning and remote-collaboration among researchers and teachers associated with our projects and programs (Lesh, Crider, & Gummer, 2001).

Interactions Among eMedia. When considering the use of eMedia in instruction, it has proven to be useful to sort our issues related to their multiple, parallel, and interacting characteristics. The meaning of these characteristics is illustrated in greater detail in the second section of this chapter. In brief,

- *Multiple* refers to the fact that more than a single medium (the external component of a model) tends to be needed to express the meaning of a given construct or conceptual system; and, more than a single medium also tends to be needed to describe a given learning or problem solving situation
- *Parallel* refers to the fact that several media tend to be used simultaneously as a person thinks about a given learning or problem solving situation – or as she communicates with herself or with others.
- *Interacting* relates to the fact that eMedia often are linked in such a way that when one medium is changed, the linked medium changes in corresponding ways.

Familiar examples of multiple and parallel functioning occur when a teacher wants to teach the idea of fractions using concrete manipulate-able materials, while simultaneously using some diagrams and written notational systems as well as spoken language. Examples of interacting functioning occur in software that we've designed, called Fraction Town (see the Web Appendix to this chapter), where multiple dynamically linked representation are used to enable students to manipulate one notation system (e.g., pies) while observing the reflection of these actions in systems that involve other types of representations (e.g., written symbols).

Multiple, parallel, and interacting functioning also can be seen in many of the transcribed solutions that we've collected for model-eliciting activities. For example, for the Summer Jobs Problem that follows, the products that students produce often include graphs, diagrams, or lists; and, a series of modeling cycles often are required to develop these products.

Characteristics of Media in Modeling

A transcribed solution to the Summer Jobs problem is given in the web Appendix to this chapter. The researchers' narrative that accompanies this transcript identified 19 distinct instances where the students externalized their

15. TECHNOLOGY-BASED REPRESENTATIONAL MEDIA

current ways of thinking using some combination of spoken language, or written symbols, or spreadsheet tables or graphs. Then, after these externalizations were expressed, the results led directly to changes in the students' ways of thinking. In this way, the ways of thinking of a student who created a graph often was significantly different than the ways of thinking of the (same) student who looked at the graph after it has been drawn. In fact, in some cases, the very process of creating a graph induces significant changes in the constructs and conceptual systems that are used to produce it. In other words, the students not only used the media to communicate with one another, they also used the media to communicate with themselves. Consequently, as the students went through multiple modeling cycles, they repeatedly expressed, examined, tested, and revised, or refined their current ways of thinking—so that their ways of thinking were increasingly powerful, sharable, and useful.

Here is an example from the transcript that shows how student's used eMedia to externalize their current ways of thinking thereby communicating with themselves and also with others.

Alan:	Look at old Willy. He's really catching on {at the end of the Summer}... Look, back here {in June} he only made a hundred and eighty-nine bucks; but, out here {in August} he was really humming.
Barb:	I think August should count most. Then July. ... I don't think June should count much. They were just learning.
Alan:	How we going to do that.
Barb:	I don't know. Just look at them {the numbers in the table} I guess.
	{pause}
Barb:	Let's see, out here {in August} Kim was best. ... Then Robin, no Tony. ... Then Robin. ... I think they're the top three. Kim, Robin, and Tony. ... How'd they do in July?
Barb:	Wow! Look at Kim. She's still the best. ... But, uh oh, look at Cheri. She was real good in July.
Alan:	Let's line them up in July. Who's first.
Barb:	Kim. ... {pause} Then Tony, and Cheri, and Robin. ... {long pause} ... Then Willy, Chad, and Jose. ... {long pause} ... And, these guys weren't very good {referring to Maria, Terry}.
Note:	While Barb was doing most of the talking and overt work, Alan was watching and listening closely. But, Carla was off on her own playing with the computer's spreadsheet, and entering lists of numbers. ... But, at this point, Carla re-enters the conversation
Carla	Look you guys, I can make a graph of this stuff. Look.

Note:	EXCEL software made this possible - would they have thought to graph It had It had been paper and pencil? For the next four minutes, Carla used the computer to flip back and forth, showing the following three graphs that she had made, explaining how she made the graphs, and pointing out who was the top money earner each month.
Barb:	Ok, let's, like, line them up for each month.
Alan:	You started doing that.
Barb:	Ok, you {Alan} read them off and I'll write them down.
Alan:	Ok, here's August. ... We got Kim, Tony, Robin ... Cheri ... Willy, Chad, and Jose. Then, Maria and Terry.
Note:	For approximately five minutes, Alan, Barb, and Carla worked together to get a list of "top money makers" each month. Then, they noticed that the rankings were somewhat different each month; so, the "trends" shown here were used as an early attempt to reduce this information to a single list.
Alan:	Look, Kim was top in July and August; and, so was Tony. ... Robin was next in August; but, she wasn't as good in July. ... {pause} ... But, she {Robin} was really good in June. ... {pause} ... I think August is most important because some of them were just learning. ... August is how they'll probably do next summer.

The "Summer Jobs" Problem

Last summer Maya started a concession business at Wild Days Amusement Park. Her vendors carry popcorn and drinks around the park, selling wherever they can find customers. Maya needs your help deciding which workers to rehire next summer. Last year Maya had nine vendors. This summer, she can have only six – three full-time and three half-time. She wants to rehire the vendors who will make the most money for her. But she doesn't know how to compare them because they worked different numbers of hours. Also, when they worked makes a big difference. After all, it is easier to sell more on a crowded Friday night than on a rainy afternoon.

Maya reviewed her records from last year. For each vendor, she totaled the number of hours worked and the money collected—when business in the park was busy (high attendance), steady, and slow (low attendance). (See the table on the following page.) Please evaluate how well the different vendors did last year for the business and decide which three she should rehire full-time and which three she should rehire half-time.

Write a letter to Maya giving your results. In your letter describe how you evaluated the vendors. Give details so Maya can check your work, and give a clear explanation so she can decide whether your method is a good one for her to use.

15. TECHNOLOGY-BASED REPRESENTATIONAL MEDIA

HOURS WORKED LAST SUMMER								
	Busy	Steady	Slow	Busy	Steady	Slow	Busy	Steady	Slow
MARIA	12.5	15	9	10	14	17.5	12.5	33.5	35
KIM	5.5	22	15.5	53.5	40	15.5	50	14	23.5
TERRY	12	17	14.5	20	25	21.5	19.5	20.5	24.5
JOSE	19.5	30.5	34	20	31	14	22	19.5	36
CHAD	19.5	26	0	36	15.5	27	30	24	4.5
CHERI	13	4.5	12	33.5	37.5	6.5	16	24	16.5
ROBIN	26.5	43.5	27	67	26	3	41.5	58	5.5
TONY	7.5	16	25	16	45.5	51	7.5	42	84
WILLY	0	3	4.5	38	17.5	39	37	22	12
MONEY COLLECTED LAST SUMMER (IN DOLLARS)								
	J U N E			*J U L Y*			*A U G U S T*		
	Busy	Steady	Slow	Busy	Steady	Slow	Busy	Steady	Slow
MARIA	690	780	452	699	758	835	788	1732	1462
KIM	474	874	406	4612	2032	477	4500	834	712
TERRY	1047	667	284	1389	804	450	1062	806	491
JOSE	1263	1188	765	1584	1668	449	1822	1276	1358
CHAD	1264	1172	0	2477	681	548	1923	1130	89
CHERI	1115	278	574	2972	2399	231	1322	1594	577
ROBIN	2253	1702	610	4470	993	75	2754	2327	87
TONY	550	903	928	1296	2360	2610	615	2184	2518
WILLY	0	125	64	3073	767	768	3005	1253	253
Data are given for times when park attendance was high (busy), medium (steady), and low (slow).								

One fact that is clear from the transcript to the Summer Jobs Problem is that, at any given moment during the time that they are working on the problem, several distinct ways of thinking may exist in the minds of the team members (or even in the minds of individual team members); and, each of these ways of thinking may be expressed using a variety of media—ranging from spoken language, to diagrams, to written notations, and to computer-generated graphs. Furthermore, because different ways of thinking, and different media, often emphasize or ignore different aspects of the situations they are used to describe, they often lead to interpretations of the situation that do not fit together nicely. Sometimes, these mismatches lead to significant conceptual reorganizations. Sometimes, two or more media (or underlying ways of thinking) are integrated. Or, sometimes, they are differentiated—with one being emphasized and refined, while others are discarded. Thus, when problem solvers are going though a given cycle during their attempts to develop a productive description (or interpretation) of the problem solving situation, a description of their thinking tends to correspond to Fig. 15.3b more than Fig. 15.3a.

FIG. 15.3a. One medium is used to express a single conceptual system which is used to interpret a given problem solving situation.

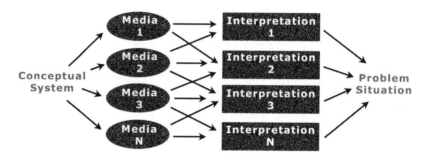

FIG. 15.3b. Several media are used to express several conceptual systems which yields several distinct interpretations of a given problem solving situation.

Every representational media has some characteristics that the system being described does not; and, every representational media also has some characteristics that the system being described does have. If this were not the case then the representational media would be identical to (and indistinguishable from) the system it is to be used to describe. Figure 15.3b is no exception to this rule. For example, even if it were possible to completely separate the various relevant representational media, the interpretations that they generate would continue to be partly overlapping. So, the boxes that we have labeled "interpretation 1" or "interpretation N" are likely to share certain objects, relationships, operations, or patterns and regularities.

Characteristics of Media in Communication

In fields such as engineering or business, where successful practitioners are increasingly becoming heavy users of mathematics, science, and technology, it's clear that, when problem solvers work on complex problems, they typically work within teams of specialists who make use of a variety of technology-based tools which in turn involve a variety of related eMedia. For example, when students use calculators or spreadsheets to respond to the Summer Jobs Problem, their solutions often illustrate a second important fact about situations where eMedia are used.

Another fact that the transcript the Summer Jobs Problem makes clear is that, at the same time that technology-based tools make it easier for problem solvers to carry out certain computations (or procedures), they usually make other tasks more difficult. For example, in order to use a calculator or a spreadsheet, problem solvers must first code and organize relevant information in forms that are recognized by the tools. Or, when the calculator or spreadsheet generates graphs (or other products), significant difficulties often occur when the students must make sense of these result. In other words, description and communication and interpretation become important mathematical activities. Consequently, expressing, describing, and communicating tend to involve nontrivial forms of mathematical thinking, when problem solvers use eMedia to develop sharable and reusable products in response to model-eliciting activities, they broaden their understandings of the relevant constructs by developing new meanings, understandings, and abilities. Consequently, because eMedia tend to lead to an explosion of productive descriptions and modes of communication, a broader range of abilities tend to contribute to success in learning and problem solving situations. For example, the transcript to the Summer Jobs Problem includes many instances where eMedia provide powerful conceptual tools for thinking about important mathematical constructs in ways that are simple and yet sophisticated. But, these same eMedia also radically increase the importance of abilities such as those related to representational fluency—and abilities associated with mathematization and communication.

PART TWO: THEORETICAL PERSPECTIVES

In the first part of this chapter, we emphasize the role of eMedia as expressive tools for communicating and modeling. We consider a models to have two parts: (a) internal constructs (or conceptual systems), and (b) a variety of external representational media. The purpose of these models is to interpret or construct other systems. And, during the process of constructing, describing, and explaining other systems, we emphasize the plural, parallel, and interacting functioning of these media. In this section, we give more details about modeling, modeling processes, and technology-based representational media; and, we emphasize understandings and abilities that become salient when learners and problem solvers work with technology-based tools.

We have argued that many of the most important functions of technology-based representational media (eMedia) are: (a) to describe or explain complex systems, and (b) to express complex constructs by providing new ways for people to communicate with both others and with themselves. These perspectives are not difficult to understand or accept. Yet, they are quite different than the traditional views where eMedia are thought of as objects to look at more than windows to look through (or communicate through).

Modeling Issues/Misconceptions. In the process of working with colleagues in mathematics education, the first author has found it necessary to clarify and resolve a variety of his preconceptions about learning as it relates to a representational media. Key ideas included the following points. First, mathematical thinking, even at elementary levels, tends to focus on seeing as much as on doing; and, what needs to be emphasized are structural characteristics (patterns and regularities) of systems that are investigated. Second, representations are not neutral; when one characteristic is highlighted, others tend to be disguised or dismissed. Third, both modeling and communication involves iterative cycles in which a variety of constructs and media interact.

Mathematical thinking focuses on seeing as much as on doing. One main idea that models & modeling perspectives adopt from Piaget is his emphasis on holistic conceptual systems that underlie people's mathematical interpretations of their experiences—and their mathematical judgments about those experiences (Beth & Piaget, 1966). In particular, models & modeling perspectives agree with Piaget that many of the most important properties of mathematical constructs and conceptual systems cannot be derived from properties of their constituent elements but instead are emergent characteristics that only become meaningful in the context of the system-as-a-whole (see chap. 3 of this book where Piaget's conservation tasks are discussed). For example, in the transcript for the Summer Jobs problem, the students developed a way to think about trends in productivity for workers using observations (data) about: (a) the amount of money earned during various conditions and periods of time, and (b) the amount of time spent working during the preceding conditions and periods of time.

Notice that productivity is not a characteristic that can be observed and measured directly. It needs to be thought of in terms of relationships that involve other characteristic—such as money earned and time spent - that can be observed directly. Similarly, trends can only be seen by investigating properties of the entire collection of data that were given. So, what we see in the transcript for the Summer Jobs problem is an iterative series of ways that students gradually developed conceptual systems that gradually evolved beyond information that was given directly toward information that could only be seen by looking at relationships and trends.

Models and modeling perspectives share with Piaget the belief that systemic characteristics of mathematical systems cannot be abstracted from external objects or events, nor can they be pure fabrications generated from internal resources of the mind. The development of these mathematical properties of holistic systems must involve more than simply assembling (or constructing, or piecing together) the elements, relations, operations, and principles. According to Piaget, systemic properties of conceptual systems-as-a-whole involve stages of reflective abstractions in which emergent properties evolve from systems of interactions at more primitive/concrete/enactive levels. Some ways that these reflective abstractions occur, and some ways that representational media play a role in these reflective abstractions, can be seen in the transcript of the Summer Jobs problem. In general, as students moved through a series of modeling cycles that were based on different ways of thinking about the information that was given, what the students did repeatedly was: (a) to express their current ways of thinking in forms that they shared, examined, and tested; and (b) modifying or adapting their ways of thinking based on results of the preceding sharing, examining, and testing activities.

Representations are Not Neutral. Bruner (Bruner, 1966) extended Piaget's perspectives by focusing on the representational tools that students use to express their ways of thinking. He emphasized that different representational media have different characteristics in terms of their power to simplify complex patterns and relationships, their power to go beyond the information given, and their generative power for developing predictions and explanations of observed information.

Representational media can be thought of as conceptual amplifiers in the sense that any media facilitates some ways of thinking and restricts other ways of thinking (Lesh, 1987). This is an idea that models and modeling perspectives adopt from Vygotsky's concept of zones of proximal development (Vygotsky, 1962; Wertch, 1985). That is, in nontrivial learning or decision-making situations, we assume that a student generally has available a variety of potential conceptual systems that have the potential to be engaged to interpret current experiences. Furthermore, each of these conceptual systems tends to be at an intermediate stage of development—not completely unfamiliar (so that it needs to be constructed beginning with a "blank slate") and not completely mastered (so it simply needs to be retrieved and applied). Consequently, the challenge for students is to extend, revise, reorganize, refine, modifying, or adapt constructs (or conceptual systems) that they DO have—not simply to find or create constructs that they do NOT have (or that are not immediately available). For this reason, choices that influence which systems actually get engaged depend on a variety of factors that include: (a) guidance provided by an adult or peer, or (b) conceptual tools (such as powerful, economical, and generative representational systems) that are readily available by luck or by design—perhaps, due to interventions from an adult or peers.

Again, instances of the preceding phenomena are obvious in the transcript to the Summer Jobs problem. In particular, especially in cases where eMedia are used, it's clear that the ways of thinking that students put into a given diagram or graph often leads to new ways of thinking that are read out of these diagrams or graphs. In fact, the student who is responsible for generating a graph is often as surprised anybody about the new ways of thinking that it generates

Representations have tremendous instructional value to promote conceptual development (Hiebert & Behr, 1988; Lesh, Behr, & Post, 1987). This is evident in Dienes (1969) principle of multiple embodiments; and, it is apparent in Bruner's emphasis on the fact that representations have the power to mold, shape, amplify, and generate ideas. However, the power of representational media cannot be divorced from the development of the conceptual systems that they are used to express. This is why models and modeling perspectives emphasize that models involve both (internal) conceptual systems and (external) media in which they are expressed; and, it is why we emphasize the plural and interacting nature of the media that accompany any given conceptual system.

Modeling Processes Involve Multiple Media and Multiple Representational Mappings. One hallmark of a mature theory is specialists in the field tend to express increasing amounts of it's ideas and ways of thinking using a single (usually written) symbolic media. However, even at advanced stages of development, specialists often continues to think using graphics, experience-based metaphors, and other media in addition to the formal symbolic tools of their trade.

In transcripts associated with chapters throughout this book, it is clear that both modeling and communication involves iterative cycles in which a variety of constructs and media interact as students move from one modeling cycle to another in attempts to develop increasingly useful ways of thinking about problem solving situations. For example, in the transcript to the Summer Jobs Problem in the web Appendix for this chapter, it is clear that, even if all of the students are thinking in identically the same way about the nature of relevant quantities, relationships, and patterns, they tend to be using a variety of interacting representational media to express these ideas. For example, when looking at a table of data, they may be thinking about the associated graphs in a variety of quite different ways. They may be using quite different language to express these ideas; and, they may invoke quite different past experiences as metaphors to explain what they believe the graphs are showing.

The availability of eMedia tends to radically expand the varieties of media that students draw upon to express their ways of thinking; and, they also tend to radically increase the importance of description abilities, planning abilities, and monitoring and assessing abilities—all of which draw upon multi-media thinking and representational fluency. Whereas, artificial intelligence models of thinking often collapse all thinking into a single representational system, there are strong reasons to believe that humans are inherently multi-media processors.

SUMMARY

Effective instructional uses of eMedia should involve more than getting people to do things better, but it is about seeing things differently. It is more than seeing the same ideas in the same ways (old topic with new technology) or new ideas in new ways (new topics with new technology), but it is about seeing the new ideas even when the technology is no longer available.

This chapter has focused on the roles of eMedia that emphasize communication, description, and explanation. Specifically, we've emphasized ways that eMedia can be used in an "expressive" role to facilitate learning and problem solving. This is seen not only in the business sector of learning, but also in mathematics education. As such, we focused on student-generated expressions rather than focusing on teacher-generate or textbook-generated expressions.

The models and modeling perspectives we have adopted assume that the meaning of a given construct tends to be distributed across a variety of interacting media—each of which clarify and de-emphasize somewhat different aspects of the underlying conceptual system. The meanings that learners or problem solvers use to create a given description of a situation often are quite different than the meanings that they are able to read out of these descriptions. Also, to develop sufficiently useful descriptions and explanations of a given learning or problem solving situation, students usually need to go through a series of iterative model development cycles. Thus, throughout the development of model (or purposeful description or explanation), as students extend, revise,

15. TECHNOLOGY-BASED REPRESENTATIONAL MEDIA

or refine relevant constructs and conceptual systems, the meanings of their representational media develop significantly—even during the process of working on a specific model-eliciting activity. For example, in solutions to the Summer Jobs problem, the meaning of a given graph may change significantly over the course of a series of modeling cycles. So, this capacity to evolve (Fig. 15.4) is an especially important characteristic of media in general and eMedia in particular.

Evolving Capacity Levels

Low Evolving Capacity	Medium Evolving Capacity	High Evolving Capacity
Explore	Adapt	Create
Modeler's Use—General		
Explore a pre-built technology-based model	Adapt a pre-built technology-based model	Create a technology-based model
Modeler's Use—Spreadsheets		
Explore a pre-built spreadsheet	Edit a spreadsheet	Design a spreadsheet

FIG. 15.4. Evolving capacity levels.

Some of the most important strategic strengths of eMedia result from their dynamic and manipulatable character. Dynamic characteristics may involve: (a) showing a process of change, or (b) having the ability to be changed by the user. Manipulatable characteristics include the idea of linked media. This new characteristic of technology-based enables the student to make changes to one representational media and the corresponding linked representational media is altered to reflect the changed state.

Chapter 16

A Models and Modeling Perspective on Skills for the High Performance Workplace

Melissa J. Dark
Purdue University

The impetus for this chapter is the increased concern about, reports on, and initiatives to address the skill shortage in the United States. For several years I have been involved in curriculum development, occupational analyses, and competency-based education, human performance, skill standards initiatives in postsecondary and professional education in the technical, engineering, and science-related disciplines. Skills in these disciplines are valued because of the role they play in research, innovation, technology, technology transfer, and economic development. And although discussing skills in these disciplines is important, it is also very difficult to do for a variety of reasons. In this chapter I consider several of the reasons why it is difficult to discuss skills for the high performance workplace including various connotations of the word skill and some of the associated problems. I then highlight some current trends and factors contributing to the groundswell of activity surrounding the skill shortage. These trends and factors are discussed in greater detail with regard to their impact on skills needed for today's high performance workplace, as well as skills that are emerging as critical for success in the future high performance workplace. Special focus is placed on skills that are needed which up-to-now have been neglected or were nonexistent. Finally, I present a framework for thinking about these new skills. The framework is an outcome of the failed attempts to meaningfully define skills and is intended to be a tool that can be used by curriculum developers, instructional designers, teachers, trainers, and other professionals responsible for educating the future workforce to think about skills in a more meaningful manner. There are several reasons why it is important to talk about skills that are emerging as critical for success, which up-to-now have been neglected, or were nonexistent. At present globalization and technological innovation are having a profound effect on our national health, prosperity, welfare, and security (Black & Lynch, 2000; Bresnahan, Brynjolfsson, & Hitt, 1999; National Research Council, 1999a, 1999b).

Globalization and technology have dramatically changed the workplace, forcing businesses to reevaluate their efficiency and ability to compete in the new economy. In an attempt to improve efficiency, businesses have sought to reengineer products, services, markets, work methods, processes, materials, machines, organizational structures, and the workforce. Among the many aspects of a business that can be reengineered, repeatedly a more skilled workforce is identified as essential to innovation and transformation (Black & Lynch, 2000). Why? Because economies and organizations are human constructed systems, whose ability to innovate and adapt depends upon the abilities of the workforce responsible for the products, services, markets, work methods, processes, materials, machines, and organizational structures that need to be reengineered. In short, innovation and transformation are the result of people.

In addition to the many reasons why it is important to identify skills emerging as critical for success in the high performance workplace, there are also many reasons why it is difficult. First of all, doing so requires linking fundamental changes in the economy to evolving workplace needs and practices, specifically with regard to skills needed for individuals and organizations to succeed. It also requires concurrently linking economic changes and workplace practices to educational practices. This is difficult for many reasons including: a) the changing and complex nature of the economy, b) the lack of a common definition of what skill is, c) the lack of a common vernacular for describing skills, and d) the lack of a useful tool or method for meaningfully talking about skills when definitions and vernacular differ.

The dynamic and heterogeneous nature of the economy and the workforce make timely, accurate, and comprehensive characterizations difficult, and characterizations of their interactions even more difficult. In this chapter, I concentrate on characterizing globalization, technological change, and high performance because they are salient trends prevalent in much of the other literature on workforce development. However, I am not going to offer an exhaustive description of globalization, the technology age, and high performance for two reasons; no such definition exists, and even if it did, I am interested only in the salient features that are making certain skills emerge as critical for high performance, which up-to-now have been neglected or nonexistent.

Another reason for the difficulty stems from the lack of a common definition and vernacular for skills. Skill has been operationalized narrowly, broadly, and everywhere in between. In this chapter, I do not construct a definition applicable and acceptable to all. I provide a working definition of skill, and distinguish between skills and skillfulness so that I can present the framework and demonstrate how it can be used for thinking about skills that are emerging as critical for success in the high performance workplace.

Finally, I present a framework for thinking about skills, which up-to-now have been neglected or nonexistent, that are emerging as critical to success in the high performance workplace in the global, technological age. I also provide examples to show how I have used the framework to identify such skills. I hope that the framework is useful as a reference point for researchers and practitioners in education and business to think about, identify and organize other neglected or previously nonexistent skills that are emerging as critical for high performance in the global, technological age.

SKILL: WHAT IS IT?

Skill is not a unitary concept (Cuban, 1997; Pearlman, 1997) that is easily operationalized. There are a variety of reasons for this. First, there are several definitions of what skill is. In a very narrow sense, a skill is a condition-action rule that specifies an observable behavior, which is very limited in scope and nature. Following are a couple of skill statements that exemplify the narrow definition of what a skill is: a) given a ruler, the student is able to describe the relationship of inch, foot, and yard, and b) given a set of numbers, the student can identify even and odd numbers up to 100.

Using this definition, a given subset of things that a person should or could know are skills, but because it is not possible to express all knowledge as action rules, this definition means that a great deal of knowing is about something more than acquiring skills. The advantage to using this definition is that it is possible to specify knowing in descriptive terms and identify specific behaviors that demonstrate skill acquisition. Descriptive terminology and specified behaviors are frequently endorsed because they provide clear guidelines for instruction and assessment. However, there are also disadvantages to this approach. Too often, the outcome is a laundry list of trivialized skills. The implication is that the sum of many smaller skills is equal to the acquisition of larger and more important learning outcomes that cannot be directly observed. Taking basketball as an example, I might be a good dribbler, passer, or shooter, but that does not mean that I am a good basketball player.

At the other extreme, skill has been used to define a wide range of personal characteristics, aptitudes, abilities, competencies, preferences, and specialized knowledge. Using this definition suggests that everything is a skill. The following are examples of skill statements that exemplify a fairly broad definition of what a skill is: a) look for patterns, and b) explain a problem situation. Even broader examples of skills are: spatial sense, and problem solving. The advantage to the broader approach is that it is usually nontrivial. However, the all-encompassing definition is a disadvantage when curriculum developers, teachers, and trainers attempt to create instruction or assess mastery using these vague definitions. Labeling everything as a skill is so broad that it is meaningless to anyone responsible for designing, developing, delivering, or evaluating instruction and learning.

Lists of skills that fail to distinguish between broad and narrow definitions are also a problem. For instance, although I might have basketball skill in the sense that I am a good passer, that is different from being a good offensive player. Being a skilled offensive player requires the integration of several skills just as looking for patterns embodies a set of skills and not a single skill as defined as a condition-action rule. These could be called different types or levels of skills, and I have no problem with that terminology. But failing to get clear about what a skill is problematic for teachers, learners, instructional designers, curriculum developers, and so forth. Although many people answer the rally cry for addressing "the skills gap," it is easy to get bogged down in personal preferences and beliefs without even knowing it.

The most difficult problem, in my opinion, is the lack of tools, methods, or procedures for bridging the gap between narrow, restrictive definitions and broad, all-inclusive definitions. I do not believe that a common definition of skill will solve the problem of communicating effectively about skills. So, without forcing a definition acceptable to all, I decided to address the problem of communicating about skills by creating a tool (the framework) to conceptualize and communicate about skills emerging as critical for success. A main function of the tool is to enable conceptualization of skills in a meaningful manner, yet in a manner specific enough to be informative for purposes of instructional design, development, delivery, and evaluation.

Because the framework was developed to conceptualize skills emerging as critical for the high-performance workplace, it is important to discuss connotations of skill that are germane to the workplace.

A Workplace Definition

Employers, businesses, and organizations seek employees who can perform competently. The nuances of skill in the workplace are aligned with and derived from expectations of competent performance, so much so that in the workplace skill is often equated with competent performance. Competent performance in the workplace includes ability and motivation. Ability is a function of aptitude and training, education, or both and motivation is a function of desire and commitment. From the employer's perspective, skill often denotes fairly broad and far-reaching abilities, such as understanding a wide variety of systematic relationships among variables for competent decision-making and performance. Competence in the workplace is far less about condition-action rules and far more about skillful performance(s). When leaders from business and industry talk about skills, they are referring more to an entity that I call skillfulness than specific skills. What is skillfulness?

Skillfulness denotes proficiency and the ability to perform with facility, adeptness, and expertise. Skillfulness also denotes continuous adaptation and dynamism, as opposed to a static skill acquired once and for all. For example, the mention of a skilled teacher, a skilled craftsman, a skilled manager, a skilled, machinist, a skilled librarian, a skilled researcher, or a skilled dancer, suggests expertise and adeptness. Characterize skillfulness also requires thinking about how one acquires the ability to perform with facility, adeptness, and expertise.

Skillfulness is acquired through the totality of experience. Totality occurs on different dimensions and includes: (a) the totality of events in a series of events, (b) the extent of participation or how active the individual was in the experience,

(c) the totality of the senses and the mind in the experience, and (d) the totality of the object (i.e., dance, research, or teaching).

As a former dancer, let me extend this example. Dance class usually starts at the barre and proceeds to the floor. Roughly one third of class is spent at the barre and the remaining two thirds is spent on the floor. There are two objectives for starting dance class at the barre: to prevent injury by warming up the muscles, and to practice specific skills including positions of the feet, posture, balance, and certain steps. If the dancer never leaves the barre, then she or he would never dance, because dance is the totality of a series of dance movements. Dance is not the mastery of the skills in isolation at the barre, but the synthesis of skills on the floor. There is a parts-to-whole approach when learning ballet. However, there is also an important whole-to-parts approach in that floor work provides a context and the need to master specific skills. Floor work provides the opportunity for dancers to practice individual steps (dance events) as a totality of events. Furthermore, if the dancer only danced on the ballet studio floor, then she or he would not have experienced the totality of the object of dance. To experience dance totally requires performing on a stage because the totality of dance as an object is a combination of many factors including practice, performance, performer(s), costumes, music, lighting, scenery, and props.

Much like dance, skills in the workforce really mean skillfulness, and are derived from the totality of experience. However, this brings us back to the problem of having skills defined so broadly that meaning for educational purposes is difficult. Let us consider the relationship between skills and skillfulness.

Skills and Skillfulness

I have defined skill more narrowly than skillfulness to operationalize it. Skill is doing; for something to be a skill, the learner has to be doing. Although I decided to use a more narrow definition, there are a few traps in using this approach that I want to avoid.

First, skill does not denote prerequisite. I have found this connotation of skill to be too restrictive. Used this way, skills are often operationalized as hierarchies of tasks, whereby subordinate tasks must be performed to specified criterion before the learner is deemed ready for and capable of acquiring the next skill in the hierarchy. For example, basketball requires skills such as dribbling, shooting, passing, and blocking, to name a few. However, it is not required that individuals learning to play basketball become proficient dribblers before they can learn passing and shooting, or play basketball. In fact, this would be a short-sighted approach when trying to help someone acquire basketball skill.

The other thing that I maintain is that skill cannot be put into a linear equation whereby the sum of the parts (skills) equals the whole (skillfulness). Skillfulness, as the whole, is more than the sum of several smaller skills put together. Again basketball is a good example. An individual skilled at dribbling, shooting, passing, and blocking, is not necessarily skillful in basketball. But because skillfulness is hard to conceptualize and operationalize, the flawed "sum of the parts equals the whole" approach is perpetuated.

These are some of the concerns and difficulties that led me to the development of a framework for conceptualizing skills emerging as critical for high performance in the global, technological age. I have built on the sum of the parts equals the whole approach by viewing skills through a lens that provides information regarding (a) the totality of the object, and (b) the totality of events, which enables contextualization of skillfulness and allows one to move back and forth between narrow, discrete descriptions of skills, broad definitions of skillfulness, and relationships thereof. The framework is a model for thinking about skillfulness.

Earlier I discussed skillfulness as being acquired through experience and more specifically through a totality of events, senses, mind or both participation, and the object. In this chapter I do not address the totality of senses, mind or both and participation. The framework I developed provides a lens for thinking about the totality of events, processes or both associated with skillfulness, as well as the totality of an object associated with skillfulness. Several chapters in this book describe how models and modeling can be useful for developing skillfulness by providing students with an opportunity to acquire experience through a totality of senses, mind or both, and participation. I leave it to those chapters to describe how models and modeling engage the totality of learners' senses in the experience, and the totality of their participation. Together, the framework and the models and modeling approach to instruction represent a models and modeling perspective on skills for the high performance workplace.

HIGH PERFORMANCE IN THE GLOBAL, TECHNOLOGICAL AGE

I promised to discuss a framework that I think is useful for conceptualizing skills emerging as critical for the high performance workplace in the global, technological age. However, to discuss a framework for conceptualizing skills that are emerging as critical for high performance in the global, technological age, I first want to highlight salient aspects of globalization, the technological age, and the high performance workplace. This discussion is necessary to understand how these trends are making certain skills emerge as critical, which up-to-now have been neglected or nonexistent.

In this section I also reference some skills frequently mentioned in the literature on high performance workplaces. It is important to note that the skills I am referencing in this next section and the two following sections do not necessarily fit the characterization of skillfulness or skill that I presented earlier. I present them here because they serve as good examples of the fuzzy descriptions that led me to develop the framework. In addition, they provide additional preface for making sense of the framework. You can see some of these skills again in the framework. I hope that the fuzzy descriptions become

16. SKILLS FOR THE HIGH PERFORMANCE WORKPLACE

clearer once the framework is used to show the relationship of these skills and skillfulness.

Globalization

Globalization is having a profound effect on individual and organizational health, prosperity, welfare, and security. In the global economy, organizations are facing increased competition against foreign companies to sell their products and services. Productivity is critical. The most productive business(es) capture market shares, generate greater profits and dividends, and pay higher wages. Productivity is a measure of how efficiently an organization transforms inputs into outputs (Russell & Taylor, 2000). Inputs commonly include materials, machines, methods, and the workforce. Transformation is the conversion of inputs into valuable outputs, and outputs are the product(s) or service(s) provided. Productivity can be improved by reducing input while holding output constant, increasing output while holding input constant, reducing inputs at a rate greater than outputs, or increasing outputs at a rate greater than inputs. Productivity improvements are essential to remain competitive in business.

The increased competition associated with globalization has increased the pressure on businesses to compete in terms of speed, innovation, variety, and customization (National Research Council, 1999). The global economy has brought an increase in the variety of products and services (output) that organizations provide. Diversification of the product portfolio is a frequently used strategy for stabilizing and or increasing output in volatile markets or both. In addition, today more than ever before, products and services are being customized to client-driven requirements rather than producer-imposed constraints. Producer-imposed constraints shaped the mass production model of yesterday. Mass production is a high volume, low variety business model, where productivity improvements were realized through standardization and redundancy of production. This model called for highly standardized and redundant jobs that utilized a restricted range of skills. The increase in customization and variety are reducing the need for standardized, redundant jobs and increasing the need for a flexible workforce capable of increased decision-making, problem solving, and communication. All in all, businesses that rely upon speed, innovation, variety, and customization to compete need a workforce capable of innovating and customizing a variety of products, services, and processes quickly.

Technology

In a broad sense, technology can be defined "as the systems and objects or artifacts that are created using knowledge from the physical and social spheres of activity" (Kozak & Robb, 1991, p. 32). Furthermore, technological innovation refers to human constructed systems that provide increased ability and capacity to do things that previously were not possible. Past technological innovations that have dramatically influenced ability and capacity include examples such as electricity, the telephone, and dedicated machine tools. Although technology has always been important, it is the breadth and speed of technological innovation that has made previously neglected or nonexistent skills emerge as critical (Berman, Bound, & Machin, 1997).

Technological innovation has profoundly affected innovation and customization, as well as the speed at which innovation and customization occur. Technology is fundamentally changing the way that humans transform inputs into outputs. Technology has three types of effects on work and occupations. It increases the skills required on some jobs, decreases the skills required for others, and creates the need for new skills (National Research Council, 1999b). However, whereas "skill requirements for some jobs may be reduced, the net effects of changing technologies are more likely to raise skill requirements and change them in ways that give greater emphasis to cognitive, communications, and interactive skills" (National Research Council, 1999, p. 71).

At the core of much technological innovation is digitization. Digitization refers to the conversion of physical phenomena and meaningful symbols like words and numbers into digital signals to create or manipulate information (National Research Council, 1999). A major effect of technology on work is that physical activities are being replaced with more abstract and analytical forms of work. Employees must understand systems that do not exhibit any physical manifestations of their functions. Organizations are increasingly more reliant upon a workforce capable of producing sharable, modifiable, and transportable tools, and adeptly encoding and decoding data and information to customize a variety of client-driven solutions, products and services quickly. As technology increasingly performs processes for us, what becomes important is to be able to put information in forms so that the technological tools can be used, to interpret the results that are produced with these tools, and to transfer the principle of doing so appropriately from one situation to the next. The symbolic interface is a "substantively complex activity that requires people to have technical skills, to conceptualize transformation processes abstractly, and to analyze, interpret, and act on abstractions" (National Research Council, 1999, p. 38). Success in the global, technological age depends primarily upon a workforce that can adapt to constant change and adopt continuously emerging technologies to make production and commerce more efficient and policy more effective (Weinstein, et al., 1999, p. 37).

High Performance

As a result of and in response to external influences such as globalization and technological innovation, organizations frequently reengineer their organizational strategies, structures, and processes in an effort to improve performance. Organizations that have improved their performance in the increasingly competitive marketplace as gauged by improvements in performance indicators such as productivity, market share, profit, dividends, and return on investments have been termed high performance. Business strategy, structure, and process reengineering initiatives are of great interest to policy makers and business leaders as benchmarks for others seeking to improve organizational performance. This is true for workplace skills as well. Whereas globalization and technological innovation certainly influence the skills critical for performance, skill gaps or needs are manifested at the organizational level. What are the characteristics of the high performance organization and the skills emerging as critical for success in the high performance workplace?

The high performance workplace is a complex system that is dynamic, continually adapting, and symbiotically linked to the economy. The high-performance workplace survives based on its ability to innovate and customize a variety of client-driven products and services quickly. The ability to innovate and customize a variety of solutions quickly requires a workplace that includes streamlined, decentralized decision-making that maximizes communication. Flexibility is critical. In the high performance workplace, employees have access to strategic goals, priorities and operating information, and are expected to positively contribute to attainment of goals by productive decision-making. Innovative work practices and teamwork are common. Diverse skills, knowledge and abilities of employees are leveraged to improve work processes. New skills are required as workers play a bigger role in improving quality, solving problems, troubleshooting systems, serving clients and cutting costs. The skills and knowledge of all members of the workforce are continuously updated and aligned with those needed to accomplish the work of the organization. As this trend continues; more and more high-performance jobs require workers with complex, up-to-date skills.

Today, high-performance goes beyond intellectual specialization to improving efficiency. Employees are no longer just expected to do things right, they are expected to do the right things right and do them at the right time. This means that organizational effectiveness cannot be accomplished only by the vertical integration of skills in an organization. (Vertical integration—the traditional way of organizing in businesses—refers to organizing departments and people within departments by functional specialization, i.e., engineering, accounting, sales, etc.) To do the right things right, organizations are recognizing that horizontal integration is required. (Horizontal integration refers to cross-functionalization, e.g., engineers who understand sales and the organizational impact of engineering decisions on sales and vice versa, and service technicians who understand finance and the organizational impact of service quality on the bottom line.) Employees must know more about everything. The power of horizontal integration comes not from employees knowing more specialized facts and skills, but consistently knowing more specialized knowledge and seeing how that knowledge applies across a variety of situations.

More and more employees are engaged in an alternating career in which they go back and forth between applying what they know in a team context (horizontally) and taking time out to learn new skills in a functional setting (vertically) (Womack & Jones, 1996). These shifts require workers to possess conceptual, analytical, communication, interpersonal and self-management skillfulness in addition to basic academic knowledge and technical skills (National Science and Technology Council and the Office of Science and Technology Policy, 1996).

Skillfulness Required for Success in the High Performance Workplace

The characterizations of skills emerging as critical for high performance that are woven throughout the previous sections are important. However, they are more characterizations of skillfulness than they are characterizations of skills, and we are back to the problem of descriptions that are so broad that they do not lend themselves to helping educators teach different things, or teach differently. I identified the following areas of skillfulness as important for success in the high performance workplace: complex systems, communication, representational fluency, group functioning, and self-management. These areas were used as I developed the framework because they appear repeatedly across several occupations and organizations. These areas of skillfulness are not exhaustive: users of the framework might choose to add others that are relevant for them. However, because of their relevancy to many occupations and organization, it is worth discussing them in further detail.

Complex Systems. Skillfulness in complex systems includes working with a multitude and variety of dynamic variables to solve problems that frequently have more than one right answer and more than one right solution path. It requires defining and clarifying problems; sorting and classifying information; evaluating the relevancy, reliability, validity, and sufficiency of information given; understanding complex situations from multiple viewpoints; breaking systems into parts; understanding systematic relationships among variables in a system; prioritizing needs, goals, problems and root causes; planning, designing, executing and evaluating a solution using divergent schemes to untangle the problem; analyzing results; recognizing when a solution is adequate and appropriate and when it is not; identifying causes for failure in solutions; and making necessary adjustments. In addition, proficiency in problem solving includes developing procedures for monitoring, inspecting, testing, verifying, revising, and documenting the problem solving process (Manufacturing Skill Standards Council, 2000; National Research Council, 1999a, 1999b; Pearlman, 1997; Weinstein et al, 1999).

Communication. Skillfulness in communication includes communicating problem statements, goals, needs, priorities, root causes, and outcomes to other audiences; communicating with others who have different skills; giving suggestions and criticisms; and accepting suggestions and criticisms (Manufacturing Skill Standards Council, 2000; National Research Council, 1999a, 1999b; Weinstein et al. 1999).

Representational Fluency. Skillfulness in representational fluency requires abstraction. Abstraction and representation includes visualizing and conceptualizing transformation processes abstractly; understanding systems that do not exhibit any physical manifestations of their functions; transforming physical sensory data to symbolic representations and vice versa; quantifying

16. SKILLS FOR THE HIGH PERFORMANCE WORKPLACE

qualitative data; qualifying quantitative data; working with patterns; working with continuously changing quantities and trends; and transferring principles appropriately from one situation to the next (National Research Council, 1999a, 199b; Pearlman, 1997).

Group Functioning. Skillfulness in group functioning includes creating strategies for sharing work tasks with the team based on the structure of the problem and the structure of the team; interacting with people in a broad array of functional roles; building team spirit; involving all group members; keeping one another on task; and adapting, cooperating, negotiating, and coordinating in the team structure for problem resolution (National Research Council, 1999a, 1999b).

Self-management. Skillfulness in self-management includes using independent judgment to make decisions; self-assessing one's own work for revision, refinement, and elaboration; and adapting quickly to new tools, new tasks, new jobs, new audiences, new teams, and new problems (National Research Council, 1999b; Pearlman, 1997).

A FRAMEWORK FOR CONCEPTUALIZING SKILLS EMERGING AS CRITICAL

Although I believe that discussing emerging skills and skillfulness for high performance is a worthwhile pursuit, my experiences led me to the conclusion that it is a fairly tough job for the many reasons already mentioned. I have seen many people stymied in their attempts to address this topic, and I have been stymied in my own attempts. I have seen the itemized lists of skills and I have constructed itemized lists of skills, only to be dissatisfied with this overly simplistic approach to conceptualizing skills and skillfulness. I have seen all-encompassing definitions of skills and I have constructed all encompassing definitions of skills, only to be dissatisfied with the inability to use my work to create instruction and assessment. This is what led me to develop the framework that I present here. I will first present the framework and describe how it is useful for conceptualizing skills and skillfulness emerging as critical. I give some examples of how to use the framework by revisiting the themes described in the previous section using the framework, concentrating specifically on skills that are emerging as critical, which have previously been neglected or nonexistent. Finally, Appendix A includes an example of a model-eliciting activity called the Summer Jobs problem. The summer jobs problem was written for middle school math students. I recommend that upon reading the problem in Appendix A, you try to use the framework to assess how and the extent to which the Summer Jobs problem develops skillfulness in complex systems, representation fluency, group-functioning, communication, and self-management. You might specifically try to identify the processes that students are required to go through while working on this problem, as well as products of thought that are a result of those processes. Finally, Appendix B includes notes from other instructional designers and educators who have worked with this problem and framework to identify specific skills and their relationship to the development of skillfulness.

This framework is based upon my work and research cutting across blue-collar, professional, technical, and managerial occupations in manufacturing, service, and government sectors. Although it was not my goal to provide a framework that I think applies to all organizations and occupations, or predicts the future (because that simply is not possible), I hope that deriving a framework from a broad cross-section will increase the applicability and usefulness of this tool.

The framework (Fig. 16.1) is a tool for thinking about skillfulness as acquired through a totality of experience, specifically the totality of the object and the totality of events. The X axis in the framework includes examples of objects or products one might produce in an occupation or discipline that together might represent the totality of that occupation or discipline. For example, an engineer who designs automobiles is responsible for a variety of products or objects of engineering. Among the many objects that they might be required to produce are: specification, descriptions or both of materials to be used in automobiles, functional and form designs of automobiles, a proof, or an explanation or both of why one type of engine is preferred over another, or assembly diagrams that show procedures for assembling the automobiles.

The objects of performance that I included are examples of objects commonly required in a wide variety of occupations, organizations, and disciplines. However, I am not suggesting that these are the only examples of objects that could be included. I encourage others who use this tool to identify objects of performance they deem important and use those in the framework.

To assess skillfulness across a totality of objects, individual skills can be organized across the X axis. This shows the extent to which students are being asked to produce a variety of objects that collectively represent the totality of the object or if students are consistently being asked to produce a limited number of objects, thereby reducing the likelihood the skillfulness being developed. For purposes of instructional design, delivery, and evaluation, this tool can be useful for providing learning activities and assessment tools to ensure that learning activities engage students in a totality of objects.

Another aspect of totality required for skillfulness is the totality of events. The Y axis in the framework includes examples of processes (events of performance) that are frequently experienced by employees in the workplace, and should be experienced by students in the classroom. Again, using the design engineer as an example, in order to produce a functional design for an automobile, the engineer would need to (a) assess customers' needs, (b) gather all of the necessary information to set up the design, (c) perform design work to produce the design, (d) monitor the design process to ensure it is going as planned, (e) inspect the design to ensure that it meets necessary criteria, (f) analyze the results of the inspection, and (g) make necessary adjustments.

16. SKILLS FOR THE HIGH PERFORMANCE WORKPLACE

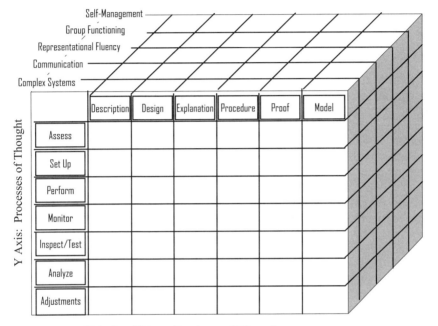

FIG. 16.1. Framework for conceptualizing skills and skillfulness.

Again, the processes I have included represent processes that I have observed as important to various occupations, organizations, and disciplines, and users of this tool should identify and utilize processes pertinent to the given situation. Furthermore, it is possible for cells to contain more than one skill. Likewise, it is possible that some of the cells can not be easily filled or require slight modification. For example, how does someone perform a description? The act of performing a description is usually expressed in terms of the form that the description is to take, that is, a set of blueprints, a term paper, and so on.

To analyze skillfulness across a totality of processes, skills should span a variety of processes represented on the Y axis. This can show the extent to which students are being asked to engage in a variety of processes that collectively represent the totality of events or if students are consistently being asked to engage in limited aspects of processing. Again, from an instructional design perspective, this would be useful for providing learning activities and assessment tools to ensure that learning activities engage students in a variety of processes.

The Z axis describes types of skillfulness that are needed for success in the high performance workplace. Again, this is not exhaustive; my criteria for selecting themes for the Z axis were universality and importance. The Z axis is a lens for conceptualizing emerging product and process specific skills within that theme. The key themes have been especially helpful to me for identifying skills that are emerging as critical for success in the high performance workplace, which until now have been neglected or were nonexistent. Let us look at a few examples.

Organizations are increasing group functioning in an attempt to improve performance. Skillfulness in group functioning includes a broad array of product and process skills. One example of a skill where learners must produce a description (product) of needs (process) might be asking learners to produce a description of the nature of the problem and the roles required of the team for problem resolution. This requires students to engage in assessing needs and to demonstrate the outcomes of this process by producing a description. Another skill required for group functioning skillfulness might be setting up the organizational design of a team for solving a given problem. This requires students to engage in setting up the group structure and to demonstrate the process by producing an organizational design of the team structure. There are numerous products that students can be required to produce. Likewise there are multiple processes that they should experience to produce those products. If we want students skillful in group functioning, then it is important that students experience a totality of processes and products for group functioning.

Increased digitization is leading to increased need for skillfulness in representational fluency. More and more, students and workers must be able to transform physical sensory data into symbols and vice versa, and transform symbols into other types of symbols. Furthermore, they must be skillful in transforming unlike data into information that can be used for decision-making. Skillfulness in representational fluency includes a broad array of product and process skills. One example might be asking learners to set up (process) procedures (product) for quantifying qualitative information. This requires students to analyze the structural form of different kinds of information and to create a tool for shifting back and forth between dissimilar representational systems. Another skill required for skillfulness in representational fluency might be to adjust (process) the variables used to predict sales (product). This requires students to reprioritize dissimilar information in a representational system. For students to become skillful in representational fluency, they must have experience with a totality of the products and processes of representational fluency.

SUMMARY

The framework described in this chapter was designed for assessing skills emerging as important for the high performance workplace. The approach I have taken to thinking about skills is unlike approaches taken in the past. I

16. SKILLS FOR THE HIGH PERFORMANCE WORKPLACE

believe that it is most meaningful to talk about skills in relationship to skillfulness. From a workforce perspective, skillfulness has been and will continue to be the expected outcome of educational efforts. However, unique economic and technological forces are exerting considerable influence on the skillfulness needed in the workforce, making this issue more critical today than it has been in the past few generations. The framework does not point out areas of skillfulness that are required for success in the high performance workplace. However, given broadly defined trends of skills needed for success in the twenty-first century, such as abstraction, complex reasoning, group functioning, and so on, this tool can be helpful for thinking about these emerging areas of skillfulness, and specific skills that attribute to developing these types of skillfulness. In addition, this tool can be used to clarify the relationship between skillfulness and skills. Specifically, the tool can help others think about the extent to which their curriculum is promoting skillfulness by including a totality of products and processes. And vice versa, this tool can help others think about how skillfulness can be further defined as specific skills for use by instructional designers, deliverers, and evaluators.

IV
MODELS AND MODELING
IN PROBLEM SOLVING
AND LEARNING

Chapter 17

Ends-In-View Problems

Lyn English
Queensland University
of Technology

Richard Lesh
Purdue University

Exercise! Do we get enough of this dreaded activity!

Dan is always complaining about how tired he gets each day. His mom said "why not exercise more to get fitter and so get less tired?" But he would always answer by saying he gets enough at school and most of his friends don't like exercise. So, does this mean most students surveyed don't get enough exercise through sport in pastime activities?

State your view.
Use the information on student pastime activities from all the countries surveyed to get the answer. Suzie and Jerry, 10th grade students

This problem is one of many that were posed by 9th and 10th grade students during a data-handling program where the mathematical products were unique student creations (English, Charles, & Cudmore, 2000). That is, students were given criteria for generating particular mathematical products, such as a problem, a persuasive case, or a data-gathering tool. What was different about these products from those of the usual classroom tasks was that students did not know the nature of the products they were to develop; they only knew the criteria they had to satisfy (e.g., Their problems had to be based on data they had gathered in a previous task and their problem was to be challenging and appealing to their overseas peers). There was more than one way of satisfying these criteria and therefore multiple products were possible. Furthermore, these products were far more complex than the usual responses demanded of students (where they produce a simple numeric answer by manipulating the appropriate information in a given problem).

Tasks that present students with particular criteria for generating purposeful, complex, and multifaceted products that go beyond the given information are referred to here as ends-in-view problems. We have adopted this term from the work of John Dewey (Archambault, 1964), who highlighted the importance of evaluating the means for accomplishing a task:

> It is simply impossible to have an end-in-view or to anticipate the consequences of any proposed line of action save upon the basis of some consideration of the means by which it can be brought into existence. Propositions in which things (acts and materials) are appraised as means enter necessarily into desires and interests that determine end-values. Hence the importance of the inquiries that result in the appraisal of things as means. (pp. 91–92).

Problem posing is an example of an ends-in-view task. The "Exercise" problem created by Suzie and Jerry represents the product they created. Suzie and Jerry had an end-in-view of what they had to develop; they were aware of the criteria for generating their problems and for judging whether one problem was better or worse than another. However, as we indicate later, the actual creation of these problems was the challenging part for the students. Suzie and Jerry generated their problem after numerous interpretations of the task goal, several deliberations over how to tackle the task, some arguments accompanied by much decision making, repeated revisions of intermediate products, and not surprisingly, some displays of frustration.

The purpose of this chapter is to examine these ends-in-view problems and to demonstrate how they can provide powerful learning experiences for students. The problems that are described in this chapter are classified according to the type of product that is to be developed and the nature of the ends-in-view that apply to each. Such products include tools (e.g., models, plans, and designs), constructions (e.g., complex artifacts, cases, and assessments), and problems (i.e., problems that are posed, questions that are asked, or hypotheses that are to be tested). We first consider some general features of these problems and then take a closer look at the various types of products. In the second part of the chapter we examine some different examples of ends-in-view problems, ones that are based on the domain of data handling. The data-handling examples illustrate how the important ideas and processes within a domain can be fostered through ends-in-view problems.

FEATURES OF ENDS-IN-VIEW PROBLEMS

Ends-in-view problems differ from the typical problems that students meet in class in two ways. First, the products that need to be developed are complex (e.g., they may comprise constructions, descriptions, or explanations). Second, criteria for testing the product are specified in addition to features that describe the nature of the product. A junior sporting team, for example, might be determined to win the grand final in the coming year. The players have an end-in-view, which is improving their performance so that they win. However, the

17. ENDS-IN-VIEW PROBLEMS

actual product they need to work towards must consist of more than just scoring lots of points. They must develop as a team, and each team must develop an intricate system of abilities. The players need to be consistently good at their game, over a long period of time. That is, achieving their goal requires long-term strategic actions that need to be carefully planned, implemented, and monitored. It is the nature of these strategic actions that is problematic for them.

Another key feature of ends-in-view problems is their relevance to today's world. In traditional textbooks and tests it is easy to find problems that refer to real situations, that is, their contexts are real world. However, the mathematics involved in solving these problems is often not real world. In contrast, ends-in-view problems not only have meaningful contexts but also address the kinds of mathematical knowledge and processes that are fundamental to dealing with the increasingly sophisticated systems of our society. As emphasized in the first chapter of this book, many of the most powerful goals of education involve helping our students, and our future leaders develop powerful conceptual tools for constructing and working with complex systems such as traffic control programs or weather forecasting systems. Students must develop the rudiments of these skills before they enter the workplace; indeed, this development needs to begin in the middle school years, if not before.

Both the products of ends-in-view problems and the processes involved in their generation differ from typical textbook problems. The products of textbook problems often would not make sense if the problem were acted out in an everyday situation, in contrast to the products of ends-in-view problems. Furthermore, in these textbook examples the criteria used to assess the quality of alternative answers often would not make sense in the real world. The textbook criteria tend to emphasize political correctness rather than practical usefulness.

When it comes to solving these textbook problems, it is generally assumed that the givens, the goal, and the legal solution steps are specified unambiguously—that is, they can be interpreted in one and only one way. This means that the interpretation process for the solver has been minimalized or eliminated. The difficulty for the solver is simply working out how to get from the given state to the goal state. With ends-in-view problems, however, it is not just reaching the goal state that is problematic, but also interpreting both the goal and the given information, as well as the permissible solution steps. Each of these components might be incomplete, ambiguous, or undefined; there might be too much data, or there might be visual representations that are difficult to interpret. When presented with information of this nature, students might make unwarranted assumptions or might impose inappropriate constraints on the products they are to develop. Thus an important aspect of ends-in-view problems is diagnosing the given situation, just as a physician would when faced with an ill patient—the physician is given a few indicators of what is wrong (i.e., some symptoms of an underlying illness), but needs to analyze these symptoms further so that the underlying disease (not the symptoms) can be treated. The physician's end-in-view is curing the patient, but exactly what this cure comprises is not yet known. In other words, the end-in-view is not identical with the product (cure) that has to be developed. Furthermore, precisely because the product that is needed is complex, there is usually more than one possible product, and achieving one that meets the given requirements is problematic. Numerous questions, issues, conflicts, and misinterpretations are inevitable in the course of product development, as are trade-offs involving opposing factors.

Although solvers do not know the exact nature of the product that is required of an ends-in-view problem, they know when they have developed one. This is because the given criteria or design specifications serve not only as a guide for product development but also as a means of product assessment. That is, the specifications enable learners to judge the suitability of their final products as well as enabling them to assess their intermediate products. With respect to these intermediate products, students can progressively evaluate their own work, identify any deficiencies and then revise and refine their work. Or, if several alternative products are being considered at the same time, then the students are able to assess the strengths and weaknesses of each.

As noted earlier, the solutions to ends-in-view problems are more than a single, simple product; rather they are comprehensive, complex products that display all or some of the following characteristics:

- Differentiation of factors, that is, different kinds of factors might be included in the finished product with trade-offs among the factors noted;
- Coordination of ideas that form a cohesive product,
- Conditional and flexible statements (e.g., If the store is to be located in a new neighborhood, then do X; if it is to be located in an established area, then do Y), and
- Cautionary statements (e.g., pointing out the limitations of sample size, alerting the client to possible sources of difficulty, error, or false assumptions. [e.g., "In making the recommendation to do X, it should not be assumed that factors A and B are unimportant. Their potential impact needs to be taken into consideration."]).

In addition to these features, the products of ends-in-view problems are designed to be shared with others and should thus be user-friendly, informative, and should capitalize on mathematical representations that can convey effectively the ideas presented. Furthermore, the ends-in-view problems that provide the most productive learning experiences for students should be reusable, particularly beyond the immediate situation. For example, a description of how to improve the performance of a school's sporting team could be applied to the improvement of other competitive school groups. This reusability feature is a significant one, given the well-documented difficulties students experience in transferring their learning to new situations (English, 1997).

TYPES OF ENDS-IN-VIEW PROBLEMS

Although ends-in-view problems that are described in this section share a number of features, they do differ in the type of end products they request. We classify these products under three broad categories: tools, constructions, and problems, each of which comprises several different examples.

Tool as a Product

Tools as products include models, mathematical descriptions, explanations, designs, plans, and assessment instruments. In general terms, tools are products that fulfil a functional or operational role.

Models

The numerous examples of models in the chapters of this book attest to their significance in the classroom, in the work place, and in daily living. Models are used for ranking items, people, and places (e.g., a wine-tasting panel develops a model to rate various wines); other types of models are used in determining loan payments on a home; and models also form the basis of complex systems such as a company's financial operations. A model always has a purpose and it must be sharable and reusable; otherwise there is not much point in creating a model.

Consider models that serve in a ranking capacity, a very common tool. Because ranking occurs frequently in daily life, there are many authentic contexts in which such modelling problems can be couched. For example, consumer guidebooks rate a range of items from cars to restaurants to holiday destinations. Sports teams are rated in newspapers, popular film stars and media personnel are rated in magazines and other consumer outlets, and people wishing to invest in the stock market have various rating guides to assist them. The Sneaker Problem and the Summer Camp Problem (created by Helen Doerr, http://soeweb.syr.edu/mathed/HMDproject/Main.html) are good classroom examples of models serving as rating tools. These problems focus on the core mathematical ideas of ranking, weighting ranks, and selecting ranked quantities. The problems are designed to encourage students to develop powerful ways of thinking about situations in which information about items (in this case, sneakers) can be assigned weights and then aggregated in some way in order to assign "quality ratings" for a specified purpose.

Descriptions and Explanations

Mathematical descriptions and explanations, which may also constitute models, can be used to illustrate and verify the results of an experiment or investigation, or to describe why something that appears superficially correct is mathematically incorrect. These descriptions and explanations tend to focus on the structural aspects of the item in question rather than the surface features and can involve multiple representations, such as graphs, diagrams, or tables. At the same time, mathematical descriptions and explanations need to be communicated to and used by others, and therefore must be couched in socially and culturally appropriate discourse.

Designs and Plans

Designs and plans are used in all walks of life, ranging from basic household schedules to complex plans for a new city underpass. Frequently, designs and plans must meet detailed and complex criteria and must incorporate appropriate mathematical and representational systems. In the Mini-golf activity (created by Michelle Heger, see Appendix A), for example, students are to develop a layout for a miniature golf course. The criteria to be satisfied are given (e.g., the course must fit within the confines of a particular building [56 feet by 65 feet] and between certain displays and demonstrations; the end of each miniature golf hole must line up with the beginning of the next hole). Students are given the mathematical shapes and dimensions of the 18 holes and must work with these representations to design the course.

Assessment Instruments

Assessment instruments are another common example of tools that are purposeful, sharable, and reusable. Such instruments are used in a wide range of contexts such as assessing the progress of learners, evaluating a company's (or a nation's) economic growth, selecting staff, and implementing quality control measures. These assessment tools normally undergo rigorous development that incorporates cycles of testing, refining, and applying. The testing instruments used in the Third International Mathematics and Science Study (U.S. National Research Center, 1996) are a case in point.

Construction as a Product

A construction normally requires students to use given criteria to develop a mathematical item, which can take many forms including spatial constructions, complex artifacts, persuasive cases, and assessments (i.e., the products of applying an assessment instrument). The criteria for a construction do not specify the exact nature of the product to be developed, rather, they set certain parameters for the design of the product. Unlike tools, constructions do not necessarily need to be reusable.

Spatial Constructions

Several examples of spatial constructions appear in the chapters of this book. For example, the Shadow Box problem requires students to construct an optical illusion viewing box in which various shapes, when viewed through a viewing hole, appear to be a square (see chapter 20 for an analysis of students'

17. ENDS-IN-VIEW PROBLEMS 303

construction of this box). Other examples of spatial constructions include the Quilt Problem (see chap. 4, this volume for an analysis) and the Toy-Train Problem (developed by Michelle Heger, see Appendix B). For the latter problem, students are provided background information about railroad gradients in a newspaper article. After completing a number of readiness questions, the students are presented with the problem about a model train for which a boy and his father are creating a train route. The model train can only handle a certain gradient, and the father and the boy need help creating a route that does not violate this maximum slope. The students are given a topographical map and miniature pieces of the model train track pieces. They are to construct a route on this topographical map such that the train never exceeds climbing the given gradient.

Complex Artifacts

Inventions are good examples of complex artifacts. The criteria for their design frequently focus on deficits in existing artifacts or on perceived societal needs. For example, Ralph Sarich, the inventor of the orbital engine in the early 1970s, believed that the conventional car engine of the time was too big, too heavy, and too expensive and thus decided to do something about it. His initial plan was to design a small rotary unit, prove that it could work, and then sell the idea to a major car manufacturer. He was not sure exactly what type of engine to build. Nevertheless, he sold his family home to pay for the venture and within two years had come up with his orbital engine, one that was lighter, cheaper, smaller, and less costly than the traditional piston design. However, concerned that his engine would take his clients (the car manufacturers) too long to tool up for building the engine, Sarich shelved the project and turned his attention to a new injection and combustion process, but with a less radical engine. He consequently established (but later sold) the Orbital Engine Corporation, which is presently a leading developer of advanced engine technologies.

Assessments

Assessments are the products of applying an assessment tool. Such products can serve a number of purposes and usually suggest or imply courses of action. For example, the results of the Third International Mathematics and Science Study provide significant insights into students' achievements in each of the participating nations. Comparative analysis of these assessments has led to a major rethinking of mathematics education in several countries (e.g., Judson, 1999).

Critical assessments of articles submitted for publication are essential products for journal editors (who have established scoring rubrics for these assessments). Reviewers who make such assessments require knowledge and understanding of the subject matter, as well as higher-order thinking processes such as critical and analytical thinking, logical reasoning, and the ability to interpret a variety of representational systems. The editors' decision on whether or not to publish a particular article stems largely from the contents of these assessments.

At the classroom level, students' assessments of their own work and that of their peers play an important role in their learning. Such assessments take students beyond thinking with a mathematical item to thinking about it (e.g., thinking about a problem that they have created themselves). In producing such assessments, students need to consider the extent to which the item meets certain criteria (e.g., whether the mathematical content is appropriate and sufficiently challenging; whether the context is appealing, etc.). These assessments allow students to identify the strengths and weaknesses of the item, provide directions for addressing any deficiencies, and offer suggestions for improvement and extension.

To complement students' own assessments, teachers can develop portfolios of rich assessment instruments such as scoring rubrics, observation forms, and ways of thinking frameworks. The assessments that result from applying these instruments provide insights into students' achievements and abilities that are seldom gained from standardized tests.

Cases

Being able to develop a persuasive case is a powerful skill in both the workplace and daily life. These cases make use of persuasive discourse to adopt a stance on an issue, to recommend one course of action over another, or to highlight an issue in need of attention, to name a few instances. Such cases are made especially effective when they draw upon mathematical data to support their claims. For example, a city council might wish to build a sports stadium at a certain site but realizes it has to convince the public that the site is the best of several possible sites. In building its case, the council might argue that the roads to the site could carry the expected volume of traffic by showing the mean number of cars using these roads during peak-hour traffic. The council might also state that the ratio of industries to home dwellers is higher than at other sites, arguing that fewer residents would be affected by the associated noise.

PROBLEM AS A PRODUCT

The ability to pose problems (in addition to solving them) is becoming increasingly important in academic and vocational contexts (Brown & Walter, 1993; English, 1998; Silver, Mamona-Downs, Leung, & Kenny, 1996). Problem generation occurs in many walks of life, whether it be a mathematician posing tentative conjectures for working hypotheses, a government agency faced with resolving a major national issue, or a finance company planning a complex merger where potential problems need to be considered carefully. Even during the modelling cycles involved in working model-eliciting activities (see chapter 1), students are engaged in problem posing, that is, they are repeatedly revising or refining their conception of the given problem.

17. ENDS-IN-VIEW PROBLEMS

As noted at the beginning of this chapter, the problems that we are able to pose are strongly influenced by our ends-in-view, including our criteria for judging the appropriateness and effectiveness of one problem over another. If our ends-in-view are limited, then our products (problems) are likely to be equally limited. And the potential for conceptual advances through problem posing is also restricted. In the classroom, for example, significant conceptual development is unlikely if learners are not engaged mathematically with the desired product and are not encouraged to consider alternative problem situations. Posing contextual problems to illustrate numerical statements (e.g., $23 - 5 = 18$) is a case in point. Students can simply create a problem by mapping text directly onto the given symbols (e.g., "There were 23 birds. Five flew away. There are 18 left."). Students who give thought to other problem situations would recognize that more than one interpretation of the statement is possible and would utilize text that supported these different interpretations (e.g., "I have saved $5 so far. But I need $23 to buy a new CD. How much more do I have to save?" We have seen some students interpret such statements in terms of identification numbers, e.g., $31 - 25$, as "Rooms 31 to 25 are on this floor.")

Obviously, the learning contexts in which problems are posed and the criteria for their creation have a significant impact on the quality of the problems produced. For example, the problem-posing activities addressed in the next section engage students in posing investigative type problems using a data base that they had generated themselves. Such problems are purposeful, sharable, and highly motivating because the students create problems for their international peers to solve, the students are given constructive feedback from the solvers of the problems (i.e., peer critiques), and in solving their own problems, the students can find answers to questions and issues that they have wondered about as they explored the data (e.g., "Why do the Canadian students prefer shopping over sporting activities?"). Such problem creations are also reusable in that many students from around the globe continue to try the problems that have been solved by numerous students before them, and the processes the students used in generating their problems can be applied to other related problem-posing situations.

We revisit these problems as products in the next section, where we also consider other types of products of ends-in-view problems that are based on the domain of data handling. These examples illustrate how such problems can foster the important ideas and processes within a particular domain.

ENDS-IN-VIEW PROBLEMS INVOLVING DATA HANDLING

Data handling is an approach to dealing with data, a frame of mind, an environment within which one can explore data ; it is not just a body of statistical content. Shaughnessy, Garfield, and Greer (1996, p. 205)

Data handling is particularly important in today's information age where organizing, describing, representing, and analyzing data together with using visual representations (e.g., diagrams, tables, graphs, charts) are fundamental life skills. Ends-in-view problems involving data handling provide students with rich and motivating learning experiences, in part because they encourage an "exploration of controversy" (Shaughnessy et al., 1996, p. 206). Furthermore, these problems take students beyond just computing with numbers to making sense of large volumes of data, quantifying qualitative information, identifying patterns and trends, producing convincing arguments supported by appropriate data, and assessing the products developed by their class peers. The problems cater for a range of achievement levels by providing manageable challenges that build on important mathematical content and processes, and allow for various approaches to solution.

The examples that follow are from a cross-cultural, longitudinal study in which 9th and 10th grade students participated in two data-handling programs. The students were from Australia, Canada, and Zambia (English et al., 2000). The ends-in-view problems described in this section draw upon the results of an international survey that was created by all the participating students. The survey itself was the product of an ends-in-view task described next.

Tool as a Product: Creating an International Survey. After preliminary discussion, the students worked in groups of 3 to 4 to develop some questions for inclusion in their survey. The students were presented with the instructions shown in Fig. 17.1. This task involved the students in extensive posing and refining of questions, taking into account the nature of the data that their questions would yield.

Think of 20 questions that you would like to include in the international survey we are constructing. Include a mix of questions that will produce nominal data, ordinal data, and interval data. Write your questions on the back of this sheet as well as creating a Word document. Remember that your questions must be suitable for the students in the other countries.

FIG. 17.1. Instructions for creating survey questions.

The final survey was posted on our website and completed online by the students in each class. The survey responses were collated and displayed on a spreadsheet for the students to access from the website. Data exploration followed, with the students wondering, conjecturing, posing questions, and searching for patterns and trends in the data. This exploration paved the way for the next ends-in-view task, namely, the posing of interesting and challenging problems for their peers to solve.

Problem as a Product: Posing Problems and Asking Questions. Prior to posing problems, the students developed their own criteria for determining:

17. ENDS-IN-VIEW PROBLEMS 307

- What constitutes a mathematical problem?
- What makes a good mathematical problem?
- What makes a mathematical problem challenging?
- What makes a problem appealing to the solver?

Appendix C contains part of a transcript of one group's discussion on these issues. The criteria that the students developed served several purposes. They guided the students in:

- Posing their initial problems,
- Testing whether one problem creation was better than another as they refined their problems, and
- Critically assessing one another's completed problems.

When presented with the task of posing problems using selected data from the survey results, the students did not know exactly what their problem would be like. However, they knew the specifications they had to meet: their problems had to be mathematical ones, they had to draw upon the survey data, they had to be appealing (set within an interesting context) and appropriate for their overseas peers, and the problems were to be challenging yet solvable. The students were also advised to give thought to the type of questions they might pose, the nature of the answers required (including use of representations and questions that called for some critical thought), and any cross-cultural issues that should be taken into account. Once students had posed their problems, they first shared them with their class peers and then made improvements based on their feedback (which used the assessment criteria the students had established previously). Following this, the students subsequently published their problems in the "Problem Gallery" on our website:
(e.g., see http://ourquestions.com/test/mitchelton1.html). The students in each of the countries then chose problems to solve from the gallery and subsequently provided feedback to the author of the problem (English & Cudmore, 2000).

Construction as a Product: Developing a Persuasive Case. The students completed two cases during the data-handling programs, one that was directed towards the school council or local state politician and the second, to the readers of a newspaper. For the former case, the students were to take the data from selected sets of survey questions (e.g., school uniform issues, participation in school sports, computer use in schools) and present a persuasive case that argued for a particular point of view (e.g., that students should have a greater say in the design of their school uniforms). For the newspaper task, the students were to undertake a comparison of data gathered from three of the participating countries and write a newspaper article based on their findings. The students were to support their cases with reference to appropriate findings from their analysis of the survey data and were to include suitable representations to illustrate their arguments (e.g., tables, graphs).

As a preliminary activity for both cases, the students were given a statistically flawed newspaper article that they had to examine critically, that is, they had to read it with a "mathematical eye." The students were to pose a number of questions that they considered should be asked of the information reported in the article.

Construction as a Product: Producing Assessments. Producing assessments of their own and their peers' creations was an important component of the students' learning. As noted previously, the students developed criteria for assessing the problems they had posed, and applied these criteria during problem generation as well as on problem completion. When critiquing their peers' problems from the Problem Gallery, the students completed their assessments online. An important aspect of these critical assessments is the assessor's feedback on how the problem might be improved and extended. The students also produced assessments of their peers' cases as they were being presented in class.

For the newspaper cases that the students constructed, they were required to produce a critical reflection and self-assessment that addressed the following questions:

1. When you were given the task of constructing a newspaper article from the data contained in the survey, what issues did you consider?
2. What questions did you ask yourself when you were trying to clarify what you were doing?
3. What obstacles did you encounter during the construction of your newspaper article?
4. How did you overcome these obstacles?
5. Does your newspaper article present a convincing argument? Why?

STUDENTS' DEVELOPMENTS IN SOLVING DATA-HANDLING ENDS-IN-VIEW PROBLEMS

In this section, we analyze some of the students' responses as they worked a couple of the data-handling ends-in-view problems. As background reading for this analysis, we recommend that chapter 19, this volume, by Zawojewski, Lesh, and English be consulted.

Clarifying the Problem

In their initial dealings with the ends-in-view problems, the students displayed many of the characteristics of an unstable product. They spent considerable time trying to interpret the nature of the end product they were to develop and identifying the nature of their clients (e.g., their international peers). In the excerpt below, a group of students is trying to clarify these aspects as they commence the task of constructing questions for the international survey:

Brian: What age is it going to be for?

17. ENDS-IN-VIEW PROBLEMS 309

Gretta: This is year 9....like, this is the readers?
Teacher: Well, your population will be...yeah. Year 9s.
Carol: Okay. A relative question to today. Do you guys want to start with numerical questions or ...how are you going to do it?
Carol: Okay. What do you want to start with? Do you want to start with numerical or categorical...
Brian: I suppose it doesn't really matter.
Carol: It doesn't matter? Just any questions...

Other students recalled a previous way of thinking to help them interpret the product they were to construct. For example, Laurel's group is trying to determine the types of questions they might pose as part of their problem construction. To help them, they are recalling the types of questions they had addressed in their science lessons:
Laurel: (Reading form the instruction sheet) "What types of questions might you ask in your problems?

Mary: In our problems. (as opposed to "your" problems as read out by Laurel)
Laurel: (Continuing to read from the instruction sheet) "In responding to these questions, you might consider what kind of data you expect in the answer, and whether you require the solver to use a statistical procedure to arrive at the answer."
Mary: Right.
Mary: I think it says, like, you know how we did all that data on our three questions---
Laurel: Yes. I know what it means.
Mary: ...and I think it says, like, what type of question we can ask about those data.
Cindy: That can be subjective or objective response? It can be, like, subjective...
Laurel: Tested. We did that in science: testable or untestable. Someone's opinion or someone's actual research.
Cindy: Yeah, whether it's opinion or whether it's fact.
Mary: It should be fact because we had data for them to use.
Laurel: Yeah.
Mary: Like, they're supposed to use that data.
Laurel: Yeah. An answer shouldn't be opinionated, because therefore you wouldn't get a proper answer...All right. So we have to think of what kind of data we expect in the things...

Reconciling Individual Interpretations With Those of Others

Individual students brought their own understandings to the group setting and thus had to reconcile their interpretations and views with those of their peers. This reconciliation process continued throughout the working of a given problem. In the following excerpt, the members of Laurel's group are modifying their own interpretations of the types of questions they might ask in their problem.

Mary: The questions we might ask would be stuff like "Which is the most preferred artist in so-and-so country?"—that type of question.
Laurel: What's our problem? Do we have to write down a problem?
Mary: I think we have to make up a problem eventually of course, but like...
Laurel: When they say questions do you mean...
Mary: Questions as in like, for them to find the answer. Like, it's a mathematical problem.
Laurel: Oh, so it's not questions like "Why is this happening?" or...it's how to find the answer.
Mary: It could be. I think it's more to the answer.
Laurel: Right, so...
Laurel: What? No, questions as in—but isn't the problem. like, you could be saying why do you like this or like that...

Internalizing and Coordinating Ideas

As students progressed on the tasks, their ideas usually became more coordinated and they were better able to detect weaknesses in the intermediate products they were developing. For example, in designing their questions for the international survey, Carol's group initially posed questions that were open-ended, without giving consideration to the nature of the data that such questions would generate. After constructing many of these questions, Carol realized that categories of response would be needed and alerted the others to this. It was interesting to see how Brian was internalizing her comments as he progressed from simply agreeing with others to becoming an important group monitor:

Brian: I still think we should do the music one.
Carol: Yeah, with another group.
Brian: And then write: "Name your favorite band," or something like that, to go with that sort of thing.
Carol: Yeah, we could do that, yeah.
Miranda: Yeah, that'd be interesting.
Carol: Writes: " Who is your favorite band?"
Brian: Yeah.
Carol: It's going to be like a huge...I mean, think of it as like, they've got to write down—how many groups are we going to get across the page.
Brian: Yeah.
Carol: Writes: "Out of these, who is your favorite group?"
Brian: Oh yeah.
Carol: Do you want to just put down five groups?

17. ENDS-IN-VIEW PROBLEMS

(Considerable discussion followed on popular bands, and which bands would be known in the United Kingdom, Canada, and the United States.).

Carol: Okay...another question.
Brian: Oh, got to put another one down there. "Don't know," because if they don't like any of them—you know, with the Spice
Carol: Other?
Brian: Yeah. And then, like "Other—please specify".
(Further creation of questions yielding nominal data followed.)
Brian: Oh, we need some numerical ones there (pointing to the students' list of questions).
Carol: Yeah...How many pets do you have?
Brian: Yeah.
Carol: (Writes.) "How many sports do you play?"
Brian: "How many days a year do you think you'd go to church?" or something like that because American people are very religious. So put that down there.
(Carol records.)
Brian: "How far is it...
Carol: "How far do you travel to school every day?"
Brian: Yeah.
Carol: Records: "How often do you use public transport?"
Brian: Yeah, that'd be a good one.
Carol: "...in a week."
Carol: (Records.) "What's your favorite movie?"—I don't know, something like that?
Brian: Or "Have you seen..."
Carol: "What is the most recent movie you've seen?"
Brian: Yeah, but like saying that there's that whole thing again.
Carol: True...Still, we're allowed to have...you know....
Brian: "How many movies have you seen out of the following list?"
Carol: Hang on—even better: "Which of these movies have you seen?"

Repeated Cycling of Refinements to Intermediate Products

A noticeable feature of the grade 10 students' discussions, particularly as they developed their problems and their cases, was their attention to refining their intermediate products. The students repeatedly cycled through refining their contextual language, refining their mathematical language (the structure of the mathematical questions or issues posed), coordinating their contextual and mathematical language, and improving any open-ended questions they wished to include (the open-ended question asked the solver to think in a critical or philosophical manner). The students frequently became bogged down in contextual concerns (e.g., which characters they should include in their problem setting), until one of the group members alerted them to the need to return to the task directions. These refinement cycles were also evident when the students acted on feedback from their peer's critiques of their finished products.

In the following discussion, Laurel's group is trying to improve the mathematical questions they wish to pose in the problem they are constructing. Notice how Mary alerts the group to the inaccuracy of Laurel's suggested question.

Laurel: Alright. We have to find the mean, median and mode? Or which one do you think? What do you think gives a more accurate answer to the most preferred musical artist: mean, median or mode? We could ask them that...
(No response from others.)
Laurel: That's a question we could ask.
Cindy: (to Laurel): I wouldn't want to be asked that.
Laurel: You don't want to be asked, but we're going to ask it. They're going to give us an answer because we don't have to think of it ourselves.
Cindy: But we do. We have to.
Mary: Well, we'd better get through this, you know?
Laurel: How about "Find the mean of the most popular musical artist...
Mary: You can't find the mean. (Laughs)
Laurel: Can we find the mode or...
Mary: Cardinal. Okay, for the cardinal question...
Laurel: So it's cardinal data?
Cindy: You could ask them to find the mean.
Mary: Slash mode...median.
Laurel: "...and median."
Mary: And you might ask...
Laurel: And how about—can we ask them their opinion on which do you think give the most accurate...
Mary: Yes, that's a good one.
Laurel: "Which of these do you think give the most accurate...
Mary: "View of the general time spent on the...
Cindy: "Shows the trend in the data."
Mary: Yeah, "trend."
Laurel: "Which of these...(writing)...do you think...give a true reflection...?
Cindy: "Gives the most accurate."
Laurel: "Gives the most accurate...description of the trends?" or...
Mary: View or....I don't know. View?
Laurel: "View of the trends presented in the data."
Mary: Why does one question take us so long?

The group then had considerable discussion on the wording of their problem context. Next, they returned to the nature of the questions they would pose in their problem. Laurel posed an open-ended question that asked the solver to

17. ENDS-IN-VIEW PROBLEMS

suggest reasons for the findings. Again, notice how Mary is acting as group monitor, reminding them of the need to move on.

> Laurel: Joe...I think we should make it a girl cause then we can ask it is there any...what factors do you think might influence the people's choice of listening and music and we say whether they're male or female because as you can see the Canadian girls like the Backstreet Boys?
> Cindy: Yeah, but there's so many answers to that.
> Laurel: Yeah, well that's what we need to make them find out, we have to ask them some questions.
> Cindy: They're only meant to be simple ones.
> Laurel: Well we're meant to ask stuff like is there any...
> Mary: People we have to hurry up you know.

Next, the group revisited the mathematical question they had posed earlier and tried to incorporate it within their chosen story context. The students found this rather difficult to do:

> Laurel: Yes it is...look is there any correlation between the people who are groovy and the people, what music they listen to. Please help Groovy Greg discover...find out...what was the kid's language?
> Mary: Kid's language...are you trying to insult us?
> Laurel: Find out whether...
> Cindy: His fears are justified.
> Laurel: Is there is a link between the people who are groovy and what music they listen to?
> Cindy: Yeah.
> Laurel: Is there a link between the people who think they are groovy and should I say like Austin Powers?
> Mary: We could say—Austin Powers wants to find out how many people think that they're groovy like him?
> Mary: Or we could say—How many people who think they're groovy like their music as well.
> Laurel: Yeah is there any correlation...
> Mary: Between...
> Laurel: Between the people who are groovy and the music that they like...
> Mary: No, but...
> Laurel: We need a story.
> Mary: Yes, we need a story.
> Laurel: With a question like um the correlation, you really can't um turn it into a story very well.
> (Brief discussion)
> Laurel: Well, we were just...we were going to try and see if there was any correlation between the people who were groovy...who think they're groovy and the people who and what music they like.

Returning to their mathematical question, the group realized that further refinement was needed:

Cindy: Copy those...have you found out whether there's any correlation?
Laurel: No, we haven't found out...we have to solve our own question before we can put it on....the thing is we have to...are we going to do that over all the countries or just over one of them or...like is there correlation in one country or is there a correlation between all of them?

The group subsequently revisited their construction of an open-ended question to be included in their problem:

Laurel: What was the question?
Mary: Like, think about the reasons why there's this difference between the...
Laurel: Why is this happening?
Mary: Yes, why is this happening.
Cindy: Why are there differences in the (inaudible word—overtalking)
Laurel: Just say...I just have to say "And why?" for that question and they can point out the reason why. Not much to that. And...now, why what?
Mary: Why there are differences between the data of the different countries. Does that make sense?
Cindy: "Why are there differences between each set of data."
Mary: Yes, each set of data. That's it.
Cindy: We're not being specific to one (problem?)
Mary: Oh, okay.
Laurel: "Differences between the data...between the...
Cindy: "Sets of data."

Laurel's group finally submitted their problem, as shown in Fig. 17.2.

Musical Madness
One day Groovy Greg was dancing down the street and he spotted his friend, Jess. As he was passing by, he noticed that Jess was listening to a different band than he was. Greg wondered if all groovy people listened to the same type of music.
Please help Groovy Greg find out if there is a link between groovy people and the music that they listen to. You can start by finding the groovy people from each country and the musical artist that they prefer. When looking at this data can you see any correlation between groovy people and the music that they listen to. What factors might affect the outcome of a person's musical preference?

FIG. 17.2. Problem created by Laurel's group.

CONCLUDING POINTS

This chapter has introduced an important problem type that receives limited attention in existing mathematics curricula. Ends-in-view problems encompass a range of authentic tasks that require the solver to go beyond the given information to generate a comprehensive and multidimensional product that meets certain specified criteria. Furthermore, because the product that is needed is complex, there is usually more than one possible product, and achieving one that meets the given requirements is problematic.

Generating a product, however, is not the only challenging aspect of these problems. Interpreting the goal, the given information, and admissible solution steps is also problematic if these components are incomplete, ambiguous, or undefined (as they often are in real life). Thus an important feature of ends-in-view problems is diagnosing the given situation to determine just what steps should be taken in working towards an appropriate product. The solver has an end-in-view of what is to be developed but the exact nature of this product is not known. In other words, the end-in-view of the solver is not identical with the product that she or he has to develop.

However, even though solvers do not know the exact nature of the product that is required of an ends-in-view problem, they know when they have developed one. This is because the given criteria or design specifications serve not only as a guide for product development but also as a means of product assessment. That is, the specifications enable learners to judge the suitability of their final products as well as enabling them to assess their intermediate products.

The authentic nature of ends-in-view problems means there are numerous different types of products that can be called for. In this chapter, we reviewed the following product types:

- Tools, which fulfill a functional or operational role, including models, mathematical descriptions, explanations, designs, plans, and assessment instruments,
- Constructions, such as complex artifacts, cases, and assessments,
- Problems, that is, problems that are posed, questions that are asked, or hypotheses that are to be tested.

The mathematical topics students study in school could be made more meaningful and more powerful if some of the standard textbook examples were replaced by ends-in-view problems. We have illustrated how this can be achieved within the domain of data handling. Instead of the teacher providing students with ready-made statistical experiences, the students generated their own. That is, they created the data base for their product developments, they developed their own criteria for judging whether one product was superior to another, and they took on the role of assessor in determining the quality of their achievements.

How students learn with ends-in-view problems requires a good deal more research. The richness of the experiences provided by these problems necessitates exploring students' learning through multiple lenses. We need to investigate how students, both individually and as a group, develop the important mathematical understandings that these problems foster. At the same time, we need to analyze students' developments in working as team members to create an agreed-upon problem solution.

ACKNOWLEDGMENTS

We wish to thank Kathy Charles and Leone Harris for their expert assistance in data collection during the data-handling program. Donald Cudmore's development and maintenance of our website is gratefully acknowledged. We also thank Michelle Heger and Margret Hjalmarson for their invaluable feedback on numerous drafts of this chapter.

Chapter 18

A Models and Modeling Perspective on Problem Solving

Judith S. Zawojewski
Purdue University

Richard Lesh
Purdue University

Problem-solving strategies taught in conventional school mathematics programs include heuristics such as "draw a picture," "try a simpler version of the problem," "use a similar problem," "act it out," and "identify the givens and goals." These strategies, originally based on Polya's (1945) work, are intended to help students connect the situation at hand to previously learned mathematical procedures, which in turn lead to a solution. The strategies are intended to provide a response to the question: "What can I do when I'm stuck?" But, for the kind of model-eliciting activities that are emphasized throughout this book, the transcripts of students' solutions reveal very few instances when the problem solvers are stuck. Most of the time, during the solutions of most problems, the students have a variety of relevant ideas and procedures available. Their problem is not that they have temporarily lost a tool or an idea. Rather, their problem is that current interpretations (descriptions, explanations) of the situation need to be extended, refined, or modified in order to meet some specified goal. Thus, the purpose of this chapter is to recast conventional problem-solving strategies with this major difference in mind. First, we examine the difference between a conventional definition of problem solving and model-eliciting activities. Then, by taking a developmental and social perspective on conventional problem-solving strategies, we recommend a revision of the strategies to meet the needs of students engaged in model-eliciting activities.

WHEN IS A PROBLEM A "PROBLEM"?

Wickelgren (1974) described problem solving in mathematics and science courses, saying "I was enormously irritated by the hundreds of hours that I wasted staring at problems without any good ideas about what approach to try next in attempting to solve them" (p. ix). His perspective of problem solving was one of searching for a lost tool (or procedure) for solving the problem. Similarly, Lester (1983) defined "a problem [as] a task for which: (1) the individual or group confronting it wants or needs to find a solution: (2) there is not a readily accessible procedure that guarantees or completely determines the solution; and (3) the individual or group must make and attempt to find a "solution" (pp. 231–232). Following this definition, problem solving would become the activity one engages in when working on a problem. Schoenfeld (1992), in his review of research on mathematical problem solving, indicated that the term problem solving has had "multiple and often contradictory meanings through the years" (p. 337). Definitions have ranged from information processing-based definitions[1] (like Wickelgren and Lester), which tend to persist in school mathematics today, to Lesh's (Lesh & Doerr, 1998) description that "the most important criteria that distinguishes 'non-routine problems' from 'exercises' is that the students must refine/transform/extend initially inadequate (but dynamically evolving) conceptual models in order to create 'successful' problem interpretations" (p. 9). The difference between these two perspectives is that the first views problem solving as the search for a powerful procedure that links well-specified givens to well-specified goals (see Fig. 18.1), whereas Lesh's model-eliciting perspective views the interpretation of the givens and goals as the major challenge, making selection and application of procedures a cyclical process integrated into the interpretation phases of problem solving (see Fig. 18.2). In a model-eliciting perspective, rather than using a fixed interpretation or procedure to process data, students are operating primarily on their own interpretations. In the model-eliciting sessions we have observed, much of what students are trying to do is to figure out what the problem is, and once accomplished the selecting and carrying out of the procedure is not very hard.

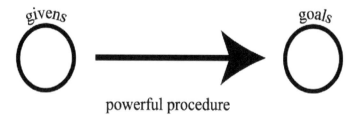

FIG. 18.1. Information-processing perspective of problem solving.

---

[1] For more about an information-processing perspective on problem solving, the reader may want to see, for example, Mayer (1985), Rissland (1985), Sowder (1985), and Schoenfeld (1985a).

18. PROBLEM-SOLVING STRATEGIES

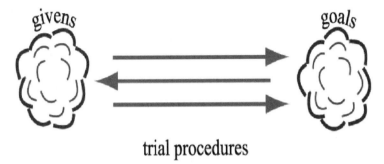

FIG. 18. 2. A modeling perspective of problem solving.

The essential characteristics of model-eliciting activities are quite different from the givens and goals types of problem solving represented in Fig. 18.1. For example, in a teacher sourcebook on problem solving developed by Charles, Mason, Nofsinger, and White (1985), the 150 givens and goals type of problems were organized into categories that are still prevalent in school mathematics texts today. Problem-solving skill activities are described as problems that promote the development of thinking processes involved in problem solving, such as: given a picture, write a story problem. One-step problems and multiple-step problems are described as "choose the operation(s)" problems. Finally, process problems are presented in a sequence that helps students choose an appropriate general problem-solving strategy. The essential characteristic of these types of problems is that the students are required to search for an appropriate tool (e.g., operation, strategy) to get from the givens to the goals, and the product that students are asked to produce is a definitive response to a question or a situation; or, where the mathematical constructs that are relevant consist of rules for computing answers to such questions. The essential characteristics of model-eliciting activities, on the other hand, are quite different, because students are required to develop, extend, and/or revise a model (e.g., interpretation, description, explanation) that is useful for accomplishing some specified purpose.[2] What needs to be produced is a model to make sense of the situation for which students' currently available interpretations of the givens and goals lack enough detail, elaboration, precision, and development.

Thinking of problem solving as situations in which a tool is lost, we believe, is overly simplistic. For example, suppose that we simply handed a relevant mathematical tool to a problem solver (e.g., a preprogrammed calculator that automatically calculates relevant answers when the given data are punched in).

---

[2] This purpose is sometimes called the "end in view" for the problem. It is what enables the problem solver to know when they have produced a construct (i.e., model, description, explanation) that is good enough. (See chapter 17 in this book by English and Lesh for further details about "ends in view"). In model-eliciting activities, it is only in a superficial sense that the production of a specified "end in view" is what is problematic about the situation.

Is there any mathematical thinking involved in simply using this tool (or calculator)? Most would agree—very little. Is there any mathematical thinking involved in finding this particular tool for oneself? If any mathematical thinking is involved in selecting such a calculator, it probably occurs when decisions are made about which tool to use—and when. Or, mathematical thinking may occur when decisions are made about how to put information into forms so that the tool can be used. But the activities of finding and using the tools are only a small part of what is required in a model-eliciting activity, and by themselves are often not problematic.

If an activity only involves tool using, then it is an exercise rather than a problem. If the activity involves tool finding, and even if the tool seems to be somehow mathematical in nature, very little deep mathematical thinking may be involved. This is one reason why some consider tool building to be the key characteristic that distinguishes problems from exercises (Gagne, 1965). According to Gagne's view, the distinguishing characteristic of problem solving is that it requires the problem solver to combine several smaller or less complex tools (or principles, or procedures, or condition-action rules) to form a larger or more complex tool.

This tool-building point of view continues to have several shortcomings from a models and modeling perspective. First, as many other chapters in this book emphasize, thinking mathematically is about seeing at least as much as it is about doing. In particular, it is about interpreting situations mathematically—by quantifying them, by dimensionalizing them, and by describing them using systems of mathematical relationships—at least as much as it is about compiling and executing a series of computations or other rule-based procedures. Second, as transcripts throughout this book illustrate, the way that interpretations get extended, refined, and revised generally involves sorting out and integrating unstable interpretations—at least as much as it involves assembling previously mastered systems to form more complex systems that are completely new. These models, or interpretations, generally evolve along a variety of dimensions. Further, at the beginning of a problem-solving episode, relevant models tend to be at intermediate stages of development.

This chapter focuses on problems that appear to be similar to traditional textbook word problems in the sense that the product that students are required to produce appears to be a traditional type of answer to a precisely specified question. Nonetheless, we will show that, even in the context of such problems, what is problematic for students often has more to do with the development of a useful way to describe the situation than it does with simply choosing an operation (procedure, strategy) and computing an answer. In fact, solutions to the activities in this chapter and transcripts of model-eliciting activities throughout this book show that students in successful modeling episodes seldom spend more than a few minutes on issues related to computation. Most of their time tends to focus on trying to make sense of the situation.

Some points that are emphasized about problem solving strategies in this chapter include the following.

18. PROBLEM-SOLVING STRATEGIES

1. The most useful strategies and procedures are not content or context independent. That is, they are learned during the solution development process for the specific problem. Which strategies and procedures a student uses in a given problem-solving situation depends on how the situation is interpreted. In other words, the strategies and procedures learned within the problem-solving episode are important components of the students' interpretations (or models) for describing or explaining the situation.
2. In model-eliciting activities, the most important purpose of most problem solving strategies and procedures is to help students refine, revise, or extend ideas that they already have (which are initially in some primitive form). The main purpose is not simply to help students find previously mastered tools (or ideas) that are lost, nor to assemble completely new tools using pieces that have been mastered in the past.
3. Any given strategy or procedure can be counter-productive in some situations, and appropriate in others, depending on the purpose for which it is used. For example, in model-eliciting activities, the general purposes often have less to do with helping students make effective use of existing ways of thinking, and have more to do with helping students overcome debilitating characteristics associated with their current (inadequate) ways of thinking.
4. The productivity of a given strategy or procedure depends on the purpose for which it is used. Consequently, in model-eliciting activities where solutions generally involve multiple modeling cycles, strategies that are productive at one stage of problem solving may be counter-productive at other stages. Similarly, strategies that are productive in one context may be counter-productive in others. Thus, knowing when and why to use a given strategy or procedure is a large part of what it means to understand that strategy or procedure.
5. Problem solving strategies should be expected to be similar to other types of mathematical knowledge or abilities in the sense that they develop along a variety of dimensions. Therefore, to inform instruction, it should be useful to investigate the development of these strategies and procedures—and to understand the nature of early levels of development.
6. Early levels of understandings often emphasize social dimensions of development. For example, in model-eliciting activities, effective problem solvers eventually learn to function as if they were, within themselves, a team of students working collaboratively. In doing so, they are able to think and also think about thinking; and, they are able to construct a way of thinking (embodied in diagrams, descriptions, etc.) and also critique that way of thinking.

An Example: The Carpentry Problem

Even though The Carpentry Problem (Fig. 18.3)[3] does not ask students to explicitly produce a model, or some other type of conceptual tool as a final product, it tends to be model eliciting during the solution attempt when students' ways of thinking become apparent. Even though the statement of the problem asks for a definitive answer, what is problematic is not the production of this answer. Instead, what is problematic is making sense of the situation so that the students' many answer-producing routines can be tested and evaluated. In other words, the real problem for students is to describe the relationships among available information.

Typically, initial solution attempts to this problem have involved the direct application of the formulas for area or perimeter of a rectangle. Or, more generally, first interpretations tend to involve a mapping to procedures with which individuals in the group are already familiar. Even though these initial interpretations are often overly simplistic or even incorrect, individuals tend to center on their first way of thinking. In our observations, it appears that group interaction is often critical for moving beyond these initially incorrect or overly simplistic interpretations. For example, the excerpt of a transcript in Fig. 18.4 reveals the beginning of a conversation that took place among two seventh grade girls and one boy as they began to work on the Carpentry Problem. Linda, Dora, and Randy (fictional names) were students in a north suburban Chicago middle school. As this excerpt illustrates, Dora immediately thinks that the problem has asked for area; so, Linda suggests a memorized procedure. But, after doing the calculations, and after having some doubt about the resulting product, Randy (who is initially drawn into the interpretation being used by Linda and Dora) suggests that the problem is about perimeter, and not area. His new interpretation was a conceptual shift, beneficial to the solution process. Then, the group proceeds to apply a standard procedure for finding perimeter without further reflection on the relationship between the problem statement and their knowledge of perimeter.

OBSERVER: Reads each task statement aloud one-by-one, waiting until the students complete the current problem before reading the next task.
1. John is constructing a recreation room in his basement. He has put up the walls and put down a floor. He needs to buy baseboard to put along the walls. The room is 21 feet by 28 feet. The baseboards come in 10-foot and 16-foot lengths. How many of each kind should he buy?
2. If John wants to have as few seams as possible, how many of each size baseboards should he buy?
3. If John wants to have as little waste as possible, how many of each size should he buy?
4. If the 16-foot boards cost $1.25 per foot and the 10-foot baseboards cost $1.10 per foot, how many of each kind should he buy if he wants to spend the least amount of money?

---

[3] This task was developed in the Applied Problem Solving Project, under the direction of Richard Lesh, Northwestern University, 1980–1984.

18. PROBLEM-SOLVING STRATEGIES

> 5. There is a sale on the 16-foot baseboards. They now cost $.85 per foot whereas the 10-foot baseboards still cost $1.10 per foot. How many of each should he buy if he wants to spend the least amount of money?

FIG. 18. 3. The carpentry problem.

Linda (0 min.)	We have to find out the wall -	Linda alludes to the wall—a flat surface
Dora	The area.	Dora also alludes to the measure of a flat surface.
Linda	Yeah, thank you. The area of the room.	Linda draws a rectangle to represent the room.
Dora	Okay. So you times 21 times 28. You guys have a calculator to do that? I'm not gonna do it.	It appears that they are searching for a procedure to link to the problem—area is length times width.
Linda (1 min.)	(After using the calculator.)The room is 588 feet.	
Randy	Ah, I don't think that's right . . . (calculating to check) . . . I, I have 588—	
Dora	500 and 80 what?	
Randy	88. 8.	
Dora	88? Okay.	
Linda	(Looking at her picture.) I don't think that was right. Let me try another longer way. (Mumbling while re-calculating 21 times 28.)	Linda calculates using a pencil and paper. She senses that something is wrong with their answer, but attributes the error to a computation difficulty rather than to a conceptual misunderstanding. She continues her focus on doing computations associated with the procedure rather than rethinking the validity of the procedure.
Randy	Wait! We only do the edges, not the whole thing.	Randy means, "We only have to find *the distance around*, not the area." This was the group's second distinct interpretation of the problem. While this was a conceptual shift it is apparent that the procedure for "finding the distance around" replaced "finding the area" as the procedure that would produce *the* answer to *the* problem.
Dora	We need to write it down this time . . . (Dora and Linda are mumbling while adding on a piece of paper.)	
Linda (2 min.)	98 feet is the . . .	Linda correctly calculates 21+ 21+28+28 (a standard procedure for finding perimeter of a rectangle.)

| Dora | That's not what I get. | Dora was having trouble applying the standard procedure correctly, adding 21 + 28. She used only the two "given" numbers available, and did not think about how the numbers could be used to describe all sides of the room. |

FIG. 18. 4. An early interpretation of the first carpentry problem.

This excerpt illustrates how a group of students initially sought and then applied procedures with which they were already familiar. In this case they initially pursued an incorrect path, which resulted in an answer of 588 feet. The unreasonable amount prompted doubts on the part of Randy and Linda. Whereas Linda thought there was a problem in her computation, Randy came up with a new insight—that the distance around was important. Dora and Linda quickly abandoned their initial interpretation and joined Randy in pursuing the perimeter interpretation.

Notable in the full transcript (Appendix A) is that the students continued to work together, with little intervention from the observer, to finally coordinate their problem interpretation and mathematical knowledge to reach viable solutions to the first three questions. Also notable was that during the session, the students appeared to use strategies such as consider a similar problem (e.g., find the area of the rectangle, find the perimeter of the rectangle), draw a picture (described next in more detail), and look back (when Linda and Randy reflected on the 588 feet). But, unlike the way in which the strategies are used as explicitly stated hints in conventional school mathematics, in this episode these strategies, instead, can be used to describe what the students were doing. Similarly, Schoenfeld (1992) found that the use of general problem solving strategies as hints during specific problem solving episodes is not particularly helpful for improving performance or transfer, but he also noted that the conventional problem solving strategies are useful in describing what students do. We have used these strategies to describe groups' behaviors during modeling sessions, and have found that social and developmental perspectives are helpful in understanding how and under what conditions students use of these problem solving strategies is productive.

So, what is the role of problem solving strategies in solving problems? Our main concern is that in conventional school mathematics, problem solving strategies (e.g., draw a picture, find a similar problem) which were intended to be means-to-ends, tend to be treated as though they are ends-in-themselves. When used as ends-in-themselves, such strategies can be counter-productive when the prompted behavior serves to lock students into premature or inadequate interpretations of the problem. Instead, we want to consider how problem-solving strategies can function to facilitate refinements in problem interpretations and help students avoid being deceived by their own inadequate-but-evolving early problem interpretations. Thus, we propose taking social and developmental perspectives on problem solving strategies.

SOCIAL AND DEVELOPMENTAL PERSPECTIVES ON PROBLEM-SOLVING STRATEGIES

The use of problem-solving strategies such as draw a picture, think of a similar problem, and identify the givens and goals evolved from Polya's suggested heuristics for mathematical problem solving (1945). Polya's heuristics were developed for university students who were well versed in various classes of mathematics problems and conventional models for thinking about mathematical situations. However, a number of studies have found that the prescriptive use of these types of general strategies is not particularly helpful for mathematically unsophisticated students (e.g., Bell, Swan & Taylor, 1981). In Schoenfeld's (1992) review of the literature on heuristics, he suggested that even with good material (e.g., Mason, Burton, & Stacey, 1982; Shell Centre, 1984) "the task of teaching heuristics with the goal of developing the kinds of flexible skills as Polya describes is sometimes a daunting task" (p. 354).

In his own work, Schoenfeld (1985b) gradually moved from a focus on general problem solving heuristics to a focus on content-dependent capabilities and on three metacognitive questions:

1. What (exactly) are you doing?
   (Can you describe it precisely?)
2. Why are you doing it?
   (How does it fit into the solution?)
3. How does it help you?
   (What will you do with the outcome when you obtain it?)
   (p. 374)

These questions were intended to help groups reflect on their current understanding of a problem. Schoenfeld indicated that initially students were unable to respond to these questions, but over time they began to ask the questions among themselves in order to prepare for their instructor's inevitable questions. Lesh (1986b) has described Schoenfeld's global problem-solving cues as considerably different from heuristics of the type described by Polya. Whereas Polya's heuristics were relatively specific activities intended to amplify students' positive conceptual capabilities, these cues were general reminders intended to help students compensate for (or minimize) the effects of negative conceptual characteristics associated with primitive interpretations, and also to go beyond first-order thinking to thinking intelligently (i.e., think about their thinking). Schoenfeld also concluded that problem-solving performance depends heavily on the content-dependent capabilities that are available to the problem solver to use for selecting, filtering, interpretation, and organizing information that is given. From a models and modeling perspective, we go beyond recognition of the importance of content knowledge in problem solving, to also understanding how "knowledge tailoring" is perhaps the most significant characterization of successful modeling sessions. Further, we are interested in the strategies that facilitate this process.

All of the foregoing suggests that we need to consider students' problem solving from a developmental perspective—as their interpretation of the problem develops, their solution develops, and the knowledge they bring to bear develops. Further, the social interaction among group members appears to be a critical component in the evolution of these various dimensions. We believe it is useful to consider how conventional problem-solving strategies can be recast when considering them from developmental and social perspectives. To illustrate, we consider three commonly taught problem-solving strategies: draw a picture, look for a similar problem, and identify the givens and goals.

Draw a Picture

Draw a picture is one of the most common problem-solving strategies taught in schools today. In a typical lesson the teacher demonstrates a problem, draws a picture of the situation, and then has the students practice drawing pictures for carefully selected tasks. They pick tasks for which the pictorial representation is likely to reveal the procedure needed to solve the problem. For example: "What is the cost of 15 oranges if 5 oranges are sold for 39¢?" The teacher may demonstrate, and then ask the students to practice drawing pictures and writing the accompanying symbolic sentences such as in Fig. 18.5. The question is—If a student can immediately translate between such a picture and a symbolic representation for picture, is this really a problem for the student?

OOOOO	OOOOO	OOOOO
39¢	39¢	39¢

39¢ + 39 ¢ + 39¢ = $1.17   (or 3 x 39¢ = $1.17)

FIG. 18.5. A correct drawing and interpretation.

Suppose a student did not have the practice provided in the foregoing lesson. Rather, the task was unfamiliar to the student. Bell, Swan, and Taylor (1981) found that individuals who were not already skilled at drawing pictures tended to draw ineffective diagrams when prompted by a teacher's suggestion. It may be that students try to draw the picture that they think the teacher wants, rather than focusing on the relationship between the situation posed and the representation drawn.

Suppose a group of three young students (for whom this *is* a problem) was asked to solve the given problem, but no diagram was given or suggested. In attempt make sense of the situation, an individual might draw the picture in Fig. 18.6. In trying to communicate his or her initial understanding to peers, it would be natural for group members to ask questions about the representation, challenge the picture's relationship to the question posed, and propose alternative ways to represent the situation.

18. PROBLEM-SOLVING STRATEGIES

```
O O O O O              O O O O O O O O O O O O O O
O O O O O   O O O O O  O O O O O O O O O O O O O O
O O O O O              O O O O O O O O O O O O O O
15 oranges   5 oranges          39¢
```

FIG. 18. 6. A possible initial drawing.

In group model-eliciting episodes we have observed, pictures tend to be drawn for the purposes of communication: with other group members to explain or challenging points of view; with oneself for reflection; and with oneself for reference at a later time. Although it is common for early representations to be oversimplified or have misrepresentations embedded in them, the pictures (or other representations) are neither good nor bad in and of themselves. Instead whether the picture is beneficial depends on the function that it serves. In fact, one picture may serve both good and bad functions within the same session, depending on the communication purposes it fulfills. In the third line of the transcript in Fig. 18.4 Linda drew a picture similar to the one seen in Fig. 18.7.[4] As Linda was drawing this picture, the labeling was sensible because it indicated the length of the top side from left to right as 28, with the length marked at the at the end of segment (as it would appear as if reading off a ruler). She similarly recorded 21 at the lower right hand corner, and so forth. This initial drawing supported the communication among group members early in the session as they worked to develop an understanding of the problem situation. However, because her rectangle was drawn as a near square, later in the problem solving session, when they returned to the picture, there was confusion about which side was 21 and which side was 28. This confusion prevented Linda and Dora from recognizing that Randy had devised an effective way of thinking about the problem. Thus, whereas the picture served as a good communication device in early phases of the problem, during later phases it was not effective as a communication tool. The same picture was both good and bad at different points throughout the solution process.

FIG. 18.7. Linda's first diagram.

---

[4] This example was used in Lesh and Zawojewski (1992).

The social discourse (which can be seen in Appendix A) led to a number of small refinements on Linda's first diagram. After 9 minutes, however, a major conceptual shift occurred in which she began to use "hash marks" and "middle labeling" on the diagram (see Fig. 18.8). This major revision was associated with a critical turning point in the group's understanding of the problem, which in turn led directly to a solution with which the group was happy. The interaction of group members during the process was critical to the evolution of the solution. Further, the pictures served as a window to the group's thinking as their collaborative solution interpretation developed over the course of the problem-solving episode.

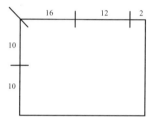

FIG. 18. 8. Linda's reconceptualized diagram.

This illustration shows how the first representation served a productive purpose for the initial communication among students. At a later point, the picture was no longer helpful for thinking about the problem, and in fact was counter productive. However, a new representation emerged which reflected the developing problem interpretation of the students. Thus, is was not that an individual picture was good or bad, or right or wrong. Instead, drawing of pictures contributed to the problem solving process, when they served communication purposes among the group members. We suggest, therefore, that the problem solving strategy "draw a picture" be recast to emphasize the use of representations for the purpose of communication. From a social perspective, groups naturally provide the opportunity for communication. From a developmental perspective, as the individuals in the group probe, question and challenge each other's interpretations and representations, the group's problem interpretation and solution develops into a more powerful, successful model.

Look for a Similar Problem

When students are first introduced to the problem-solving strategy, look for a similar problem, it usually follows the introduction of a particular class (or classes) of mathematics tasks. For example, consider the "train problems" that most readers have encountered in high school algebra. A student can be taught a class of "one train overtakes another problems", and also learn about two subclasses: *when* one train overtakes another and *where* one train overtakes another. Unless students have previous familiarity with this class of problems, it

18. PROBLEM-SOLVING STRATEGIES

is unlikely that they can identify appropriate similar problems that could lead them to the correct procedure needed. Model-eliciting activities are not in a class of problems that call on particular procedures. Rather, creating the procedure for solving a problem is often what is required as a solution. Thus, premature identification of a similar (or familiar) problem may lead students to solve the wrong problem, because during early phases of problem solving, students tend to center on superficial features or inappropriate approaches to the problem. However, the problem selected as similar is not necessarily a good or bad choice. Its usefulness can only be judged in terms of the function it serves during the problem solving process. For example, in the Carpentry Problem, the group's first two interpretations of the problem were: area procedure, and perimeter procedure. Although the first was an incorrect interpretation and the second was an oversimplification, the group's use of these two similar problems seemed to be natural steps that quickly led the group to thinking about the relationships between the length of the walls and the baseboards. Interestingly, their third interpretation was the closest to the given problem, but was quite problematic. The group began to test a series of trial solutions, using combinations of 10-foot and 16-foot boards to exactly cover (i.e., with no waste) the distance around, although the task only required that they minimize the number of seams. At this point, the no waste interpretation almost derailed the solution process. The following excerpt (Fig. 18.9), illustrates how this restriction led students to ignore or distort known facts—illustrating the egocentric tendency of problem solvers during early and intermediate stages of problem solving sessions. Because the three students expected to find an answer that did not have much waste, a correct solution to the 21 foot side involving a 10-foot and 16-foot board was implicitly rejected, presumably because of the 5 foot waste.

Linda (5 min.)	Okay. Now for the 28 length, for the 28 length walls all you need is 2, 16 boards. So that's 4, 16 boards. And then, with the other walls.	Linda was concerned about "waste," so perhaps she calculated 16 + 16 = 28 to subconsciously accommodate to the "minimize waste" restriction.
Dora	Ah, I follow you so far.	
Linda	It's 21 feet and you'd only buy 2 10-foot lengths of whatever it's called, and you'd have 1 foot left over. I'd just buy four of each and you'd have 2 feet left over.	Linda was confused about which of the two quantities was "left over." In this case the wall was 21' and she had purchased only two 10' baseboards. Neither or Dora nor Randy noticed the error even though they were monitoring carefully at the time. They all seemed to expect that there would be some combination of boards that would fit perfectly on each side, or at the very least that the small "left over" pieces would result in minimal waste.
Linda (5	Okay. Now for the 28 length, for the 28 length	Linda was concerned about "waste," so perhaps she calculated 16 + 16 = 28 to subconsciously

| min.) | walls all you need is 2, 16 boards. So that's 4, 16 boards. And then, with the other walls. | accommodate to the "minimize waste" restriction. |

FIG. 18.9. Distorting known facts to fit expectations.

The struggle continued, with the group switching back and forth between the minimizing waste and minimizing seams. After about 15 minutes, when the group was plagued by frustration, they began to distance themselves from their current ways of thinking (about the current similar problem) and entered a brainstorming phase (see Fig. 18.10). In this excerpt, it appears that Randy suspected their lack of success might not have been due to using a correct procedure incorrectly, but that there might be a better way to think about the problem.

Randy	We goofed. I know we did.	
Dora	What if you added (i.e., substituted) something? Just bought one big long one instead of having all these little broken up parts?	Dora was brainstorming with some "wild" suggestions apparently trying to help the group get out of its "rut" concerning the way it was thinking about the problem. Here she suggested, "What if we could just buy one long board, rather than 10' or 16' lengths?" She seemed to be aware of the fact that she was changing the intended conditions of the problem - perhaps in hope of striking on a new idea.
Obs.	Well, you get two choices. At a lumberyard they have only two sizes. There are 10-foot and 16-foot lengths. There aren't any that are bigger than 16's -	
Dora	Oh, m'God.	
Obs. (17 min.)	So, those are the only two options you have . . . Your goal is still to try to have, eliminate these ugly cut marks on the side of your wall. So -	
Linda	Okay.	
Dora	Why chop up half of it?	Dora was pointing to one of her pictures when she asked this question. Her gestures indicated that she was suggesting a new way to think about solving the problem: "If we always try to avoid making cuts in *any* of the 10-foot or 16-

18. PROBLEM-SOLVING STRATEGIES

		foot boards, wouldn't this produce an answer with the fewest number of seams?"[5]
Linda	(Recounts seams in her most recent picture.) 1, 2, 3, 4...	
Obs.	If you could avoid it, probably you wouldn't want to.	
Linda	(Continues counting.) 5, 6, 7, . . .	
Dora	What do you mean?	
Linda	(Continues counting.) 8, 9, . . .	
Obs.	If you don't have to chop a piece in half -	
Linda	(Continues counting.)	
Dora	Yeah, I know . . . unless you couldn't avoid it?	

About half a minute later:

Linda	Why don't we just use all small pieces so the lines you see are supposed to be there? (Pause.) What's the matter with seams? They're not ugly. (Pause.)	This brainstorming tactic is: if you can't solve the problem given, change the parameters.
Randy	No, I . . .	
Dora	Okay, do something like this*	See Dora's diagram below (*).
Linda	But that's only 20 feet. We need 28 feet.	
Randy	Hey, wait a minute. Wait a minute, wait a minute.	
Dora	Geez, you're right. Okay, I got it. I got it. No, the only thing that can times and equal 28 feet?	Linda's many little pieces suggestion reminded Dora of her earlier "equal sized pieces solution."
Linda	Is there—	
Randy	Hey wait a minute.	
Dora	Yes, there is. 7 times 4.	So, Dora has suggested using four 7 foot 3 x boards to cover the 28 side, and perhaps was thinking ahead to three more 7 foot 3x boards to cover the 21 foot side.

---

[5] The conjecture is not entirely implausible, but neither is it completely correct. Cuts occur not only at "cuts," but also at ends of boards. Perhaps Dora's idea could have spawned a useful idea, but at the time it was made, neither Linda nor Randy realized what Dora was suggesting.

* Dora drew this diagram:

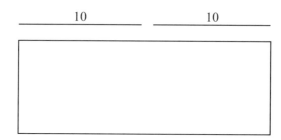

FIG. 18.10. Distancing and brainstorming.

This brainstorming phase of the session seems to illustrate that the group as a whole implicitly recognized that it was centering on an interpretation of the problem that was a dead end. The group seemed to be trying to distance itself from its current solution attempts, and explore other ways of thinking about the problem. The power of this brainstorming stage was that it helped the group look at the same problem from different points of view, rather than continue a search for similar problems. Brainstorming for alternative perspectives is a powerful strategy because it helps a group focus on how it is thinking about the problem. The social setting of group problem solving allows the group members to consider the problem from different points of view. Further, understanding the problem-solving process as developmental emphasizes to students (and their teacher) that successful solution attempts are likely to include a few dead end paths, and the way out of these dead ends is to consider the problem from different perspectives.

Our suggestion is that the problem-solving strategy "look for a similar problem" be recast to consider the given problem from different points of view. From a social perspective, groups naturally provide the opportunity for perspective taking. From a developmental perspective, as the individuals in the come to grips with each other's points of view, their group's problem interpretation and solution evolves to one that increasingly takes into account more details and relationships.

Identify the Givens and Goals

Identify the givens and goals, as a problem-solving strategy, has embedded in it an assumption that the givens and goals can be identified. In many of the model-eliciting activities in this book, the information given and the solution goals are not immediately clear, calling for problem formulation on the part of the students. In the Carpentry Problem, however, the givens and goals were explicit in the problem statement. The transcript excerpt found in Fig. 18.11 reveals that this group, in particular, was quite good in identifying the givens. They gave an unusual amount of attention to their understanding of the givens and goals— unlike many groups we observed who proceeded to try to solve problems even

18. PROBLEM-SOLVING STRATEGIES

when they knew they were unsure about the exact meaning of some parts of the problem.

Obs. (4+ min.)	Question 2. If John wants to have as few seams as possible, how many of each size should he buy?	
Linda	That's the problem?	
Dora	If John wants to have as few seams as possible, how many—	Dora was verbally paraphrasing the problem after the observer's reading.
Linda	What are seams?	
Dora	Oh, I know. Seams are, um	
Linda	Oooooh.	
Dora	Shut up! I do. Okay. All right, they're um, they're like um. They're seams.	
Randy	Yeah, what are seams?	
Dora	Okay, um.	
Randy	Now we know what those things are - what are seams?	
Dora	You know, you know, um, like the end of your pants or something. Or the end of a skirt.	
Obs.	Um-hm.	
Dora	How you can crease it over and—	
Randy	Oh.	
Obs.	What's a seam on that thing? (Pointing to the baseboard in the classroom.)	
Dora	On what thing?	
Obs.	(Again, points to the baseboard in the classroom.)	
Linda	The corner…stairs.	There were stairs in the classroom, so Linda was correct.
Dora	Corners.	
Randy	Ooooh.	
Dora	Yeah, the corners.	
Obs.	Okay, it's where they made a cut. Like here. (Pointing to a seam that was not in the corner.) And here point to a corner) and here (pointing to a noncorner).	

FIG. 18. 11. Clarifying the "givens."

Although this group of students carefully examined their understanding of the givens, they had difficulty understanding the goals of the problem, particularly when they transformed the minimizing seams goal to minimizing waste. The struggle that this group encountered concerning the goals was evident even 12 minutes into the session, at which point the students seemed to be unsure about what problem they had solved. In the excerpt in Fig. 18.12 the observer was not clear about which problem, #1 or #2, the students felt they knew how to solve. Linda stated, and Dora and Randy nodded in agreement, that they had not yet dealt with their "screw up" on the first problem. So, presumably, they were claiming to have solved the second problem—that is, the minimizing seams problem.

Obs.	You bought different wood this time than you did last time, right?	It wasn't clear which problem the group was claiming they understood how to solve, so the observer attempted to elicit an explanation.
Randy (12 min.)	Right.	
Dora	Yeah.	
Linda	Yeah. Except for this one, we screwed up.	
Obs.	In this case, you were supposed to find a solution that has as few seams as possible.	
Linda	Right. This has—	The picture that Linda pointed to was one that Dora had been drawing - which presumably showed the solution that Linda was trying to describe.
Obs.	Where are the seam cuts? Are there any cuts in the wood that you can avoid?	It was still not clear what problem the group was claiming they knew how to solve. So, before moving on to problem #3, the observer attempted to re-confront Dora, Linda and Randy with problem #2.
Linda	1, 2, 3. Three. Three seams.	
Dora	You can't avoid 'em all.	
Obs.	You can't avoid them? (Pause.)	
Randy	We can or we can't?	Randy was asking the observer, "Are you saying that we *can* avoid some?"
Linda	I don't know.	

FIG. 18.12. Difficulty identifying the "goals."

The suggestion, "identify the givens and goals," is not necessarily helpful when the task requires problem formulation, or when students have interpreted the givens or goals inappropriately. Rather, it is more productive to allow the social process of dialogue and discussion to lead to the clarification of the problem in a holistic way (where the givens, goals and solution are all considered and reconsidered at the same time). Illustrated in the previous excerpt, the givens gained clarity as the students discussed the meaning of terms such as "seams" (in Fig. 18.12), with some input from the observer. The group began to sort out the goals through a complex process of sharing problem interpretations, and with some prompting on the part of the observer. The interesting thing was that the group did not need to understand the givens and goals before making progress. Rather, they developed a problem solution as they were making sense of the givens and goals.

From a social and developmental perspective, identification of the givens and goals is integrated into the on-going problem-solving process. Clarification comes from the interaction of the group members, and evolves as the group's understanding of the problem develops. The more students revise their interpretations, explain their new interpretations to each other, and reconcile their views, the clearer the givens and goals become. Thus, asking students to identify the givens and goals too early might solidify their interpretation prematurely. We suggest that this problem solving strategy be recast to provide opportunities for students to explain the problem situation to someone else. In small groups, this happens naturally as students explain their interpretations to each other, take on different points of view, and use picture to communicate—simultaneously and gradually clarifying their understanding of the givens and goals in the problem.

SUMMARY

The conventional problem solving strategies that students often encounter in school mathematics are taught as suggestions for techniques to try when they are "stuck." In model-eliciting problems, we seldom saw student invoke these strategies consciously. We found them, instead, to be reasonable ways to describe what students were doing. We also found that use of the strategies, in and of themselves, did not necessarily lead to good or bad results. However, we saw how students may be helped to move beyond their initially egocentric interpretations and their tendency to center on certain features of the problem by the group interactions. If we want to consider problem-solving strategies for instructional purposes, it may be useful to recast them from a social and developmental perspective. In the Carpentry transcript the pictures drawn developed into more complete, detailed, and coordinated representations as the group's interpretation of the problem was socially negotiated. The series of similar problems that were driving the group's solution processes became increasingly similar to the actual problem as the students cycled back and forth between the problem statement and their diverse interpretations. The givens in the problem were fairly clear to the group at the outset, but clarity about the goals of the problem developed as their problem interpretations evolved through their interactions. Thus is seems that we need to think about the conventional problem-solving strategies with respect to the function they serve in the problem-solving process, and it also seems that those functions are largely social. A summary of the problem solving strategies from a social perspective can be found in Table 18.1. It includes the most salient features of the three problem solving strategies described above when considered from a developmental and a social perspective.

TABLE 18.1
Reconceptualizing "Problem-Solving Strategies"

Conventional Problem Solving Suggestions	Social Perspective on Problem Solving Strategies
Draw a picture.	**Use representations for communication purposes.**
When a student does not fully understand a problem, and is prompted to draw a picture, he or she may draw a wrong picture and act as though the picture is correct.	When a student doesn't fully understand a problem, he or she may draw a wrong picture, but the picture will be revised once communication has taken place among peers and the student develops a deeper understanding of the problem.
Identify a similar problem.	**Look at the problem from various points of view.**
When a student does not fully understand a problem, and is prompted to identify a similar problem, he or she may identify an irrelevant problem and then solve the irrelevant problem.	When a student doesn't fully understand a problem, he or she may identify an irrelevant problem as similar, but abandons or modifies the "similar" problem as a result of challenges by peers or of public testing of the current problem interpretation.
Identify the givens and goals	**Explain the problem to someone else.**
When a student does not fully understand a problem, and is prompted to identify the givens and goals, the student may identify the wrong givens and goals and then solve the "wrong" problem.	When a student doesn't fully understand a problem, the student may initially identify the wrong givens and goals, but revises the givens and goals as the problem interpretation and solution evolve as a result of explaining current interpretations to each other.

Chapter 19

A Models and Modeling Perspective on the Role of Small Group Learning Activities

Judith S. Zawojewski
Purdue University

Richard Lesh
Purdue University

Lyn English
Queensland University of Technology

The problem-solving process in many of the most important 21st century contexts involves teamwork, as illustrated in chapter 12 by Oakes and Rud. However, this point of view is not typically reflected in school mathematics, where students generally work in isolation, especially at the middle and high school levels. It is difficult for schools to change this pattern because parents, administrators, and even teachers find small group learning activities suspect. They ask question such as: What do the "smart" kids learn if all they do is teach the other kids what they already know? What do the "not so smart" kids learn if the smart kids do all of the work? The purpose of this chapter is to examine the role of small group work with model-eliciting activities in order to understand its potential value, and to provide initial guidance for implementing these types of activities. The type of tasks that form the basis of the problem solving experiences discussed are model-eliciting activities, which we believe are most like the kinds of problems that students of today will be solving in the workplace tomorrow. This chapter begins by describing how the design of the tasks embeds social aspects into the activities. We then explore the mathematical power that students as a group have when working on such problems. Cognitive development is then addressed from two different perspectives. First we examine how the group's mathematical model, which underlies their final product, develops during a problem-solving episode. We assume that groups' interpretations of a problem evolve from unstable to increasingly coordinated and stable models. Based on this assumption, we then discuss the potential for small group interactions to ameliorate the problems associated with unstable models, which are common during early phases of model-eliciting activities. The chapter ends with helpful hints for implementing model-eliciting activities with small groups, based on their experiences and observations of the authors.

SOCIAL ASPECTS IN THE DESIGN OF MODEL-ELICITING ACTIVITIES

The model-eliciting activities described throughout this book are of the type that we believe will be pervasive in the 21st century. As machines increasingly take over procedural and computational work, humans are increasingly designing new mathematical models to meet the needs of new and unfamiliar problem situations over procedural and computational work, humans are increasingly designing new mathematical models to meet the needs of new and unfamiliar problem situations. Model-eliciting activities have embedded in them social dimensions that reflect the teamwork and communication required in designing and using new technologies. This section describes the ways in which social dimensions are embedded in the these tasks based on the design principles described by Lesh, Hoover, Hole, Kelly, Post (2000).[1]

Two of the task design principles, the model construction principle and the model generalizability principle, work hand-in-hand to ensure that a task requires the development of a mathematical model that can be shared and reused. The model-eliciting activities are designed to have goals that serve social functions, such as producing explicit constructions, descriptions, explanations, justified predictions, assessed situations, and/or generalized procedures/plans. These types of products are meant to be shared or communicated with others who are interested in this problem, or other problems of the same type. Shareable products are in contrast to conventional problem-solving solutions in which expected responses are mathematical answers to mathematical problems, which usually do not need to be shared. With no external audience in mind, students tend to write the solution for themselves, and are often willing to use untested heuristics, or base decisions on single-case successes. Given the responsibility to develop a shareable tool creates the need for a mathematical model that will holds up under the scrutiny of others, much like the work of a mathematician or scientist who must publish papers for his or her professional society. To illustrate the social aspects of the model-construction and generalizability principles, consider the Big Foot Problem in Chapter 1 (Lesh and Doerr, this volume). Students are given an indicator of a person's height—a very large footprint. The solution to the problem requires the creation of a product—a "How To" Tool Kit for police detectives to use to figure out the height of a person given only their footprint. By requiring the creation of a mathematical model that can be used by a specific audience (the police detectives) for a particular purpose (for use in solving crimes) successful problem solving groups develop their products by posing and testing various models on each other, and by imagining how the police detectives would use the final product.

The personal meaningfulness principle for task design means that students are asked a "real" question rather than a "mathematics class" question. For example, coaches of sports teams do not compute the performance statistics of

---

[1] Lesh, Cramer, Doerr, Post and Zawojewski review the design principles in Chapter 2.

19. THE ROLE OF SMALL GROUP LEARNING ACTIVITIES

players for the sake of finding probabilities. Rather, the "real" problems are about making decisions, such as "Who should I bring to bat third in the roster?" Statistics and mathematical procedures are used to assist in formulating the final answer (i.e., the decision). Whether or not a task embodies the personal meaningfulness principal depends, in part, on the students' personal cultures and experiences. It may be that individual students in a small group have never encountered the given context (e.g., softball), but working collaboratively provides a diversity of personal cultures and experiences from which the group can draw to make sense of the situation. The social process of making sense of the problem posed can increase the interest and motivation of individuals in the group, and create a momentum that contributes to the group's perseverance. Developing a classroom culture that supports this level of group activity is not trivial, but model-eliciting activities are carefully designed and field tested to meet the personal meaningfulness principle, and thus facilitate interest and perseverance on a task.

The self-evaluation principle is served when the criteria for "goodness of response" are embedded within the task. Model-eliciting activities provide such criteria by giving students a clear audience and purpose for the model they are constructing—in other words, the criteria are embedded in the social context of the task. A self-assessment rubric has been developed for these activities (see chapter 7, this volume) that describes the quality of a product with respect to the needs of the given audience. The rubric essentially turns students back to the task statement to evaluate their progress. Model-eliciting activities that meet the self-evaluation principle are unique from most mathematics tasks that are used as problem solving activities. For example, consider the open-ended task, "explore the patterns of sums of consecutive numbers." The difficulty with self-evaluation in this type of problem is that even when groups of students are working together, their collaboration and conversation do not help them answer questions such as, "Are we finding the right patterns?" or "Have we done enough to be finished?" They must turn to the teacher to evaluate their progress or to decide if the work is complete. On the other hand, when criteria for "goodness of response" are embedded in a task, students can potentially return to the task statement of to revisit the audience and purpose, providing criteria by which to assess their own progress. Shifting students away from using the teacher as the only source of evaluation to using each other and the task statement is challenging, but potentially rewarding work. The work of self-evaluation is essentially social as students learn to take on the perspectives of the client, and listen to and value each other's points of view.

The model-externalization principle for task design requires that a group's solution is an external documentation of their mathematical model. The externalization of models is critical to cyclic modeling process (expressing, testing and revising ways of thinking) and for assessing intermediate and final products. External documentation of models is socially motivated in three ways in model-eliciting activities. The first occurs during the modeling session, as individual students naturally begin to use various forms of representation to pose and test their problem interpretations. As illustrated in chapter 18 (Zawojewski and Lesh) individuals eagerly explained their points of view to each other while also using diagrams to make their points. As they did so, they enriched their verbal descriptions while simultaneously leaving a trail of evidence that the group and/or the teacher could use at a later time to trace the development of the group's model.

The second form of documentation results from the task design, which requires the group to produce a final product that can be shared or reused. This final product reveals the group's thinking because it the explanation, procedure, etc., has been produced for a specific client and purpose. Thus, the final solution fully reveals the group's way of thinking about the problem. A third motivation for external documentation occurs when a teacher requires groups to present their final products to each other. The listening groups naturally find themselves comparing and contrasting their own model to the model presented, leading potentially to an additional round of revision.[2]

Teachers and researchers find the external model documentation very helpful in their assessment of student progress. However, it is important to keep in mind that external representations produced during the problem-solving episode do not provide definitive windows into individual students' thinking, because the very act of creating the representation can lead to a reinterpretation of the problem situation. Similarly, a group's final product is not necessarily a window into the individual's thinking, for each student has developed his or her own interpretation of the groups' model. The final product is the model posed as the solution by the group, and provides opportunities to interpret the construct produced by the group as a unit.

The simple prototype principle steps outside the immediate solution process to the use of the shared experience as a way to talk about subsequent mathematical activity with peers, oneself and/or the teacher. For example, one model-eliciting activity asks students to assess whether or not a bank robber could have escaped with a million dollars in small bills using a duffel bag. The client is a responsible TV news reporter who has requested a written assessment of the situation be produced prior to broadcasting the story on the news. In our work we have seen this problem serve as a shared experience to which teachers and students refer back at a later time. The socially shared history is evident when students recognize that the model they developed for the original activity is related to, and can be adapted to new and unfamiliar problems, such as estimating the number of breaths one takes in a life time, or estimating the number of baseballs that can fit into a classroom. Further, we have had some teachers report that students recognize specific mathematics concepts or procedures that they learned in model-eliciting activities when they reappear in more traditional lessons.

---

[2] The public sharing of models also provides students with opportunities to articulate their model. In our work, this has been an important part of the process because students' verbal description of their models often lag behind the sophistication of the model they have produced. Students need opportunities to explain their thinking clearly.

Model-eliciting activities also serve as prototypes for understanding the processes needed to solve complex modeling problems. For example, as discussed in chapter 18 (Zawojewski and Lesh), students learn the power of representing their ideas externally (e.g., drawing a picture) to communicate with their peers, and the power of taking different points of view as they experience the process of comparing, contrasting, and reconciling their initial interpretations of a modeling task.

The social dimensions that underlie model-eliciting activities put into place four mechanisms that help to ensure that the group will develop an increasingly effective model (Lesh, et al. 2000). First, the tasks are designed to prompt diverse ways of thinking, making alternative interpretations and approaches salient to the group. Second, a selection process naturally begins from the social interaction, bringing to the surface unproductive ways of thinking about the problem, which can then be refined, modified or rejected. Third, the group problem solving leads to propagation, in which productive ways of thinking are spread and integrated throughout the development of the model. Finally, the use of model-eliciting activities with small groups leads to conservation, in which productive ways of thinking are preserved over time. Thus, we use model-eliciting activities as the circumstances in which we consider the role of small groups in modeling activity in the school mathematics classroom.

THE POTENTIAL FOR MATHEMATICAL POWER

Slavin's review of the literature on cooperative groups (1989–1990) established that there are circumstances under which cooperative learning does indeed increase student learning. However, teasing out the circumstances under which collaborative work leads to increased mathematical power has been the challenge. For example, Good, Mulryan, and McCaslin (1992) have identified many factors that need to be addressed in research on small group work. Watson and Chick (in press) and Noddings (1989) have raised concerns about how the various factors that might contribute to, or detract from, students' learning, or influence levels of functioning during the group problem-solving process. In this section, we begin by describing how the task design has addressed some critical factors that contribute to successful small group work. Then, we discuss the potential power that small groups bring to the problem-solving situation.

Task Design and Factors Affecting Small Group Problem-Solving Success

Through our experience we have learned that one of the most critical factors for ensuring productive small group activity is that an appropriate task be selected, which is why we are focusing on model-eliciting activities for this discussion. Beyond the design principles described above, Cohen (1994) has suggested that the mathematical power of collaboration can only develop when no single individual is likely to solve the given problem on his or her own. Most of the model-eliciting activities we use have been ones that no individual in a middle-school group could solve independently. Thus, by placing our discussion of small group work in the context of model-eliciting activities, we can focus on understanding the processes that lead to the potential for mathematical power in collaboration.

A second factor we have found to be of critical importance in productive small group modeling is the teacher's role during the episode. Keeping the cognitive demand of a task high during the implementation phase is as important as selecting tasks that have a high cognitive demand in the first place (Stein & Smith, 1998). We encourage teachers to turn over mathematical authority (in other words, the mathematical power) to the groups, and to refrain from stepping in to help groups get on the right track. Refraining from assisting students is uncomfortable for most teachers, as well as the students. Thus, over the years, certain types of supports have been built into many of the model-eliciting activities we have used. For example, readiness activities are provided to establish familiarity with the context of the model-eliciting activity, and to provide an opportunity for the teacher to assess prerequisite knowledge (see Lesh, Cramer, Doerr, Post, & Zawojewski, chapter 2). When teachers feel that the students have been given an opportunity to understand the context and have demonstrated the prerequisite knowledge, they are more willing to turn mathematical authority over to students. Another way in which teachers are supported is by providing model-eliciting activities that are extensively field tested. Given information that the tasks are age appropriate, and that other middle school students have withstood the struggle and found solutions, helps teachers leave the problem solving to the students. The careful design and extensive field testing also results in tasks that have clear problem statements with the criteria for goodness of response built into them. This feature allows teachers to turn the students back to the problem statement when they ask questions like, "Are we on the right track?" or "Is this what you want?"

The Potential Power of the Group

The skills of the group are potentially broader and more accurate than those of the individuals. The information brought to the task is potentially greater, because each individual is tapped for his or her relevant knowledge. Tool use is often enhanced, for example, when one student who is quite adept at using a spreadsheet brings that skill to the collaborative process. Watson and Chick (in press) describe how collaboration amplifies (i.e., "lifts") the skills and capabilities of the group and indicate that "the synergy of the group can be conducive to lifting each of the participants beyond the levels they could have attained on their own, by providing a stimulus for their contributions and through encouraging justification for particular approaches." However, they also argue that the amount of lift observed in a collaborative event is very rarely above the capability of its most capable member. Our perspective is different, we have seen group products that represent more power than would be possible for any one individual in the group (see Middleton, Lesh, & Heger, chapter 22). We have also observed so-called "lower ability" and "higher ability" students

19. THE ROLE OF SMALL GROUP LEARNING ACTIVITIES

each gain access to activities that they otherwise would not have had on their own.

Peer interaction has the potential to amplify the interest and motivation of the students involved, increasing the potential mathematical power. Perseverance, a benchmark of success in real work situations, can be sustained in group situations whereas individuals may give up. Although a waxing and waning of interest happens on an individual level within the episodes we have observed, the momentum of the group often recaptures the individual's interest. Finally, satisfaction with the group's progress can be cumulative, particularly because the model-eliciting problems do not have a unique "correct" solution path and answer. As the model is refined, the groups test and assess their new ways of thinking, and thus can feel a growing satisfaction and finally a sense of completion.

We have also found that successful groups develop mathematical power by flexibly and fluidly taking on different roles (e.g., leadership, monitoring, and calculating) throughout the session (see Middleton, Lesh and Hegar chapter 22). This is in contrast to others, such as Cohen (1996), who suggests that assigned roles are desirable because they take over much of the managerial function of the teacher. But, we have seen students who are not typically described as leaders in mathematics class emerge as leaders at various points in times during group modeling activities. We have also seen the so-called leaders naturally take on roles as monitors and processors at other times during successful modeling episodes, such as in the transcript described in Middleton, Lesh, and Heger (chapter 22, Appendix B). Conversely, we have also observed that students' unwillingness to give up initially acquired roles can lead to unproductive results, as illustrated in a transcript (Appendix 1) of a small group solving the Wallpaper Problem (Fig. 19.1).

You have a roll of wallpaper. Each roll costs $17.35. The wallpaper store does not sell parts of rolls and they will only return half the cost of any unused rolls. Determine how much it will cost to wallpaper this room.

FIG. 19. 1. The wallpaper problem.

The following transcript excerpt was obtained in a research setting, in which the task was administered to three seventh grade students in a north suburban Chicago middle school. They were in a small room that was usually used as a pullout room for primary grade special education students. There were a few small chairs, tables, cabinets, and a dollhouse. The full transcript reveals that two models, or ways of approaching the problem, emerged during the session. The first model was to compute the area in standard units of measure and compare that with the area of a roll of wallpaper. The second model was to use floor-to-ceiling strips of wallpaper as a unit of measure for area.

The full transcript reveals that the students spent most of the session measuring and calculating the area of the walls, trying to figure out how the information about area could help them decide the number of rolls needed. Jim frequently tried to initiate a rethinking of the problem, but Barb's leadership was unyielding. Hope consistently followed Barb's lead. The excerpt in Fig. 19.2 illustrates how the lack of flexibility in role taking kept this group from coordinating their collective knowledge. Their inability to act as a unit derailed a successful solution. In this excerpt, which is after 30 minutes of computing areas and trying to figure out how to use the results from the calculations, Jim has an insight about a different way to solve the problem.

Jim	All we have to do is find out how many rolls we have to buy to do it, right? We don't have to do it - - -	He intends to finish the sentences with "find the area" which is what his teammates have been doing through the first 30 minutes of the session.
Barb	Right, Jim . . .	She uses a condescending tone.
Jim	Oh, so . . .	
Barb	Oh, so . . .	
Hope	(Laughs)	
For the next few minutes, Barb continues to work alone, calculating areas using length measurements from various walls. Jim goes back to reinterpret the previous information given the measurement, in light of his new idea— which he has not articulated explicitly. He then puts the roll of wallpaper to the wall and during the next segment describes and demonstrates to Barb, Hope, and the Observer a clear, simple, and nearly perfect solution to the problem. In the process he regenerates all of the information that he needs to solve the problem.		
Jim	1....2....3..that takes care of another roll ....1....	
Barb	What are you doing, Jim?	
Jim	2....3... that takes care of another roll, that's 3 rolls so far	
Barb	Jim!	
Jim	. . . 1 . . .	
Barb	Jim!	
Jim	What?.....2.....	
Barb	What do you mean "takes care of another role?"	
Hope	(Laughs)	
Barb	There are not—Jim!	
Jim	....3, that's another role . . . 4 what? Well, we have to go back to the store and buy some more roles.	
Barb	Jim, you don't know what you're doing.	
Jim	Yes I do.	

Barb	No, because you keep going 1, 2, 3, that takes care of another roll.	
Hope	5 rolls, 5 rolls?	
Jim	No, now, you see . . . no, wait, then let me explain, let me explain. OK, each - - this thing is 33 feet - - 30–33 feet long, right? Right! And how long is that going down like that 9 feet, 6 inches.	*When Jim says, "down like that", he indicates the height of the wall.*
Barb	9 feet, 6 inches.	
Jim	How many times will 9 feet, 6 inches go into 33 feet? 3 times, right? Right, so then if you go up one . . .	
Barb	(Laughs)	
Jim	Once up again, that's twice, up again, three times, then you're all out!	*Jim is making large gestures, acting out the procedures at the wall.*
Barb	(Laughs)	
Jim	Then you're all out of one roll. What's so funny? Isn't that right? (He faces Barb and Hope). Isn't that right? (He faces the observer.) Isn't that right? (He faces the heavens.)	
Obs.	Convince them. If they - - if they don't think you're right.	
The session continues until the end with Barb and Hope laughing at Jim, as he tries to explain again. Barb and Hope get all concerned about "left over" pieces from Jim's method, and continue to refuse to consider him as possibly leading a new solution method. The group never reaches a consensus about a good response.		

FIG. 19.2. Excerpt from wallpaper problem—lack of role change.

It was noteworthy that throughout the session Jim's insights were nearly always reformulations of the problem or its goals, rather than tinkering with the way Barb and Hope were doing the problem. He attempted to take a leadership role when he was explained his method to his teammates, but they rebuffed him, and laughed at his approach. He appealed to the observer, and the observer even encouraged him to try to convince them, yet he made no headway. The headstrong Barb continued to lead Hope down the original dead-end path. This excerpt illustrates what can happen when roles in the problem solving process are static and inflexible. What does this suggest for classroom practice? Perhaps mathematical power of a group may be curtailed if we assign static group roles to students. In this case, Barb naturally took on the role (it was not assigned), and simply did not let it go. Although it is not easy to change the personalities of students, we can help them become aware of their own tendencies and the potentially counter-productive effects of inflexibility (see Middleton, Heger and Lesh, chapter 22). If Barb could become aware of her own tendencies to be possessive of the leadership role, and could also become aware that maintaining the role of leader throughout a modeling episode is not always productive, she may begin to monitor herself. Perhaps she would become increasingly able to let go, and seriously consider other's points of view. Perhaps she could learn to consciously tell herself to give others the change to take on leadership roles when they have different ideas than her. Questions about how individuals learn about their own tendencies, monitor those tendencies, and change their behavior in modeling activities are questions open to research.

In our experience, when a small group of students is given an appropriate modeling activity, the group has the potential to bring more mathematical power to a task than any of the individuals involved. During their joint effort, two types of cognitive development can potentially take place. First, the solution product itself evolves from an initial interpretation, which is often over simplified and perhaps even incorrect, to one that has accounted for and coordinated the details of the problem posed. Second, the individuals in the group learn about themselves as problem solvers while acquiring knowledge and skills from one other as they participate in the collaborative effort.

COGNITIVE DEVELOPMENT IN SMALL GROUP MODELING ACTIVITIES

Researchers such as Yackel, Cobb, and Wood (1991) have used social perspectives to make sense of the learning and problem-solving behaviors of individual students. This perspective is important because teachers, parents, and administrators are interested in the achievement of individuals in the group. A different perspective that needs consideration, is one that is consistent with most workplace environments—that the final group product is the most important aspect of performance. While the contributions of each individual are assumed to be critical, the evaluation of the product occurs at the team level. Individuals' growth and contributions are in the background instead of the foreground. In this section, we consider cognitive development from this group perspective by exploring the role of small groups in the cognitive development of the models underlying the group product. We then consider the ways in which the group setting has the potential to ameliorate the unstable characteristics of models during early phases of the problem-solving activity.

Development of the Models

To understand how social dimensions of understanding are involved in the model development process, it is useful to think of a group's initial model as being similar to an uncoordinated system of physical actions—and to think of

19. THE ROLE OF SMALL GROUP LEARNING ACTIVITIES

the process of developing a system of mathematical relationships and operations as being similar to the process of coordinating a system of physical actions. For example, good tennis tutors can be thought of as using a physical version of Dienes' constructive principle. In chapter 2 Lesh, Cramer, Doerr, Post, and Zawojewski describe a tennis tutor helping a beginner to hit balls in relatively simple situations. The tutor starts by hitting the ball to the beginner waist height with no spin, so that little movement is needed to "meet the ball" on the forehand side. Over time, as the tutor provides progressively more complex volleys for the blossoming athlete to return, the athlete's actions gradually become more stable and better coordinated. Specifically:

- more information is taken into account (such as backspin or forespin on the ball),
- crude activities are differentiated into refined activities (such as when a straight forehand shot is modified slightly to produce top spin or chop shots), and
- simple activities are integrated into progressively more complex and flexible systems.

Similarly, the development of a group's mathematical model has the potential to evolve from a collection of individual interpretations and perspectives to an efficient and flexible coordinated group model. We observed this evolution of mathematical models as groups have worked on the O'Hare Problem (Fig. 19.3). This was a task that involved only straightforward uses of elementary mathematics. During the typically 40-minute solution attempts, groups spent no more than 5 to 10 minutes engaged in number crunching activities. Rather, they spent most of their time refining their interpretations of givens and goals. Final solutions tended to involve coordinating, differentiating, and integrating several initially unstable conceptual models concerning time, distance, cost, and so on. Because the task involved both too much and not enough information, models developed by the group had to be constructed so that meaningful patterns could be used to filter and/or simplify the situation, to fill in holes, or to go beyond the information given[3] (Lesh, 1981).

---

[3] Problem solving capabilities, which go beyond the information given, are described by Bruner (1973) as those needed to maximize the transferability of learning to new situations. From our perspective, it is from the use of model-eliciting activities that students learn to model and use modeling as a way to grapple with everyday problems.

> Orally posed: "What's the best way to get from here (a north suburb of Chicago) to O'Hare Airport?"
>
> NOTE: The term "best" was left undefined in this problem because a goal was to have students investigate processes involved in problem formulation, information interpretation and trial solution evaluation. A map was available, but no other suggestions were given about whether "best" was intended to mean shortest, quickest, safest, simplest, least confusing, most convenient, least expensive, or some other possibility. Also, no suggestions were made about whether a car was available, - or whether a bus, taxi, limousine, or train might be needed. However, these kinds of information were available if students requested it.

FIG. 19.3. The O'Hare Problem

During our research, this problem was implemented a number of times with small groups of students, revealing some common patterns in ways that students thought about the situation.[4] Typically, when students first began to work on the O'Hare problem, alternative ways to think about the situation were not explicitly recognized or sorted out. Students used the map to try to identify a route that was an unconscious mixture of assumed goals: shortest, quickest, easiest, and simplest. At first they initially noticed only a few salient features; for example, when considering the quickest route, they often did not consider any known traffic patterns near shopping malls or factories at either starting or quitting time. At the same time, they frequently made unwarranted assumptions about the problem situation or imposed unnecessary constraints. This was illustrated when students who assumed that a car was available also assumed that paying for parking was not an important consideration, or that leaving the car for an extended period of time presented no difficulties. Gradually, as one or more of the initial interpretations became better organized and more refined, more details began to be noticed with respect to that interpretation, and several interpretations began to get sorted out and integrated. Consequently, by the end of the problem solving session the students, who initially functioned as individuals, generated group products that often were:

> *detailed*—for example, for any given route, a great many details were noticed related to factors such as time, convenience, or possibilities of traffic jams;

---

[4] In Appendix 2 is a full transcript for the O'Hare Problem. In this session, a team of three seventh-grade students were in a suburban school north of Chicago. Earlier in the school year these students already had completed more than half a dozen model-eliciting activities of the type described throughout this book. The students were used to working as groups without teacher intervention during the problem-solving sessions.

differentiated—for example, different kinds of factors were sorted out, and trade-offs were noted—such as the fact that potentially fast routes were often high risks for accidents or traffic jams;
integrated—for example, "If you care about factors A, B, and C, then you should choose a way that deals in a combined way with these factors;"
conditional and flexible—for example, "If it is rush hour, then do X, if it is not, do Y or Z;"
aware of assumptions and possible sources of difficulties or errors—for example, "If your car is in the lot for a week, you will pay a lot, and someone may steal your radio."

Transcripts of small groups solving the O'Hare Problem have illustrated to us how initially a group's model is a disorganized bundle of individual actions and perspectives. Individuals in the group each form their own interpretation of the problem situation, and each acts as if their peers not only understand one's own perspective, but that their peers are also operating from the same point of view. Over time, in successful groups, students begin to communicate, compare, and contrast their perspectives. They notice that individual group members have different interpretations and they finally resolve the differences. When the group begins to function as a sufficiently well-coordinated system, a conceptual reorganization often occurs, and is often associated with a simpler, more elegant way to think about the problem situation. Although the new model is simpler, it still maintains an accounting of all the details and relationships originally discovered. The reorganization of the ideas allows the new, more integrated conceptual structure to require less conceptual space to "run the system." This happened for Jim in the Wallpaper transcript. He was initially inundated with the facts and figures of measuring the wall, finding the area, and trying to figure out how the area would help them decide on the number of wallpaper roles. Once he developed a different way of thinking, using the roll of wallpaper as a unit of measure, and imagining rolling the paper along the walls to find out how many rolls were needed, he had a much simpler model which still accounted for all of the details and relationships in the problem statements. Once a model has reached conceptual reorganization and simplification, new facts, model-reality mismatches, and internal inconsistencies within the model can be noticed, and the model refined accordingly. When a group begins to act as a unit, they continue this process until they together decide that the model-based predictions meets the purpose of the client in the problem.

Unstable Models in Early Phases of Problem Solving

When engaged in a model-eliciting activity, members of a group usually begin by functioning individually, each noticing certain, salient features of a given situation, ignoring other relevant features. Simultaneously individuals distort their initial interpretations to fit their prior conceptions and tend not to even notice that peers have different ways of thinking about the situation. These tendencies to center and exhibit egocentrism, described in depth in chapter 2

(Lesh, Cramer, Doerr, Post, and Zawojewski), are associated with immature models, and inevitably occur during early stages of a group modeling activity. These tendencies to center and exhibit egocentric tendencies do not appear to be age related. We have seen children who tend to display cognitive characteristics which seem quite adult-like, particularly after extensive experiences with modeling activities. We have seen adults who tend to display cognitive characteristics that seem quite child-like, especially when first encountering model-eliciting problems. What we inevitably see is that successful small groups always go through cycles of development from relatively unstable, uncoordinated individual interpretations to more coordinated, and increasingly stable group models. Based on our observations we believe that there are two mechanisms that seem to foster development from unstable to stable models. The first mechanism is perspective taking on the part of individuals in the group, and the second is when group members begin to internalize what were once external group behaviors.

Perspective Taking. Piaget and Beth (1966) describe early stages of cognitive development in which centering and egocentricism are prevalent. While these characteristics of immature cognitive development have traditionally been associated with particular age ranges, our work with model-eliciting activities reveal that these traits are more closely associated with immature models—regardless of the age of the members of the group. This is illustrated in the Wallpaper excerpt, when Barb and Hope centered on finding the area and displaying their egocentric tendency by sticking with their dead-end approach to the problem. Jim (who was of the same age), on the other hand, could be seen to have his personal model evolve as he listened to, tested, and questioned what his peers were doing. Barb and Hope, on the other hand, were unable (or unwilling) to seriously attend to Jim's new way of thinking about the problem. Thus, the group's model remained unstable and never led to the development of an effective response.

Members of the group who solved the O'Hare Problem (in Appendix 2) seemed to be consciously recognizing that different perspectives exist, and working to understand and reconcile their various ways of thinking. Perspective taking occurs when individuals encounter the interpretations of others in problem situations and are motivated to understand and reconcile the different points of view. Social processes that lead to perspective taking happen externally in small group settings when students challenge each other, monitor each other's procedures, and test their own and each other's hypotheses. These external processes have the potential to provide images and experiences which individuals can eventually internalize. The internalization of perspective-taking behaviors has the potential to contribute to the individual's own metacognitive abilities.[5]

External-to-Internal Processes. The social context of group problem solving provides opportunities for the group members to think *about* a proposed model.

---

[5] See Lesh, Lester, & Hjalmarson, chapter 21 for further discussion on metacognition and model-eliciting activities.

19. THE ROLE OF SMALL GROUP LEARNING ACTIVITIES

This is in contrast to thinking *with* a model, where the emphasis is on using the model to produce an answer. Thinking *about* a model means that the model itself is the object of critique; it is tested, compared to alternative models, and as a result is refined. Thinking about models is similar to what Vygotsky (1978) referred to as "higher order thinking," which "originates as actual relations between human individuals" (p. 57). He described the internalization of external processes as consisting of a series of transformations (pages 56–57). The first two are relevant to our discussion:

1. An operation that initially represents an external activity is reconstructed and begins to occur internally.
2. An interpersonal process is transformed into an intrapersonal one. "[E]very function in the child's cultural development occurs twice: first on the social level, and later on the individual level; first, between people, (interpsychological) and then inside the child (intrapsychological)" (p. 57).

We have observed these two transformations across successful small group model-eliciting activity. It appears that over time, students begin to expect that others will have alternative perspectives. They begin to anticipate what others' perspectives might be. Further, they seem to more readily notice hallmarks of early, unstable models in need of refinement. The following chart (Fig. 19.4) indicates how some characteristics of unstable models are naturally ameliorated in small group situations, and how they can become part of the consciousness of individuals who have internalized the external behaviors of small groups.

External Behavior Evident in Small Groups	Internalized Behavior as a Result of Working in Small Groups
Characteristic of Unstable Models: When a situation is thought about in one way, other alternative ways often tend to be temporarily forgotten	
In small groups one member may recall or have recorded a previous way of thinking and bring that perspective back into the conversation when the group becomes "lost" in a new perspective.	*As a student internalizes group modeling processes* he or she maintains an awareness that when one thinks about a situation in one way, other alternatives tend to be temporarily forgotten. From the group experience the individual learns to keep track of previous ideas and to consciously go back to review previous perspectives, as was experienced in small group sessions.
Characteristic of Unstable Models: Within a given way of thinking about the situation, the "big picture" is often lost when attention is focused on details—or, details are neglected when attention is focused on the "big picture."	
In small groups different members see things from different perspectives,	*As a student internalizes group modeling processes* she or he

thus there is the potential for a group member to revive the "big picture" perspective when the focus on details bogs down the process (or that a group member will call attention to details when the "big picture" becomes too unwieldy).	maintains an awareness that within a given way of thinking about the situation, the "big picture" is often lost when attention is focused on details – or, details are neglected when attention is focused on the "big picture." From the group experience an individual learns to regularly revisit the "big picture" and the "details" throughout the problem solving process.
Unstable Model Characteristic: Only a small number of details can be kept in mind at one time.	
In a small group the various members of the group tend to remember, and keep track of, different sets of details. When group members try to coordinate the details, they find that recording their information to share with the group is an effective means to manage the details all at once.	*As a student internalizes group modeling processes* he or she maintains an awareness that only a small number of details can be kept in mind at one time. From the group experience, the individual has learned that recording the details is an effective means to keep track of them all and to eventually coordinate the information.
Unstable Model Characteristic: Some of the information is misinterpreted, usually because the group members try to force the information into previously conceived models.	
In small groups, when some of the information is misinterpreted, it is often because the group members are trying to force the information into previously conceived models. The group members have the potential to act as independent monitors, noticing different interpretations and prompting explicit examination of each others' interpretations.	*As a student internalizes group modeling processes* she or he becomes aware that information is often misinterpreted, especially in early phases of the problem solving process. Thus the individual has learned from the group process to explicitly monitor interpretations, checking and rechecking them as the problem solving session progresses. The individual acts as if various members of a group were explicitly sharing interpretations and subsequently revising their interpretations.

FIG. 19.4. External-to-internalized ways of dealing with characteristics of unstable models.

19. THE ROLE OF SMALL GROUP LEARNING ACTIVITIES

MANAGING AND IMPLEMENTING SMALL GROUP MODELING-ELICITING ACTIVITIES

A models and modeling perspective on learning and problem solving requires a shift in the view of the learner, as well as a shift in the view of the teacher. One major difference is that the teacher's role is to provide opportunities for groups to engage in multiple cycles of expressing, testing and refining their problem interpretations (i.e., models). This is in contrast to the role most teachers associate with their role in teaching problem solving: to help students identify the given information and the solution goals for a problem, and then to help students search for a procedure to link the givens to the goals. The most important teacher roles in small group modeling is to create the need for students to create significant models, tools and representation; to set the stage for the activity; and to listen carefully during the solution process.

When using model-eliciting activities over time, many of the common classroom management problems decrease, largely due to the nature of the tasks. Given the design principles for model-eliciting activities, and the extensive field testing, students usually find the tasks interesting and motivating. Because the problems are substantive and in need of group effort, students tend to work together productively, with individuals naturally shifting among various roles (e.g., leader, monitor, calculator). However, when first introducing model-eliciting activities, teachers commonly have many questions about planning and managing the sessions. We share some suggestions which are based on our own experience with these types of tasks.

How Should the Groups Be Formed?

The two most common questions asked are, "How many students should be in a group?" and "How should I decide which students to put together in a group?" The number of students in a group is guided by our goal to maximize opportunity for all students to be engaged at all times, which leads to the use of small groups. However, if the number is too small, then the amplification powers of the group may be lost. In our work, we have tried using two, three, four, and five students in a group. We found that when there are four or five students, subgroups of two or three tend to form within the group. If subgroups work separately for long periods, and complete a number of modeling cycles prior to comparing notes, their ways of thinking may drift quite far apart. In this case, trying to reconcile two different, yet potentially successful, interpretations is often difficult and can lead to frustration as the group tries to formulate one final product. On the other hand, we find that groups of two students often do not have the combined mathematical power to adequately grapple with the substantive problems. Furthermore, when one student in a pair is perceived as strong and the other weak, the so-called strong student simply tells the so-called weak student what to do, and the solution is often weak, representing the work of only one student. We have had most success when three students work together. Their combined skills and perspectives tend to be enough to deal with the problem complexity, and the roles taken by the members tend to shift naturally and effectively. While there are times when two of the three are actively engaged, and the third is thinking or working independently, this is usually followed by the third expressing a new idea to the other two members, which often leads to testing, refinement and reconciliation of the ideas at hand. Our recommendation is to aim for groups of three, and when necessary use a fourth person.

Deciding who should be in the groups requires consideration of multiple factors. Keeping in mind that the goal is to provide opportunities for students to take on different perspectives, and to have skill levels amplified, it behooves the teacher to combine students who have different ways of thinking and different types of skills. For example, a teacher might want to select one student who is good at computational procedures, accompanied by a student who understands mathematical concepts, but might demonstrate inaccuracies in computation. A third student, who might be considered a weaker student by his peers, may be taught to use a tool that is likely to be helpful in the solution process. For example, one teacher introduced the use of spreadsheets (EXCEL) to a small group of students, who were then distributed into the different groups. By adding this level of proficiency to a select group of student, those who are typically thought of as liabilities in group problem solving have the potential to become assets.

Of course, a teacher needs to assess students' personalities and anticipated behavior problems for particular groupings. By trying out different combinations of students, and juggling both the cognitive and management goals, individual teachers can learn about the idiosyncrasies of their particular students, which can in turn inform the decisions about who to combine into groups.

How Should the Problem-Solving Episode Be Started?

In our recent work with middle school students and their teachers, we have included readiness activities that are implemented prior to the actual model-eliciting activity. The readiness activities are based on a newspaper article, that introduces the students to the context, and some questions that review the content of the article. Occasionally the readiness questions will address a few prerequisite computational skills, but they never scaffold the any particular approach to the model-eliciting activity. These activities have been effective in helping students and teachers feel more comfortable as they get ready to encounter the challenging model-eliciting activities. There are alternative ways to implement the readiness activities. Students can read the article and do the questions as homework, and then review their responses in class. Or the students can be asked to read the newspaper article out loud in class, followed by individual, group and/or class work to respond to the questions about the article. Some mathematics teachers team with language arts teachers, so students can complete the reading in their English classes. The first time a model-eliciting activity is used, the readiness activities may need quite a bit of teacher direction.

19. THE ROLE OF SMALL GROUP LEARNING ACTIVITIES

However, in the long run, readiness activities can be turned over to the small groups or to individuals as homework.

Upon completion of the readiness activities, copies of the model-eliciting activity can be distributed. Some teachers provide only one copy per group, and ask one student to read the task statement to the others. The teammates are encouraged to listen carefully and ask questions for clarification after the problem is read. Other teachers prefer to give each student a copy, so individuals can begin to read and formulate initial ideas independently. After about five minutes, they tell the groups that they can begin to work together. For the initial experience with model-eliciting activities, a teacher may want to preface the activity with some introductory remarks about how the task is unlike most mathematics problems they have done before. Further, initially the teacher may want to have a student read the task statement aloud to the class, and provide time for questions. Since the solutions to model-eliciting activities are explanations, assessments, or decisions to be made, the products that students are to produce are not like solutions to mathematics problems they have previously completed. Therefore, after reading the problem as a class, teachers can be encouraged to have the students clearly identify the client, the product, and the purpose for the product that they are to produce. This discussion can help students focus on the nature of the expected response

What Is the Teacher's Role During the Modeling Session?

Frustration is often a first reaction on the part of students. For many, their past mathematics classroom experiences have led them to believe that when given a problem, they are supposed to be able to immediately search for, identify and apply the correct procedure. Thus, when they are unable to identify a particular procedure right away, they feel the problem is unfair, or that the teacher has poorly prepared the students for the task. This initial frustration is difficult for many teachers to tolerate. Rather than giving students hints about how to approach the problem, high cognitive demand is retained by reassuring students that this problem is like a real world problem where they get to identify the mathematical ideas they find relevant, and where they have to figure out the mathematical questions they need to pose in order to solve the problem. Teachers can suggest to the students that they can ask their peers for ideas, and that brainstorming about ways of thinking might be helpful.

Doerr (personal conversation) has indicated that one of the best things a teacher can do during model-eliciting activities is to listen. It gives teachers something to do and provides them with an opportunity to learn a great deal about their students' thinking. Sometimes teachers develop their own observation forms, taking notes by listening and watching during a modeling session, and later sharing observations with other teachers who also implemented the same problem. Giving a focus, or purpose, for listening and observing is an effective way to help teachers acquire a new role while simultaneously reflecting on what their students are learning. Teachers can be asked to identify different interpretations students used during the solution process, or to identify the different representations students used and for what purpose. By understanding their students' thinking, teachers are in a better position to choose subsequent activities, ranging from skill-based lessons to additional modeling activities.

Students often ask teachers for help, especially during the first few model-eliciting activities. It is difficult for teachers to refuse to respond, yet the goal is to leave the problem solving to the students. Further, many students and teachers find it difficult to tolerate the inefficient approaches and wrong directions that typically surface early in the modeling episodes. A teacher may find that in the beginning, it is better to stay physically away from the groups, because students try to draw the teacher into telling them how to do the problem. Interventions that do not remove the cognitive demand from the task are possible. For example, teachers may reassure students that it is normal to begin with primitive ideas, and that more stable and effective ideas will evolve naturally as they continue to work on the activity together. If students ask for clarification about what they are to do, the teacher can ask the students to return to the task statement, and to (again) clearly identify the client, the product they are to produce and the purpose for the product. As a last resort, the teacher may suggest that each group member observe a different student group for about five minutes, to get different ideas about ways to grapple with the task. When the group members reconvene, they will bring different perspectives to the activity at hand, which may help "jump start" their own modeling process.

Preparation Prior to the Modeling Session

To prepare for the classroom implementation of model-eliciting activities, teachers need to have opportunities to solve the problems themselves, also working in small groups of peers. Completing the student-level activities provides teachers with opportunities to experience how initial interpretations of a problem are likely to be primitive and unstable, and how perspective taking facilitates the multiple cycle of expressing, testing and revising individual and group interpretations of the given problem, of the approach to use, and of the final product. Given this experience, a teacher may become more comfortable with the initial discomfort of students, and find the words to reassure the students that their initially primitive ideas will develop into more stable models, and that this cyclic process is natural and to be expected. On the other hand, having solved the problem, some teachers have an urge to tell students how to go about finding a solution! This illuminates the importance of having the teacher groups share their different responses with the whole group, just as they will with students. By developing an appreciation for how different products (based on different models) can satisfy the client's needs, teachers may become committed to allowing students to develop their own mathematical models for solving the problems.

A teacher also needs to select the task to use with the students. There are two important considerations. First, the teacher needs to select tasks that challenge all students in a group, making the task is truly a problem-solving experience for all involved. On the other hand, the problem also has to be within reach of the group as a whole. This usually is not a problem when using model-eliciting activities that have been field-tested with similar-age groups. The field testing is important because most teachers tend to underestimate what their students can do, and thus eliminate some model-eliciting activities that "look too hard." The second aspect of preparation is that the teacher needs to recognize that for many of the students, the initial small group modeling experience is breaking an implicit contract that students have formed with their teachers. Students think that every problem should be scaffolded in a way that tells them exactly what to do (Shannon & Zawojewski, 1995). Model-eliciting activities are often the first time that students are expected to figure out the intermediate mathematics questions to pose in the process of forming an explanation, prediction, justification, or assessment. Students think that if they do not immediately know what to do that either the teacher posed a poor problem or that the teacher did not teach them the "right stuff" in previous lessons. Teachers sometimes feel the same way, too! Over time and multiple experiences, we have found that both the students and teachers come to terms with new expectations and greater satisfaction with the products produced by small groups of students.

REFLECTIONS

A teacher entering the world of model-eliciting activities with small groups needs to view the process as a model-eliciting task in and of itself. The first implementation is likely to be unstable, with both students and teachers wondering whether it is going to work. But, as groups of students learn that together they are quite powerful, that they need to listen to each other, and that they can indeed solve these complex problems, they develop a clearer, more stable idea of how modeling activities can be accomplished. Similarly, teachers initially struggle with their students' feelings of frustration. They struggle to find ways to keep the cognitive demand of the modeling activity high, and to allow the students to do the problem solving embedded in the task. As teachers learn to listen, and turn the mathematical authority over to the students, they begin to notice how the small groups provide the mathematical power needed to accomplish the tasks. They begin to see how perspective taking is an important part of the problem solving process. Over time, the teachers begin to see the ways in which students internalize the external small group problem modeling processes. Increasingly, the teachers' model of students' modeling becomes more stable, and routines for implementing these types of problems become established.

Sustained experience with model-eliciting activities becomes increasingly satisfying to both the teacher and the students. Further, the processes students learn contribute to their skills for the workplace of 21st century. As we look to the future, the social dimensions of problem solving reaches well beyond face-to-face collaboration—to problems and challenges that are emerging in the fast-pace changes in technology. Members of problem-solving teams in business, industry, and academics today require quick adaptation to new methods of interacting and working together. Collaborating partners are often scattered across the country or throughout the world, requiring the use of communication tools that were unavailable until very recently. Working with colleagues across the country via Internet video conferencing, for example, requires new protocols for interaction. Even using the computer as a tool for carrying out procedures requires problem solvers to learn computer-based dialects in order to tell the computer what data to use and what procedures to conduct on the data. These dialects change with each new version of a software tool, and the tools needed to solve a problem often change as the team moves from problem to problem. The pervasive role of various types of social interactions in the 21st century work place makes the case for incorporating small group problem solving compelling for school mathematics.

Chapter 20

Local Conceptual Development of Proof Schemes in a Cooperative Learning Setting

Guershon Harel
University of California, San Diego

Richard Lesh
Purdue University

Three interrelated areas comprise the theoretical perspective of this chapter: *abstraction* (Piaget, Inhelder, & Szeminska, 1960), *proof scheme* (Harel, 2001; Harel & Sowder, 1998;), and *local conceptual development* (Lesh & Harel, in press; Lesh & Kaput, 1988). This theoretical perspective has guided our investigation into three main questions:

1. What proof schemes does a small community of learners (three middle school classmates) share?
2. What cognitive disequilibria occur in this small community of learners who are working collaboratively to solve open-ended problems, and how do such disequilibria bring about reconceptualization of existing proof schemes?
3. What are the characteristics of local conceptual developments in which transitions and reconceptualizations of proof schemes occur during relatively short problem-solving episodes?

This chapter provides insights—rather than complete answers—to these questions. We are particularly interested in local conceptual development in the context of students' construction, retrieval, and application of proof schemes.

The chapter is organized in four sections. The first section focuses on a brief description of Harel and Sowder's (1998) conceptual framework of proof schemes and on Lesh and Kaput's (1988) notion of local conceptual development, as well as on Piaget's notion of abstraction. The Piagetian notion of abstraction is, in our view, fundamental to any conceptual development in mathematics, local or long term. Although Piaget's theoretical perspectives developed exclusively with individual learners to trace their long-term intellectual growth, we find the theory equally suitable—in fact indispensable—for analyzing local conceptual developments occurring when a small community of learners working collaboratively. Local conceptual development—theorized first in Lesh and Kaput (1988) and elaborated in Lesh & Harel (in press)—is a new concept that has arisen in mathematics education research, as mathematics educators extend basic Piagetian concepts beyond generally development in naturally occurring environments to the development of powerful constructs in artificially rich mathematical learning environments. Thus, to the previous list of questions, we add:

4. What parallels exits between local conceptual development and long-term conceptual development?
5. How can knowledge of local conceptual development inform us about important roles for teachers in cooperative learning settings?
6. In what ways might insights about local conceptual developments provide us with an understanding of the principles that govern long-term conceptual developments?

In the second section, the theoretical perspective resulting from these primary ideas is used to analyze a sequence of interactions among three eight graders working collaboratively to generate a useful response to a model-eliciting activity called the Smart Shadows Problem (see Appendix A). It is important to remark that we do not argue that the proof schemes employed by these students were constructed in a vacuum during the single session—namely, that what the students developed was entirely absent at the start of the session. On the other hand, anyone reading what the students accomplish in this session is likely to recognize that they developed something that was (for them) new. It is this something, together with the path of its development, that are the focus of this chapter. We conclude with a summary section depicting the local conceptual developments that emerged from the session, and describing possible implications for research and instruction.

THEORETICAL PERSPECTIVE

Lesh and Kaput's Notion of Local Conceptual Development

Models are constructed to describe situations mathematically. In some cases, the construction of such models demand sophisticated representational media, such as systems of differential equations and sophisticated software program. In other cases, the models involve merely ordinary spoken language, drawings, metaphors, and so forth. Lesh and Doerr (1998) described a generic four-step modeling cycle that seems to underlie solutions to a wide range of problem-solving situations:

1. description that establishes a mapping to the model world from the real (or imagined) world,

20. DEVELOPMENT OF PROOF SCHEMES

2. manipulation of the model in order to generate predictions or actions related to the original problem solving situation,
3. translation carrying relevant results back into the real (or imagined) world, and
4. verification concerning the usefulness of actions and predictions.

The notion of a modeling cycle has been investigated in the context of students' solutions to more realistic problems than those given in most traditional curriculum materials. For example, Behr, Lesh, and Post's investigations into rational number concepts and proportional reasoning and Lesh and his colleague's (Lesh & Akerstrom, 1982; Lesh, Landau, & Hamilton, 1983) research on the use of mathematics in everyday situations illustrate a variety of situations in which problem solving is characterized by progressions through the preceding kinds of cycles. However, after observing hundreds of solutions to such everyday problem-solving situations, one of the main conclusions that has been formed is that multiple modeling cycles tend to be required, not just a single cycle. Furthermore, in cases where the relevant construct focuses on a single easy-to-identify foundation-level ideas, and when the development of this idea has been investigated by developmental psychologists or mathematics educators, the modeling cycles that problem solvers go through often are remarkably similar to the stages identified in developmental research. For instance, proportional reasoning is an example of such a foundation-level idea—it is perhaps the single most important construct in middle-school mathematics curricula—and it also is a construct that has been investigated extensively by both developmental psychologists (e.g., Piaget & Inhelder, 1958) and mathematics educators (e.g., Lesh, Post, & Behr, 1989). Consequently, it is striking when, during a single 90-minute problem-solving episode, students go through a series of modeling cycles in which the ways of thinking are remarkably similar to the stages of development identified by Piaget and others. One example of such a transcript is described in the first chapter in this book. Others are given in Lesh & Harel (in press); and, another example is given in Appendix B of this chapter.

Harel and Sowder's Conceptual Framework of Proof Schemes

The phrase "proof scheme" refers to what convinces a student about the validity of a claim, and to what the student offers to convince others. Harel and Sowder (1998) provide a psychological framework, informed by historical, philosophical, and cultural analyses, for examining students' conceptions of mathematical proof. Additional data analyses and historical and epistemological considerations have led to a refinement of this framework (Harel, in press). According to this theoretical framework, proving is defined as the process employed by a person to remove or create doubts about the truth of an observation[1] (Harel & Sowder, 1998). A distinction is made between two processes of proving: *ascertaining* and *persuading*: "Ascertaining is a process an individual employs to remove her or his own doubts about the truth of an observation. Persuading is a process an individual employs to remove others' doubts about the truth of an observation." (p. 241). Thus, "a person's proof scheme consists of what constitutes ascertaining and persuading for that person" (p. 244).[2]

The framework consists of three main classes of proof schemes (Fig. 20.1), the *external conviction* proof schemes class, the *empirical* proof scheme class, and the *deductive* proof schemes class. In what follows, only proof schemes that are relevant to this paper are described:

Proving under the external conviction proof schemes class depends on either (a) an authority such as a teacher, a book, or a classmate; (b) on strictly the appearance of the argument; or (c) on symbol manipulations, with the symbols or the manipulations having no potential quantitative referents in the eyes of the student. These behaviors correspond to the *authoritarian proof scheme*, the *ritual proof scheme*, and the *non-quantitative symbolic proof scheme*.

Proving under the empirical proof schemes class depends either on evidence from examples of direct measurements of quantities, substitutions of specific numbers in algebraic expressions, and so forth, or perceptions. Accordingly, these behaviors correspond to two proof schemes: the *inductive proof scheme*, and the *perceptual proof scheme*.

All the proof schemes in the deductive proof schemes class share three characteristics: the *generality* characteristic, the *operational thought* characteristic, and the *logical inference* characteristic. The first characteristic means that the subject understands that the goal is to produce a convincing argument which applies to all (not just most) of the cases. The second characteristic means that the subject forms goals and subgoals and attempts to anticipate their outcomes. The third characteristic means that the subject understands that evidence in mathematics ultimately must be based on logical inference rules. However, it should be noted that, although evidence in mathematics rests on logical inferences rules, it is almost never the case that one develops proof by reasoning with these rules alone. Inductive and abductive reasoning (Pierce, 1931-1958) are almost always integral parts of the search for and construction of evidence.

Of particular importance in this chapter is the *transformational proof scheme*. The term *transformation* entails changes in images, where an image is what Thompson (1994), based on Piaget (1967), characterizes as a cognitive entity that "support[s] thought experiments and support[s] reasoning by way of quantitative relationships" (p. 230). Transformations are goal oriented. They may be carried out for the purpose of leaving certain relationships unchanged;

---

[1] Terms such as *observation* and *conjecture* are used in Harel and Sowder (1998) with specific meanings.

[2] Proof schemes can include conceptual systems for interpreting situations, transforming descriptions to generate predictions, and so on.

but, when a change occurs, the learner needs to anticipate it and apply appropriate operations that would compensate for the change.

For the session of small group cooperative learning that we describe later in this chapter, the path of local conceptual development that emerged includes clear marks of transformational proof schemes emerging from perceptual proof scheme activities.

The three classes of proof schemes that we emphasize might be classified around Brousseau's (1997) distinction between didactical obstacles and epistemological obstacles. The former refers to the impact of narrow instruction, whereas the latter, inevitable in nature, refers to the impact of the meaning of the concept. Harel (in press) suggests that the formation of the external conviction proof schemes is due to narrow instruction more than to natural conceptual development; whereas, the formation of the deductive proof schemes is due to natural conceptual development more than to narrow instruction. The formation of the empirical proof schemes is the result of both impacts. That is, although these schemes are inevitable, current instruction has failed to help students recognize the role of empirical reasoning in proof production, understanding, and appreciation.

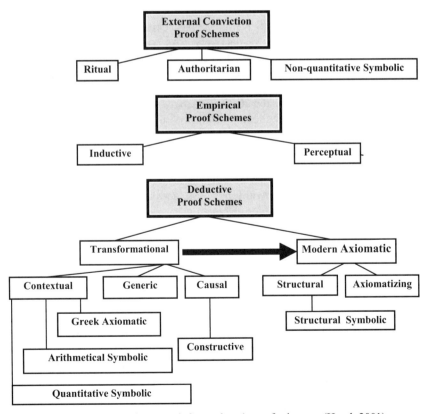

FIG. 20.1. A framework for students' proof schemes. (Harel, 2001).

ABSTRACTION

The proof scheme framework we have just presented are used to illustrate the fundamentals of Piaget's concept of abstraction. However, to demonstrate the uniform and broad applicability of this concept, we accompany our discussion with examples of elementary, as well as advanced, learning processes. Other related examples are given in Lesh and Carmona chapter 4 of this book on "Piagetian Conceptual Systems & Models for Mathematizing Everyday Experiences."

Piaget distinguished between two main categories of abstraction: empirical and reflective. To describe Piaget's notion of empirical abstraction, Steffe, Cobb, and von Glasersfeld (1988) defined two cognitive structures: *template* and *concept;* whereas a template serves to recognize things as repetitions of past experience, a concept serves to re-present (i.e., play back) prior experience:

Empirical abstraction concerns a perceptual (sensory) experience and results in a template that serves to recognize further experiences as similar or equivalent to the past one; eventually the abstracted template turns into concept and can be re-presented as an internalized item without the presence of the sensory material in actual perception. (p. 333)

As templates, the cognitive structures that underlie judgments about color, sound, and taste, enable one to recognize past experiences of blue, sound of thunderstorm, and sweet taste, respectively. As concepts, on the other hand, they enable one to think of past experiences of blue, sound of thunderstorm, and sweet taste in absence of their occurrence by re-presenting (i.e., recreating) these experiences.

In mathematics learning, the distinction between template and concept is significant. For example, a student, presented with the equation $y = 2x - 3$, may recognize the equation as an instantiation of a function—that is, $2x - 3$ is the output for an input x—but may not be able to think of such a function in the absence of one. For this student, function is a template rather than a concept. With "function" as a concept, on the other hand, the student is able to describe the input-output process in general terms and, more important, to independently create examples and non-examples of function.

The main distinction between empirical abstraction and reflective abstraction is that whereas the former extracts properties by perceptual interaction with objects (e.g., seeing, touching, lifting, and the coordination of those), the latter extracts properties from actions or operations on objects. This distinction entails deep consequences: A property that has been abstracted from actions or operations

> must in addition be transferred somewhere, that is, on a different plane of action or operation. In the case of abstraction proper [i.e., empirical abstraction] this question does not appear since we are dealing with a property of an object, which is assimilated by the subject. In the case of reflective abstraction, however, when the subject extracts a property or a form from actions or operations on a plane P1, he must then

20. DEVELOPMENT OF PROOF SCHEMES

transfer it to higher plane P2, and this is thus a reflection in quasi-physical sense (as in the reflection of a light beam). But for this form or property to be assimilated on this new plane P2, it must be reconstructed on this new plane and therefore subjected to a new thought process that will this time mean "reflection" in a cognitive sense . . . . reflective abstraction is the general constructive process of mathematics: it has served, for example, to evolve algebra out of arithmetic, as a set of operations on operations. Cantor constructed transfinite arithmetic in the same manner: he put into one-to-one correspondence the sequence 1, 2, 3, 4, . . ., with the sequence 2, 4, 6, 8 . . . This generates a new number (№) which expresses the "power (a number) of the denumbrable," but it is an element of neither sequence. (Piaget, 1983, p. 125).

Piaget further distinguishes among three categories of reflective abstraction: *pseudo empirical abstraction, reflecting abstraction, and reflected abstraction* (Vuyk, 1981),[3] which Steffe et al. (1988) formulated as follows:

> Pseudo-empirical abstraction concerns a motor (kinesthetic or attentional) action, something the subject does, and results in a pattern that serves to recognize further kinesthetic or attentional actions as similar or equivalent to the past one. (p. 333)

By *pattern* is meant "a collection of elements that can be characterized (and recognized) by the spatial and/or temporal relations that connect them" (p. 340).

> Reflecting abstraction derives from pseudo empirical abstraction and produces an interiorized action or operation that can now be carried out "in thought," i.e., with re-presented sensory material when that material is not perceptually available.

> Reflected abstraction derives a rule from several pseudo-empirical or reflecting abstractions and produces an "operation with operations" that no longer requires any specific perceptual or re-presented sensory material to be carried out in thought.

---

[3] It should be noted that the definitions used for these types of abstraction in the literature is inconsistent and in some cases even conflicting. As was pointed out by Vuyk (1988), Piaget himself was unclear on these terms. For example, she indicated that "reflected abstraction and reflexive [reflecting] abstraction are sometimes used as synonyms and sometimes distinguished by Piaget" (p. 121). Her interpretation of these terms conflicts with that by Steffe, Cobb, and von Glasersfeld (1988). Although Vuyk views reflected abstraction as inferior to reflecting abstraction, Steffe at al. see them in opposite order of sophistication. Dubinsky and Lewin (1986), on the other hand, do not distinguish between reflecting abstraction and reflected abstraction, and talk only of reflective abstraction.

To illustrate relationships among the three kinds of reflective abstractions in the process of knowledge construction, consider the following scenario. As was mentioned earlier, empirical abstraction leads to the extraction of color as a common feature of objects. Once color has been abstracted as a template, it becomes a raw material for further actions—using it as a criterion to classify objects, for instance. A child who empirically constructs a template of color may use it to classify a collection of blue objects and white objects. He or she might then place the two collections in alignment (kinesthetically or attentionally) and observe that the collection of the white objects is in one-to-one correspondence with the collection of the blue objects—an action resulting in the construction of a pattern. This process is a pseudo-empirical abstraction because the child has extracted a property—one-to-one correspondence—from her or his action on the objects, rather than from the objects themselves.

Reflecting abstraction is signified by the production of an interiorized action or operation that can be carried out in thought, with re-presented sensory material when that material is not perceptually available. Thus, continuing with the aforementioned scenario, the child, after constructing the one-to-one correspondence between the two collections of objects, may realize that a certain spatial configuration of the objects would not change the one-to-one correspondence state. At this stage, he or she is able to conceive different hypothetical situations in which the collections of objects are in different spatial configurations and yet are in a one-to-one correspondence state. This process of reflecting abstraction is a significant step toward conservation of number and other quantities. More detailed examples are given in Lesh & Carmona, chap.4 of this book, Piagetian Conceptual Systems and Models for Mathematizing Everyday Experiences.

Reflected abstraction is reflection on reflection. It signifies the ability to produce operations with no specific perceptual or re-presented sensory material. At this stage, the child understands that once two collections are in one-to-one correspondence, they continue to be in this state under any special configuration. This general conclusion is derived from logical necessity rather than from observations of imagined one-to-one spatial configurations. With this ability, the child is likely to be a quantity-conserver. He or she is able to realize that collections comprised of objects of any kind, any size, and in any spatial configuration that are in a state of one-to-one correspondence are of the same numerical quantity.

Collaborative Problem Solving Episodes

The three students (Al, Bev, Candy) whose solution to the *Smart Shadows Problem* is given in detail in Appendix C were inner-city African-American students who were in a remedial math course for eighth graders. The session occurred in the late fall after the students had gained experience working on four other model-eliciting activities (Lesh & Harel, in press). In particular, these students were experienced at working in groups, and at working independently for a full class period. They also were experienced at having their work

20. DEVELOPMENT OF PROOF SCHEMES

videotaped. Furthermore, because their school used block scheduling of classes, the class periods were twice as long as in many schools. On problem-solving days the typical classroom routine was for students to work on a problem during one full class period; then, during the following class period, they would make presentations about their work.

The *Smart Shadows Problem* is described in detail in Appendix C.

Just as in the case of most other model-eliciting activities described in this book, a math rich newspaper article was discussed on the day before the problem solving session was intended to begin. In the case of the Smart Shadows Problem, the newspaper article described an upcoming school science fair that was going to focus on the theme: perception and illusion. One of the exhibits that the article described was about similarities between photographs, shadows, and slices of a cone or a pyramid.

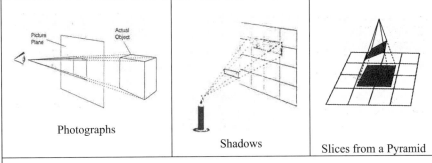

Photographs   Shadows   Slices from a Pyramid

For the Smart Shadows Problem itself, the client was a group of students who were planning to set up an exhibit in which a point source of light would be used, and non-square shapes (like the ones shown below) would be used to make square shadows. The problem statement asked the students to write a brief two-page letter to the client describing: which of the following figures can be used to make a square shadow, and exactly how should the figure be held, relative to the light and the wall, in order to make a square shadow.

Note: The light source was a small penlight flashlight with the reflector painted black so that it will come close to providing a "point source" of light.

FIG. 20.2. A general description of the Smart Shadows Problem.

The students were also given the six cardboard shapes in Fig. 20.3, which they scattered on the table.

FIG. 20. 3. The above six cardboard shapes were scattered on the table.

The transcript of the students' solution process can be partitioned into eight episodes that occurred within a single 90-minute class period. The transcript begins shortly after the students read the assignment.

Episode 1: Perceptual Proof Scheme.

1. Al: So ... Which of these make a square [pointing to the six shapes on the table]?
2. Bev: A square shadow. Ya!
3. Al: Where's the light. ...How d'you work this thing.
4. Bev: Gimme it.
5. Candy: Here, I'll do it. ...Look.

Candy picks up the rectangle without much thoughtfulness and makes a shadow on the wall. The rectangle is held parallel to the wall. So, it looks like a large version of the cardboard in Fig. 20.4.

FIG. 20.4. The rectangle is held parallel to the wall.

For several minutes the students were experimenting in making shadows with several of the shapes. One of the issues they discussed was how to make shadows without blocking the light with their hands. They solved the problem using the forked split in a broken pencil. This allows them to clip the pencil to a cardboard shape in a way that looks like a lollypop.

6. Al: Yeah, but it's not square. See.

After a long pause during which the students were looking over the other cardboard shapes, somewhat more thoughtfully, Candy picked up the diamond—apparently because its sides are all the same length, like the square, and then she said:

7. Candy: OK, let's try another one. ...Nope, it doesn't work.... (pause)... None of 'em will work.
8. Bev: Nope. None of 'em.

Despite Bev's declaration, the students continued—without exchange of words among them—their shadow manipulation activity by picking up different shapes and using them to make shadows.

The 1 through 8 response episode lasted for about 10 minutes. During this time the students were not working well as a team. Rather, they were working in parallel—as three individuals working in the same place. There was little collaboration and little communication was occurring to coordinate their efforts. At this point the students had little raw material to act upon; it merely consisted

20. DEVELOPMENT OF PROOF SCHEMES

of limited empirical observations, mainly perceptual and devoid of any transformational reasoning.

The three students, however, were experienced in collaborative work—an invaluable experience in processes of proof and refutation. Students have learned how to use pauses and intermissions to engage effectively in ascertainment and persuasion—two essential processes of proving. The silence signifies individual ascertainment pauses—a time interval during which the participants shift from group communication—used for collaborative exploration and chiefly for persuading each other—to individual reflection—used for introspection and ascertainment.

At this point in their interaction, the perceptual proof scheme seems to have dominated their ascertainment and persuasion ventures. They perceived shapes as a whole rather than analyzing them in terms of their properties. The only shapes that were being considered seriously were those that seemed to look something like squares (i.e., the rectangle and the diamond). The students seemed to identify the features of the target object—square, whose apparent features are equal sides and all of its angles are right—and accordingly chose quadrilaterals with those features (rectangles and diamonds) as the source object. The correspondence between the target object and the source object resulted, for these students, in a pattern—a product of pseudo-empirical abstraction, as we have explained earlier—and is alone governed students' judgment. Transformations and their possible consequences were not present at this point of the iteration. For example, the shapes were not tilted relative to the wall, and the light was held constantly perpendicular to the wall.

Episode 2: Early Emergence of Transformational Proof Scheme. Until this point in the session, the students have only tried to hold the shapes parallel to the wall. They have not tried tilting any of the shapes. Then Candy held the rectangle, and tilted it to make the long sides look shorter.

9. Candy: Look guys. Look when you move it. ...See.

This is a critical response that had changed the course of the subsequent events. It was the first time the students applied transformations and observed their consequence in the form of the changes in the shadows. It should be noted, however, that their observation at this point is perceptual rather than reflectively abstract; namely, it is not a thought action carried out in absence of respective perceptual cues.

Here, for the first time in the session, the students were explicitly tilting and turning the cardboard shapes, moving away from the fixed pattern they constructed previously, and carefully observing corresponding changes in the shadows they produced, whereby new patterns were constructed.

It is important to point to two critical characteristics that are absent from this episode—characteristics that reflect the dominance of the perceptual proof scheme and the absence of transformational proof schemes.

First, even though the students began to pay attention to ways the shadows were changing when the shapes were manipulated, they were not exploring what features were invariant under the transformations or how to compensate for the changes. Nor were the student thought-experimenting—applying actions or operations in thought to hypothesize conditions for the desired outcome (reflected abstraction).

Second, the transformations applied by the students were not goal oriented and were not invertible. Namely, the students merely varied the positions and observed the shadows that resulted. They attempt to find a transformation whose outcome is a desired shape.

Thus, students' activities during this episode are dominated by their empirical proof scheme—specifically by their perceptual proof scheme, and yet this signifies the beginning of a transition, for the students shifted from reasoning about properties of static objects to reasoning about properties derived from physically transformed objects—pseudo abstraction in Piaget's terms.

Episode 3: Limited Control of Transformations.

10. Al: [It's] still not square! ... Look, it's tilting in. It's skinnier up here.

The shadow created by Al looked something like the trapezoid shape shown here.

FIG. 20. 5. The shadow looks something like the trapezoid shape shown here.

11. Candy: Oh Geez!

Encouraged by outcome of Al's action, Candy tried to control the outcome of her actions by turning the rectangle around more, and she moved it in and out relative to the flashlight and the wall. But, because the shadow never looked square, she gave up—a fact Bev used as evidence for her previous claim (8):

12. Bev: I said. None of 'em will work. None of 'em are square.

At this point the verbal interaction among the students ceased—another individual ascertainment pause—and the students were just looking at the shapes, picking them up, and putting them down again without trying them. Sometimes, they were looking at the shapes through one eye, and noticing how their shapes seemed to change when they are tilted. Finally, Al got an insight from these explorations, and he tried to make a shadow with the diamond. He tilted the diamond so its shape changes as shown below.

FIG. 20. 6. Shapes seemed to change when they are tilted.

20. DEVELOPMENT OF PROOF SCHEMES

This point undoubtedly marks a significant change in students' conceptualization. First it was their first sign of success—a counterexample to their earlier conclusion of the impossibility of the task. Second, and more important, until now, their responses were based mostly on trial and error, uncontrolled transformations of positioning a shape and perceptually observing its shadow. Now the transformations began to be controlled. Al, for example, quit just looking at the shapes and he tried to make a desired shadow with the diamond. This response clearly was based on prior reflection and it was the beginning of a goal-oriented action.

This action is an indication of possible emergence of an operation, not operation per se. For the students seemed to be making the implicit assumption that a shape only makes a square shadow if it already looks something like a square, by having either all of its sides equal or all of its angles equal. In this way of understanding, the inputs for the transformation, not its possible output, were predetermined.

Without making their thinking explicit, the students were reasoning by analogy—implicitly making predictions from the geometry of vision to the geometry of shadows. But, the students did not follow up on this analogy; and, they were still thinking of shapes-as-a-whole. For example, just like the early stages of reasoning identified by Van Hiele (1986), Bev refused to recognize a diamond as being a square – simply because its sides are not vertical and horizontal. Even though Al quickly corrected her in this particular instance, similar tendencies persisted and were apparent for all three students.

Episode 4: Controlled Transformations. The new pattern noticed by Al—the correspondence between the diamond to the square—is of particular significance in another respect: It was the first occurrence of a counter example to the previous assertion expressed in Responses 7 and 8. Despite Bev's objection to the counter example, it destabilized the other members' confidence of their earlier conclusion about the impossibility of the task, and they turned to pursue a new direction. Thus, this cognitive conflict led the team to attempt controlling the transformations' outcomes:

13. Al: Look at this. ... I did it. Look.
14. Bev: That's not square. It's a diamond.
15. Al: Ya, it is. Look

But, Bev did not want to call Al's shadow a square apparently because its sides were not horizontal and vertical. To convince Bev, Al tilted the cardboard shape so the shadow had horizontal and vertical sides—an argument that convinced both.

16. Candy: Hey. Yeh! ... That it. But ...
17. Bev: But that's it! [that's all of the shapes that work]. The others aren't like that.
18. Candy: Like what.

The students seem to be making the implicit assumption that a shape only makes a square shadow if it already looks something like a square by having either all its sides equal or all its angles equal.

19. Bev: [It's] got square sides. Others don't.

When Bev said, "square sides," her gestures suggested that what she meant was that the sides must be all the same length—like they are for a square.

20. Al: I think this one'll work. --- [Long pause] ... Nope. I can't get it. ... This is close. But, I can't get it

Here, for the first time, the students clearly were investigating how continuous transformation in one figure lead to other kinds of continuous transformations in the shadow. In particular, this is the first instance that the students were paying attention to relationships between transformations. This action requires carrying out operation in thought, which is an indication of reflecting abstraction and early signs of complete transformational proof scheme.

Episode 5: A fall back.

21. Candy: Let's try some others.

Several minutes pass as all three students picked up shapes, turned them in the light, and watched how the shadows changed. Now that the students share raw material to act upon collectively, they were working together well. For example, one was holing the light, one moving the shape, and one checking the shadow—paying attention to what each other were doing and saying.

22. Bev: This one here [holding the shape in her hand] won't *ever* work. That's for sure.
23. Candy: What about this one [pointing to the shape]. It don't work either?
24. Bev: Let's see. . . . No way. ... Nothin's the same [none of the sides are the same length].

Several minutes pass. The students seemed to be getting frustrated. Bev was turning and tilting the irregular quadrilateral; but she seemed to be focusing on the fact that all four sides of the shape were different in length and so she fell back to her previous judgment about the impossibility of the task.

25. Al: Hey look. Here's a dog.

Al was using his fingers to make a shadow that looks like a dog opening and closing its mouth to notice changes in properties from one instance to another.

20. DEVELOPMENT OF PROOF SCHEMES

On the surface, this seems an off task behavior. But it is more likely that this type of interaction serves a bridging mental break toward the next round of attempts to advance their goal.

Episode 6: Coordination Within Transformation.

26. Bev: What if you put the light over here. ... Whoa! It goes off the wall.

Bev was noticing that part of the shadow was no longer visible on the wall where they have been projecting shadows. Until this point in the session, the students had not consciously manipulated the direction the light is shining. They had simply been shining the light perpendicular to the wall. Now, they were shining it at an angle. Thus, we see here how the students were taking into consideration the source and direction of the light. Rather than holding the light fixed and moving the shapes, they were now holding the shape fixed and moving the light source—and they were explicitly investigating how the shadow changes when the light source moves. So, they were thinking about the relationships between the light source, the cardboard shapes, and the shadows. In other words, the students were coordinating the transformation from within: the transformation action (manipulation of light beam), the transformation domain (the cardboard shape) and the transformation range (shadows).

27. Candy: I don't think we can do that. ...We gotta keep the shadows little. Those big ones get all fuzzy out there.

Here, the students moved the shape and light source so that part of the shadow went beyond the side of the wall, and above the ceiling. So, Candy responded that when the shadows get too big, the edges of the shadow get fuzzy, or go beyond the limits of the wall they're using as a projection screen

A long ascertainment pause had passed, during which the students were trying out, individually, different shapes and different placements of the light source. Then they began working as a team, with one moving the light, one turning and tilting the shape, and one looking carefully at the shadow.

28. Bev: Look, I can make this side big ... or little.

Bev had been focusing on the longest side of the trapezoid, and she noticed that she could make the shadow of this side as long (or as short) as she wanted by tilting it or by moving it closer to the light source. Note that here the students are focusing explicitly on transforming a single side of a shape. They are not simply transforming the shape-as-a-whole.

29. Candy: So, make this the same as this [pointing to the top and bottom sides of the trapezoid].

Again, the students were referring to the relationship between pairs of sides, rather than simply focusing on the size of individual sides, and the general shape of the whole shadow.

Episode 7: Coordination Between Transformations. In the previous two episodes the students made two important advancements: They were able to coordinate the three components of a transformation—the domain, the range, and the rule of the transformation—and were able to vary some parts of inputs while maintaining other invariant. This development is a strong indication that the students, at this point of the process, were in control of the impact of the perceptual proof scheme, for they were able to manipulate what seems perceptually fixed. These conceptualization enabled the students to move to the next critical stage: to explicitly consider relationships between transformations.

30. Al: Try this one again. ...{Long pause} ... We gotta make this the same as this.

Long pause. Several minutes pass. Al was holding up the rectangle; and the two sides he was referring to were adjacent sides.

31. Al: There I got it.

Al tilted the rectangle so that the adjacent sides of the shadow were the same length.

32. Bev: But it's all messed up here [pointing to one of the angles of the shadow]

Here Bev was referring to the fact that, by making the two sides the same length, the angles are no longer right angles. This was the first time in the session that the students have been explicit about paying attention to variations in angles, apart form the contribution that angles have to the general shapes of figures-as-a-whole.

33. Candy: Oh Geez! ... You guys going to the game tonight.

Clearly the students were getting frustrated. But, they were continuing to play with the shapes and shadows as they talked about other off task topics.
By continuing to experiment, Al made two of the adjacent sides of the rectangle almost the same size. Now, he was trying to keep these two sides the same size while also tilting the rectangle to make the size of the interior angle look more square (i.e., 90 degrees). He was not paying any attention to the other sides or angles.

34. Al: Yep.
35. Bev: We're getting' killed. ... I'm goin'. Joe [her boy friend] thinks he gotta play.

20. DEVELOPMENT OF PROOF SCHEMES

Several minutes passed. While Bev and Candy talked about topics that were mostly off task, Al continued to experiment with the rectangle.

36. Al: Look. I got it. Almost. ... It's almost square.

This was the first that the students had tried to get to the final result using a series of simpler transformations, each of which focused on a single attribute of the desired final result.

37. Candy: What about here? [pointing to the angle that's opposite to the one between the two adjacent sides]. It's still messed up.
38. Al: We can do it. We can do it. ... Just gotta do 'em all at once.
39. Bev: Here. Le'me see. ... How'd you do it?

Bev took the rectangle away from Al, and she started manipulating it the same way Al suggested.

40. Al: First, make these the same {referring to two adjacent sides). ...Like this. [He guides the other student's hands to tilt the shape so that the adjacent sides are the same length. ...Now, keep it like that, and move it so the angle is square.
41. Candy: I see. I see. ... Long pause ... That's cool. Lookit.

Candy took the rectangle away from Al and Bev and followed the instructions she heard Al give. She did so by first making the two adjacent sides the same length. Then, she kept these two sides the same while also tilting the shape to make the angle between them a right angle.

42. Al: Let's try this 'un again.

Al was holding the irregular quadrilateral that they had decided didn't work.

FIG. 20.7. Irregular quadrilateral.

And all three students were experimenting with all of the shapes again. After several minutes, Al said:

43. Al: It's the same. ... Just gotta make these [adjacent angles] the same. Then, tilt it to make the corner square. ... Then, just move it [left or fight] till this is OK too.

Al was trying to get one part of the shadow right and to keep it constant while focusing on other parts.

FIG. 20.8. Experimenting with a parallelogram.

44. Candy: Try this one. ... [experimenting with the parallelogram] ... It don't work [while holding up the parallelogram].
45. Al: Le'me have it. Here. Get these first [referring to two of the adjacent sides]. There. Now, keep 'em like that, and tilt it so the angle's square.

The students were now turning their procedure into something approaching a formula for success that works for most shapes.

46. Candy: What about these [pointing to the non-adjacent sides and the other angles].

This is the first time that the students have simultaneously paid any attention to the transformations of all four sides and all four angles.

Episode 8: Formulation. At this point in the session, the students were getting very careful about inspecting all of the properties of the shadow—not just one or two properties. Also, the students went beyond thinking that "If it's a square, then its properties A, B, and C" and they start thinking that "If it has properties A, B, and C, then it's a square. That is, a square is defined by its properties.

47. Al: Just slide it back and forth.

Here, Al was adding another rule at the end of the steps in his existing formula.

48. Bev: Hey, we did it.
49. Al: So, all of 'em work.

After a long pause, all of the students were experimenting with the various shapes.

50. Candy: This one don't work. {*She's holding up the concave quadrilateral.*}
51. Al: Why not. Le'me have it. ...{Long pause}... Nope, this don't work. It's always poked in.

By "poked in" Al meant that the concave vertex can never be made to look convex.

52. Candy: All of 'em but this one {pointing to the concave figure). We can do 'em all but this.
53. Bev: Let's write our letter.

20. DEVELOPMENT OF PROOF SCHEMES

After another a several-minute pause, Bev started writing a letter to "the client" for the smart shadows problems. She was writing down the steps Al had described. Occasionally, she asked Al and Candy what to write, while they were talking "off task"—but still playing with the shapes.

54. Candy: Is this really square? Or does it just look square?... This [top] part doesn't look right.

Candy had been playing with the trapezoid. She was pointing out that when the trapezoid is tilted to make a shadow that is close to a square, the top edge is slightly fuzzy compared to the bottom edge.

FIG. 20.9. Trapezoid tilted to make a shadow "close" to a square.

55. Al: That's just {because of} the light... Look we can make this side big or little. So, we made it just right. ... And we can make this corner big or little. ... So, all we gotta do is move it [back and forth, as he demonstrates] to make this like this.

Here we see the students recognizing the generality of their hypothesis in that they recognized the difference between looking square to actually being square.

The letter that is shown next is a reduced form of the one that Bev wrote with help from Candy and Al. While she was writing, the teacher stopped by the table and briefly helped with formatting and editing. But, this resulted in small changes in the style—but not the substance of the letter. The words were the students'. While Bev was busy writing, Candy and Al continued to talk (mainly off task) while also playing with the shapes, light, and shadows. In addition to using the cardboard shapes, they also used their hands to make more animal faces and other funny shadows. Then, Al tore the diamond in half to make two equilateral triangles; and, he asked Candy "How many different triangles do you think we can make?" After making shadows with the triangles, they soon concluded that:

1. We can make any triangle we want.
2. We can make the angles as big or as little as we want.
3. We can make the sides as big or as little as we want.

The next day after the students had finished writing their letters to their client (students in another school), they made brief 5-minute presentations to their whole class about how they had solved the problem.

When Bev, Cindy, and Al gave their presentation, Bev read the letter, and Candy and Al demonstrated with the light and cardboard shapes. When the teacher asked them to go beyond explaining how their method works to also show why it works, Al explained by tearing one of the quadrilaterals in half to make two triangles. Then, he showed how to make half of a square with this triangle. Finally, he drew a sketch, like the one in Fig. 20.10, showing; (a) how any of the convex quadrilaterals can be cut into two triangles, and (b) how any of these triangles could be used to make half of a square shadow. After all of the students in the class had made presentations about their work, Al, Bev, and Candy revised their letter to include the four steps shown below.

Whereas the students' first letter described a method that required much tinkering to make it work, their second letter constituted a concise and to them a general proof that a square shadow can be made using any convex quadrilateral. The procedure that is specified uses lines (and triangles) that were not given in the shapes that were given. Also, the students implicitly introduced movements in three dimensions. That is, they slid the shape in one direction; they turned it in a second direction; and finally, they turned it in a third direction.

Dear Students,
You can make square shadows with these shapes.

Shapes like this one don't work. ▶

Follow these three steps.
1. **Make one corner square.**
2. **Make the two sides the same.**
3. **Go on around and fix each of the sides and corners.**

It works. We showed it.
Bev, Candy, and Al

Take the shape and cut it into two triangles. You can do it two ways.

Use straws to make kite pieces.
Only the kite pieces and tilt them to make a square shadow. It is easy.

Put the paper back on the kite and the shadow will look square too.

FIG. 20.10. Solution to Smart Shadows problem.

CONCLUSIONS

In this chapter, our goal has been to apply three interrelated theoretical perspectives—the Piagetian notion of abstraction, Harel and Sowder's conceptual framework of proof scheme, Lesh and Kaput's outline of local conceptual development—to guide our investigation into the Questions 1 through 3 in the context of solutions to a model-eliciting activity that involved both deduction and justification.

A Conceptual Development Path: From Perceptual Reasoning to Transformational Reasoning

In the problem-solving session described in this chapter, the transformational proof scheme is in sharp contrast to the proof schemes in the external conviction class and the empirical proof scheme class. A case in point is the perceptual proof scheme (in the empirical proof schemes class), which was used heavily by the students: It consists of physical perceptions and a coordination of perceptions, but it ignores transformations on objects and it does not completely or accurately involve anticipating results of transformations. Such perceptions "constitute an imitation of actions that can be carried out in thought (e.g., rotations of objects)... [but they] cannot be adequately visualized all the way to [their] ultimate conclusion before [they have] actually been performed" (Piaget & Inhelder, 1967, p. 295). With a transformation proof scheme, on the other hand, one is capable of "pictorial anticipation of an action not yet performed, a reaching forward from what is presently perceived to what may be, but is not perceived" (Piaget & Inhelder, 1967, p. 294).

The session under investigation was divided into eight episodes that involved different ways of thinking and that capture the transition path from the perceptual proof scheme to the transformational proof scheme. The following are notable benchmarks along this path:

Benchmark #1: Perceptual Reasoning. At first, the perceptual proof scheme dominated the students' ascertainment and persuasion activities. The students did not attend to the properties of the shapes, but shapes as a whole. In particular, they did not attend to transformations and their possible consequences.

Benchmark #2: Quasi-Transformational Reasoning. The students began to apply transformations and observe their consequence in the form of the changes in the shadows. However, though their observations were perceptually, rather than reflectively, based in that they were not the result of actions that were carried out thought in absence of respective perceptual cues. Nor were their observations the result of operations—actions that are goal oriented and invertible. The students' activities during this episode were dominated by perceptual proof scheme, yet they reveal the beginning of a transition, because the students begin to move from reasoning about properties of static objects to reasoning about properties derived from physically transformed objects.

Benchmark #3: Controlled Transformations. From this point on, we began to see a significant change in the students' conceptualization. The change seems to be the result of a counterexample to their earlier conclusion of the impossibility of the task. They moved from trial and error attempts—uncontrolled transformations of positioning a shape and perceptually observing its shadow—to controlled transformations, or goal oriented actions, by trying to make desired shadows with the diamond.

Benchmark #4: Transformation From Within. This reconceptualization led the students to focus on coordinating the transformation from within: the transformation action (manipulation of light beam), the transformation domain (the cardboard shape), and the transformation range (shadows). They were able to coordinate the three components of a transformation, and, more important, they were able to vary some parts of inputs while maintaining other invariant—a clear indication of a conceptual control.

Benchmark #5: Transformation From Between. Here, the students moved to explicitly considering relationships between transformations, not just among parts of the same transformation. This was expressed in their ability to simultaneously pay any attention to the transformations of all four sides and all four angles.

Benchmark #6: Hypothetical Reasoning—Reflected Abstraction. The culminating point in this local conceptual development session is students' ability to formulate a general theorem (in their view) with the understanding of the meaning of a for all statement. As we have alluded to in our previous analysis, an essential element of reflected abstraction is the ability to purge a concept from restrictions. This process is essential in forming, for example, a mental representation of the for all condition in mathematics theorems that conforms to the conventional mathematical meaning of this term. Without this meaning, proof in mathematics looses its essentiality. When a for all statement in geometry, for example, is proved in reference to a figure attached to the proof description, the student must understand that the proven property is not limited to the referenced figure but continues to hold for any other figure—imagined or not—that obeys the theorem's conditions. This ability is usually taken for granted, but as it has been shown this commonly assumed understanding is absent among many students (Harel & Martin, 1989). The formulation episode (Episode 8) clearly shows students' remarkable performance in that they fully understood this indispensable aspect of proving.

Implication for Instructional Development

A remarkable characteristic of the preceding problem solving session episode is that the students utilized the empirical proof scheme—perceptual in this case—in a way that's seldom been seen in our past studies of proof. Also, it is remarkable that the external conviction proof scheme was not present in any of the episodes. Again, this is significant because, in virtually all past studies of proof schemes, external conviction proof schemes dominated students' reasoning (see, e.g., Harel & Sowder, 1998). Apparently, this shift away from

20. DEVELOPMENT OF PROOF SCHEMES

external proof schemes and toward empirical and logical proof schemes resulted from the students' experiences with model-eliciting activities in which (a) the constructs, descriptions, and explanations that needed to be developed could be based on extensions of students' personal knowledge and experiences; (b) students clearly understood why, and for whom, the relevant constructs, justifications, and explanations were needed; (c) the constructs, descriptions, and explanations were expressed in externalized forms that could be tested and refined repeatedly; (d) students' were able to judge the relative strengths and weaknesses of alternative responses, without relying on opinions from other authorities; and (e) in particular, responses were not accepted as being "correct" (politically) based on the opinions of teachers or other authorities—even in cases where these responses might not be recognized as being sensible to give in everyday situations beyond school.

The preceding observation supports our earlier claims that the formation of the external conviction proof schemes is due to narrow instruction more than to natural conceptual development—and that the formation of the deductive proof scheme is based on natural conceptual development more than to narrow instruction.

Judgments about whether a given construct/description/justification is correct (from the point of view of some unspecified expert) are much different than judgments about whether the construct/description/justification is useful (in a situation where the criteria for success are clearly understood by students). Also, students' performance on proof-related tasks tends to be very different depending on whether: (a) the students use assumptions that they consider to be obvious in order to formulate conclusions about issues that they consider to be problematic; or (b) students use assumptions and procedures that they do not understand in attempts to prove things that they do not consider to be problematic. Yet, from the perspective of students, most of their school experiences with proof fall into the latter class.

When proof-related tasks are designed to satisfy principles of design for all model-eliciting activities (Lesh, Hole, Hoover, Kelly & Post, 2000), the transcripts that are given throughout this book suggest that extraordinary achievements can be expected from surprisingly young students—even in activities involving geometric proof, and even among underprivileged students who have been enrolled in remedial mathematics courses based on records of low performance in past situations that emphasized traditional textbooks, teaching, and standardized tests.

Implication for Theory Development

For model-eliciting activities, the sequences of modeling cycles that students go through during 90-minute problem solving sessions problems often appear to be compact versions of the developmental sequences that psychologists and educators have observed (over time periods of several years) concerning the natural evolution of relevant constructs and capabilities. For example, for the transcript that is given in this chapter, the sequences of modeling cycles are very similar to stages that have been described by Van Hiele (1986), Piaget (e.g., Piaget & Inhelder, 1958), and many mathematics educators who have investigated the development of spatial-geometric knowledge, or the development of abilities related to argumentation and justification. Furthermore:

- Mechanisms that developmental psychologists have shown to contribute to general conceptual development often can be used to help clarify the kinds of problem solving processes (or heuristics, or strategies) which should facilitate students' abilities to use (and even create) substantive mathematical ideas in everyday situations.
- Mechanisms that contribute to progress in local conceptual development sessions often can be used to help explain general conceptual development in areas such as proportional reasoning.

Proof can be considered to be the name for a whole category of capabilities; but, it also can be considered to be the name of a specific explanatory system that is applied to a specific problematic situation. In this chapter, we have focused on local conceptual developments that refer to the latter kinds of cases. Nonetheless, because appropriately designed model-eliciting activities typically require students to produce results that are sharable and reusable, it tends to be appropriate to use the term "proof schemes" to refer to the constructs and explanations that students develop.

Chapter 21

A Models and Modeling Perspective on Metacognitive Functioning in Everyday Situations Where Problem Solvers Develop Mathematical Constructs

Richard Lesh	Frank K. Lester, Jr.	Margret Hjalmarson
Purdue University	*Indiana University*	*Purdue University*

The models and modeling perspective described throughout this book has several distinctive characteristics applicable to metacognition and higher order thinking. First, whereas traditional perspectives generally treat metacognition and higher order thinking as if they consist mainly of content-independent processes, a models and modeling perspective assumes that the meanings of specific metacognitive and higher order abilities are closely associated with particular concepts and situations. That is, students' interpretations (or models) of learning and problem-solving situations are based on holistic conceptual systems including both metacognitive and cognitive components. In fact, it is assumed that these conceptual systems also include components such as those that involve beliefs, awarenesses, attitudes, identity, group functioning, and how students feel about a problem solving (Garofalo & Lester, 1985; Lester, Garofalo, & Kroll, 1989; Middleton, Lesh & Heger, (chap. 22, this volume). Consequently, metacognition is assumed to include the control, regulation, and monitoring of these and other components of students' models.

Because metacognitive and cognitive components are included in conceptual systems, a second distinctive characteristic of a models and modeling perspective is that both lower order and higher order abilities are assumed to function in parallel and interactively—not just unidirectionally and sequentially (one after the other; see Fig. 21.1). For example, whereas traditional conceptions of metacognition recognize that higher order thinking influences lower order understandings and abilities (perhaps by determining which is used when and how in a given situation), a models and modeling perspective also recognizes that lower order understandings and abilities often influence metacognition and higher order thinking (because, in any given situation, the amount of conceptual energy that needs to be dedicated to lower order thinking is one of the factors that influences the extent to which relevant metacognitive processes are engaged). Thus, a models and modeling perspective assumes that, as students increase their ability to use relevant lower order constructs or processes, related higher order and metacognitive processes are more likely to function effectively.

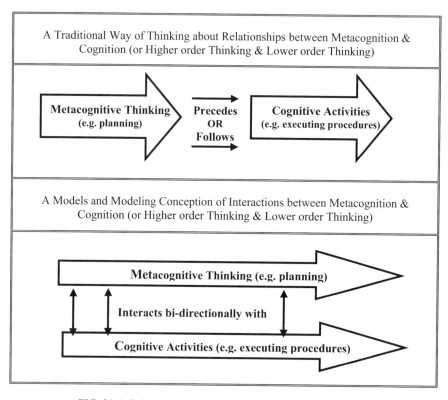

FIG. 21.1. Relationships between metacognition and cognition.

A third distinctive characteristic of a models and modeling perspective assumes that metacognitive and higher order abilities should be thought about developmentally. Moreover, the development of these abilities should be expected to occur along dimensions similar to those that apply to lower order understandings and abilities. In particular, development is expected to be similar to what Zawojewski and Lesh (chap. 18, this volume), describe for students' developing understandings of problem-solving strategies. For example, when we observe students' primitive metacognitive abilities, we expect to observe the gradual internalization of external processes—as well as the gradual abstraction of concrete abilities. Internalization may occur when experienced problem solvers begin to function as if they were, within themselves, a team of individuals who flexibly shift among a variety of alternative ways of thinking about a problem-solving situation—or among a variety of strategic roles that

21. METACOGNITIVE FUNCTIONING

range from planner to worker to recorder to monitor to assessor to communicator. This is one reason why, in our research on metacognitive functioning, the problem solvers that we investigate often are teams working collaboratively, as well as individuals-working-in-isolation. It also is why we have found it useful to view team learning and problem solving through the lens of theories about individual cognition (Kelly & Lesh, 2000) as well as viewing individual learning and problem solving through the lens of social cognition theories (Cobb, 1994, 1995).

A fourth distinctive characteristic of a models and modeling perspective emphasizes that the productivity of specific metacognitive functions often varies from one problem to another and from one stage of problem solving to another. This is in contrast to traditional perspectives that often advocate teaching metacognitive processes as if they were always equally productive across all problems and all stages of problem solving. Therefore, because solution processes in model-eliciting activities typically involve a series of modeling cycles and the productivity of metacognitive functions varies during different modeling cycles, some of the most important components of metacognitive understandings include decisions about when and why to employ specific metacognitive processes. For instance, during early modeling cycles, relatively unmonitored brainstorming may be productive in order to develop a variety of possible solution methods. Whereas, during later cycles, progress may depend on detailed monitoring and assessment.

In this chapter, beyond describing important aspects of our models and modeling perspective about metacognitive and higher order thinking, another goal is to describe several productive types of problem-solving experiences that encourage the development of metacognitive abilities and higher order thinking. To begin, we describe several commonly used definitions of metacognition, higher order thinking, and habits of mind and suggest how these definitions change under a models and modeling perspective.[1] Second, we discuss characteristics of metacognitive abilities that naturally occur during model-eliciting activities. Third, we examine several implications for classroom instruction by focusing on several different types of "reflection tools" or "debriefing tools" that are designed to help students reflect on (and improve) their metacognitive practices.

---

[1] Several researchers (e.g., Cuoco, Goldenberg, & Mark, 1996; Schoenfeld, 1992) have argued that a goal of mathematics instruction is to help students think like mathematicians do. Thus, a goal of instruction for them is to instill in students those "habits of mind" that come naturally to mathematicians. Because these habits of mind include metacognitive habits, we included habits of mind in our discussion.

COMMON DEFINITIONS OF METACOGNITION, HIGHER ORDER THINKING, AND HABITS OF MIND

Metacognition

An early and influential definition of metacognition was given by Flavell (1976): "Metacognition refers to one's knowledge concerning one's own cognitive processes and products or anything related to them." (p. 232). This definition has been used frequently by mathematics educators, but has invited criticism as being too vague and broad to be useful. Still, such criticisms have not prevented Flavell's definition from being applied in different venues (e.g., Brown, 1987; Campione, Brown, & Connell, 1989; Garofalo & Lester, 1985; Kilpatrick, 1985; Schoenfeld, 1985). For example, Schoenfeld (1985) obviously was concerned with what students know about their own cognition when he asked them questions such as: "What are you doing?, Why are you doing it?, and Is it helpful?" And Kuhn (1999) emphasized situations in which "Metacognitive knowing operates on one's base of declarative knowledge... What do I know, and how do I know it?" (p. 18).

For Flavell's definition to be useful, two basic questions arise: "What is meta? What is cognitive?" (Brown, 1987, p. 66,). For the purposes of this book, our answer to the second question is that the cognitive level includes: (a) students' models (constructs, conceptual systems); and (b) facts, skills, understandings, and processes that directly contribute to the development or use of these models. Consequently, our answer to the first question is that thinking becomes metacognitive when students shift from thinking with the preceding cognitive elements to also think about them—by managing them, by monitoring them, by modifying them, and so on.

What difference does it make to shift attention toward metacognition that focuses on models and modeling, rather than (for example) information processing procedures, at the cognitive level? According to information processing theories of cognition, mathematical knowledge and abilities are generally thought of as consisting mainly of condition-action rules. Similarly, a models and modeling perspective on cognition also recognizes that a great deal of mathematical knowledge can be thought of as condition-action rules. But, as Fig. 21.2 suggests, information-processing theories generally give the impression that the condition-action rules resemble finishing nails (where the conditions are rather small and insignificant, and the procedures are large and important). However, according to our models and modeling perspective, the condition-action rules resemble thumbtacks (where the conditions involve complex conceptual systems within which associated actions are comparatively small).

21. METACOGNITIVE FUNCTIONING 387

Information Processing Theories: A "finishing nail" metaphor

Models and Modeling Theories: A "thumbtack" metaphor

FIG. 21. 2. Metaphors to describe condition-action rules for information-processing theories versus theories about models & modeling.
Note: The "heads" of the nails and tacks represent the conditional components of the rules. The "necks" of the nails and tacks represent the procedural components of the rules.

According to an information-processing perspective, progress tends to be visualized as movement along a path from givens to goals. Information is presented and then processed along a path to arrive at a solution. For example, given a rowboat, students could be asked to row across the Wabash River from Point A to Point B so the students' problem is getting across the river. They need to use the oars properly and efficiently. In contrast, according to a models and modeling perspective, goals are visualized not so much as places where one arrives, or short answers that one gives, but as products that one designs and produces. Thus, even though the end in view (see English & Lesh, chap. 17, this volume) is clear for judging the quality of products that need to be produced, the exact nature of the products is not known. To continue the river-crossing metaphor, a models and modeling problem might ask students to design a method for getting across the Wabash River from Point A to Point B. In this case, the problem is to design something with very specific criteria for success and quality. The focus of the problem shifts from getting to Point B to how to get to Point B most effectively. In addition, the product the students come up with is useful not only for the problem at hand, but can be generalized and extended to other situations as well.

Suppose the primary function of problem solving is to link together procedures for processing information (that is assumed to be given in an immediately useful form at the beginning of the session) to make progress toward a goal whose nature is clearly understood. Suppose further that what is problematic is that one of the relevant procedural tools is temporarily lost. Under these conditions, metacognition is needed mainly to help problem solvers maximize positive characteristics within initially (adequate) ways of thinking. What this means is that the usefulness of current information needs to be maximized because only the precise amount of information needed to solve the problem is available and the students need only to find the right tool to solve their problem. From this perspective, students need to spend more time finding the right way than sorting out the wrong ways. But, if problem solving is thought of as being primarily about developing models (or conceptual systems) for making sense of givens, goals, and possible actions, and if the product that needs to be developed is a complex artifact whose nature is not precisely known, then metacognition is needed mainly to help problem solvers minimize negative characteristics of current (inadequate) ways of thinking and develop beyond them. For example, a negative characteristic of a group of problem solvers' current way of thinking about a problem might be that they have not considered all the relevant information and are focusing on only one piece of the data. This means their solution is probably insufficient unless they are given a reason to move beyond their current incomplete way of thinking and consider other pieces of information. In contrast, each individual in the group may have too many possible alternatives for thinking about the problem and need to choose between them. This is opposed to traditional problem-solving situations where the students' problem is to find a solution path rather than choose between paths.

One purpose of minimizing negative characteristics is that students go through a series of modeling cycles to produce products for the kind of model-eliciting activities that are emphasized throughout this book. Initial conceptions of the problem may not be used throughout the problem-solving process. Metacognition tends to focus on very different kinds of functions than when initial conceptions of givens and goals do not change significantly from the start to the end of problem-solving sessions. Namely, the purpose of metacognitive functions is to help students move from one way of thinking to an improved way of thinking or from cycle to cycle. This is different from traditional notions of problem solving where metacognition serves to help students stay on the right path or remove obstacles in the path.

Higher Order Thinking Skills

Traditionally, higher order thinking has been associated with "the capacity to engage in the processes of mathematical thinking, in essence doing what makers and users of mathematics do" (Stein, Grover, & Henningsen, 1996, p. 456). However, among mathematics educators who focus on K–12 schooling, it has been common to imagine that mathematicians spend most of their time doing tasks that are similar to solving puzzles, or answering textbook-style word problems. By contrast, a models and modeling perspective assumes that a more accurate portrayal of what mathematicians spend most of their time doing is constructing and investigating structurally interesting systems—or using them to make sense of real-life situations. Thus, according to a models and modeling perspective, mathematicians are seen as designers as much as problem solvers in the usual sense of the term; and, only a small portion of their higher order abilities are seen as being reducible to complex skills. Again, just as in other chapters throughout this book, mathematical thinking is considered to be about seeing at least as much as it is about doing and, powerful models for making sense of complex systems are not seen as reducible to skills or condition—action rules.

Because many of the same processes tend to be cited as examples of both metacognition and higher order thinking, the following questions arise: How is higher order thinking different than metacognitive functioning? What is the relationship between higher order versus first order thinking? Historically, one difference between metacognition and higher order thinking is that metacognitive functions generally refer to functions that act on cognitive functions, whereas higher order thinking has been more skill oriented—in the sense that (a) higher order thinking tends to be divided into a discrete list of skills, and (b) the main function of higher order thinking skills are to operate on lower order facts and skills. In fact, especially in cases where attention shifts from instruction to assessment, the term "higher order thinking skills" tends to be emphasized—presumably in an attempt to sound less vague (and more measurable) than simply referring to higher order thinking. Thus, the following sort of definition of higher order thinking skills often is invoked. "In simplest terms, higher order thinking measures include all intellectual tasks that call for more than information retrieval" (Baker, 1990, p. 7).

One shortcoming of attempts to reduce higher order thinking to lists of skills is that the "skills" that are described actually tend to be names for classes of skills or categories of skillfulness as discussed in Dark's chapter 16, in this volume. There is another shortcoming of this view. If higher order thinking is reduced to a list of discrete skills, the tacit assumption is that such abilities can be taught in the same way that lower order skills tend to be taught—that is, in a one-at-a-time, isolated, and decontextualized manner (separated from one another and from the lower order knowledge and processes they are intended to act). In addition, implicit in considering higher order thinking as a list of skills is the assumption that all skills are equally useful across all problems and across all stages of problem solving when, in fact, different skills may be counterproductive at different stages of problem solving and with different problems (as discussed previously).

A third shortcoming of reducing everything to skills is that the terms "higher order" and "lower order" invite the value judgment that one or the other is relatively unimportant. For example, lower order skills may be thought of as being less significant because they are viewed as trivial and less powerful. Or, lower order skills may be thought of as being more important because they are considered to be prerequisites to higher order thinking—as though complete mastery of lower order processes must be achieved before higher order processes can have any meaning or usefulness. According to a models and modeling perspective, neither of the preceding points of view is considered to represent more than a half-truth. That is, both views are expected to have serious negative consequences if they are allowed to influence instruction too strongly.

A models and modeling perspective assumes that higher order thinking is as necessary for mastery of lower order skills as the converse. But, this perspective goes farther than this by also assuming other kinds of relationships between higher order and lower order thinking. These relationships are summarized in Table 21.1. The categories listed in the left most column of Table 21.1 were identified by examining different categories of abilities that are commonly referred to as involving some type of higher order thinking in standards documents produced by the National Council of Teachers of Mathematics (NCTM, 1989, 2000) and other organizations. To sum up the message presented in Table 1, traditional perspectives about mathematics instruction and assessment focus on moving from givens to goals when the path is blocked, whereas models and modeling perspectives focus on the development of powerful conceptual models.

TABLE 21.1
Six Categories of Higher Order Objectives Commonly Emphasized in Standards for Instruction and Assessment

	TRADITIONAL PERSPECTIVES	MODELS & MODELING PERSPECTIVES
Mathematics as Problem Solving	Focus on *general (Polya-style) problem-solving strategies and heuristics:* Answers to "What can I do when I'm stuck?"	Students go beyond thinking WITH these strategies and heuristics to also think ABOUT them (e.g., under what conditions, and for what purposes, are they useful).
Mathematics as Reasoning	Focus on *meta-cognitive processes* (thinking about thinking) ... when students are planning, monitoring, and assessing solution steps or choosing among alternatives.	Students go beyond BLIND thinking, and also think ABOUT thinking so that the processes can be used more effectively (e.g., under what conditions, and for what purposes are they useful).
Mathematics as Communication	Focus on *mathematizing realistic problem-solving situations* using: spoken language, written notation, concrete objects, pictures, diagrams, or experience-based metaphors.	Students focus on representations that have the greatest power and usefulness, and they go beyond thinking WITH models and representation systems to also think ABOUT similarities and differences, and strengths and weaknesses, for a variety of purposes.

21. METACOGNITIVE FUNCTIONING   391

Mathematical Structure	Focus on *analyzing strengths and weaknesses of alternative conceptual systems* (i.e., mathematical models) which are used to interpret, describe, explain, or make sense of a given problem solving situation.	Students focus on constructs that have the greatest power and usefulness, and they go beyond thinking WITH these interpretation schemes to also think ABOUT them—their strengths and weaknesses of alternative ways of thinking and the underlying patterns and regularities that they emphasize.
Mathematical Connections	Focus on *mathematics as a complex system* of interrelated concepts and processes.	Students go beyond isolated PIECES of information to construct well-organized SYSTEMS of knowledge, and, they go beyond thinking WITHIN isolated topic areas to also integrate and differentiate ideas BETWEEN topic areas, subject matter areas, or domains of experience.
Mathematical Dispositions	Focus on *beliefs and attitudes* about: (i) the nature of mathematics, (ii) the ways mathematics is useful in real life situations, and (iii) the nature of their own mathematical abilities.	Students go BEYOND doing mathematics to also think ABOUT the nature of mathematics, ABOUT ways that mathematics is useful beyond school, ABOUT the nature of their own personal mathematical capabilities, and ABOUT the effects of their beliefs and attitudes on all the above.

According to Table 21.1, a models and modeling perspective does not distinguish between metacognition and higher order thinking. The important claim is that, regardless of whether we refer to metacognitive functions or to higher order thinking, the relevant understandings and abilities are parts of the models that students use to interpret learning and problem-solving situations. Thus, they interact with other components of these models, and, like other components of the models, they develop over time, in cycles, and along a variety of dimensions.

Habits of Mind

Like others, Cuoco, Goldenberg, and Mark (1996) stated: "We would like students to think about mathematics the way mathematicians do" (p. 377). They characterize mathematicians as people with good metacognitive habits; and, they define "habits of mind" to be "ways of thinking that one acquires so well, makes so natural, and incorporates so fully into one's repertoire, that they become mental habits—not only can one draw upon them easily, one is likely to do so" (p. 14). Thus, habits of mind are viewed as something students need to develop in order to think mathematically and to be successful problem solvers.

We agree, of course, that many desirable metacognitive behaviors seem to function habitually for problem solvers who are consistently successful. But, just as for many other half truths that occur when attempts are made to use simple metaphors to explain complex understandings and abilities, the habit-of-mind metaphor has many shortcomings when it is used to characterize most of the metacognitive functions that we find to be most useful in activities that focus on models and modeling. Many of these shortcomings occur because habits are generally thought of as behaviors that function without conscious reflection whenever a situation occurs that involves the conditions that stimulate them, whereas conscious control is an essential characteristic associated with most of the metacognitive behaviors that are of greatest interest to us. One reason this is true is that in modeling activities where multiple modeling cycles are needed to develop adequate ways of thinking, the productivity or nonproductivity of most metacognitive behaviors varies from one situation to another—within tasks, between tasks, and over time. For example, sometimes it is useful to give a great deal of attention to detailed planning and monitoring. But, at other times, brainstorming may be important, where criticism and details need to be temporarily avoided. (For example, at the beginning of a problem-solving session when students are generating ideas about how to solve the problem.) Similar statements apply to many other metacognitive functions.

Other shortcomings of the habit-of-mind metaphor occur because there are well-established ways that people make and break habits such as brushing their teeth, smoking, or overeating, and, in general, these methods are not the kind that we advocate for developing productive metacognitive abilities—often because the most important metacognitive abilities tend to involve a great deal more conscious decision making than the term "habit" implies.

For the purposes of this chapter, a final set of shortcomings that we want to emphasize has to do with the way mathematicians or scientists are characterized. Like diSessa, we are impressed with the fact that scientists do not all think alike, work alone, work from hypothesis to proof, or work linearly. Also like diSessa, when we try to develop productive learning activities for problem-solvers-in-training, we believe that designers may provide a more productive metaphor than "scientists." (diSessa, Hammer, Sherin, & Kolpakowski, 1991) One reason this is true is that designers generally judge the adequacy of their work based on specific and clear criteria that are close to the task. That is, the client is clear, the purposes for which the product is needed are clear, relevant assumptions are clear; and, the tools that are available are clear. But, for scientists, the criteria often are quite remote from the immediate task and they depend on the scientist's membership in a scientific community that has taken many years to develop assumptions, procedures, and goals that often are not obvious in specific problem-solving tasks. For example, in the formal study of

21. METACOGNITIVE FUNCTIONING   393

geometry, Euclid's postulates are generally considered reasonable starting points from which other facts must be deduced. Why not begin with facts that are obvious if paper folding is used—rather than a compass and straightedge—to construct geometric objects? Or, why not use primitive commands that are understood by certain computer programs? Answers to such questions seldom are obvious to problem-solvers-in-training. Yet, they directly impact many metacognitive functions such as those that emphasize planning, monitoring, and assessing.

A models and modeling perspective assumes that designers work with models that develop over time, in cycles, along multiple dimensions, with specific criteria for success, and clients and teams of collaborators whose identities, purposes, and perspectives are known. A modeling perspective also assumes that different higher order thinking processes (such as regulation, monitoring, and planning) are useful at different points in problem solving, and that it is important for the problem solver to understand when, where, and why a given process is useful. Even processes such as reflection, or characteristics such as intense engagement, are not useful all of the time! Sometimes it is useful to take a break or to temporarily step away from intense problem- solving activities. So, knowing when, where, why, and how long are important components of what it means to understand relevant metacognitive functions.

CHARACTERISTICS OF METACOGNITIVE ACTIVITIES THAT OCCUR NATURALLY DURING MODEL-ELICITING ACTIVITIES

This section refers to examples given in transcripts in the appendixes to this chapter. All transcripts report responses that average-ability middle school mathematics students generated for model-eliciting activities of the type described throughout this book The first transcript involves a problem called Reaction Time. It asks students to catch a dropped object, and to compare their own reaction times using the hand they use for writing, and their other hand. The second transcript involves a problem called Paper Airplanes. It asks students to develop a system for scoring a paper airplane contest including three different flight paths. In both transcripts, we identify instances where the students' comments or actions strongly suggest that they are engaging in some sort of metacognitive activity. Then, interpretations added to the transcript discuss the impact (or the lack of impact) of these apparent metacognitive activities. For example:

- A student may suggest a plan, or request the generation of a plan. But, the plan or suggestion may be ignored by others—as well as being ignored by the student in his or her follow-up activities.
- A student may notice errors, or may identify current activities that seem to be unproductive, nonsensical, or lacking in direction. But, these

observations may be ignored by others—as well as being ignored by the student in his or her follow-up activities.

Other issues that are highlighted in the transcripts focus on evidence regarding the nature of developmentally primitive versions of students' apparent metacognitive activities. For example:

- A student may monitor or assess others' behavior before being effective at monitoring or assessing his or her own behaviors.
- A student may use pictures or diagrams to communicate with others before being effective at using such representations to examine his or her own thinking.

Discussions about the preceding kinds of issues are organized into four sections corresponding to the distinctive characteristics of a models and modeling perspective that were described in the introduction of this chapter. That is: (a) metacognitive abilities are content- and context-dependent, (b) lower order and higher order abilities function in parallel and interactively, (c) metacognitive abilities should be expected to develop along dimensions similar to other cognitive understandings and abilities, and (d) the productivity of specific metacognitive functions varies from one situation to another—and (in model-eliciting activities) from one modeling cycle to another, or from one stage of problem solving to another.

Metacognitive Abilities Develop

For mathematical topic areas such as those dealing with early number concepts, rational numbers and proportional reasoning, or early algebraic or geometric thinking, mathematics educators are familiar with the process of investigating the nature of students' developing understandings and abilities. But, when attention shifts toward problem-solving strategies, or toward metacognitive abilities, developmental perspectives are rarely adopted. For example, few have investigated developmentally primitive versions of problem-solving strategies such as "look for a similar problem." Notable exceptions occur when Vygotskian researchers (e.g., Lambdin, 1993) have investigated external-to-internal or intrapersonal-to-interpersonal dimensions of cognitive development for metacognitive abilities such as those that involving monitoring or assessing one's own thinking. For example, the ability to monitor and assess the behaviors of others may contribute to the development of the ability to monitor and assess one's own behaviors. Or, diSessa (diSessa et al, 1991) has investigated the development of several important kinds of metarepresentational abilities such as those that are involved in developing and interpreting graphs. What about other kinds of metacognitive processes or strategies that contribute to expertise in learning or problem solving? Could interpersonal-to-intrapersonal (or concrete-to-abstract, or intuitive-to-analytic) dimensions of cognitive development also apply to strategies such as: (a) look for a similar problem, (b) clearly identify the

21. METACOGNITIVE FUNCTIONING

givens and goals, (c) work backwards, or (d) draw a diagram or picture? In this book, Zawojewski and Lesh (chap. 18, this volume) argue that the answer is yes. They argue that similar dimesions apply to such heuristics. For example, in many of the transcripts that are appended to chapters throughout this book, the following phenomena occur quite frequently.

- A developmentally primitive version of look for a similar problem appears to be look at the same problem from a different point of view (such as the point of view of another person in your group).
- A developmentally primitive version of draw a picture to examine your own thinking is draw a picture to explain the problem to a friend.

In general, productive problem solvers eventually appear to function as if they were, within themselves, several people sitting around a table working collaboratively. When they adopt one way of thinking, they do not lose sight of alternative ways of thinking. When they focus on one kind of information, they do not forget other possibilities. When they focus on details, they do not lose sight of the big picture (or vice versa). And, generally, when they are busy thinking in one way, they also are able to think about thinking. This is one reason why, in our research on model-eliciting activities, we often investigate problem solving in sessions in which the problem solver is a team of 3 to 4 students working in collaboration—rather than being a single student working in isolation. It also is why our research often compares teams with individuals in much the same way that other researchers have compared experts with novices, or gifted students with average ability students. That is, investigations focusing on one type of problem solver (e.g., a team) often provides useful ways to think about another type of problem solver (e.g., an individual).

Another point that is illustrated repeatedly in transcripts of model-eliciting activities is that, when solutions involve multiple modeling cycles, and when we focus on discussions and actions that take place during the early cycles, students' ways of thinking often sound like parallel monologues rather than collaborative discussions. That is, even though students may be sitting at the same table and talking to each other about the same problem, the students are not really working together, and they are not really communicating with one another. Instead, they are often working in parallel, without recognizing that they are thinking about the situation in completely different ways; and, they are often talking at one another rather than talking to or with one another. For example,

- Each student may have underlying assumptions or interpretations about the problem situation that they have not shared explicitly with their group members, but are affecting how they are working and thinking about the problem.
- One student may begin working individually on one aspect of the problem while the other 2 to 3 discuss a different aspect of the problem.

- One student may make a suggestion to help the solution process that is ignored by the other group members.

The preceding kinds of conceptually egocentric behaviors are especially likely to occur for relatively inexperienced problem solvers—or for experienced problem solvers who are in relatively unfamiliar situations. But, even in such circumstances, when students gradually progress through several modeling cycles in which their thinking becomes clearer and better organized, individuals not only refine and extend their own ways of thinking, but they also coordinate, differentiate, and integrate their thinking with the thinking of others.

Lower Order and Higher Order Abilities Function in Parallel and Interactively

In other chapters of this book (see chapter 22 by Middleton, Lesh, & Heger), we have described how, when students go through the process of developing increasingly useful conceptual systems for making sense of a model-eliciting activity, the modeling cycles that they go through often are strikingly similar to the cyclic processes that appear when students gradually coordinate systems of physical activities—in fields ranging from sports, to performing arts, to crafts involving pottery, carpentry, or cooking. That is, when physical abilities are poorly coordinated: (a) students typically find it difficult to notice more than a small amount of relevant information about the situations that they encounter, and (b) they often distort what they see to fit preconceptions that are not completely appropriate. (For example, in learning to ride a bicycle, a child typically notices very little about her or his environment, some obstacles seem huge that later become small, and some hazards go unnoticed that later are paid close attention.)

In general, as mental or physical activities become better organized, and as they begin to function more smoothly, students' interpretations of relevant experiences tend to become less barren and less distorted. Then, as a second-order result of noticing more and better information, increasingly higher levels of organization tend to be required. So, additional coordination cycles are needed.

Similar phenomena appear to occur for conceptual systems. For example, because better organized and smoother functioning conceptual systems require less attention, they appear to create conceptual space so that students are able to deal with more information and to engage in additional activities—these latter activities include metacognitive activities in which students simultaneously think and think about thinking.

To provide a simple metaphor for thinking about the preceding kinds of parallel processing, we have found it useful to refer to familiar experiences doing things like riding bicycles. For example, during the period of time when we were writing this chapter, one of the authors had the following experience. One morning, while he was riding his bicycle to work, he was continually (without ever stopping) whistling a new bluegrass tune that he and a group of friends had tried to play the preceding evening. Also, he needed to show up at work with a clear plan for the lecture that he was scheduled to give immediately after arriving at work. So, during this trip from home to work, his main attention was focused on planning the lecture. But, he also rode the bicycle through dangerous traffic without ever stopping and without getting killed, and he never quit whistling a tune that had been difficult to remember just one evening earlier. So, it is in the sense of successfully carrying out multiple complex tasks simultaneously and continuously that we speak of parallel processing.

A point that we want to emphasize is that the preceding kinds of parallel processing also appear to apply to activities in which students simultaneously think and think about thinking. Clearly, metacognitive and higher order thinking both influence and are influenced by lower order cognitive processes. That is, as lower order conceptual systems become better organized, and as they function more smoothly, more conceptual space appears to be available for metacognitive functioning. This is why, in the transcripts that accompany this chapter, and in most other transcripts appended to chapters throughout this book, we see many instances where metacognitive activities are only effective after students' current ways of thinking have achieved a critical level of organization and coordination. The appendices include situations similar to the following when metacognitive activity is abandoned in favor of lower-level thinking:

- One student carrying out complex computations, ignores or is not able to answer the question "Why are you doing that?" posed by another student.
- A group of students draws a picture that is useful in the moment, but later impedes understanding of the problem situation.
- A group makes a computational error, but ignores it because their result simplifies the problem.

Metacognitive Abilities Are Both Content- and Context-Dependent

For researchers who have investigated the thinking of people who use mathematics in everyday situations (Lave, 1988; Nunes, Schliemann, & Carraher, 1993; Saxe, 1988, 1991), or people who are heavy users of mathematical ways of thinking in their jobs (Hoyles, Noss, & Pozzi, 2001; Pozzi, Noss, & Hoyles, 1998), it is clear that these peoples' ideas and abilities tend to be organized around experiences at least as much as they are organized around abstractions (Greeno, 1998; Greeno et al., 1997; Schank, 1990). For example, a person may consider a collection of ideas to be related, not because they are logically connected, but because they have been used together in familiar problem-solving situations. Furthermore, this is especially true for people who are at early stages of development in the relevant fields.

Situated cognition, of the preceding type, tends to be emphasized naturally in model-eliciting activities. This is because models (unlike purely abstract conceptual systems) tend to be developed for specific purposes in specific contexts. Consequently, even after the relevant conceptual systems have been constructed, their meanings generally retain nuances that reflect the situations that provided the context for their development. Yet, the purposes that create the need for models generally include the need to be sharable (with others), reusable (in other situations), and modifiable (for a variety purposes). Therefore, the models that are developed tend to be explicitly designed to be useful in a variety of either structurally or mathematically similar situations or both. That is, they are designed both to generalize and to transfer.

The preceding point is significant because "situated learning" often is portrayed as being highly context specific—and as not transferring to other situations whose similarities are based on abstract generalizations. In fact, even for learning theories that do not emphasize the situated nature of learning, lack of transferability and generalizability are difficulties that many theories have not addressed well or completely. For example, such difficulties have been especially troublesome for learning theories that view mathematical knowledge as if it consisted of nothing more than lists of condition-action rules (or procedures)— especially if they assume that:

- Understanding is defined by doing the rules correctly, not on recognizing relevant conditions that determine when and where the rule should be used.
- The conditions are thought of as depending mainly on information that is objectively given, rather than information that involves relationships and patterns that are inferred—but that are not objectively given.

A models and modeling perspective avoids transferability and generalizability difficulties because of the following assumptions: models for making sense of complex systems are considered to be some of the most important components of mathematical knowledge and understanding, and the purpose of these models is precisely to provide meaningful ways for students to construct, describe, explain, manipulate, or predict patterns and regularities associated with complex systems. Consequently, in the transcripts that are appended to chapters throughout this book, we see that, when students work on model-eliciting activities during 1- to 2-hour problem-solving sessions, they typically spend only small portions of their time doing calculations or executing other kinds of data-processing procedures. They spend most of their time developing ways to think about the information, relationships, patterns, and regularities that are relevant. What they are transforming (and changing) is not data as much as it is their own ways of thinking about the data. This characteristic implies that students move beyond thinking with procedures to thinking metacognitively about procedures when carrying out modeling activities.

If performing procedures is emphasized, and if procedures are viewed as inert entities that lie dormant until something happens to wake them up, then it seems obvious to expect that procedures that are mastered in one situation are not likely to be transferred to other situations particularly from a situated cognition perspective. But, if attention shifts toward dynamic conceptual systems (i.e., models) that actively seek out situations where they can function,

21. METACOGNITIVE FUNCTIONING

then it seems equally obvious that, to use Abraham Maslow's observation, "when the only tool you use is a hammer, every problem begins to resemble a nail." So, transcripts throughout this book illustrate that over-use (over-generalization) often replaces under-use (lack of transfer) as the main learning difficulty that needs to be addressed. In model-eliciting activities, some of the most obvious impediments to progress often occur when: (a) problem solvers lock-in too quickly on "canned procedures" (and officially sanctioned ways of thinking) that need to be modified substantially in order to produce results that are most useful, and (b) problem solvers assume that, when errors occur, or when results are produced that do not make sense, procedural mistakes are the cause rather than flawed ways of thinking. In both cases, the problem solvers difficulty is not that they do not have a procedure, cannot decide which procedure to use, or have not employed a procedure, but that they may have over-generalized a procedure incorrectly to the present situation.

According to a models and modeling perspective, students' interpretations (or models) of learning and problem-solving situations are assumed to be based on holistic conceptual systems that include both metacognitive and cognitive components—as well as a variety of beyond-cognitive factors such as feelings, attitudes, values, and understandings related to personal identities or roles needed to promote productive group functioning. Thus, the transcripts that are appended to chapters throughout this book show that, during the solution of model-eliciting problems, problem solvers not only develop models for making sense of their experiences, they also (simultaneously and in a similar manner) develop metacognitive and beyond-cognitive understandings and abilities that are useful when these conceptual systems are employed.

Productivity of Metacognitive Functions Varies From One Situation to Another

Many researchers and teachers have observed that, to be successful, problem solvers generally find it useful to plan their work carefully, monitor and assess their progress continuously, and work diligently (without getting off task). Nonetheless, in the model-eliciting activities described in the appendixes to this chapter, just as in many other kinds of situations where people work on complex issues over significant periods of time, it is easy to find instances showing that:

- Sometimes it is important to "jump in and mess around" before attempting to formulate explicit plans. In fact, during early stages of grappling with a problem, when the problem solver is likely to have developed only primitive ways of thinking about the situation (including relationships, patterns, regularities), it may be counter-productive to get locked into a single premature plan.
- Sometimes brief periods of brainstorming are useful, where participants explicitly avoid monitoring and assessing ideas or procedures that are suggested in detail. Much like the story of the centipede that became paralyzed when it was asked to describe the order in which it moved its

legs, when people work on complex problems in fields like business and engineering, paralysis by analysis sometimes derails progress.
- Sometimes breaks are useful, to avoid getting too close to the situation by stepping back, and to get energized for critical periods where breakthrough efforts are needed. In fact, during investigations of students' behaviors during model-eliciting activities, exceptionally productive problem solvers (or teams) often are not those who always have their nose to the grindstone and never engage in off task behaviors. Instead, productive groups and individuals develop a sense of when it is time to take a break and when to become intensely engaged. Furthermore, if teachers introduce interventions that prohibit such breaks, productivity often appears to suffer.

One reason why the preceding facts emerge so clearly in many model-eliciting activities is that the conceptual systems that are relevant to the solutions of such problems nearly always are at intermediate stages of development. That is, they are not completely unknown; yet, they are not sufficiently developed to deal with the problem at hand. Consequently, appended transcripts show that solutions to such problems generally involve a sequence of several modeling cycles in which relevant conceptual systems are extended, modified, or refined, and they also show that different metacognitive functions tend to be productive as problem solvers move from primitive to better developed ways of thinking and from cycle to cycle.

Single-cycle problem-solving situations can be characterized as getting from givens to goals when the path is blocked or when a critical procedure is not immediately available. By contrast, for model-eliciting activities, the goal of most metacognitive functions is not to help students find a lost tool, and it is not to help students think of some relevant ideas when they have none. In fact, when the goal of a problem is to design an artifact (or tool) for accomplishing some specified purpose, solution processes seldom fit the metaphor of proceeding from one point to another along a path or eliminating obstacles, and the goal of productive metacognitive functions seldom fits the metaphor of helping students do something sensible when they are stuck (i.e., when they have no relevant ideas to bring to bear on the situation). For example, in the transcripts that are appended to this chapter, just as in transcripts that accompany other chapters throughout this book, the students are seldom without an ample supply of relevant ideas.

In model-eliciting activities, what is problematic tends to be the fact that students must develop beyond their own early (and often barren and distorted) ways of thinking about the situation. Consequently, especially during early modeling cycles, many of the most useful metacognitive functions are those that help students overcome negative characteristics (such as conceptual centering or conceptual egocentrism) that are associated with primitive, unstable, or otherwise inadequate ways of thinking. They are not those that are intended to provide hints about positive steps to take or what to do next. In fact, a counterintuitive fact about model-eliciting activities is that, during early modeling cycles, when teachers try to give hints about correct ways of thinking about the problem, these hints often derail progress. Why? Because even correct ways of thinking often need to be developed further and students may not realize what makes the correct way of thinking correct for their problem or fit assumptions they may have about the problem. So, when a current way of thinking is identified as being correct, students often cease to think about further refinements, elaborations, or revisions that are needed. In fact, as Zawojewski and Lesh show in chapter 18 on problem solving, if strategies such as draw a picture lock students into current ways of thinking, rather than providing ways to examine or revise current ways of thinking, then they may inhibit rather than facilitate progress. For example, in transcripts for model-eliciting activities, students may draw a picture (or look for a similar problem, or clearly identify givens and goals) and, rather than providing a way to help students examine and revise their current ways of thinking, the result may effectively replace the problem that was given and lock students into their initial conceptualizations.

IMPLICATIONS FOR INSTRUCTION: DEBRIEFING TOOLS FOR STUDENTS AND INTERVENTION TOOLS FOR TEACHERS

This section describes a variety of reflection tools that we have been developing to help us investigate the nature of students' developing metacognitive abilities—as they are used in model-eliciting activities in elementary mathematics. However, these same reflection tools also are used to help students develop more effective metacognitive abilities; and, they also are useful to help teachers develop productive roles (e.g., beyond being a hint giver) when their students are working on model-eliciting activities.

The effectiveness of these reflection tools depends on the facts that, when students work on model-eliciting activities: (a) their solution processes usually involve a series of distinct modeling cycles in which different metacognitive understandings and abilities tend to be productive, (b) following their problem-solving activities, it is not difficult to teach students to be able to describe some of the most obvious stages that they went through, and (c) these students can also be taught to give insightful answers to brief questionnaires (reflection tools) about changes in their own metacognitive roles and behaviors as they progressed from one stage of problem solving to another.

For the preceding approach to be effective, it is not necessary for students to give complete and accurate descriptions of the stages and modeling cycles they have gone through. In fact, even for researchers and teachers who review videotapes of the sessions, it generally requires many viewing cycles before relevant patterns in behavior emerge clearly. What is needed from students is simply that they become able to detect instances where changes occurred in their profiles of metacognitive behaviors—or where changes occurred in group functioning, or in personal roles, or in feelings or attitudes toward their experiences. We are also more interested in the development of their thinking and how they reflect over time than in the accuracy of one particular instance. Because their metacognitive abilities are continuously changing and simultaneously changing as their understanding of the related conceptual systems change, accurate pictures from instant to instant would be almost impossible to generate or assess. However, a developmental picture showing cycles of development over time is more instructive.

Ideally, some of the reflection tools for students should be completed in a very short period of time. The purpose of this is to minimize interruptions to the problem-solving process if they are given while the students are still working on the problem. Also, because we have seen the benefits of multiple representations and technology-based representational media on conceptual development (Johnson & Lesh, chapter 15, this volume) and we assume that metacognitive abilities develop in a similar manner to conceptual systems, the tools can have a graphical or pictorial component available either in the analysis of the students' responses or in the generation of students' responses. This does not prohibit written as well as quantitative data from being used in the reflection tools.

Reflection tools for teachers should help them understand how their students thinking developed metacognitively. In the case of teachers, their reflection about their students' thinking develops in a similar manner to the way their students' metacognitive abilities develop. Namely, their interpretations develop along multiple dimensions, in cycles, in a context-dependent manner, and in parallel. In fact, teachers monitoring and assessing students as they work on model-eliciting activities are observing many groups working in parallel and are processing information about what each group is doing and their progress on the solution. In order to help teachers assess their students' thinking, they need tools that allow for these characteristics. Their tools should also employ multiple representations, technological components to facilitate data analysis and recording, and require a minimum of time to complete because teachers are collecting so much data simultaneously. The teacher's role shifts from hint-giver to way-of-thinking-improver meaning that more of the teacher's responsibility lies in providing contexts where students' ways of thinking improve naturally than in giving hints for correct solutions. Typical interventions we have used to help students improve their way of thinking include asking students to describe their current way of thinking, explain current assumptions about the problem, and describe their solution process thus far. Similar to reflection tools for students, teacher reflection tools should include multiple representations and both quantitative and qualitative data. After responding to the questions emphasized on the preceding kinds of "reflection tools," students often are able to participate in productive discussions about issues such as: How did you know when to get back "on task"? Why did you switch roles (at this point)? When were you frustrated and what did you do to deal with this situation?

CONCLUSION

Our purpose in this chapter has been to describe a models and modeling perspective on metacognition designed to facilitate the understanding of a concept that has (in the past) been very difficult to understand because notions of metacognition and higher order thinking were not well-grounded or very complete. This has included reexamining what is meant by problem solving in general and then interpreting what that means for metacognition. In addition, we examined the strengths and weaknesses of a variety of notions related to metacognition and higher order thinking and presented our viewpoint about ways to improve on those notions. A key part of the perspective that has guided our analysis is that metacognitive and higher order abilities are connected to conceptual systems and models and interact with them bi-directionally. This implies that if we view conceptual systems as developing along multiple dimensions, in cycles, in context, and in parallel, then we can make similar statements about their related metacognitive functions.

Chapter 22

Interest, Identity, and Social Functioning: Central Features of Modeling Activity

James A. Middleton	Richard Lesh	Michelle Heger
Arizona State University	*Purdue University*	*Purdue University*

As usual, Vygotsky has spurred Western psychologists to take a radically different perspective on the process of academic development...

> ... their separation [intellect and affect] as subjects of study is a major weakness of traditional psychology since it makes the thought process appear as an autonomous flow of "thoughts thinking themselves," segregated from the fullness of life, from the personal needs and interests, the inclinations and impulses of the thinker. (Vygotsky, 1962, p. 8)

Research on mathematical teaching, learning, and knowing has a long and impressive history, investigating cognitive processes, developmental trajectories, and pedagogical techniques that explain and predict mathematical achievement (Grouws, 1992). In a parallel history, the research on beliefs, affect, and motivation has seen considerable recent development, especially in explaining why, despite our best efforts, these processes, trajectories, and techniques seem to have a broad impact on only a narrow and privileged portion of our society, and leave the rest out (Middleton & Spanias, 1999). Perhaps we have ignored a key purpose of education in our desire to create the next generation of scientists and mathematicians, and, in our attempts to enhance intellectual development, perhaps we have ignored Vygotsky's notion of the "fullness of life." If mathematics learning is portrayed as "thoughts thinking themselves," and if curriculum activities are thought of as "deeds doing themselves," then we are not likely to attend to the place of the child as both an intellectual and affective being, situated within a community of intellectual and affective beings, attempting to coordinate their actions to enhance their personal success and happiness. The purpose of this chapter is to describe a way of thinking about the activity of doing mathematics that treats the affective and motivational (i.e., dispositional) state of the child as integral, and indeed inseparable, components of children's mathematical interpretations of learning and problem solving experiences (i.e., models). This perspective is crucial, we think, to understanding why children do what they do, what the nature of their knowledge is given a particular set of processes that may impinge upon their learning experience, and why they sometimes (despite all of our good intentions) fail to do what we want them to do. The route through which we came to this particular stance reflects the history of the authors as psychologists, our gradual conversion by innovations in the learning sciences, and especially by means of our experiences as designers of mathematics tasks and curriculum. Through successive phases in our own learning histories we have come to view learning as constituting more than change in behavior potential, knowledge as more than configurations of neural firings, and motivation as more than either incentive or coercion. In brief, our chapter sums up this history and projects a research agenda on modeling that allows students to take an active role in behaving intelligently in complex quantitative and spatial tasks.

We begin with a brief review of several of the most critical relevant features of intelligent behavior in complex situations, highlighting recent findings which suggest that mere knowledge of content is not sufficient for acting smart. Next, we discuss the distinct possibility that children do not have an epistemology that compartmentalizes mathematical activity into one category and self knowledge into a separate category, but that both disciplinary knowledge and knowledge of the self co-evolve, and reflexively impact each other in their development. In the second half of the chapter, attention shifts toward exploratory research that all three authors have been conducting to focus on instructional activities and reflection tools that are designed to help students take charge of their dispositions, which we characterize as profiles of interacting cognitive, social, and emotional components. These activities and tools focus on helping students control their characteristics to their own advantage as they move from one situation to another, or as they move from one phase of problem solving to another during the process of grappling with a model-eliciting activity of the type emphasized throughout this book (see also de Corte, Verschafel, & Op'T Eynde, 2000).

It Takes More Than Just Ability to Be Stupid

In his recent work, Perkins and Tishman (1998) do an admirable and convincing job of explaining many reasons why intelligent behavior is not merely a byproduct of ability, whether ability is conceived as hard-wired programming or intellectual capacities that can be developed to differential levels through education. Rather, before a person initiates a sequence of intelligent behaviors, he or she must first be sensitive to the potentialities of information (e.g., by being attentive or alert to occasion for utilizing information), and be disposed to act upon that information (e.g., by valuing the domain sufficiently to create a problem space). In essence, the implications of Perkins and Tishman's research represents a move from thinking about knowledge of a domain as purely "cold" cognition toward knowledge as socially indexed and value-laden. Rather than thinking of a child's (or a class's) understanding of mathematics as those aspects of memory that can be mapped onto the structure of mathematics as a set of axioms, we begin to see knowledge as entailing facets of students' whole experiences that have taken on the label "Mathematics." As such, the label not only points to, say, ratio tables as models for connecting concepts of rational numbers, but also to the problem sets, social norms, and peculiarities of students' personal learning situations that accompanied rational number instruction.

In the United States at least, we see children for whom the label mathematics conjures up images of failure, arbitrary rules set by others, and academic ostracism. Ostensibly, these children understand that mathematics is both useful and important, but fail to engage in mathematics in any depth, subsequently electing not to continue taking mathematics after their compulsory education is ended (Carpenter, Corbitt, Kepner, Lindquist, & Reys, 1981). Even for children who seem to perform adequately in their classes, the connection seldom is clear between this performance and their goals and aspirations for the future. Consequently, even among these students who are generally more successful in mathematics, who are less fearful of mathematics, and who have an academic self-concept that is strong in relation to mathematics, they too frequently drop out of higher level mathematics courses in high school or college (Amit, 1988).

There has been a long and rich debate about the degree to which achievement that children exhibit as a result of academic activities is impacted by *expectancies*—those feelings of personal agency in performing a task—and *values*—the degree to which the task itself and its outcomes and contingencies are considered worthwhile to pursue. In general, research suggests that both expectancies and values are important in mathematics achievement. Students use expectations of success to guide their degree of involvement in immediate tasks. Nonetheless, these expectations, however much they impact any given situation in particular, are relatively unrelated to long-term engagement patterns, whereas the value students place on content seems to predict subsequent involvement and engagement in the field (Meece, Wigfield, & Eccles, 1990). This suggests that the tasks we design for students must be facilitative of the abilities students bring to the situation, but also, inasmuch as the experiences within which students experience the epistemology and content of mathematics constitute mathematics as a field, they also should be facilitative of the development of value for mathematics and generally positive mathematical identities (e.g., Lemke, 1997).

Moreover, a point that has been stressed throughout this book is that the degree to which mathematics is considered useful and important for immediate and subsequent application should be a focal problem for research in mathematics education. Although there is general consensus that mathematics is both useful and important, few studies have seriously examined the level and topical distribution of mathematics across its numerous applications for the purpose of informing curriculum and instruction. For excellent examples of serious treatments of real-world mathematics and its relation to curriculum and standards, see Smith (1999).

Long-term mathematical engagement, therefore, beyond that which is coerced, is dependent the value students place on the mathematics and the context in which it is applied. It follows then that an understanding of the development of mathematical values and dispositions is necessary for the development of a modeling environment that transcends the school environment and manifest itself throughout students' lifespans in the world of work.

Structuring a Mathematical Self in a Modeling Environment

Models can be thought of as representational systems for structuring experience (Hestenes, 1992). They leave off the superfluous features of a domain while retaining the structure that defines it. But, whereas the majority of chapters in this book address the question of how students construct the mathematical components of their models, this chapter emphasizes that students' interpretations of a mathematical experience are not simply what have traditionally been termed "cognitive" in nature. They also include aspects of affect, beliefs, individual differences, individual roles while working in a group, group functioning, and numerous other self-related constructs. That is, the models that students use to interpret experiences include not only components focused on mathematical content, but also components focused on mathematical practice and on the self in relation to content and practice (Debellis & Goldin, 1998). Furthermore, even though it is possible to isolate these components for analytic purposes, they relate to each other in a highly interdependent causal fashion. Whereas content knowledge can be thought of as a component that structures the information gleaned in carrying out a task, knowledge of the self-in-the-task (such as "motivation") can be understood as components that structure the social roles one has played, the success one has enjoyed, and the systems of negotiation entered into to retain a positive sense of agency while carrying out a task. Together, the content and self-components contribute to the ways in which mathematical histories and practices occasion differential interpretations in different children for what is ostensibly considered the same task.[1]

Although different mathematical and self-related components that students construct are of interest in research related to the preceding issues, it is really modeling in the sense of in situ structuring of experience that we want to emphasize in this chapter. How do students simultaneously construct both a framework for understanding classes of mathematical tasks as embodiments of mathematical structure and also a framework for determining their identity in relation to mathematics?

One way that a models and modeling perspective helps us examine this structuring of experience is that model-eliciting activities focus attention directly on students' evolving interpretations of learning and problem solving situations.

---

[1] For a similar discussion related to the notion of multiple embodiments of mathematics, see Simmt and Kieren (2000).

Second, because interpretations of a model-eliciting activity often change significantly from one stage of problem solving to the next, we can use modeling problems to more directly observe the relationship between students' modeling of mathematics and of the self in relation to mathematics over a reasonably brief period of time. For example, as transcripts given throughout this book show when students engage in modeling problems whose solutions involve multiple modeling cycles, they are often able to develop significant models (by which we mean useful, mathematically sophisticated, and affectively demonstrative) over relatively brief periods of time (e.g., one or two class periods). Consequently, it should be possible to design modeling activities (and accompanying reflection tools) explicitly to facilitate change in interest, identity, and other self-related aspects.

Implications of a modeling perspective suggest that dispositional constructs, especially those that serve to direct behavior, are dynamic and not constant. Children exhibit different motivational patterns, affective reactions, and cognitive and social engagements in different circumstances. The research is strongest and most easily discernible across tasks (situations), but because the information these factors provide are integral to the modulation of effort, they are also variable within a task (Middleton & Toluk, 1999). For example, transcripts throughout this book illustrate that many different profiles of noncognitive abilities can lead to success—by a profile we mean the peaks and valleys of thoughts, feelings, arousal states, and social interactions as they play out across the lifespan of a problem-solving activity. While working on a modeling problem, students' profiles often change as they progress through the various modeling cycles as depicted in Fig. 22.1.

Because interpretations of problem situations are dynamic, in the long run, a person who is successful is the one who learns how to either manipulate his/her own profile to suit the situation, or who alters the situation to some degree to fit his or her profile. Or, when the preceding perspectives are applied to a social situation, the successful community tends to be one that is able to coordinate the individual contributions of its members (the profile peaks) in such a way that the activity goals are realized. This kind of modulation requires a certain amount of flexibility in students' willingness to adopt different personae (to use the Greek metaphor for personalities as masks—personae—that an actor could don so that a play could be performed with fewer players than parts). But they must come to know when to adopt a certain persona and why.

Lastly, we argue that the self-components of the model—that is, personae—that emerge from modeling activity have a slightly different manifestation than the content-components. At some point the two are separable; that is, a graph of a quadratic function reveals little of the person or community that created it, save that they have some affinity for visual displays. Likewise, the student who smiles in an algebra class reveals little of his or her understanding of quadratic functions. At a larger grain size, however, the two can be seen as one: A single "model" of the mathematical experience. The student who smiles while drawing the graph reveals something about how the task has mapped onto her self system.

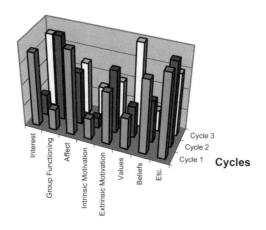

FIG. 22.1. A student's changing profile across modeling cycles.

It is this level of analysis that we have failed to address in our research on thinking. When does the content become one with the child? What does the smile signify about the graph, and what does the graph signify about the smile? If, as Perkins and Tishman assert, ability to solve problems does not determine intelligent behavior, but serves to assist volition, then these become crucial questions for research on modeling and symbolization.

THE FIT BETWEEN THE ACTIVITY AND THE SELF

Because mathematical self concept and self efficacy in mathematics tasks are learned knowledge (Middleton, 1999), an examination of students' experiences that contribute to the development of self-statements should shed light on the knowledge of, and about, the mathematics students are constructing.

In general, the perception that mathematics is difficult and that their ability to do mathematics is poor leads students to avoid mathematics given the opportunity (Hilton, 1981). Reasons for these kinds of feelings include perceived nonsupportiveness of the teacher, and the changes in the rules of engagement in mathematics from the lower elementary to the middle school grades (Eccles, Wigfield, & Reuman, 1987; Midgley, Feldlaufer, & Eccles, 1989). By the middle grades, most students begin to perceive mathematics to be a special case, where smart students succeed and other students merely get by or fail. They begin to believe that success and failure are attributable to ability, and that effort rarely results in a significant change in their success patterns (Kloosterman & Gorman, 1990). Moreover, there is evidence that a significant proportion of students recall bad experiences in mathematics class as significant influences on their learning (Hoyles, 1982).

Although the sources of satisfaction and dissatisfaction are similar for mathematics compared to other subjects, the ways in which students internalize their experiences in mathematics are markedly different. In mathematics, students tend to have strong feelings about what they are capable of doing, and they tend to internalize these feelings into their self-concept more so than in other subjects. Anxiety and feelings of inadequacy and shame are common interpretations of bad experiences in mathematics (Hoyles, 1982).

Students, especially young students, rarely have larger curricular goals for engaging in mathematics tasks. Their learning situations have been designed and selected for them. So, there are two possible evaluation methods that these students use: ascertaining the fit of an activity to their personal dispositional profile, and to some extent, socially identifying with their teachers and other students (Middleton, Littlefield, & Lehrer, 1992). These task evaluations can then be consolidated into a task-specific self-concept, the degree to which the parameters for involvement in a task fit the personal beliefs a student has about their abilities, desires, and aspirations (e.g., Wigfield & Eccles, 1992).

At the same time, however, the research on optimal experience in academic endeavors suggests that during involvement in tasks, the self ceases to be an independent entity and becomes part of a larger system (Cziksentmihalyi, 1990). In fact, when individuals report sustained involvement in difficult tasks, they also tend to report a broadening of self beyond the limits of individuality to a point where the self exists in a higher plane of complexity, being redefined in terms of the activity within which it is situated. (Note the remarkable similarities in this explanation of loss of self in the activity and the descriptions of identity cf. Lemke, 1997; Wenger, 1998). This is quite different from seeing the task as a good fit for a student's particular dispositional profile; instead, because the profile is dynamic, and because the task itself is dynamic, it suggests a coevolution of both task and child in such a way that neither can be seen as distinct from the other.

Mathematical Modeling and Self-Modeling Are Reflexive

Another ramification to our work on the nature and function of modeling suggests that students' mathematical modeling and their self-modeling is reflexive. That is, neither can be understood entirely without the other, and understanding the contribution of both is to understand the child better than merely knowing each in isolation. If this is indeed true, then the link between mathematical understanding, contextual referent, social role-taking, and mediation extends beyond a means-end approach to get children to learn mathematics, to an approach that favors the child becoming a mathematical thinker, including the structure of content and the dispositions of a lover of math (see also Wenger, 1998).

This process of becoming must involve modeling of mathematical situations, where situations are defined more broadly than activities. As Cobb and colleagues argue, mathematical learning is partly a process of intellectual inheritance by which individuals appropriate the mathematical ways of knowing of their larger society (Cobb, Gravemeijer, Yackel, McClain, & Whitenack, 1997). We are arguing that mathematical self-knowledge is a mathematical way of knowing, enculturated through sociomathematical norms, and manifesting itself as individual and communal beliefs and values. According to this perspective, mathematical situations are not seen as moments in time, nor as activities leading up to the satisfaction of a goal state (what might normally be termed "episodes"), but instead are better thought of as threaded conversations that evolve, branch, and converge as individuals negotiate the role their self plays in the practice of the larger community. The meanings these conversations take on are both personal and social, where the mathematical self is only seen in relation to the mathematical community.

To be competent, activity in such a community of practice must embody the following three criteria: mutuality of engagement—the coordination of social acts (e.g., Mead, 1910) of an individual with other members of the community (in the pursuit of a common goal); accountability to the enterprise—the ability to understand the goals of a practice, taking responsibility for one's actions in the community, and contributing to the community's pursuit of its goals; and negotiability of the repertoire—the ability to take a repertoire of skills and ways of behaving, and apply them flexibly (but appropriately) to novel situations the community may face. For competence to become optimal experience, then, the coordination of social acts in the classroom must be on focused, challenging activity, where the students' individual contributions to the group's practices embody their unique histories and abilities, but where those unique contributions are mutually constitutive in the construction of a coordinated model.

So, while children are learning about fractions with unlike denominators, they also are learning the rules for engagement in tasks that ask for knowledge of fractions with unlike denominators. They learn who is considered smart and who is not in relation to the task. They learn how the classroom community respects those with mathematical knowledge and those without. They learn where each fits within these parameters. It is unlikely that, in the future, given the opportunity to engage in similar activity, a student will forget all of this contextual baggage and apply his or her "pure" fractions with unlike denominators scheme to the purely mathematical information given. More likely, a child will utilize this knowledge to choose whether or not to involve himself, to monitor the fit of the task to his goals, and to regulate his interactions with others in the community (Middleton & Toluk, 1999). This process, in turn, will determine the degree to which he will be able to engage his understanding of the content proper (e.g., Perkins, Jay, & Tishman, 1993; Perkins & Tishman, 1998). In the case of involvement in complex, interesting tasks, a new kind of self-awareness or self-concept may emerge: One that is inseparable from the content. What elements of design, then, can be abstracted from the research on tasks that engender both high motivation to learn and quality content understanding?

First, it suggests that couch potatoes do not have more fun. Sustained intellectual and often times even physical effort must be an integral feature of modeling activity. Second, the ability to locate one's self in a nexus of potential communities of practice or fields of inquiry seems to be critical for navigating the social landscape of a task. Third, the ability to utilize the local expertise of other learners (e.g., personae, which may include particular content expertise, representational fluency, or communication skills) in a collaborative effort allows students of diverse abilities and backgrounds to be successful in complex mathematical situations. Lastly, as we allude to throughout this chapter, tasks that are designed to elicit and transform students' models of mathematically rich situations, are likely candidates for curricular improvement along these lines of inquiry. These points are developed in depth under the following three themes: Interest/Motivation, Identity, and Social Functioning.

INTEREST, IDENTITY, AND SOCIAL FUNCTIONING

To conclude our discussion, we have chosen to address three specific dispositional constructs: Interest/motivation, identity, and social functioning. It is useful, though somewhat artificial, to separate these three constructs for our discussion. In general, the research literature on each is distinct epistemologically, focusing on different facets of behavior, different units of analysis, and utilizing different methods of inquiry. Moreover, the ways in which these aspects of learning contribute to the functioning of the individual versus the collective is at different levels as they are conceptualized. However, it must be made clear that when one speaks of having an interest, one is also to a great extent, expressing who he or she is. Similarly, when an identity is mapped out for an individual within a conglomeration of possible communities of practice, the result is in effect, a statement of relative interests. Likewise, when one is having an interesting experience, he or she is, by nature redefining one's self in terms of social norms and parameters.

Interest/Motivation

Motivation is a term that is often used to describe a general state of attitude[2] either for or against some task or course of action. When we say a child is motivated, we mean that he or she displays an attitude of compliance with the requirements for participation in a task, generally accompanied by overt expressions of positive affect, directed attention, and persistence. When we call a child unmotivated, we mean that he or she does not follow the rules, frowns or acts out, or cannot stay focused (Middleton & Spanias, 1999). Motivations,

---

[2] We use the term *attitude* here in the same sense as the old nautical term. As in the nautical context, an attitude is a way of behaving as a person charts a course through some conceptual and/or practical terrain. Attitude is not a concept in and of itself, but represents a coordination of thoughts, feelings, and practices that have some stability over time, but that manifests itself interactively with the task parameters.

therefore, are inextricably bound to the child-in-experience. To say that a child is either motivated or unmotivated is to compare a task or field of practice with the participatory actions of an individual.

The case has been made elsewhere that there is no such thing as an unmotivated child (Middleton & Toluk, 1999). The term unmotivated, rather than meaning an absence of will or desire, actually refers to the actual attitude of the child to do something else than we, as educators, desire. There are a number of reasons for this mismatch of our goals versus those of the child. For example, a child may actively engage in a visual/spatial task, whereas shying away from a task that requires algebraic manipulation because he or she feels more control when utilizing a preferred strategy. Another child may fail to see the application of a mathematical model to any scenario they might find useful or interesting. Still another may be taught that mathematics is not for them. Each of these cases of mismatch, and converse matching scenarios can be shown to be influenced significantly by a comparison of the requirements for participation in school mathematics and the self-system of the child. In general, children tend to choose to engage more in tasks that map neatly onto their self-system than tasks that are outside that self.

Generally, the kinds of motivational models we are describing are adaptive systems that interpret and restructure experience in a format that guides future engagement. As anticipatory structures, they posit tentative definitions of success, orient the individual to the probability of success in an activity, guide behavior, and monitor the outcomes of behaviors.

As evaluative structures, motivations serve as templates by which in situ experience can be mapped onto the individual's self-system. This mapping process ensures that once engaged, a student can disengage before damage to the self (both physical and psychological) can occur to any great extent.

The tradeoff of stimulation and control levels an activity affords determines the degree to which an activity has the potential to be considered worthwhile enough to initiate engagement. With moderately high levels of stimulation (e.g., challenge, fantasy, and curiosity), and moderate control over success (e.g., provision of meditational tools, appropriate task difficulty, and flexible ways of solving the problem), a student tends to engage in the activity and enjoy his or her engagement (see also the following discussion of "Flow" below). As a student monitors the level of stimulation they receive from the engagement, and the level of control they have over the task, the task value (an index that records the relative benefits of engagement) is modulated. Initially, positive affect corresponding to high stimulation and sufficient control may give way to displeasure if the task ceases to be stimulating, or if the control is taken away (for a detailed model of this evaluative process, refer to Middleton & Toluk, 1999).

These evaluative processes allow the individual to get the most learning out of a situation. By monitoring the levels of stimulation and control afforded by an experience, a student can continue to engage up to the point where the cost of engagement (fatigue, time, etc.) creates a situation of diminishing returns over the benefits of engagement (grades, enjoyment, and feelings of success). This

22. CENTRAL FEATURES OF MODELING ACTIVITY

maximizes the level and amount of learning (both academic and self-related) that can take place in a given activity which, in turn, increases the adaptivity of the system.

Note that learning in terms of being able to produce recognizable mathematical terms, notations, and arguments is not particularly important except as the student perceives that retaining such facets of the situation increases the adaptivity of his or her system to the conditions under which they exist. Many disposition-related choices we witness in schools do not lead students to engage in the kinds of content-related learning we might desire. Nevertheless, the reasons behind their choices generally reveal remarkably adaptive thought. A student who continually experiences failure when presented with mathematical tasks would be foolish indeed to continue engagement when presented an alternative. Avoidance behaviors such as checking out, feeling sick, or deferring to others are creative, goal-oriented, and extend the adaptivity of the self-system where continued involvement in the same old mathematics might not.

Although self-perceptions of agency in a domain become more differentiated and intractable with age, it appears that even very young children construct models of a domain, and their competence in relation to it (Byrne, 1984).

Interestingly, on a societal level, it seems that mathematics is one of the few basic subject areas where it is okay to be incompetent. Few people seem to be overly embarrassed at being innumerate whereas the same thing cannot be said about being illiterate. This tendency can partly be explained in that people tend to discount poor performance in a domain when the domain is less central to their self concept than other domains (Harter, 1986). That is, "if it is not important to me, then it really does not matter if I am good at it or not." Moreover, mathematics seems to be a slightly different case societally in the level to which students develop strong feelings about their own competence and their role in the classroom environment. So, on both an individual and a societal level, the value accorded mathematics is lower than other subject matter, and this valuation is tied to people's beliefs about their own competence and the subsequent interest (or lack thereof) these beliefs engender.

Interest is a special conceptual domain within which activities that have provided continually optimal levels of arousal and control are labeled. Cziksentmihalyi (1990) would call these "Flow Activities." The difference between the Flow perspective and the modeling perspective taken in this chapter is that Flow activities are those that are classified by the researcher as those that have historically provided optimal experience. Interest, as we are defining it includes activities and fields of practice as they have been classified by the individual. To say an activity affords "Flow" or "Interest" is saying roughly the same thing—it is a statement of the probability that a person experiences optimal arousal and control. To say, however that an activity is an interest is to study the attitude of a child, and to place the child, his or her self, squarely within the activity.

Another cluster of points that we need to emphasize here is that: Many children who think they like math (i.e., find it an interest), like what seems to be a narrow and distorted view of what mathematics is (or should be) beyond schools; conversely, many children who think they hate math hate what seems to be a narrow and distorted view of what mathematics is (or should be) beyond schools. So, the ways we can help them may focus less on changing their views of themselves and instead focus more on changing their views about the nature of mathematics, and the nature of situations in which mathematics is useful—that is, by redefining the very nature of school mathematics, including the parameters that define what knowledge, skills, tools, and situations are important, we may be able to erode the time-honored tradition of hating math (e.g., Middleton, Leavy, & Leader, 1998).

Identity

In the mathematics classroom, there is no widely held expectation by either teachers or students that the students in the class will grow up to be mathematicians. Nor are the practices of the teacher or the teaching tools applied to mathematics instruction explicitly designed to develop an identity—either as an amateur mathematician (a lay participant in the community of mathematics), an appreciative aficionado of mathematical practices (a fan of the sport so to speak), or an informed consumer (a critical thinker in the general sense). Regardless, it is clear from the research on mathematical attitudes, emotions, and motivation (see reviews by McLeod, 1992, and Middleton & Spanias, 1999) that children do develop identities with respect to the school mathematics practices in which they are engaged for upwards of 12 years and that these identities exist in a larger network of potential identities grounded in peer interactions, cultural and ethnic affiliation, interests and predilections (Lemke, 1997). Some configurations that make up these networks do in fact lead some children to develop identities within the community of mathematics creators, consumers, and aficionados. However, the vast majority of students place themselves squarely outside these communities and actively avoid engaging in deep mathematical practices as they matriculate into high school and higher education.

These tendencies beg a central question for research: "How do identities develop within ad hoc communities such as those established within a classroom, and how do these ad hoc identities relate to or even determine, in some situated way, the development of identity within existing communities of practice such as those of scientists, engineers, venture capitalists, or teachers of mathematics?" To some extent, we expect (and research bears out) that what happens in the culture of the classroom does in fact determine future engagement patterns as measured by courses taken in college, intended majors, subsequent approaches to mathematics instruction, and so forth. (Amit, 1988; Meyer & Fennema, 1985). However, how this happens is a mystery.

Identity in Ad Hoc Communities. Identity is a layering of participatory events by which our experiences and their social interpretations mutually inform each other (Wenger, 1998). *Who one is* is constituted to a great extent, by what one does. *What one does* is interdependent with *what others do*. Identity is at once social, describing the relationship of the student working and playing in a community of students (i.e., the form of membership in a community), but also individual, describing the uniqueness of the one among many (i.e., the ownership of participation).

Identity is also an interplay between the local activity and the global community to which an identity is identified. An identity helps an individual determine the answers to such existential questions as, "How does the here and now stem from, relate to, and in part determine, location within a nexus of potential communities to which I might ascribe?" In essence, because an individual can neither go forward in time, nor backward, identity situates the individual in a temporal frame by which the appropriateness of actions can be evaluated, and by which future courses of actions can be plotted against.

We use the term *ad hoc* to describe communities that are made up out of necessity, under the auspices of particular goals, to perform particular practices, or to produce particular products. A small group activity in the mathematics classroom is one example of such a community. In classrooms, the makeup of small groups changes often over the school year. Some group structures are designed to distribute the unique characteristics of students in the classroom, some are designed to consolidate these same characteristics, and some are developed in a random fashion. As designed organizations, these groups are charged to engage in some task and to produce some product (an answer, a model, a report, etc.). Following the completion of the task, the ad hoc community may be disbanded as its immediate function is completed.

Such transitory communities, by their very nature, cannot develop mature, articulated identities among their members. However, certain roles can be fostered that, if the roles afford transportability across other social situations, can become routinized into taken-as-shared practices by the larger community of the classroom.

Identity Within Existing (Introduced) Communities. Over the long haul, how ever much such roles facilitate the social norms of the class, they do not appeal deeply to the epistemological and disciplinary norms of mathematically sophisticated communities. Even the provision of real-world contexts within which to situate mathematical activity may not overlap significantly with the practices of the kinds that mathematics creators, consumers, and aficionados would recognize as legitimately mathematical. Even solving sophisticated quantitative and spatial problems of the kind emphasized in this book, without the emphasis on structure and system that leads to models which have transportability from one situated activity to another, may detract from the capability of the student to develop a meaningful identity with respect to the field.

Mathematics, its associated practices, and its ways of thinking exist historically as a well-developed field. Such fields (or *discourses,* according to Gee, 1997) embody the kinds of criteria we ascribe to communities of practice. Moreover, because of its maturity, mathematics has many subdisciplines that constitute communities of practice among which are the traditional categories of number, algebra, geometry, and statistics. Whereas these subdisciplines overlap significantly, they have unique foci of inquiry, unique forms of inscription, and unique histories that govern their development. A critical question for the design of a modeling curriculum is, "To what extent do the tasks that engage children, afford the involvement of children in these unique foci, histories, terms, and patterns of inquiry?" For it is only through realistic engagement in such tasks that children can develop any kind of entrée into the world of mathematicians. It is only then that they can develop an identity with respect to school mathematics (perhaps the only mathematics they will seriously pursue at any time in their lives) that is consistent with a legitimate identity with respect to mathematics as a larger field.

Moreover, quantitative and spatial reasoning is not solely the purvey of mathematics as an axiomatic field. It is also the heritage of the sciences (both natural and social), engineering, and the professions. Indeed, many fundamental mathematical principles have been given birth by the need to model empirical phenomena (as the close ties between today's calculus and mechanics curricula attest). To think not only about the mathematics, but also with it is a basic design element for model eliciting tasks, but what then determines quality mathematical thinking, if it is not purely set by the community of mathematicians?

Figure 22.2 illustrates the situatedness of the practices of children in a well-designed modeling activity, the Bigfoot Problem, within the nexus of numerous potential communities of practice within which the task is designed to reside. In such a task, the child's activity is at once fully mathematical, fully journalistic and fully forensic science, affording an entrée, or potential identification with each of the communities to which the task appeals. At the same time, it is also fully pedagogical, playing out in a designed environment with roles carefully prescribed by the task writers and the teacher to afford particular discourses related to ratio and proportion (see Appendix A).

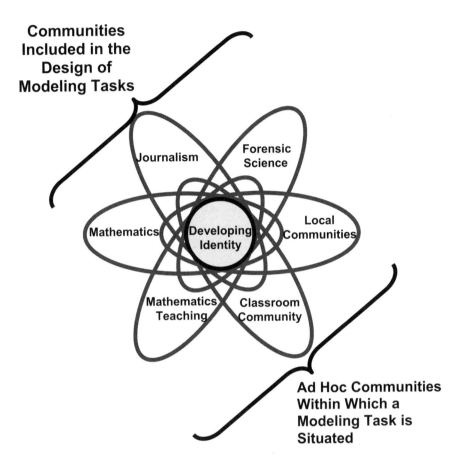

FIG. 22.2. Locating identity within a nexus of potential existing and ad hoc communities.

Social Functioning

Interestingly, the notion of identity as we have conceived of it in this chapter, as a sociocultural phenomenon, is highly related conceptually, though its epistemology is disjoint, with that of social intelligence, the act or state of knowing how to work with others. Moreover, these ideas are not new or even revolutionary. Eighty years ago, for example, E. L. Thorndike argued that social intelligence, the ability to understand and act wisely in relation with others, is also integral to a person's intelligence (Thorndike, 1920). Similarly, Gardner (1983) popularized the notion that there is not just one kind of intelligence, but many including interpersonal intelligence. Gardner's work is echoed by Sternberg's research in arguing that social intelligence is distinct from academic abilities and plays a key part in allowing people to be successful in the practicalities of life (Sternberg, 1985).

We would like to stress that this ability to discern the dynamics of the social situation into which students are thrust and use that discernment as a tool for successful functioning is especially relevant for interaction in modeling eliciting problems, working in small groups, using powerful conceptual tools (including diagrams, language, and symbols), and constructing complex products using processes that typically require telescoping sequences of testing and revising. Division of labor, ability to take the perspective of others, and effective communication all involve a sense of the social relations and unique abilities of each member of the group. Thus, students' success in social (i.e., group) problem solving relies not only on their subject matter knowledge, but also their social skills and, many of the defining characteristics of modeling problems, such as the complex solutions and the progression through development cycles, accentuate social skills (such as taking on multiple perspectives) that standard problems do not afford.

Complexity in Modeling: Products and Problem Solvers

The products that modeling problems ask students to produce are complex. For example, students may need to create descriptions, explanations, representational systems, languages, or physical models to make sense of the problem-solving situations. These complex products generate five implications for social functioning.

1. Because these constructions require more than just one-word or one-number answers, they often involve multimedia communication capabilities, which in turn, require that the students deal with divergent viewpoints, access to information, and distributed expertise—essential social competencies.
2. Because the solution paths involve more than one-way routes from the givens to the goals of the problem, students find that they need to use skills such as planning, monitoring, assessing, and communicating that again require the consideration of the larger sociocultural system within which the problem parameters are embedded.
3. Because the problem solver is often a team consisting of diverse specialists, rather than being an isolated individual, students need to be able to communicate clearly with each other and synthesize the contributions of the group members.
4. Because sophisticated tools are often used, problem solvers may be relieved of computational drudgery only if they are able to describe situations in forms that fit protocols that these tools will recognize (which are often mathematical formalisms); so, once again, communication capabilities are highlighted. In this example, the communication skills students develop are normatively accountable (e.g., Harre & Gillette, 1994) to the particular mathematical community that developed and uses the computational tool, and that thus sets the parameters for its appropriate use.

22. CENTRAL FEATURES OF MODELING ACTIVITY

5. Finally, because the products that problem solvers produce often include conceptual tools that need to be reusable and transportable, problem solvers need to be able to anticipate the needs and understandings of others to whom the conceptual products are shared. Note that by using the word *share*, we assume the "taken-as-sharedness" of the meanings that the conceptual tools take on both for the producer(s) and for the intended audience(s).

As in the real world and in school, problem solvers often are teams, not just isolated individuals working on a problem. Such teams are instances of complex systems with emergent properties that go beyond the characteristics of the individuals within the group. In Oakes and Rud, (chapter 12, this volume) they describe how in fields such as engineering, managers value most highly graduates who can work in teams, adapt quickly to sophisticated tools, and communicate to diverse groups of people ranging from managers, to colleagues, to customers. Thus, they want people that can function as parts of groups coordinated towards the fulfillment of a common purpose, not just as individuals, acting alone or in parallel.

So how can we conceptualize the group as being an emergent complex system comprised of more than just the sum of individual group members' contributions? Consider the chemical analogy presented by Asch (1952): Water is made up of the elements hydrogen and oxygen. Yet, water has properties different than those of hydrogen and oxygen. Furthermore, the molecular combination of hydrogen and oxygen can take different forms such as ice, steam, and liquid depending on how the elements are configured, and the conditions under which they react. Thus, water is not simply the aggregate of its components. It is, however, determined by the arrangement of its components. Similarly, the group construction is not just the aggregate of its individuals' developments. Its practices are determined by the ways in which individuals co-develop, and its functioning is impacted by the conditions of the situation. As Asch states, "We need to see group forces arising out of the actions of individuals and individuals whose actions are a function of the group forces that they themselves (or others) have brought into existence. We must see group phenomena as both the product and the condition of actions of individuals" (Asch, 1952, p. 251). Cobb et al. (1997) make much the same point from a sociocultural perspective when they describe an emergent perspective on classroom research as opposed to Asch's social psychological perspective.

Kauffman (1995) offers similar discussions about biological systems. As he writes, "the complex whole may exhibit properties that are not readily explained by understanding the parts. The complex whole, in a completely non-mystical sense, can often exhibit collective properties, 'emergent' features that are lawful in their own right" (p. vi). Another common example of how complex group behavior can result from simple individual behaviors occurs in Douglas Hofstadter's book, *Godel, Escher, Bach* (1980). In his book, he highlights the differences between an ant colony as a whole and the individual ants that compose it. Each individual ant is rather simple whereas the ant colony as a whole can accomplish sophisticated tasks such as an efficient collection of food or the development of a nest. Rissing, Pollock, Higgins, Hagen, and Smith, (1989) document that even ants develop complex, sharable, modifiable, and transportable communicative acts (which interestingly include the development of chemical inscriptions) to guide fellow ants to quality food sites.

As teachers and educators are aware however, merely placing students within groups does not guarantee that they produce complex solutions. Numerous educators have verified the benefits of small collaborative groups (Davidson, 1985; Good & Biddle, 1988; Good, Reys, Grouws, & Mulryan, 1989-1990; Johnson & Johnson, 1974; Noddings, 1989; Sharan, 1980; Slavin, 1989; Slavin, Sharan, Kagan, Lazarowitz, Webb, & Schmuck, 1985). However, they all preface these conclusions with statements regarding the situations or conditions necessary to produce productively functioning groups. In fact, in fields like management and engineering, it is common to engage in team building activities and to receive feedback about group functioning in the hope of producing more functional groups. Most of the feedback from such team building activities and many of the conditions found to be conducive to group functioning by the aforementioned researchers center upon social interactions. For example, Schmuck and Schmuck (1983) found that students need to be able to paraphrase, describe each others' behaviors, check impressions, make clear statements, describe their own behaviors and feelings, and provide and receive feedback to fully optimize the capabilities of the group. Similarly, Hatch and Gardner's interpersonal intelligence contains four components that enhance the abilities of students to work together:

1. Being able to initiate and coordinate the efforts of a network of people,
2. Negotiating solutions by serving as a mediator, preventing conflicts, or resolving conflicts,
3. Providing a personal connection through empathy, teambuilding, or by responding appropriately to people's feelings, and
4. Analyzing social environments by being able to detect and have insights about people's feelings, motives, and concerns (as cited in Goleman, 1995, p. 118).

Goleman (1995) describes the following skills as rudiments of social intelligence: teamwork, open lines of communication, cooperation, listening, and speaking one's mind.

Whereas solving a modeling problem, these interpersonal skills tend to be important because students must coordinate, manage, and communicate their actions, thoughts, and desires to each other to produce the associated complex product. Furthermore, because they are asked to create complex products, their interactions are even more multifaceted than with standard problems. Thus, we see that in a modeling context, the problem solver and the respective interactions are also complex.

22. CENTRAL FEATURES OF MODELING ACTIVITY

Progression Through Model Development Cycles

When groups of students complete modeling problems, they progress through a series of development cycles in which they revise, refine, or extend their thinking about the problem. These development cycles introduce new considerations into the functioning of small groups that standard problems do not invoke. As students revise, refine, or extend their thinking in a modeling problem, they also alter the functioning of the group and the roles of the individuals in relation to the group. For example, a group may find it beneficial in the beginning of a modeling problem to brainstorm ways to approach the problem. To be successful however, the group will at some point need to terminate their brainstorming and select a procedure or idea to follow. Typically, when students first make this choice of what idea or approach to pursue in a modeling problem, they pursue a rather barren or distorted idea.

From a transcript of students working on the Big Foot Problem (see Appendix B for the full transcript), we see that one of their original attempts to find the height of the do-gooder distorts the proportional relationship between height and footprint.

Ben: "Let's see." {Ben puts his foot next to the footprint. Then, he uses his thumb and pointing finger to mark the difference between his shoe and the footprint. Finally, he uses his fingers to imagine moving the distance between his fingers to the top of his head.} "He's this tall." {indicating the height where his finger is pointing.}

Amy: "Hmmmm! I guess so. Let me try it. - - - Yeah, I guess that's about it. He's about this big."

As students continue to work with a weak idea, they begin to realize ways to strengthen their idea or approach. Sometimes they recognize that they wish to completely discard their idea. In the Bigfoot transcript, the students completely discard their additive idea when Frank, another group member, comments:

Frank: "No way, man. I'm that big," {pointing to the height that Ben was indicating} "and my feet aren't THAT big. - - - He's a lot bigger than that."

If students do discard an initial idea, they may return to another cycle of brainstorming. Ben, Amy, and Frank go through such a process of brainstorming after discarding their additive idea.

Amy: "So, how big is he?"
Frank: {pause} - - - "I don't know. - - - Six foot maybe."
Ben: "Yeah, about six feet."
Ben: {pause} - - - "That's still not big enough. Look Amy, his feet {pointing to the footprint} are twice as big as yours. If he's twice as big as you, he's ten feet tall."

Amy: "Uh huh! Mmm!"
Frank: "So, what're we gonna do?" {Silence and mumbling among the group}

A tradeoff between investigation and verification continues until a group feels satisfied with their solution. Following their break and brainstorming, Ben, Amy, and Frank begin another cycle, but this time, they use a proportional relationship that leads them to a satisfactory answer (see Appendix B). Similar cycles occur for the group with respect to several other dimensions including impulsivity versus reflexivity, time-on-task versus time-off-task, and making goals versus completing tasks (please see the fourth column of the transcript for comments about these noncognitive aspects). Thus, group functioning varies throughout the problem. Groups do not remain fixed upon one goal. Furthermore, remaining fixed would likely severely limit the group's ability to solve the problem.

Parallel to these changes in the group's functioning, the individual is also likely to alter their individual roles to complement the group's activities. Students may shift from listening to talking, from directing tasks to carrying out tasks, from receiving feedback to giving feedback, from recording the group's work to overseeing another group member's recordings, and so forth. As an example, in the Bigfoot transcript Amy alters between leadership roles and listening roles. Thus, individuals within the group do not remain fixed upon one role either.

This propensity to change contradicts the common belief that the group's functioning and an individual's role within a group must remain constant throughout an activity. Furthermore, within a modeling problem, it may prove unproductive for the instructor to assign group roles such as director, recorder, listener, questioner, or materials manager because it is more natural for the students to alter their own roles. In the article, "Grouping for Instruction in Mathematics: A Call for Programmatic Research on Small-Group Processes," Good, Mulryan, and McCaslin describe some of the group processes associated with problem solving activities, "(1) maintaining the intention to learn (2) while enacting alternative task strategies (3) in the face of uncertainty" (1992, p. 173). They state that such characteristics "assume, indeed require, that the problem-solving tasks encourage, or at least allow, the student to take a flexible approach" (p. 174). Thus, because modeling problems include these characteristics, teachers should allow and encourage students to be flexible in their roles and processes. Furthermore, Rohrkemper and Corno (as cited in Good, Mulryan, & McCaslin, 1992) argue, "teachers often engineer tasks to enhance success and, in so doing, design tasks that are so predictable that they impede student opportunity to explore alternative task- or self-regulation strategies" (p. 174). We are emphasizing that labeling fixed student roles or group roles proves to be antithetical to group success; and furthermore, teachers and students should make a conscious attempt at role malleability. The Zawojewski chapter on small group modeling (chap. 19) provides a transcript example in which two group members retain control of the group's work

22. CENTRAL FEATURES OF MODELING ACTIVITY

throughout the entire problem. Their control prevents the third group member from expressing his idea, a particularly productive approach to solving the problem. As a result, the group's final solution suffers. This represents the modulation of an appropriate tradeoff between exploiting the distributed expertise extant in any group, and the immediate needs, the here and now functioning required for resolution of the group's goals.

If we look further at the development of relevant modeling skills, it becomes clear that an internalization dimension results from the external social dimensions. As discussed in the Zawojewski chapter, modeling problems serve as an avenue for students to internalize some external group interactions. As Good, Mulryan, and McCaslin wrote, "the capacity for self-regulation through adaptive inner speech begins in the social world" (1992, p. 174). Students have to observe the monitoring behaviors of the group before they can emulate such monitoring behaviors within themselves. A Vygotskian perspective calls for challenging conceptual and problem solving tasks that can provide students with the opportunity to enhance their self-involved and strategy task-involved inner speech. In addition, chapter 1 as well as chapter 20 by Harel describe why we often refer to model-eliciting activities as "local conceptual development sessions".

Ordinarily, students complete a series of modeling problems, and their ability to navigate through the shifts in their group functioning and individual roles occurs naturally. Appropriately with respect to the argument just presented, the teacher does not prescribe roles for the students. With such an approach, some groups are more successful than others. In particular, successful groups know when and how to adapt their group functioning and individual roles. They recognize when it is fruitful to jump into the problem and "get their hands dirty" or when to proceed with thoroughness; they recognize when it is productive to take a short social break and when it is time to return to work; they recognize when planning their progression proves valuable and when too much planning hinders task accomplishments. To assist students with improving their abilities to make such judgments, we hope to develop a student reflection tool that would allow students to investigate their group functioning and individual roles. Furthermore, we hope to develop student reflection tools to address the two previous dispositional constructs, interest/motivation and identity.

REFLECTION TOOLS

The goal for understanding the role of affect and motivation in modeling activity is perhaps best stated as having the student influence affect instead of having affect influence the student.

Students' interpretations of mathematical experiences include both disciplinary content and dispositional aspects. Thus, we can expect that students' dispositional components such as "identity" develop along similar lines as their components of mathematical concepts such as how to add fractions. Their models progress from being barren and distorted to becoming more refined, as students recognize essential structures, and omit irrelevant details. Usually, a modeling problem is used as a classroom impetus for the refinement of students' mathematical components. However, we are now striving to develop an impetus to challenge students to refine their dispositional components. As Lemke (1997, p. 46) points out,

> Reasoning *is* an affective state or process, and an enjoyable one for the mathematician. All cognition, because it is embodied, is necessarily also affective. We do not think without feeling. When a kind of thinking is a good-feeling, we tend to become good at doing. Feeling tone can be a guide to the quality of our cognition or it can signal to us a conflict of identity.

To remarry students' feelings about mathematics with their knowledge of mathematics, we have attempted to create student reflection tools, for example, metacognitive prompts that engage students in actively using their affective information as legitimate for sustaining optimal engagement in model-eliciting tasks.

The notion of metacognitive prompting during problem solving is not new, nor is the use of computing technologies as a medium for presenting prompts at opportune moments in the process of solving problems (Koedinger, 1998). What is new is the application of the notion of metacognitive prompts to the development of positive dispositional knowledge during problem solving.

Heretofore, feedback on dispositional aspects of mathematics activity has been either primarily tacit, consisting of automatized comparison of current arousal states against a dynamic tolerance threshold (e.g., levels of stimulation and control versus levels of anxiety and/or boredom), or contingency-related focusing on the outcomes of dispositional thinking rather than on their causes (e.g., praise for "good" behavior, punishment for "bad," (Middleton & Toluk, 1999). What if a system were designed to prompt students directly about what aspects of their involvement with a task facilitated their interest, identity, and social functioning in the task, and then provided feedback on potential strategies for students to alter their involvement to positively impact both content and disposition in these three areas? What would such a feedback system look like?

As described throughout this book, students progress through various development cycles as they complete a modeling problem. While the students progress through these cycles, their profile of dispositional constructs changes. The student reflection tools we are designing assist students' in identifying such changes. The first step in such a process, before answering questions specifically about dispositions during a period of time, is for students to define meaningful units of activity to which dispositional labels can be attributed. To coordinate these profiles with those more closely related to content modeling, students must be directed to explicate what they believe their development cycles were for a particular modeling problem (e.g., when they started, what aspects of their activity characterized each cycle, etc.). After identifying their cycles, the students can then engage in thinking that helps them to explicate the nature of their interest, identity, and social functioning in each cycle. The tools are designed ultimately to be answered on a technological device (likely a graphing calculator) that allows the students responses to be tallied and displayed as interactive feedback through a visual image of their profile.

In general, the system we envision needs four classes of prompts. First, students need to understand that they may be able to change the level of stimulation and control afforded by their involvement in the activity to be in their favor for optimal learning. This requires a set of diagnostic prompts to ascertain whether or not stimulation and control tradeoffs are optimal. If levels are not optimal, the next few prompts should refine the diagnosis to allow the student to state what aspects of the situation are under their control, which are not, and to project how to gain control over their involvement.

Second, students need to understand that their modeling activity appeals to real communities of practice. Unfamiliarity with certain fields in the sciences, social sciences, or business realms within which tasks are designed violates the reality principle of instructional design which affirms that tasks must be meaningful in terms of what students bring to the context (cognitively and socially), and in terms of how they perceive themselves in relation to the context (Lesh, Hoover, Hole, Kelly, & Post, 2000). A set of questions that allow students to voice their level of understanding of the practices of the various communities, including the epistemology of the domain of mathematics, will enable them to compare their own work with the kind of work that would be legitimized by those communities (i.e., what part they are playing with respect to mathematically sophisticated fields). This allows students a way of ascertaining the level of legitimacy of their peripheral participation (e.g., Lave & Wenger, 1991), even when the teacher may not be a practicing member of a community. In short, such prompts must allow students to address the clichés, "When would anyone ever use this?" and "That's not how they do it in the real world?"

Third, students need a nontrivial way to understand the dynamics of their group and assess the effectiveness of their participation in relation to others. Understanding that any conglomeration of individuals contains distributed expertise and how this expertise can be exploited to produce quality work is a vital feature of design teams in engineering, business partnerships, and academic research projects. Prompts that enable students to see their unique (potential) contribution to the collective knowledge building, and acknowledge the contributions of others, should allow them to adopt more meaningful and efficient personae than providing only generic roles.

Fourth, because all of this is new stuff, such a system needs to provide a small bank of suggested strategies to get kids started in taking charge of their own learning. How does one gain control over their engagement in a model-eliciting activity? How might I learn more about forensic medicine so that my report is of high quality? How do I as a visual thinker communicate with the human computer across from me? These are real dilemmas that, as of yet, we have little empirical work to go about providing justifiable answers.

A Reflection Tool Example: Interest

Interest can be shown to be both categorical, storing information about the accumulated value of tasks and pursuits for the purpose of making deciding whether or not to engage in a task in the first place (e.g., labeling a pursuit as an interest), and adaptive as it can be modified in the process of engagement to reflect whether or not one would want to continue one's engagement (Middleton & Toluk, 1999). The first feature of Interest is often termed interestingness. The second feature is often labeled *situation-specific interest*. The degree to which a pursuit is deemed interesting (i.e., has Interestingness) is determined in large part by consistently experiencing situation-specific interest (Hidi & Anderson, 1992). Clearly to impact the former, feedback must focus on the latter.

There are two primary task-specific determinants of interest: *stimulation* and *control*. These two factors exist in a tradeoff where too much of one equates with not enough of the other. However, the relative levels of each considered optimal varies across students: What may be considered highly stimulating and possible for one child may be anxiety-producing in a second, or boring for a third. How then can a child get a handle on their own involvement in a task such that when levels of stimulation and control are not optimal, they can alter them in such a way as to make their engagement worthwhile? The following flowchart (Fig. 22.3) depicts our first attempt at the general structure of an interest tool.

Notice that each prompt moves the student closer toward a set of appropriate suggested strategies for optimizing their levels of stimulation and control. In the case where the levels are currently optimized but the student still dislikes the task, the student is referred to the teacher to diagnose whether the conditions eliciting the response were task-related or related to identity or social functioning.

Ideally, students will utilize this tool immediately after they have finished working through one modeling cycle. Pragmatically, it may be possible only following completion of several cycles or a complete model eliciting activity. To reconstruct their learning history, students begin by identifying how many cycles and the types of cycles that they progressed through in addressing the modeling problem. Then, for each of these cycles, they would answer the questions that appear in the Interest Tool. To answer each of the questions, the students will either move a slider to reflect the level of their response or select from a list of visual graphics that best characterize their disposition.

22. CENTRAL FEATURES OF MODELING ACTIVITY 429

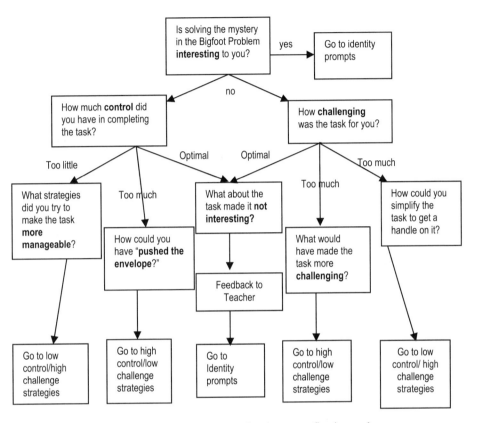

FIG. 22.3. Initial structure of an interest reflection tool.

Finally, as a functional part of classroom discussion, students are asked to reflect publicly about how to recognize positive patterns of engagement, diagnose difficulties with dispositions to learn mathematics, and project together specific strategies that can bring the interest, identity, and social functioning of the class up to optimal levels. Questions that address these issues might look like the following thought questions:

1. What did I notice about the way I spent my time during the different cycles that might be helpful in solving the next problem?
2. What did I notice about my reasons for working on the different cycles that might be helpful in solving the next problem?
3. How did the way I spent my time differ in the various cycles?
4. Would it have been more helpful to spend more or less time in Cycle 1 (or 2 or 3)?
5. Would it have been more helpful to spread out the time that I was focused on the problem in Cycle 1 (or 2 or 3)?

6. Were most of my reasons for working on the problem in Cycle 1 (or 2 or 3) more extrinsic or intrinsic?
7. What was my main reason for working on the problem in Cycle 1 (or 2 or 3)?
8. How did my reasons for working on the problem change from cycle to cycle?
9. How could I use these questions to be more successful on the next problem?

By examining students' interpretations of such feedback through observations, interviews, and classroom discussions, we hope to examine how students can refine their dispositional knowledge components in much the same way as they refine their content components. Ideally, as the students complete the reflection tools over a series of modeling problems, the tools allow the students to go beyond merely thinking about how to solve a problem, to consciously thinking about how to grow in both knowledge and disposition as they struggle to generate viable transportable models of complex mathematical phenomenon.

CONCLUDING REMARKS

In the early stages of education research, intelligence was thought of as a single, general, ability that sometimes was called general intelligence (G). It was an attribute (i.e., a label) that was associated with a given kid. A child could be considered intelligent to some degree, and instruments like those devised by Binet and Terman could measure the degree to which the child held the intelligence attribute.

Next, Howard Gardner (1983), Guilford (1967), and Sternberg (1985), pointed out that the myriad ways in which people could perform intelligently cannot be collapsed into a single factor. So, the state of the art evolved from a single-number characterization of a person to a vector-valued characterization of that person (e.g., G $\rightarrow$ (A, B, C, D, ...)). This innovation also pointed out that many of the most important attributes of intelligent behavior involved social and emotional intelligences as well as cognitive intelligences. However, even though Gardner himself claimed that these intelligences cannot be treated as simple hardwiring, because they were called "intelligences," and because they still focused on general abilities within a limited number of conceptual domains, they continued to be treated as labels for a given kid.

Next, Perkins, Jay, and Tishman (1993), emphasized that the multiple possible intelligences people could exhibit were more a function of knowledge than genetics. This perspective asserts that intelligence develops over time, and that intelligence can be influenced through instruction (or experiences). Therefore, intelligences are no longer considered fixed and invariant labels to be attached to individual kids. Rather, they are components of knowledge. Moreover, the perception that one's intelligence is indeed malleable, and that one can become smarter through hard work in fact leads people to develop healthy and productive attitudes towards academic tasks, and in turn this engenders greater success for people holding this perspective than those who believe that intelligence is fixed (e.g., Dweck, 1986).

Our own innovation to this history developed over the past three years as we contemplated writing this chapter. In particular, we extend the preceding point of view by explicitly treating the concept of multidimensional vectors as constituting components of the models that students use to interpret learning and problem solving situations. The principal difference between our view and the general view of the literature is that we shift attention beyond stable knowledge components toward interpretations that vary from situation to situation and from one phase of problem solving to another within a given situation. So, we are no longer labeling kids, or even a kid's knowledge, we are describing vector values and variable functions that mark out a trajectory of development within a complex domain like mathematics.

This development is all well and good for the field. However, there is a more practical aim to our current work. Namely, we wish to utilize our understanding of the ways in which interest, identity, and social functioning drive the development of mathematical content learning. The development of reflection tools should help students recognize patterns of behavior and affect that impact the ways in which their goals and aspirations play out in the field of mathematics, and how these behaviors and affect can be used as a tool for achieving and maintaining a sense of present and future efficacy.

To return to the statement quoted from Vygotsky (1962), if mathematics education, and its accompanying research agenda is divorced from the fullness of life that mathematics and its applications purport to enhance, then wherein lies the motivation for students to continue their relationship with mathematics and mathematics-related fields beyond those courses that are compulsory? Moreover, to what extent can mathematics educators make the ethical stance that our field has value either for the hand-to-mouth motivation of solving a particular problem, or for larger goals of improving society? If the problem is not meaningful, or the goals too abstract, then we cannot.

However, if we take seriously the evidence that shows us that children are both intellectual and affective beings, and that these two facets of intelligence coemerge, then, perhaps, we can design learning environments that are meaningful, that have value, and that ultimately tap into the fullness of life.

V
MODELS AND MODELING BEFORE AND AFTER MIDDLE SCHOOL

Chapter 23

Beyond Constructivism: An Improved Fitness Metaphor for the Acquisition of Mathematical Knowledge

Susan J. Lamon
Marquette University

One of the most compelling needs in mathematics education is ensuring that all students have access to high quality, engaging mathematics. In 1985, when mathematics education was focused on mathematical problem solving, there was a consensus among researchers that ATI (aptitude–treatment–interaction), cognitive correlate research, and work from other research paradigms using psychometrics (the measurement of mental abilities)—however useful these were to the understanding of human cognition—had been misused in education. Particularly in mathematics education, the assumption that some people have the ability to learn mathematics, whereas others have less or none, was used to sort and track students. Mathematics educators were called to promote a "heightened sensitivity to individual differences" (Silver, 1985, p. 259), but 15 years later, Kilpatrick and Silver (2000) reported that the notion of innate ability was so entrenched that little progress has been made in addressing individual differences and equity issues.

The equity principle in the new mathematics standards states that "reasonable and appropriate accommodations must be made as needed to promote access and attainment for all students" (National Council of Teachers of Mathematics, 2000). None of the suggestions for promoting equity, however, are under the control of the individual classroom teacher. Sufficient material, human resources, and access to technology may be important, but how, specifically, does the individual teacher promote equity and diversity? The history of mathematics education has taught us that when rhetoric lacks guidelines for translation into practice, that translation may be made in simplistic and unproductive, hand-waving ways. Almost any existing practice will be rationalized to a good fit. When I talk to teachers about constructivism and ask them how they have modified their instruction to accommodate individual construction of knowledge, the response almost always, is that they have their students work in groups with hands-on activities, and then they facilitate a post-investigation discussion. The class reaches a consensus and the students have constructed for themselves the mathematics that took thousands of years to produce. Wait a minute! How does this take into account the different knowledge structures with which students come into instruction? Does the construction of knowledge happen merely because of the classroom organization? The manipulatives? What role does content play here? How do we know that we are giving all students access to mathematics? How does the teacher build a model of the students' conceptual structures and track changes? In this chapter, I argue that classroom interpretations of constructivism are not necessarily headed in a useful direction and that before the pendulum of reform sweeps too far to the right, it may be time to consider alternative, but not necessarily competing, perspectives on the development of mathematical knowledge. As I study children's construction of content knowledge in a complex domain, I find that it is counterproductive to ignore students' individual differences before and during instruction. They may be ignored, but they do not go away. A new longitudinal study focused on children's development of rational number meanings and operations (Lamon, 1999, 2001a, 2001b, in press) encouraged diversity by allowing students to build fraction meaning on the strengths and ways of thinking that they showed before fraction instruction. For 4 years, in grades 3 through 6, five cohorts of students each built their understanding of rational numbers on a different initial interpretation of the symbols a/b: as a part–whole comparison, as a measure, as a quotient, as an operator, and as a ratio. This study provided a unique opportunity to compare unconventional approaches with traditional instruction and to document the sequencing and growth of ideas, the breadth and depth of the understanding that developed, and the connections children were able to make across the domain of rational numbers. The ability to track learning in the same group of children for an extended time gave a much different picture than we may have gotten from short-term intervention studies or by using cross-section snapshots of students in different grade levels. Children's work from this study supplies supporting evidence for the assertion that eliciting and managing diverse interpretations and ways of thinking is the proper goal of mathematics education. The microevolutionary process of producing and abandoning conceptual tools in favor of more useful ones reflected the students' multidimensional growth in the rational number terrain and suggested an improved fitness metaphor for the development of mathematical knowledge.

ENCOURAGING AND MANAGING DIVERSITY

Instruction that focuses on the forest but not the trees, instruction that emphasizes what children's minds have in common and deemphasizes the ways in which they are different, is incongruous with the rest of today's world. The culture of our technological age invites, embraces and manages diversity. The Internet forces us to develop systems for accommodating all of the information

23. FITNESS METAPHOR FOR KNOWLEDGE ACQUISITION 437

with which we are bombarded. We make audio and video contact with people from all around the world, and almost daily updates force us to be flexible and comfortable with diversity and change. Nonetheless, conformity remains the hallmark of many aspects of mathematics education and of mathematics education research. In most teaching experiments, children are assigned randomly to different treatments, or treatments are applied to intact classes. The goal is to show dramatic, short-term gains, when all children, regardless of their backgrounds, have been fit into the same mold. And those gains always show up! How many times is it reported that children who initially understood X ended up understanding Y, and those whose preinstructional understanding was W ended up understanding Z? Similarly, in fraction curricula, the part-whole interpretation remains the chief interpretation used in introductory fraction instruction, with no other rationale except that it is the way we have been doing it since we started teaching shopkeeper arithmetic.

Suppose children were screened before instruction and then provided different inroads to a complex mathematical domain, say rational numbers, based on the thinking styles they had already developed up to that point? Suppose that students were provided the time to express their thinking, test it, and refine it—not for just one class period, not for just a year, but for 4 years—until they grew into more powerful ways of thinking? Suppose that students were not taught any rules. What would develop? Suppose students were not compared to one another, but only encouraged to develop unique personal landscapes, rich in rational number understandings and connections? For 4 years, a radical teaching experiment on rational numbers encouraged such diversity and multiple interpretations.

By the start of third grade, children are already diverse in their mechanical abilities, their perceptual abilities, their allocation of attention to different aspects of a situation, their criteria for making decisions, their expenditure of effort depending on purpose or perceived usefulness; and the degrees of confidence with which they make statements about knowing. They rely on their past experience—sufficient or not—for making probabilistic decisions and arbitrary and convenient explanations are proxies for necessary and sufficient conditions when asked for proof. Nonmathematical concepts and practices affect their approach to mathematics, among these family and cultural practices, fairness, social customs, and vocabulary for which they have developed meaning elsewhere.

On what basis might we go about matching children's existing ways of thinking to one of the five fraction interpretations? Beginning with the mathematical ways of thinking that were critical to each of the interpretations, questions were devised to probe the children's thinking and all of the children were interviewed just before the start of third grade to determine what kind of ideas they already had about each interpretation. For example, the fraction as an operator is intimately tied to composite functions, fractions as quotients are more closely related to partitioning than to comparing, whereas ratios are more closely related to comparing than to partitioning. In fact, it was possible to determine a match between each interpretation and the children who had a predisposition to think in a certain way that would aide their understanding of that interpretation. Two strong preferences were noted. Some of the children showed a distinct ability to think of multiple steps in a composite way, whereas most children merely concatenated steps. These children generally fell into two groups—concatenators or composers—according to the way they answered the questions such as those in Appendix A. Likewise, children answered the questions in Appendix B quite differently, distinguishing themselves as sharers or comparers. Most were able to partition fairly, but some were clearly better at comparing.[1]

The point is that already by the start of third grade, children were predisposed to thinking in different ways and their individual differences were taken into consideration by matching existing ways of thinking with the demands of the different fraction interpretations. It is a patently wrong-minded notion that instruction can offer a single meaning for the symbol $\frac{a}{b}$ that would make the same sense to every child. Instead, one of five rational number interpretations was matched to each child's ways of knowing and thinking so that instruction could build on their strengths.

Admittedly, it is not going to be practical for a teacher to interview every student in every classroom. However, there has to be some viable way for a teacher to determine what goes on in each student's head and to build up a model of the student's conceptual structures, or else there is no basis for deciding whether change is occurring or not, and whether it is occurring in a desired direction. One way to do that is to devise questions that admit multiple inroads to a mathematical system.

Many years ago, Jerome Bruner (1963) was highly criticized and misinterpreted when he claimed that "Any subject can be taught effectively in some intellectually honest form to any child at any stage of development" (p. 33). Essentially he distinguished mathematical sophistication from level of representation. In my own research, I have found his statement to be true. Most of the big ideas in mathematics may be understood at first, on a simplistic level, but over time, there occurs a broadening and a deepening of the idea; it is always possible to know it in a more sophisticated way. Take, for example, the problem in Appendix C, CROWS! Third grade students can solve the problem simply by using plastic gears and a counting strategy. But because a rate is such a complex mathematical idea, it develops over time, building connections to many other mathematical ideas and gaining in sophistication. There are levels of understanding accessible to very young students, as well as more advanced levels appropriate for college-age students. Appendix D gives a partial list of

---

[1] The children who were better at comparing tended to be younger siblings in their respective families, rather than first-born children. My theory is that the need for younger siblings to hold their own in their respective families might make them better comparers. Mothers frequently give snacks to the older child with the directions to share them with a younger brother or sister. After being shortchanged a few times, the younger child begins to monitor the division of the goods.

the kinds of solutions that students produce to CROWS!, ranging from counting solutions to calculus solutions.

An important feature of the problem is the second question that pushes students beyond the first answer-giving stage to adopt a higher level perspective, often a more generalized approach. These questions work very well in group investigations, but the third feature, the report, is completed by each student independently. Like adults who have witnessed the same crime scene and then give substantially different reports to the police, different children report different results from the investigations in which they have participated together. In the reports, the teacher can determine, on an individual basis, how sophisticated a student's understanding of rate really is.

Other chapters in this book elaborate on the benefits and construction of this type of model eliciting, model developing, and model assessing problem. What is significant about these problems is that because they can be solved at many different levels of sophistication, every child has access to the mathematical content. Children's diverse backgrounds are all accommodated, as many different children solve the problem, the teacher gains perspective on the aspects of improvement. Even more important than the kinds of mathematics that may be applied to the situation (these may be predicted even before giving the problem to children), is that teachers gain the insight to become psychologists of the subject matter. How do children think about mathematics? How does that thinking develop? And how can that information be used to better design instruction?

TOOLS AND MENTAL SYSTEMS

The longitudinal fraction study provides interesting insights into the nature of tools and what they can tell us about the underlying mental models that govern their use. A mental model is a system that a person uses to explain real phenomena. It consists of mathematical elements or objects (e.g., quantities, ratios, rates, functions, percentages, averages, probabilities, or derivatives), ways of thinking about, reasoning with, and operating with those objects, and some information about when to bring them into play (patterns, regularities, contexts, etc.). A mental model usually functions at an unconscious level, but nonetheless, underlies the manner in which a person thinks and acts in any given situation.

Tools come into play when a person has a purposeful task to accomplish. A tool may be a widely used instrument or implement, such as a ruler. A tool may be a shell, such as computer software, that is imbued with meaning by the user. A tool may be a unique contrivance, a "Rube Goldberg," designed to get some job done. Another kind of tool is a constrained result, one commissioned by someone else, or designed to fulfill some specific need. Clearly, there is no all-purpose tool.

The usefulness or value of a tool depends in part on the user. We typically do not call something useful in the hands of an unskilled person. Thus, a set of manipulatives, say, pattern blocks, may serve as a tool for the teacher, but not necessarily for the students. Algorithms are tools for some people and not for others. We typically do not use tools until we have a need for them—a need to describe, to explain, to build, to justify our thinking, to solve a problem, or to make a prediction—and we intend to find a way to meet that need.

Tools may be used incidentally in getting to a result and then forgotten, or they may be used promote reflection about the system to which they are attached. A conceptual tool is a device that captures a student's thinking, thus making imagination, categories, and all sorts of mental routines and connections sharable and reusable. We accept the risk that some details are lost in the translation. In the classroom, a conceptual tool serves the dual purpose of providing the teacher a lens for viewing part of what is going on inside the student's head, and provides the student a means to consciously access part of his thinking so that reflection and revision are possible.

Diverse as we are, there seems to be a universal force among us to constantly better the tools we are using. As pleased as students are to share their tools, they develop no attachment to them. They are further able to embrace the mathematical modeling perspective that tools and solutions that work very well under one set of conditions do not work as well under different circumstances. Good ideas and workable tools are used again, perhaps with some modification, and the "Rubes" just fade away.

What kinds of powerful mental systems do students have at the start of third grade? Most people would be surprised to learn that such young children have some highly developed systems by which they interpret their world. Take, for example, the comparers identified in the fraction study. Their work and interviews before instruction suggested that they were predisposed to looking at the world in a comparative way and throughout instruction in the ratio interpretation of the symbol $\frac{a}{b}$, which was designed to build upon their predisposition for comparing, they showed some remarkable thinking. The student thinking recounted in Appendix E is a fifth grader's response to a rate question: Who grew faster? Asking children challenging questions often reveals their fledgling status with more complex mathematical ideas, and with this information the teacher or researcher can plan the next step to push the student into new territory. He used a ruler to help convey his point as he struggled to explain himself. The ruler was a tool but it was not measuring inches or centimeters. It was, in this instance, a handy devise for demonstrating steepness, and for providing others a window to the student's thinking. Although he did not have the words, the student's representation of the situation revealed that he saw some connection between slope and rate of change. In subsequent instruction, the teacher introduced the conventional vocabulary—rate of change, slope, and so forth—to help this student talk about the situation, and because the mental framework was already present, the words became useful tools for him.

Evidence of a mental system or model consists of the meaningful and appropriate use of tools, a language, actions, and reasoning (ways of thinking). One characteristic of the actions and language and ways of reasoning is that they are not mechanical. They are clearly not repetitions of other people's words or actions. Appendix F demonstrates these features in the work of the ratio class in the longitudinal study. It is curious that many adults who study these examples cannot understand what the children were doing. This, I claim, is further evidence for mental models. Those who do not interpret the world in a comparative way may not have constructed a model similar to the student models, or their ratio model may not be developed to the same level of sophistication.

THE DIMENSIONS OF MATHEMATICAL KNOWLEDGE

To facilitate students' development of powerful ways of thinking in mathematics, teachers need to know the dimensions along which improvement reveals itself. What is powerful and what is not? What is good and what is better? What constitutes movement in the "right" direction?

The general stages of human cognitive development proposed by developmental psychologists such as Piaget are not specific enough to inform instruction in content domains. Research in most of the complex mathematical domains is not advanced enough to fully support the task of psychologizing the content for instruction. How then can teachers carry on their work? Good teachers have probably always known that the real issues is not "Does Johnny know X?" but rather, "How does he know it?" Children are clearly capable of using sophisticated mathematical ideas although they may lack the vocabulary to talk about them and the symbols to capture them. This observation necessarily causes problems for those who seek to describe such abstractions as learning, understanding, and knowing. But for the purpose of instruction, what learning is, is less important than recognizing it when it occurs and noting the conditions that facilitated its growth.

How many other dimensions are there to the many ways of knowing a particular piece of content? In general, children's mental models, as reflected in their tools, action, speech, etc., change along many dimensions: from primitive to sophisticated; simple to complex; concrete to abstract; intuitive to formal. Sometimes these changes occur over long periods of time, although they may be observed in briefer problem-solving episodes. A longitudinal fraction study revealed that mathematical content may be so cognitively challenging for children, that even some of the historical, basic tenets of qualitative change in cognitive processing, such as conservation principles, may be called onto shaky ground as children rearrange what they know and the way in which they know it.

Under different rational number interpretations, children acquired meanings and processes in different sequences, to different depths of understanding, and at different rates. Ending profiles were as diverse as beginning profiles. The time-honored learning principle of transferability was robust. Not only did children transfer their knowledge to unfamiliar circumstances, but to other interpretations that they were not directly taught. After 4 years, every child had a knowledge of fractions that extended beyond the traditional part—whole interpretation. As their mental models improved, students were able to move:

- from making judgments based on empirical evidence (often, the ways things look) to making judgments based on logical connections (logical argumentation),
- from using one interpretation to freely moving among several interpretations (complexity),
- from making broad, general statements about the way things work to making nuanced judgments and interpretations (accuracy),
- from focusing on global characteristics to noticing both global and specific components of a given situation (balance),
- from stating the obvious to seeking underlying structure (depth),
- from making redundant statements to making more focused statements (precision),
- from weak conceptions of what constitutes evidence to distinguishing necessary and sufficient conditions (proving),
- from prematurely closing thought processes to engaging in long-term investigations (persistence),
- from focusing on a single all-purpose solution to recognizing conditions under which other solutions are feasible or preferable (flexibility),
- from imitating to using clever, creative twists (ownership),
- from searching for anything that works in a single situation to developing and improving techniques with broader applicability (viability).

The student work in Appendix G demonstrates many of these characteristics. Their processes for naming fractions between two given fractions are unique; they show ownership of the ideas and procedures, evidence of deep understanding. Martin's techniques show unusual flexibility and viability.

These results suggest that there is a great deal of danger in overworking the assumption that children's learning proceeds through stages that can be observed and reformulated into an instructional sequence. Implicit in associational and stage metaphors is the assumption that children's thinking is highly structured. But this perspective is remiss in that it fails to distinguish between the brain and the mind. Mechanical and stage metaphors overplay the innate machinery that humans have for learning and minimize the real beauty of the human software—that our biological makeup provides the basis for a mind that is creative, adaptable, richly structured, capable of producing new ideas, and constantly seeking to better its own ideas. Our minds are combinatorial and recursive (Pinker, 1997) and this means that with a finite number of models, there are countless possibilities for combining simple pieces of knowledge, growing and reorganizing systems, and embedding system within others. For example, the children whose work appears in Appendix H had never been taught any rules for dividing fractions, but when they were asked how they might divide 1 by $\frac{2}{3}$, their mental models proved strong enough to allow them to think about the problem. When we see such remarkable work, it is difficult to continue to second-guess students and to decide what they are capable of doing.

Many of the generalizations about stages and order of learning in mathematics education have been made by taking snapshots of children's thinking. A persistent problem in snapshot research is that a child's actions at any particular time really depend on many causes—constraints and affordances in his immediate environment. What the child might have done under other conditions, over longer periods of time, goes undiscovered. Stages are based on a desire for uniformity rather than on a tolerance for diversity. Certainly, belief in stages and in orderly learning can make the diversity inherent in every classroom seem more manageable, but research in the complex domain of fraction learning suggests that linear models of development are impoverished. They neglect the richness of the landscape traveled by students as they build mental models in complex domains. For example, consider the work of the children in Appendix I. In the longitudinal fraction study, children with highly developed notions about measurement named and renamed fractional parts using a process called unitizing, or thinking in terms of different-sized chunks. They were guided by their strong understanding of the compensatory relationship between size and number of pieces. In particular, notice Brad's comparison of $\frac{11}{12}$ and $\frac{17}{18}$. The research literature suggests that a comparison of this nature would be the most difficult for students because of the lack of integral multiples between and within the fractions (Karplus, Pulos, & Stage, 1983). Yet, for children like Brad and the others in his class, whose thinking was anchored in a strong measurement system, the presence or lack of numerical relationships did not affect their ability to compare fractions. This is just one example of the fact that recognizing the power in alternate ways of thinking may, in fact, challenge certain conventions, expectations, results, and structures associated with mathematics and education.

VACCINATIONS AND THE GROWTH OF MATHEMATICAL KNOWLEDGE

It is well known that a child who has had measles or chickenpox is immune to further attacks of the disease and, better yet, even in the absence of a disease, bacteria administered in the form of a vaccine protect the body against future attacks from that specific strain of bacterium.

What do equity, diversity, and vaccinations have in common? No, this is not the opening line of a bad joke, but rather, an invitation to step into a different world for just a few moments. As difficult as it is to read technical language outside of one's field of expertise, there are some remarkable correspondences between current literature in mathematics education and the work of scientists studying biological evolution. (See the links provided in Appendix J.) right down to some of the terminology. Furthermore, couched in the language of natural selection is a description of children's problem solving in complex domains.

Think of an antigen as a foreign invader, such as a toxin, a virus, or a bacterium. Antibodies are substances in an animal's blood that appear after the animal has been infected with a disease. Antibodies protect the animal from future infection by the same kind of invader by binding to the antigen and labeling it for destruction. Proteins then destroy the antigen by perforating its cell membrane or by completely engulfing it.

For over 100 years, scientists have been studying the response of the immune system to antigens. The dilemma is this: There has to be a precise fit between antibody and antigen. That is, antibodies are very specific; they can bind to only one type of antigen. Nevertheless, there is no apparent limit to the number of antibodies that an animal can produce. How can a finite genome possibly hold the information necessary to produce an infinite number of possibly necessary antigens? An animal can even produce antibodies when novel, artificial substances are introduced!

One major theory that was advocated for about 25 years, was the *instructionist* theory, also known as the *template* theory. Essentially, it was proposed that, rather than obtaining information from within the animal, the antigens themselves served as a kind of template for producing antibodies, much like a tailor would construct a suit of clothes using the customer as a template.

Eventually, the instructionist theory ran into many difficulties and was replaced by a more explanatory theory, sometimes called a constructivist theory. An animal cannot be taught to make specific antibodies unless it already has some capacity to make antibodies of that nature before the antigen arrives. Thus, scientists conclude that antibody formation is a selective process and that instructive theories are wrong. It is also a constructive process, because building on its innate information, the immune system responds to a vaccination with a rapid and random form of mutation in antibody genes. In short, this antibody response is a speedy and miniature form of evolution. A highly diverse population of antibodies is produced as the result of a random shuffling of genes. The diversity generated by genetic variation ensures that at least one antibody will be produced that is able to bind with an antigen and serve as the raw material for natural selection. When the mutations produce antibodies with high affinity for antigen, these antibodies begin to clone themselves and build resistance to the disease. As they remove more and more antigen from circulation, cells with less affinity die off.

What is significant is that this process entails natural selection for progressively improved biological function. Improved biological function, rather than some end result targeted by an intelligent creator, drives the process. This remarkable process occurs without the services of a benevolent, providential instructor. Furthermore, this is not a theoretical simulation, but a natural biological phenomenon, evidence of a beautifully functioning system at work.

The process begins when someone outside the system introduces a problem that needs to be attacked. Initially, there are diverse, competing reactions. Once the reaction begins, outside intervention is not needed to keep it on course. Natural selection does not steer the course, watching over the process like a teacher over children, ready to crack their hands with a ruler if they go off on a tangent. Many more antibodies are produced many of which play an active role in the immune system. The more effective ones eventually overpower the reactions that are too weak to do the job. The attack builds strength over time. The powerful connections endure.

This biological evolutionary process is a metaphor for the development of mathematical knowledge. And two of the prerequisites for natural selection are enough variation and an enough time (Pinker, 1997). This was certainly true of children in the longitudinal study. For the first few years, from grades 3 to 5, in comparison to the children who received traditional part–whole fraction instruction, the special classes appeared not to be making much progress. However, when they had finally found processes that worked and these were internalized, they had more powerful ways of thinking than they children who had been taught to execute traditional algorithms.

CONSTRUCTIVIST AND SOCIOCULTURAL PERSPECTIVES IN THE CLASSROOM

Current translations of a constructivist epistemology into mathematical pedagogy are based on the assumption that children can reconstruct mathematical ideas and processes for themselves if they can be engaged in a social enterprise driven by investigation, negotiation, and renegotiation of the meaning of concepts. Both cognitive constructivist perspectives (Davis & Maher; Noddings, 1990) the Vygotskian sociocultural perspective that views mathematics as an enculturation process (Bishop, 1988), emphasize interaction and negotiation. These theories no matter how skillfully they have been employed in research (Cobb, 1995) have been reinterpreted and abused in most classrooms to the extent that more attention is paid to the learning environment, tools, and teaching methods, than to what is being constructed and the purpose for constructing it. Teachers have no sense of the demographics of change. Which children and how many of them are doing the constructing? How quickly is the change happening? What does the group project reflect? How enduring are the ideas students produced? Do the children really own the ideas or were they spread like rumors? Perhaps the most disastrous result of the skewed emphasis on form (classroom organization and procedures) is the loss of regard for individual psychology. Thus, matters of diversity and equity are overlooked.

Negotiation assumes a certain degree of social competence. Social competence includes the ability to see yourself as an initiating and controlling agent, as opposed to being a responder to forces in your environment. This is very different from intelligence, or creativity, logical reasoning, and other qualities associated with problem solving and higher mathematics. On one hand, interaction with others may promote social competence by helping students to adopt multiple points of view and tolerance of perspectives other than their own, but on the other, large segments of a young student population may be left out of classroom negotiations, even when they are capable of mathematically sophisticated thinking, because of social immaturity.

After a certain time for investigation and group work, discussion often assumes that all of the children have had the same experience and development and are now ready to attach appropriate mathematical terminology and symbolism, but we know that different children have different experiences of the same activity or investigation. Thus, the discussion or negotiation often shows little regard for individual differences in background, perception, developmental level, or mathematical sophistication. In fact, it is assumed that the students who are confident enough to engage in the whole class discussions are speaking for the class.

Because children do not constitute a real mathematical community, the teacher is seen as the representative of the mathematical community. The teacher's dual role of intelligent designer and culture-bearer make it nearly impossible to prevent "steering" the discussion to a predestined conclusion. Most often, in whole class discussions, the teacher picks up on all of the useful germs of ideas and helps the class to assemble them into the picture that he or she holds as the ideal conception for the students. The purpose of the discussion is to smooth out the individuals differences in understanding the situation and bring them more quickly into line with conventional ideas, methods, and symbols and language. This variety of constructivist pedagogy certainly contrasts with the constructivism in the previous discussion of natural selection and immunizations.

Children, in fact, naturally produce good ideas, but often their ideas do not come "fully baked." Ideas may need debugging and often go through several rounds of revision. Furthermore, the revision does not consist of random substitutions. Each student needs to focus his or her individual brain power on improving some product in order that value-added iterations may result. In complex mathematical domains, value-added iterations of children's good ideas are time-consuming and they do not spread like contagious ideas or rumors. The fact is that if the mind were so passive that it would accept anything that is handed down the line, it would have died out in the process of natural selection. Certainly more than classroom negotiation is required to build useful mathematical knowledge.

THE BOTTOM LINE

What a person really knows—and what really matters in mathematics education—is an individual internalization of interpersonal practices and cultural knowledge. Culture may affect psychology in the sense that a child is developing mathematical ideas in coordination with others in the classroom, but

23. FITNESS METAPHOR FOR KNOWLEDGE ACQUISITION 447

emphasizing the communal nature of learning as if it some panacea and neglecting individual knowledge building is a mistake. Teachers are in the mind-building business. Unless their efforts are based on knowledge of children's thinking before and during instruction, everything they do is haphazard.

Anecdotes from classroom where model eliciting, building, and documenting problems are used are always the same. Children's work is of higher quality than the teacher ever expected it to be, and the children are intensely motivated when they are immersed in the problem. For the student, the idea that endured, and won out above all the rest because it worked better, is personally empowering.

Chapter 24

Using a Translation Model for Curriculum Development and Classroom Instruction

Kathleen Cramer
University of Minnesota

Other chapters in this book emphasize a variety of ways that representational fluency is an important component of students' models and modeling abilities—and an important part of what it means to understand basic mathematical constructs in topic areas ranging from early number concepts, to rational number concepts, to concepts in algebra, geometry, probability, statistics, or calculus. The purpose of this chapter is to report some ways that the National Science Foundation supported Rational Number Project (RNP) has used the following translation model (Lesh, 1979) to develop curriculum materials and classroom activities that are aimed at helping both students and teachers develop deeper and higher order understandings of some of the most important ideas in the school mathematics curriculum.

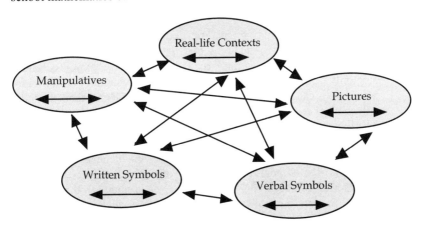

FIG 24.1. The Lesh Translation Model.

The Lesh translation model suggests that elementary mathematical ideas can be represented in five different modes: manipulatives, pictures, real-life contexts, verbal symbols, and written symbols. It stresses that understanding is reflected in the ability to represent mathematical ideas in multiple ways, plus the ability to make connections among the different embodiments; and, it emphasizes that translations within and between various modes of representation make ideas meaningful for students. Thus, the Lesh translation model extends Bruner's theory by adding to his three modes of representation real-life context and verbal symbols. The Lesh Model emphasizes interactions within and among representations. For example, the arrows connecting the different modes depict translations between modes; and, the internal arrows depict translations within modes. The model suggests that the development of deep understanding of mathematical ideas requires experience in different modes, and experience making connections between and within these modes of representation. A translation requires a reinterpretation of an idea from one mode of representation to another. This movement and its associated intellectual activity reflect a dynamic view of instruction and learning.

BACKGROUND

Confusions sometimes arise because of the multiple ways in which mathematics educators use the term "model." Therefore, it is useful to sort out the following three types of models that occur in the context of the Lesh translation model that is shown in Fig. 24. 1:

- First, the diagram as a whole often is referred to as a model. That is, the diagram describes a model that researchers (or teachers, or curriculum developers) can use to think about the nature of productive learning activities for children. It is not a model that students are expected to use.
- Second, within Fig. 24. 1, specific diagrams, concrete materials, or symbolic descriptions (equations that describe everyday experiences) also may be referred to as models. But, in this case, two possibilities commonly occur: (a) models that teachers (or curriculum developers, or researchers) can use to communicate with students about specific mathematical constructs, or (b) models that students can use to interpret (describe, explain) learning or problem solving experiences.

The preceding three uses for the term model are not inconsistent. In each case, the symbols (or words, or diagrams, or materials) are useful for constructing, describing, explaining, manipulating, or predicting other systems; and, in each case, the relevant person (student, teacher, researcher, or curriculum developer) uses the symbols (or words, or diagrams, or materials) as external embodiments of internal conceptual systems. So, in each case, the relevant models have both internal and external components. Yet, in much the same way

24. TRANSLATION MODEL

that a photograph of a chair may be referred to as a chair even though everyone knows that it is not a real chair, we may refer to equations, or diagrams, or sets of concrete materials as models even through they are only the external embodiments of the relevant internal conceptual systems.

Mathematics by its very nature is an abstraction. The requirements for effective learning do not lend themselves to the direct learning of abstractions. Rather, learners rely on embodiments or external representations of the constructs and conceptual systems to be learned. From these embodiments, students are encouraged to abstract and to generalize so that the relevant concepts to be learned exist without dependence on the representations used to foster their development and acquisition. Embodiments of mathematical ideas can vary in complexity. For example, chips can be used to embody addition as the physical union of two sets. Base 10 blocks can be used to represent whole numbers or decimals. A table of values or an algebraic equation can be used to describe quantitative relationships among variables; or, pictures of chips, Cartesian graphs, or diagrams can be used to represent other mathematical constructs.

Similarly, the Harvard Calculus Project has used tables, graphs and algebraic equations to provide learners with various models in the analysis of mathematical functions. This curricular approach has been termed the "rule of three"—or more recently, the "rule of four" to include student verbal interactions. Kaput (1989) refers to the tables, graphs, and algebraic equations as the "big three" representations in secondary school mathematics; whereas, the representation systems referred to in Fig. 24.1 are much more familiar in K through 8 classrooms. In this book, chapter 15 by Johnson and Lesh extends Fig. 24.1 to include Kaput's "big three" and adds a new layer of representations which involves technology-based counterparts to Kaput's "big three." The resulting system might be referred to as the Extended Lesh Translation Model for improving students' and teachers' understandings of important mathematical concepts.

To design the kind of curriculum units that are described in the chapters of this book on model development sequences, the Lesh translation model tends to be especially useful in model exploration activities as well as in warm-up and follow-up activities for model-eliciting activities. To understand how this is done, it is useful to give a brief description of how this translation model is grounded in earlier theories developed by Jean Piaget (1960), Zoltan Dienes (1960), and Jerome Bruner (1966).

The use of concrete materials to teach mathematics to elementary and middle-school learners has a long history in mathematics education. For example, Piaget's stages of intellectual development continue to influence mathematics educators as they consider how direct interactions with concrete embodiments of the mathematical systems can contribute to the development of underlying concepts at different ages (Post, 1988). In particular, the belief that students should be allowed to construct their own concepts in a global, intuitive manner, usually beginning with personal experiences is related to Piaget's concrete operational stage of intellectual development. Similarly, Zoltan

Dienes' Dynamic Principle emphasizes active student learning (Post, 1988), and identifies parameters for instructional sequences utilizing concrete representations to learn mathematics (see the chapter: *Model Development Sequences*, for another review of Dienes' Principles).

Dienes' theory suggests that students learn mathematics from concrete materials by moving through three stages: (a) play stage, (b) structured activities stage, and (c) the becoming aware stage (Post, 1988). During the play stage, learners have experiences with the concept in informal ways. For example, when using base-ten blocks, they characteristically play with the materials often discovering for themselves the relationships among the different blocks. This play stage is followed by more structured activities whose structure is the same as the constructs or conceptual systems to be learned. In particular, Dienes' well-known trading games are examples of structured activities that are designed to help build students' understanding of place value. The third stage of Dienes' approach is characterized by the explicit examination of the relevant mathematical concept.

Three other important features of Dienes' theory for active learning are the perceptual variability principle, the mathematics variability principle, and the constructivity principle (Post, 1988). The perceptual variability principle suggests that conceptual learning is maximized when children are exposed to a concept through a variety of physical contexts or embodiments. This multiple embodiment idea supports using several manipulative materials for any given mathematical idea to be learned. Examining the different representations, comparing differences and similarities among them facilitates mathematical abstraction. For example, students who use base-ten blocks, chip trading, and Unifix cubes to study place value can abstract the similar elements from these materials. What is common among the different materials is the mathematical abstraction that students are to construct.

The mathematical variability principle suggests that the generalization of a mathematical concept is enhanced when mathematical variables irrelevant to the concept are systematically varied while keeping relevant variables constant. For example, the mathematical variability principle suggests that, when learning about place value, students should examine different number bases. Examining similarities and differences among different bases allows students to make appropriate generalizations about place value. It is the regrouping scheme common to the different bases that is highlighted, not just the ability to compute.

Dienes' constructivity principle suggests that there are essentially two types of thinkers. The constructive thinker is roughly equivalent to Piaget's concrete operational thinker; whereas, Dienes' analytical thinker is similar to Piaget's formal operational thinker. The constructivity principle asserts that construction should precede analysis. In classroom settings, this principle is applied when children experience mathematical systems concretely before being asked to work with the mathematical ideas abstractly.

The perceptual and mathematical variability principles promote the complementary processes of abstraction and generalization. All four principles—the dynamic principle, the perceptual and mathematical variability principles, and the constructivity principle—provide educators with a framework for using physical embodiments to enhance mathematics learning.

Jerome Bruner (1966) suggested, "the structure of any domain of knowledge may be characterized in three ways" (p. 44). His three modes of representation are defined as *enactive mode* (physical representation), *iconic mode* (pictorial representation), and *symbolic mode* (written symbols). Although Bruner never explicitly stated that the optimal sequence is from enactive to iconic to symbolic modes, mathematics educators often have assumed this linear progression of learning. This idea of teaching using three different modes of representation has had a great impact on curriculum development for primary-aged children. The program known as *Mathematics Their Way* (Baratta-Lorton, 1976) is a noteworthy example.

Although the work of Piaget, Dienes, and Bruner provides a case for teaching mathematics by providing learners with a hands-on, activity-oriented environment in which students interact with a variety of manipulative materials and other representations, their work does not provide teachers with a conceptual framework to help them implement this vision of mathematics teaching and learning in the classroom. The Lesh Translation Model was designed explicitly to provide teachers with such a framework for organizing instruction to actively involve students in building mathematical understanding. What is suggested in the Lesh Translation Model is a model for using models in the mathematics classroom. It builds on the work of Piaget, Dienes, and Bruner, and, it also provides teachers with a framework for putting these theories into practice.

The rest of the chapter describes how the Lesh Translation Model was used in two different curriculum development projects. The chapter also briefly describes data documenting the development of children's fraction understandings among fourth- and fifth- graders using a fraction curriculum based on this model. Teachers' reflections on learning mathematics in an environment organized around the Lesh Model are shared. The chapter ends by considering how this model might impact classroom teachers' curriculum decisions.

USING THE LESH TRANSLATION MODEL IN CURRICULUM DEVELOPMENT

The Lesh Translation Model was used to guide the development of the Rational Number Project's *Fraction Lessons for the Middle Grades: Levels 1 and 2* (Cramer, Behr, Lesh, & Post, 1997a, 1997b). This curriculum was based on the theoretical perspectives outlined briefly in the preceding section. The effectiveness of the curriculum was studied in short and long-term teaching experiments with fourth- and fifth- graders. Results of these teaching experiments have been reported in a variety of publications and offer educators insights into the teaching and learning of fractions (Behr, Wachsmuth, Post, & Lesh, 1984; Bezuk & Cramer, 1989; Post & Cramer, 1987; Post, Wachsmuth,

Lesh, & Behr, 1985). For a discussion of insights into students' thinking about ordering fractions see Appendix A: Students' Ordering Strategies.

The RNP curriculum was field tested in a study involving some 1,600 students (Cramer, Post, delMas, in press). Initial decisions in the curriculum development process involved the identification of mathematical concepts to be learned. Rational number topics developed included the following: part-whole model, order, equivalence, concept of unit, addition, subtraction, multiplication and estimation. Learning activities to facilitate children's learning of these ideas were designed to involve children in five different representations and translations within and between these modes of representation. The manipulative models used in the lessons included fraction circles, Cuisenaire rods, paper folding, and chips. The use of several models reflects Dienes' perceptual variability principle and the Lesh model's within mode (manipulative) translations. Pictures of different manipulative models were integrated throughout the lessons. Connections to written symbols involved word names for fractions (three-fourths) and formal symbols for fractions [3/4]. Realistic contextual settings were written for each topic. This mode of representation is unique to the Lesh Translation Model, as is the identification of verbal symbols as a mode for representation. Language plays an important role in this translation model. Small group work and large group discussions facilitate students' use of language. The physical objects, pictures, and context become the objects about which students converse. Students' informal language mediates translations among different representations. Students' language can indicate readiness to translate to formal symbols. When students can easily talk about their actions with manipulatives, or describe how story problems can be solved with pictures, they are ready to record these actions with written symbols.

Consider the following example from the RNP Fraction Lessons for the Middle Grades Level 1 (Cramer, Behr, Lesh, & Post, 1997a). Students are asked to show the fraction two-thirds with paper folding while the teacher models the same fraction with colored chips. Students then describe how the two models are alike and different. Later in the same lesson, students are presented with two story problems and are asked to decide which model, paper folding or chips, would best represent the fraction in the story. These activities represent first, a manipulative to manipulative to verbal translation, and then a real-life to manipulative translation. (For the complete lesson, see Appendix B: RNP lesson from level 1).

In all 45 lessons from the two RNP fraction curriculum books, children are actively involved with manipulatives; they represent math ideas with pictures and diagrams, they solve story problems, and they connect these different representations to written symbols. Language supports translations students make.

This curriculum was studied under a variety of circumstances. Small group teaching experiments with six to eight students preceeded large classroom teaching experiments. The curriculum was revised throughout the process based on what was learned from working with children. An experimental study of the final version was conducted in a large suburban district south of Minneapolis,

24. TRANSLATION MODEL

Minnesota. Sixty-six fourth- and fifth-grade classrooms participated in the study; 33 classrooms were randomly assigned to use the RNP curriculum, whereas 33 classrooms were randomly assigned to use one of two traditional series. Instruction lasted between 28 and 30 days; classes were 50 minutes long. The questions asked were as follows:

- Do fraction related differences in student achievement exist between fourth- and fifth-grade students using the conceptually oriented curriculum (one based on the Lesh Translation Model) and students using district adopted commercial textbooks?
- If differences in student achievement do exist, what is the nature of the differences occurring in students' thinking and understanding?

Written post and retention tests were used to shed light on the first question. Interview data were used to explore the second question. Multivariate analysis of variance was used to analyze the comparison of posttests and retention tests. The MANOVA design consisted of one within subject factor (post vs. retention) and two between factors (RNP vs. Text) and grade levels (four vs. five). The dependent variables were total test scores on the two tests. A separate MANOVA was run on 15 subscales of the test using a similar design. Significance was set at $P < .0033$ to reflect Bonferroni adjustment to control for test-wise error.

Surprisingly grade level did not significantly influence achievement results. RNP students regardless of grade level had statistically higher mean scores on overall post and retention test score. RNP means were higher on 4 of the 6subscales : concepts, order, transfer and estimation.

Results showed that conceptually focused experiences based on a variety of mode translations and extensive use of several manipulative models enabled RNP students to learn more about initial fraction concepts than their counterparts using a traditional text. Interview results further supported our belief that instruction based on the translation model does in fact result in higher levels of student understanding and conceptual development (Cramer & Henry, in press).

Different groups of students were interviewed either during the 5 weeks of instruction or just at the end of instruction. A group of 14 fourth graders at the end of the study were asked to estimate the sum: $2/3 + 1/6$. None of the seven textbook students asked this question had an adequate understanding of fraction size to find a reasonable estimate to this simple problem. Their responses ranged from: *I have no idea*, to *between "1 1/2 and 2 just a guess*,*"* to finding the exact answer to use as an estimate. The seven RNP students all gave a reasonable estimate and were able to clearly explain their thinking. Their thinking depended on the mental images they constructed for fractions. This response exemplifies RNP students responses: *"Between 1/2 and 1...two-thirds is more than one-half; then add one-sixth. It's not greater than one because one-sixth is less than one-third."*

RNP students were better able to verbalize their thoughts. But then, the RNP lessons emphasized student discussions. In the lessons the manipulative models and other representations became the focal point of the discussion; students' language facilitated the translations between and within the different representations students used.

USING THE LESH TRANSLATION MODEL TO DEVELOP MATH COURSES FOR ELEMENTARY AND MIDDLE-GRADE TEACHERS

Two National Science Foundation teacher enhancement projects (PriMath and The RNP) have been involved in developing content courses for teachers and evaluating teachers' growth in content knowledge. The goal of these courses was for teachers to learn mathematics within an activity-oriented environment, which reflects the vision of the current reform movement in mathematics education. The Lesh translation model was used as a guide to develop all four content courses: Functions and proportionality, number and numeration, geometry, and data. The functions and proportionally course is the first course in the sequence and has been offered to over 250 teachers.

The Functions and Proportionality course was organized around the exploration of linear, quadratic and exponential functions. Figure 24.2 shows a diagram of the course content. This represents the mathematics to be taught. The first part of the course looks at linear, quadratic and exponential functions in a variety of ways. These are the mathematical variates of the concept of function and represent an example of Dienes' mathematical variability principle. Connections among different representations are emphasized throughout the lessons. Linear functions are then examined in detail, again with the focus on multiple representations and connections among them (Dienes' perceptual variability principle). The last part of the course looks at proportionality. Teachers use their understanding of function and in particular, linear functions, to understand what is special about proportional situations. As the mathematical characteristics of proportional situations are developed, teachers use previous learning to develop a deep understanding of this topic.

24. TRANSLATION MODEL

Functions & Proportionality:
Looking for Patterns and Regularities in Number

Represent concrete and real- world examples of functions using tables, graphs, verbal rules, & algebraic expressions: linear, quadratic and exponential.
  *Look for patterns among different representations. *Make connections among different representations based on patterns observed.
Explore Linear Functions in detail:
  • Multiple representations
  • Connections among representations
  • Slope and y-intercept
      -multiple representations
      - changes in context and rule affect graph

Proportionality as a Special Linear Function:
  • Multiple representations; math characteristics
  • Examples and non examples
  • Mathematical contexts

FIG. 24. 2. Mathematics content for functions and proportionality course.

To teach the mathematics, the Lesh translation model was used to develop the type of activities needed for teachers to learn this content. The model suggests that to develop a deep understanding of these mathematical ideas teachers would need to experience them through multiple representations (context, manipulatives, pictures, verbal symbols, and written symbols) and by translating from one representation to another. The first step in writing this course was to develop activities using concrete models. Geoboards, pattern blocks, paper folding, and string were among the concrete materials used to represent linear, quadratic, and exponential functions. Figure 24.3 shows an example of a typical problem used in the class. (For the entire lesson, see Appendix C: F & P lesson 3.)

Teachers collected data based on a concrete model. Leading questions guided teachers to use their own language to describe patterns and functional relationships (translation from manipulative to symbolic to informal language). Students than translated from their informal language to written symbols for the function rule. Graphical representations followed; this represents a translation to the pictorial mode. After three lessons, teachers interacted with three different types of functions using concrete models, tables, graphs, and written symbols; their language mediated the translations within and between modes of representation. Working through multiple examples with concrete models, and making translations to other modes of representation, are examples of operationalizing Dienes' perceptual variability principle. Looking at similarities among the examples across different modes of representation, teachers abstracted mathematical characteristics of linear, quadratic, and exponential functions.

Paper Folding Patterns
Take a piece of paper and fold in half as many times as you can. After 1 fold there will be two regions. How many regions will occur after three folds? four folds? How many folds are possible? Complete the table below. Imagine that there is no limit on the number of folds.
Questions?????

1. Describe several patterns found in the table. How does the number of regions grow?
2. Algebraically, describe how the number of regions is related to the number of folds. Use this rule to determine the number of regions given 18 folds.
3. Add another heading to the table: Area *of the smallest region.* Complete the table for this variable.
4. Describe patterns observed in the new table. How does the area of the smallest region grow? Describe this relationship algebraically.
5. Graph both relationships. How are they alike? different?

FIG. 24.3. Sample activity involving concrete models.

Concrete representations, language, pictures, and written symbols are only four of the five modes of representation students should experience. Using the Lesh model as a guide to develop this course, one other representation is still needed. Real-life contexts are important representations; learners need to experience functions within a context to develop a deep understanding of functions. Science contexts were used in the course. Figure 24.4 shows a sample problem in which students are presented with a context involving the relationship between Celsius and Fahrenheit temperatures.

Students answer questions that ask them to make translations from context to an equation (real-life to written symbols), and then from the equation to a table (written symbol to written symbol) and then to a graphical representation (written symbol to pictures). Students use the different representations to answer other questions. The lesson plan for this problem emphasizes the teacher's role in helping students see the connections among the different representations and how each representation can be used to answer the questions. Other realistic contexts were also used. Figure 24.5 shows a problem where students translated from a story problem to an equation to a graph. Questions asked students to make connections between the graph and information in the story. (For the entire lesson, see Appendix D: F & P lesson 7).

24. TRANSLATION MODEL

> A common function scientists use shows the relationship Between Celsius and Fahrenheit temperatures. To change Celsius to Fahrenheit, the Celsius temperature is multiplied by 9/5 and then 32 is added to that product.
>
> [You may want to use your graphing calculator to answer these questions.]
>
> (a) Record this rule algebraically.
>
> (b) Make a table showing the Celsius temperatures and their matching Fahrenheit temperatures.
>
> Graph the relationship between Celsius and Fahrenheit. Use the table, graph or function rule to answer these questions.
>
> (c) About what Fahrenheit temperatures match these: 24°C, -20°C, 12°C, 30°C, -5°C. [Describe your strategy].
>
> (d) About what Celsius temperatures match these: -10°F, 32°F, 41°F, 14°F, -5°C. [Describe your strategy].

FIG. 24. 4. Sample activity involving context.

> The local Telephone Company charges $5.95 plus 7¢ per minute for long distance services each month. If m is the number of minutes and c is the cost for a month, what is the functional relationship? Using the calculator, graph the function.
>
> If you do not make any long distance phone calls one month, how much do you pay?
>
> What point on the graph represents this cost? What do we call that point?
>
> Where on the graph is the rate: 7¢ per minute represented?
>
> If the monthly charge changes from $5.95 to $3.00, how will the graph change? What will stay the same?
>
> The cost per minute changes to 10¢ per minute while monthly charge stays the same. How does the graph change?
> What will stay the same?

FIG. 24. 5. Sample activity showing translations among context, symbols and picture.

The Functions and proportionality course has been well received by virtually all teachers who participated in the course. The first two groups of teachers who participated were primary teachers involved in PriMath. The course met for 3.5 hours a day for 10 days. At the end of the course teachers were asked to reflect on the following four statements and to respond to them by circling the most appropriate response. The mean for the second group of teachers is noted in parentheses.

• My knowledge and understanding of this mathematical topic has increased.
(1) strongly disagree (2) disagree (3) neutral (4) agree
(5) strongly agree   Mean = 4.8(4.8)

• I am more aware of the importance of this topic as a critical component of mathematics.
   (1) strongly disagree (2) disagree (3) neutral (4) agree
   (5) strongly agree   Mean = 4.6 (4.9)

• I appreciated the teaching style of the instructor for these sessions.
   (1) strongly disagree (2) disagree (3) neutral (4) agree
   (5) strongly agree   Mean = 5.0 (4.9)

• My interest in learning more challenging mathematics has increased.
   (1) strongly disagree (2) disagree (3) neutral (4) agree
   (5) strongly agree   Mean = 4.4 (4.7)

When asked what the most valuable part of the course was for them, teachers' responses reflected their understanding of the mathematics and an appreciation for the teaching practices. Examples are shared next:

- Seeing the patterns and understanding patterns is the key to understanding math.
- Seeing relationships between numbers and how beautifully everything fits together.
- I finally saw all the wonderful connections and interrelatedness of rules, graphs, tables, and so forth.
- I understand topics as I never have before. I remember algebra as not making any sense; now I have multiple strategies for figuring things out.
- The way the instructor moved from concrete to abstract activities and used past activities to show how it related to new ideas.
- The concrete, real-world contexts that were used.
- Constructivist learning and seeing the instructor model that for us.
- Working through the daily exercises, learning to see important relationships, and learning to communicate my understandings.

24. TRANSLATION MODEL

- The instructor's style led us to discover things and then relate them to rules.
- Learning what it was like to work in a group.
- Learning old stuff in new and better ways.
- Working in a group that had similar learning styles.
- Modeling of teaching; focus on communication, connections, reasoning all within a problem-solving framework.
- Working with others bouncing ideas and information back and forth.
- All the practice and relationships to the real world have been so helpful.
- Seeing all the connections; concrete examples.

Teachers were also asked about the least valuable aspects of the course. Only a few responded to this question. A small number of teachers in each group thought the pace was too fast.

Teachers responded in positive ways to learning mathematics from a curriculum that emphasized multiple representations and translations among the different representations. Their own reflections of the course showed insights into the power of learning mathematics organized around this model. These two cadres of teachers completed two other courses developed using the Lesh translation model. One dealt with number and numeration; the other with aspects of geometry. Their responses to these courses were as positive as for the functions class. At the end of the PriMath teacher enhancement project, which involved teachers in two summer institutes and academic year seminars, teachers were asked what they learned by participating in this project. Even though they were involved in a variety of teacher enhancement activities, the majority of the reflections dealt with the personal and positive outcomes from learning mathematics content.

USING THE LESH TRANSLATION MODEL TO GUIDE CLASSROOM DECISIONS

Although this model guided the development of a fraction curriculum and content courses for teachers, classroom teachers using commercial curriculums (traditional and reform) can also use it to help them make good instructional decisions.

No curriculum provides sufficient experiences in all of Lesh's five modes of representation with translations between and within them. Using the Lesh model, teachers can identify weaknesses in the curriculum they use, supplementing as needed. For example, consider the fifth-grade unit Patterns of Change from the *Investigations* curriculum (Tierney, Nemirovsky, & Noble, 1996). This is an excellent unit to introduce students to the idea of function and growth. A classroom teacher using the Lesh model as a guide to organize learning activities would observe that the unit involves students with manipulatives, context, pictures, written, and verbal symbols. Connections among the different representations are emphasized. By using the Lesh model as a lens to evaluate the completeness of the unit, a teacher would notice that only one manipulative model and one context are presented in the unit. The unit would be improved by supplementing it with other manipulative models. In addition to the tiles used, a teacher could provide lessons using geoboards and pattern blocksto represent functional relationship (Cramer, 2001). The teacher might also include contexts other than speed as used in "Patterns of Change." For example, students might consider realistic contexts involving taxi fares, purchasing food in quantities, or converting within the metric system. Students should reflect on these new physical models and contexts, discussing how they relate to the concrete model and context used in "Patterns and Change."

Teachers will find the need to supplement traditional series to a much greater extent than a standards-based curriculum. The addition chapter of the first-grade book by Burton and Maletsky (1998 develops meaning for addition byinvolving students primarily in picture to symbol translations. Lessons are introduced with one manipulative activity, but the student work, which follows in the textbook, does not integrate concrete models to any extent. Although options within the teacher's guide provide some activities using concrete models and story problems, these activities do not form the main focus of the lessons. A teacher would need to include more daily manipulative experiences. Looking closer, a teacher would see the need to include more story problems, particularly ones that ask students to translate from the story to a concrete model. Teachers would also see that the students are not involved in enough opportunities to talk about mathematical ideas. Viewing the chapter with the Lesh Translation Model in mind, teachers would realize that they would have to supplement the chapter a great deal.

SUMMARY

The purpose of this chapter was to present a case for considering the Lesh translation model as a guide for curriculum development. Building on the theories of Piaget, Bruner and Dienes, the Lesh model suggests that a deep understanding of mathematical ideas can be developed by involving students in activities that embed the mathematical ideas to be learned in five different modes of representation with an emphasis on translations within and between modes. The Lesh model was used in developing the RNP fraction curriculum. A large-scale study with fourth- and fifth- graders showed the effectiveness of this curriculum in developing fraction understanding over traditional curricula.

Mathematics content courses developed with this model have been successfully used with elementary and middle-school teachers. These courses, with Minneapolis teachers, have proved so popular that the district's new Urban Systemic Initiative aimed at improving mathematics instruction in Grades K through 5 will be using these courses as a foundation for its teacher enhancement activities.

Classroom teachers can use the Lesh translation model to guide how they implement district-adopted curriculum within their classrooms. The model can

24. TRANSLATION MODEL

help teachers determine the types of activities to supplement a curriculum so they can effectively meet students' instructional needs. Generally, most curriculums can benefit from more manipulative models and more problem contexts with translations from these modes to other modes of representation.

Although it has not been addressed in this chapter, the Lesh translation model can be an effective tool for developing assessment items (Cramer & Bezuk, 1991). Instruction and assessment should be aligned. Assessment tasks can be constructed around the translations within and between modes of representation. This allows teachers to assess understanding beyond procedural skill.

The power of the Lesh Translation Model can be seen in its multiple uses: a model for curriculum development, a model for classroom curriculum decisions, and a model for assessment.

Chapter 25

Integrating a Models and Modeling Perspective with Existing Research and Practice

Marilyn Carlson
Arizona State University

Sean Larsen
Arizona State University

Richard Lesh
Purdue University

This chapter provides an illustration of how the models and modeling perspective can be used in collegiate mathematics education research and instruction. The models and modeling approach (chapter 1) provides instructional designers with a well-defined structure for creating curriculum. The curricular activities for this approach, referred to as model-eliciting activities (Lesh, Hoover, Hole, Kelly, & Post, 2000), are designed to encourage students to make sense of meaningful situations, and to invent, extend, and refine their own mathematical constructs. The resultant student products reveal students' thinking and provide both teachers and researchers with a powerful lens for viewing students' reasoning and concept development. In this chapter, we discuss how this perspective has influenced both our research and instruction. For the past 5 years we have been engaged in research to investigate undergraduate students' understanding of *rate of change* (Carlson, 1998) and *covariational reasoning* (coordinating two varying quantities while attending to the ways in which they change in relation to each other; Carlson, Jacobs, & Larsen, 2001). These studies have identified aspects of covariational reasoning and have pointed to specific difficulties that students encounter when reasoning about dynamic events. Much of the data from our past research was gathered using specific mathematical tasks. Using the insights gained from this earlier research (Carlson, 1998), we modified these tasks to adhere to the six principles for developing model-eliciting activities (Lesh et al., 2000). We then utilized these new activities in a small-scale study that investigated undergraduate students' covariational reasoning abilities.

First, we present a brief sketch of Marilyn Carlson's research on covariational reasoning and the theoretical framework that resulted from that work. We then describe the process of converting a specific covariation task, discussed in Carlson (1998) to a modeling eliciting activity. This is followed by a description of a study that uses the newly developed model-eliciting activities.

Finally, we conclude this chapter by describing the insights gained into the effectiveness of model-eliciting activities in both revealing students' thinking and facilitating the research process.

REASONING ABOUT CHANGE

The rapid increase of mathematical applications requires that all citizens be fluent in modeling continuously changing phenomena, especially phenomena of dynamic situations (Kaput, 1994). Research has revealed that conventional curricula have not been successful in promoting these modeling abilities in undergraduate students. Several studies (Carlson, 1998; Kaput, 1992; Monk, 1992; Monk & Nemirovsky, 1994; Saldana & Thompson, 1998; Tall, 1992; Thompson, 1994a, 1994b) offer insights for addressing these problems by identifying and describing the complexities encountered by students when representing and interpreting dynamic events.

While investigating young children, Lesh et al. (2000), Kaput (1994), and Confrey and Smith (1995) have observed that, when provided the proper motivation and tools, middle school students are capable of creating and analyzing sophisticated mathematical models. Kaput (1994) has also advocated that with the use of powerful tools (e.g., Simcalc) young children can begin to engage in activities to explore change and accumulation of change, while building a strong conceptual foundation for the major ideas of calculus. These results suggest that similar outcomes may also be possible for undergraduate students.

Consequently, we have responded by working to improve undergraduate students' ability to create and interpret models of dynamic events. Previous research (Carlson et al., 2001) has produced a framework for describing and analyzing the cognitive activities involved in coordinating two varying quantities while attending to the ways in which they change in relation to each other (i.e., covariational reasoning). Our initial attempts to improve instruction of covariation have involved the creation of activities to promote preservice secondary teachers' covariational reasoning abilities.

As we have become more informed about the models and modeling perspective and the role of covariational reasoning in building and interpreting mathematical models of dynamic function events (i.e., a functional relationship that denotes a pattern or process of change), the following questions arose:

- Can research related to functions (Carlson, 1998) and covariation (Carlson et al., 2001) inform the design and use of model-eliciting activities?
- What effect does the use of model-eliciting activities have on the development of undergraduate students' covariational reasoning abilities?
- How can the models and modeling perspective inform the refinement of the Covariation Framework?

- How can the models and modeling perspective inform future investigations of students' covariational reasoning abilities?

Motivated by these questions, we began the process of integrating a models and modeling perspective with our covariation research and current instructional practices.

A COVARIATIONAL REASONING FRAMEWORK

The Covariation Framework (Fig. 25.1) describes six categories of mental actions that have been observed in students when applying covariational reasoning in the context of representing and interpreting a graphical model of a dynamic function event (Carlson, 1998). Our research findings have revealed that covariational reasoning does not necessarily involve all six mental actions, nor does it consist of a sequential progression from MA1 through MA6, nor do experts always begin their reasoning at MA6 (Fig. 25.1). However, we have evidence that sophisticated covariational reasoning is characterized by the ability to analyze a situation using MA6, together with the ability to "unpack" that mental action by using MA1 through MA5.

Covariation Framework

Categories of Mental Actions (MA)
- *MA1)* An image of two variables changing simultaneously;
- *MA2)* A loosely coordinated image of how the variables are changing with respect to each other (e.g., increasing, decreasing);
- *MA3)* An image of an amount of change of the output variable while considering changes in fixed amounts of the function's domain;
- *MA4)* An image of rate/slope for contiguous intervals of the function's domain;
- *MA5)* An image of continuously changing rate over the entire domain;
- *MA6)* An image of increasing and decreasing rate over the entire domain.

FIG. 25.1. Framework for covariational reasoning.

The following section provides a description of the framework categories in the context of the Bottle Problem, a problem that we used in our earlier research (Carlson, 1998; Carlson et al., 2000). In this chapter, we focus on the graphical representation system, as this was the initial context in which we observed students' difficulties in applying covariational reasoning (Carlson, 1998).

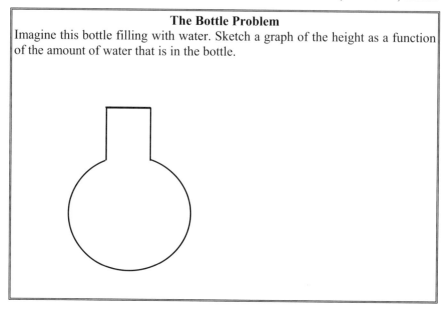

The Bottle Problem
Imagine this bottle filling with water. Sketch a graph of the height as a function of the amount of water that is in the bottle.

FIG. 25. 2. Earlier version of the bottle problem.

FRAMEWORK DESCRIPTION

MA1) An image emerges of the water level changing while imagining increasing amounts of water in the bottle.

MA2) An image emerges of the height increasing, as the amount of water in the bottle increases.

MA3) A fixed amount of water is imagined being added to the bottle, while concurrently constructing an image of the height of the water inside the bottle. This step is repeated by imagining the same fixed amount of water being added to the bottle with the construction of a new image of the amount of change in the height until the bottle is imagined as being full.

MA4) An image of the slope/rate of height change with respect to an imagined fixed amount of water is constructed. As successive amounts of water are imagined, the rate of change of the height with respect to the amount of water is imagined and adjusted. This process is repeated until the empty bottle becomes full. Representing these changing rates on a graph involves the construction of successive line segments with the slopes adjusted as each new amount of water is imagined.

MA5) An initial continuous image of the slope/rate of height change with respect to volume is formed. As the volume of water is imagined to change continuously, the rate of change of height with respect to volume (e.g., slope) is continuously adjusted. In the context of a graph, this results in the construction of a smooth curve.

MA6) An initial image of changing rate emerges. While imagining the filling of an empty bottle, distinctions are made between the decreasing rate in the bottom half of the spherical part, the increasing rate in the top half of the sphere and the constant rate in the straight neck. In the context of the graph, this results in the formation of a concave down construction, followed by a concave up construction, followed by a straight line. Inflection points are interpreted as situations where the rate changes from increasing to decreasing, or decreasing to increasing.

This framework both guided the development of covariation activities and served as a lens for analyzing and describing students' thinking. (See Appendix A for a more complete description and explanation of the Covariation Framework.)

EARLY ATTEMPTS TO TEACH STUDENTS TO REASON ABOUT DYNAMIC EVENTS

Curricular activities were designed to promote the development of 16 preservice secondary teachers' covariational reasoning abilities. Worksheets were written to assist these students in acquiring the ability to reason about dynamic events, as described in the covariation framework. This curriculum included activities that asked students to represent, in real time, various dynamic events and to produce graphs of specific dynamic behaviors. Students were prompted to verbalize the rationale for their constructions (e.g., Explain why the generated graph is smooth. In your own words, explain what this graph conveys about the changing rate of this situation. Discuss the nature of the changing shape of the graph in the context of the dynamic event.) This initial instructional intervention, although based on a research foundation, did not employ the models and modeling perspective. In particular, the activities were not designed to adhere to the six principles of model-eliciting activities. Pre- and post-instructional assessments indicated only moderate movement in these students' covariational reasoning abilities. For instance, only a few of these preservice teachers showed significant improvement in their ability to construct and represent images of slope/rate and changing slope/changing rate while imagining continuous change in the amount of water (MA5 and MA6). Although specific students appeared to demonstrate some improvements, evidence of incomplete understanding and blurred concepts (Monk, 1992) continued to persist among most of the participants in this earlier study.

The insights gained from these early instructional interventions provided both the foundation and motivation for developing covariation-based model-eliciting activities. As a first step in creating these new activities, we modified the Bottle Problem to adhere to the six principles of model-eliciting activities.

INTEGRATING A MODELS AND MODELING PERSPECTIVE

In the first chapter of this book, Lesh and Doerr describe the defining properties of model-eliciting activities (p. 6). In this section, we describe how we have used these guidelines to develop activities to promote and reveal acts of covariational reasoning.

Model-eliciting activities involve local conceptual development, a property that is especially relevant for achieving our goals. According to Lesh and Doerr (chap. 1, this volume),

> ...model-eliciting activities can be designed so that they lead to significant forms of learning (Lesh, Hole, Hoover, Kelly, & Post, 2000). Furthermore, because significant forms of conceptual development occur during relatively brief periods of time, it often is possible to observe the processes that students use to extend, differentiate, integrate, refine, or revise the relevant constructs. Consequently, to investigate cognitive development, it is possible for researchers to go beyond descriptions of successive states of knowledge to observe the processes that promote development from one state to another. (Lesh, 1983, p. 21).

If activities can be designed to promote significant development of covariational reasoning in a short period of time, they have the potential to provide valuable insights for researchers and curriculum developers. Activities that promote local conceptual development may be useful for practicing teachers as well. In addition to the instructional value of the activities, the teacher is able to observe and analyze subtle aspects of each student's mathematical development, rather than only the more obvious problems and insights of a small and vocal group. When working through a model-eliciting activity, students are asked to produce descriptions, explanations, procedures, and constructions. These types of products reveal much of the process that leads to their development. In essence, the students leave a paper trail outlining, in their own words, the reasoning they used when completing the activity.

When working with traditional curricular activities and assessments, teachers frequently find themselves trying to figure out what a particular solution tells them about a student's thinking. In most cases, they can say nothing more than the student did or did not do the problem correctly. They are often unable to say anything about what the student does know or what they can do. However, when we examine solutions to model-eliciting activities, it is easier to observe how students are thinking about a mathematical situation and to find out what they can do. The products, themselves, allow us to observe their processes (e.g., reasoning, verifying, and justifying), and not just the failure to produce an expected answer.

CONVERTING THE BOTTLE PROBLEM TO A MODEL-ELICITING ACTIVITY

The Bottle Problem has been a valuable research tool for revealing students' covariational reasoning abilities. However, by incorporating model-eliciting principles into this problem, we believe it is possible to create an activity that revealed even greater insights into the student's thinking, and one that was more effective in developing students' covariational reasoning. The new model-eliciting version of the bottle activity (Fig. 25.2) is presented in Fig. 25.3, and is followed by a discussion of the new activity in light of the six principles for developing model-eliciting activities.

Bottle Model-Eliciting Activity

Dear Math Consultants,

Dynamic Animations has just been commissioned to animate a scene in which a variety of bottles will be filled with fluid on screen. We need your help to make sure this scene appears realistic.

We need a graph that shows the height of the fluid given the amount of fluid in the bottle (a height/volume graph). Below, we have provided a drawing of one of the bottles used in the scene. Please provide a graph for this bottle and a manual that tells us how to make our own graph for any bottle that may appear in this scene.

FIG. 25. 3. The model-eliciting version of the bottle problem.

INCORPORATING THE SIX PRINCIPLES

The process of transforming the Bottle Problem into a model-eliciting activity was guided by the six principles (Lesh et al., 2000). We approached our task by discussing aspects that would be desirable in the new activity and made choices about the task-design using both these criteria and the six principles. An elaboration of this process follows.

The Reality Principle. Will students make sense of the situation by extending their own knowledge and experiences? Perhaps the most challenging part of adapting a traditional activity to a model-eliciting activity is satisfying the requirements of the reality principle. For this problem, we needed to devise a context that motivated or created a need to apply covariational reasoning. The context that we created involved a production studio that needed assistance in producing realistic animations of bottles filling with liquid.

This context required the use of covariational reasoning because such a studio would need to coordinate the fluid level of the bottle with time. Because we wanted to compare our results with earlier findings, we decided to require students to coordinate the fluid level (height) with the amount of fluid in the bottle (volume), though a truly realistic task would involve the coordination of height with time. Additionally, our research goals were focused on covariational reasoning in the graphical representation because this representation is so crucial in mathematics and has been shown to be problematic for students. Therefore, our task focused on representing the situation graphically, although it is likely that a real production studio would use a different approach.

These decisions limited the role of the realistic context of the activity. However, as is conveyed below, the activity is consistent with the other five principles, resulting in a task that is both thought revealing and model eliciting.

The Model Construction Principle. Does the task immerse students in a situation in which they are likely to confront the need to develop (or refine, modify, or extend) a mathematically significant construct? Does the task involve constructing, explaining, manipulating, predicting, or controlling a structurally significant system? As previously mentioned, the original activity partially met these criteria by requesting that students produce a graph of the situation. However, because the original activity only requested that students produce a graph for one bottle, we modified the activity by prompting students to develop a general model (i.e., a manual) for analyzing this type of dynamic situation. (This activity actually required the production of two models, the graph and the instruction manual. However, our reference to the model refers to the instruction manual because this was the primary product that we requested.)

The Self-Evaluation Principle. Does the activity promote self-evaluation on the part of the students? This principle was addressed by requesting that students produce a manual for any bottle. This request provided students criteria for assessing the quality of their model. In particular, the students needed to determine whether their instruction manual was an effective guide for producing graphs of different-shaped bottles.

The Construct Documentation Principle. Will the question require students to reveal their thinking about the situation? This was the primary motivation for developing a model-eliciting version of the Bottle Problem. In order to move our research forward we needed greater insight into the thinking involved in graphically representing a dynamic situation. The students revealed how they thought about the situation by creating an instruction manual.

The Construct Generalization Principle. Does the model provide a general model for analyzing this type of dynamic situation? The requirement that students produce an instruction manual prompted students to generalize their covariational reasoning in the context of the bottle (Fig. 25.2) to create a model that is applicable to a wide variety of different-shaped containers.

The Simplicity Principle. Is the situation simple? The situation made a specific request for students to represent graphically the filling of a bottle. As such, the situation was not particularly complex. Further analysis is necessary to determine whether this situation is simple enough to allow it to play the role of a prototypical problem.

A SMALL STUDY EMPLOYING THE MODELS AND MODELING PERSPECTIVE

Following the creation of the model-eliciting activities, a small study with 22 preservice elementary teachers, enrolled in the course "Mathematics for Elementary School Teachers" at a large public university, was conducted. The class was taught using a student-centered approach with students regularly working in groups while making their mathematical constructions. All students had completed a course equivalent to college algebra, with four students also completing introductory calculus at the university. The students were paid for their participation and were informed that their willingness to participate required that they make a strong effort to think through three model-eliciting activities. Students were videotaped outside of class and each group of two to four students spent approximately 4 hours completing the tasks. The subjects were comfortable conversing with one another because the groups were essentially the same as their class groups.

During the first session, students were asked to physically model the graphs of seven different functions using calculators and CBRs (Computer Based Rangers; these devices consist of motion detectors linked to graphing calculators. As the students walk, their motion is represented on the calculator screen in the form of a distance/time graph.) In addition to actually producing each graph, they were asked to describe the movement that would generate a specific graph. After completing this task, they were given a model-eliciting task that required them to write a strategy guide to prepare an individual to physically model any graph without experimentation.

CBR Model-Eliciting Activity
Dr. Erikson's physics class is preparing for a lab exam about motion. One of the things they will be required to do is to use a CBR to reproduce (by walking) distance-time graphs that Dr. Erikson draws on the board. They do not know what graphs they will be asked to produce in advance. Please write a strategy guide that will prepare Dr. Erickson's students to reproduce any possible graph. Below are two graphs that Dr. Erikson has used in the past. Note: the students will be allowed only one attempt to produce a graph.

FIG. 25.4. The CBR model-eliciting activity.

The second model-eliciting activity involved a distance/time graph in the context of an airplane flight. The final activity was the model-eliciting version of The Bottle Problem described earlier (Fig. 25.3).

SELECT RESULTS

Analysis of students' manuals and their discussions while writing the manuals revealed new information about students' covariational reasoning abilities, as well as valuable insights regarding the effectiveness and usefulness of the activities. Select results are presented to illustrate the ways in which these activities revealed students' thought processes. (Student solutions are available for all groups for both the CBR activity and the Bottle Problem. See Appendixes B and C.)

Were the activities thought-revealing? Consider the following solution that was produced by one group for the model-eliciting version of the Bottle Problem (Fig. 25.5).

This solution provides interesting insights into how this group thought about the situation. While their thinking (Fig. 25.5) appeared consistent with the general characterization of MA4 of the covariational framework, it differed from the actual description provided in the framework. Recall that MA4 involves the construction of an image of the slope/rate of the change in height with respect to an imagined amount of water. In this case, however, the students did not imagine the rate of change in height with respect to an imagined fixed amount of water. Instead, their image of rate appeared to depend on the width (cross-sectional area) of the bottle at a given height, and was determined by comparing the volume and height for a small disk of water. This observation was also substantiated by analyzing the videotaped interactions among the members of this group.

25. INTEGRATION WITH RESEARCH AND PRACTICE

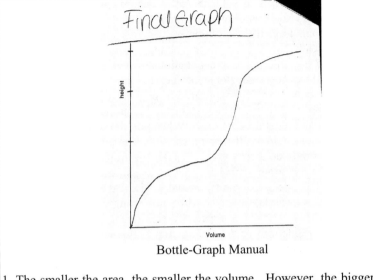

Bottle-Graph Manual

1. The smaller the area, the smaller the volume. However, the bigger the area, the bigger the volume (e.g., #1, 2, 3 on "Graph #1" (as labeled on the bottle).
2. If you don't use a lot of volume the height increases more rapidly. If you do use a lot of volume, the height will increase, just not as rapidly as the volume.
3. The slope determines how high and how far across the line will be.

FIG. 25.5. An example of the students' solutions.

Mary:	"How do you explain that?"
Chris:	"Like when you have a little bit of volume, you don't need a lot volume but you have a lot of height. And when you have a big area and less height, you are going to have more volume and its going to go flat."
Joni:	"Or as the unit goes more parallel to each other, whether if its this way or this way, it's going to be less height."
Mary:	"What if we said it this way, on a graph the greater the height, the steeper the graph will be. The wider it is, the more level it is. Do you know what I am saying? How do we say that?"
Chris:	" The reason the graph is like this is because there is so much volume needed for little area."
Joni:	"As the sides go out, it requires more fill."
Chris:	"Requires more volume verses height."
Joni:	"So if the unit is not so wide, it requires less volume and more height…"

FIG. 25. 6. Transcript excerpt.

Figure 25.6 provides excerpts of students' conversations as they responded to this task. As is revealed, the process of negotiating one final product encouraged individuals to justify and verbalize their responses (e.g., the reason for the graph is because there is so much volume) and to compare their own ideas with those of the group. These negotiations provided additional insights into each individual's concept development and reasoning patterns. As can be observed (Fig. 25.6), these students appeared to express different justifications for the graph's shape (i.e., Mary: "The greater the height, the steeper the graph the wider it is, the more level it is." Chris: "When you don't need a lot volume, you have a lot of height." Joni; "If it is not so wide, it requires less volume and more height."). Not only do these expressions deviate from the language of the framework, but like the group's manual, they express conceptualizations that also differ from the ideas of the framework (i.e., the students are not coordinating the output variable, height, while imagining changes in the input variable, volume). This example is especially significant because it provided implications for the refinement of the Covariation Framework. In particular, it suggests that the language used in describing MA4 is not general enough to describe the various ways that a student might think about rate of change of one quantity with respect to another.

The thought-revealing nature of these activities is particularly important to us because we are introducing these activities in our courses for preservice secondary teachers. We are encouraging these future teachers to use these activities to gain access to the developing understandings and reasoning patterns of their students. The implications for our students' future teaching practices are significant, because they are being introduced to an efficient and powerful means of gaining regular access to their students' thinking.

IMPACT OF THIS STUDY ON OUR FUTURE RESEARCH

In the previous section, we provided one example of the impact of this study on our ongoing refinement of the covariation framework. In this section, we describe some other implications of this study on our continuing investigation of covariation reasoning. The written solutions and the videotape transcripts enabled us to capture many instances of covariational reasoning, spanning the entire range of the Covariation Framework. However, the students' expressions of these mental actions sometimes differed qualitatively from the descriptions provided within the framework. These observed differences point to questions for further research as well as potential refinements of the covariation reasoning framework.

Students often treated height as the input variable. In the case of the bottle activity it was expected that students would treat the height as a function of volume because they were asked to create a height versus volume graph. However, the models produced by the students revealed that they frequently thought about the volume as the dependent variable. This suggested that our interpretation of the framework in situations where the function underlying the dynamic situation is invertible needs to be more flexible.

Students often treated time as the input variable. This was revealed in the students' written products and was evidenced by their use of terms and phrases like *faster, slower, quickly, longer,* and *more time to fill.* Their choice to replace the independent variable (volume) with time appeared to reduce the cognitive load for the students by allowing them to focus their attention on the changing nature of one variable or quantity. However, there was evidence that the students who used this approach did not completely understand the relationship between the height versus time graph and height versus volume graph. When asked, one group indicated that if the water were not poured at a constant rate, their *height* versus *volume* graph would be affected (a misconception). These findings suggest that the role of time in covariation reasoning needs to be further explored.

The groups did not describe a process in their instruction manuals that involved either point plotting or curve smoothing. All of the groups were able to produce correct smooth graphs for the given bottle. However, none of the groups described a process that would lead to a smooth curve by point plotting or the refinement of line segments representing average rates of change (even when their initial constructions used one of these two approaches). Instead, they gave instructions that focused on determining the rate of change at any point on the bottle and described the rate of change as an object that they could move along the domain. This observation should provide important implications for our continued investigation of covariational reasoning. In particular, it suggests a possible description of the development needed for students to form a continuous image of rate of change.

CONCLUSIONS

The students' final products, reasoning abilities, and persistence exceeded our expectations for the study. All 22 of these preservice elementary teachers provided a reasonable graph for the bottle activity. Factors that appeared to contribute to their successes include: the requirement that students verbalize their reasoning and both receive and give feedback to their peers (a feature of the model-eliciting activities); the requirement that students continue refining their solutions until a reasonable answer was produced (also a feature of the model-eliciting activities); and, the fact that they were allowed to take whatever time they needed to complete the tasks. In retrospect, these activities appeared to have the effect of placing each student in a role similar to that of a teacher. They were expected to provide a clear, logical, and defendable rationale for their solution, and like good teachers, they rose to the challenge.

The results of this study suggest that informal exploration may have promoted greater engagement and sense-making on the part of these students. Further, the unexpected successes exhibited by these preservice elementary teachers suggests a need for continued investigation of the power of these activities in transforming context-specific notions into more general models.

As a follow-up to this study, we have begun developing model-eliciting activities to promote students' development and understanding of the major conceptual strands of introductory calculus. We call for others to join us in investigating the effectiveness of model-eliciting activities for developing other mathematical concepts in undergraduate mathematics. We also call for additional investigations into the effectiveness of model-eliciting activities for promoting undergraduate students' reasoning, communication and problem solving abilities.

ACKNOWLEDGMENTS

We would like to thank Sally Jacobs for her assistance in preparing this chapter and would like to acknowledge that NSF funds, grant #9876127, supported this work.

Chapter 26

Models of Functions and Models of Situations: On the Design of Modeling-Based Learning Environments

Beba Shternberg
*Center of Educational Technology,
Tel-Aviv, Israel*

Michal Yerushalmy
University of Haifa, Israel

BRIDGING MATHEMATICAL CONCEPTS AND THE PHENOMENA THEY DESCRIBE

Modeling as both an important mathematical goal and a central pedagogical tool is currently a common part of reformed school mathematics curricula. Because of its power to support various interpretations of mathematical modeling, to represent situations that are external to mathematics and to create opportunities for constructing and manipulating models of mathematical concepts, technology can play a central role in strengthening learning.

This article focuses on two interpretations of mathematical models: *Didactical models*—models of mathematical concepts, and *Mathematical models*—models of physical phenomena. Both types of models are incorporated as a means for learning the concept of function, in a technology intensive *Learning System*, which is based on the synergy of modeling situations and constructing mathematical language to describe them.

Definitions

Following the introductory chapter (Lesh & Doerr, chap. 1, this volume), we define two meanings of the use of the term model in the context of mathematical modeling:

1. A didactical model of a mathematical concept consists of objects familiar to the learner. The operations defined on the objects correspond fully to the mathematical actions on this yet unknown mathematical

concept. Modeling in this sense uses operations on objects to support the construction of an understanding of more formal reasoning about mathematical concepts and about their essential properties. An example of such model is the Cuisenaire Rods for supporting the foundation for learning the concept of number (described by Nesher, 1989). Another example is the 'Stair model' which is often used as a model for functional variation and is intended to establish further mathematical constructs in the field of calculus (Yerushalmy, 1997).

2. A mathematical model is a mathematical construction that describes a class of phenomena external to mathematics (e.g., temporal phenomena). Modeling in this sense is concerned with using mathematical language (including the notations of the didactical models) for reasoning about the phenomenon.

Approaches for Mathematical Modeling

Three fields of knowledge are involved in mathematical modeling: One is the physical field—the *signified field* in which situations occur. These situations can be described verbally or numerically using terms of the physical field. The second is the signified *mathematical field* that consists of abstract mathematical concepts and often includes the concepts' images (as defined by Tall & Vinner, 1981). The third is also the mathematical field—*the signifier*—in which both models reside to signify two other fields: A didactical model signifies a mathematical concept, and a mathematical model describes phenomena.

A traditional application approach to modeling offers students an opportunity to analyze applied situations and mathematize them using the symbolic language that they have been taught. This pedagogical sequence starts with the mathematical signifier field where the mathematical language (symbols and operations) is taught first. The language defines and describes the signified concept (e.g. a number concept, a function concept). Later on, when interacting with the signified physical field, the mathematical signifiers—the mathematical representations of the concept—are used to model the situation. It should be noted that this happens usually only after students have mastered the language of symbolic representation required to express their models of the situation. Thus, the link between the physical field and mathematics is weak and even artificial, and the construction of formal mathematical language often remains meaningless and cannot be applied later on.

In order to close the gap between the physical and the mathematical fields, various approaches (often described as environmental approaches or situated approaches or context-based approaches) start with applications or daily life situations, assuming that learners can describe the situations in ordinary language and proceed from there to formal mathematical language. There is a wide range of alternative approaches that differ in the ways and depth they use daily life situations. In the less effective cases, the situations are used as a motivation, followed immediately by decontextualization (e.g., "Each table has 4 legs; 5 tables would have 20 legs; how many legs would have any number of

26. MODELING-BASED LEARNING ENVIRONMENTS

tables? Answer: 4x"). Research suggests that neither the traditional application sequence nor the motivation approach offer meaningful modeling experiences. Deeper use of daily life situations requires that the instructional sequence should be fully constructed as part of the students' field of experience (Boero et al., 1995), and gradually moved on to mathematical language: syntax and operations. The transition from context to mathematical model depends on the availability of classroom activities that are part of students' experience and would support the desired conception. There is a continuing debate regarding the interference between formal language and natural language, the linguistic transfer, and the role of formal language within the sequence (Nesher 1989; Pimm, 1995; Sutherland & Balacheff, 1999). An attempt to use the other type of models—the didactical models of mathematical concepts—to address the complexity of establishing a meaningful formal language is described by Nesher (1989). At the heart of Nesher's Learning System (LS) lie objects that are not just illustrative devices. The operations performed on the objects are rigorously defined and are fully and uniquely mapped onto the formal mathematical operations and syntax. In the LS approach, mathematical models are a major means toward understanding the properties of formal mathematical language. As in traditional approaches, the physical world enters LS sequence only after the formal language has been meaningfully understood. Thus, whereas the didactical modeling offers significant ways to grasp the properties of the concept of number (e.g., comparison of quantities as objects, group of binary operation) and connect them to formal language, the link between the physical field and the mathematical concept is not firmly established.

On the Need for Didactical Models: The Case of Calculus Students

In our search for an approach to algebra that would use modeling in a meaningful way, we looked at modeling actions and models constructed by high school calculus students when facing modeling problems. According to their curriculum, the students supposedly had the mathematical and physical knowledge necessary for solving these modeling problems The following description of their attempts to solve the Car Problem focuses on the relationships they demonstrated between the conceptual properties of a function and the formal use of symbolic language.

The Car Problem: Field-testing of a racing car involved measuring the speed for 10 seconds. The initial speed is 20 meters per second. During these 10 seconds, the car was monotonically slowing down and stopped at the 10th second. The following table describes how the speed changes.

Time (sec)	Speed (m/sec)
0	20
2	14
4	9
6	5
8	2
10	0

What is the distance traveled during the 10 seconds?

To model nonconstant rate temporal phenomena, one can use the set of tools provided by calculus. The Car Problem deals with decreasing speed in linearly changing acceleration. A complete and accurate solution of the Car Problem should include a symbolic model of the given speed followed by integration and computation:

$$v(t) = 0.125t^2 - 3.25t + 20 \qquad s(10) = \int_0^{10}(0.125u^2 - 3.25u + 20)du = 79.17 \text{ (km)}$$

Although it is not too difficult to identify the quadratic pattern of the given speed values, neither the exact formula of *v(t)* nor the change of the distance can be easily guessed and have to be computed.

We asked each of 30 high school calculus students to solve the Car Problem during a one-half hour interview. In their curriculum, they had been presented from time to time with similar activities either as an application or an introductory example in their calculus course, or as part of their high school physics course. We assumed that this group of highly motivated successful students, who learned about functions in several courses (algebra, precalculus, and calculus), had acquired all the necessary mathematical tools, and could be inspired by both the physics and the mathematics. Therefore, it seemed interesting to observe this group thinking about modeling and about the concepts of function. Almost all the students[1] faced a fundamental struggle with issues that lie at the heart of modeling tasks, and we would argue that the analysis of their struggles indicates an absence of conceptual connections. The following is a description of three typical complexities encountered in the attempts to solve the problem.

Math Is Not For Reality—Physics Is

One of the complexities of teaching a modeling-based curriculum is how much one has to care about the realistic details of the situation and what sort of

---

[1] There were two students who reached an almost correct mathematical model, which suffered from computation mistakes.

mathematics should be known a priori. Because modeling is usually considered to be an application, students' efforts often concentrate on finding the correct choice of a familiar formula. Students who study physics, as the students in our sample, often assume that details they have to pay attention to in a physics class can be ignored in the symbolical models dealt in the mathematics class. Yet it is not only the emphasis that is different in mathematics and physics: students who study physics often conclude that the techniques used in each subject are different as well. School physics, unlike traditional mathematics strands, often invites the use of graphs and numerical data and experimental processes with different parameters etc. (e.g. Dapueto & Parenti, 1999; Thompson, 1994).

The following is an example of how this complexity arose during the interview with Tzvi. Tzvi starts to draw a graph [speed versus time] "..to organize the ideas." While drawing a graph he comments: "It is a decreasing curve and the slope shows the de-acceleration... and I know it decreases... ." However after awhile, he muses: "I know what I would have done in physics [and here] I can't. In physics we have to do it very accurately and we would draw it on a millimetric paper and then would count the squares under the curve to compute the area. Here I can't know what the slope is... This slope [points to the line he drew] is not accurate. If it were not so steep, maybe I would do something like that [he straightens the curve] and then compute the area of the triangle using the formula.

Tzvi, who demonstrates a fine understanding of the global phenomenon, cannot make the connection between mathematical models and models in physics he used to work on. It seems that for Tzvi graphs, counting, and accuracy belong to the domain of physics where one tries to capture the reality in the most accurate way. Whereas modeling in mathematics class means for him working with formulas, algorithms and abstractions without worrying too much about the authenticity of the physical phenomenon. Later on in the interview, Tzvi does try to link the two sets of tools and to figure out a formula, but then he realizes that the only way he knows how to formalize the idea of a slope is by using a linear equation, which contradicts his analysis of the curve behavior.

"Just Having a Formula Wouldn't Help Me"

Unlike Tzvi, Adi does not feel that knowing the formula means having a correct answer. She also does not remember the "formula for such a sequence". However once the interviewer proposes to help Adi with the formulas, she answers that having a formula would not actually matter: "But it would not really help me to know the formula—I remember we do it like that—we write here x minus 1 and then it decreases. I really don't remember." Although it seemed that Adi was 'on track' in computational terms when she tried to look for the distance as a sum of a sequence, she gave up because her previous experience with sequences was not connected at all to such temporal phenomenon.

This attempt of making sense of the chosen formal language for modeling was also part of Yoni's work. Yoni graphed the given data and looked for the integral as a way to find the area under the curve. He decided that to complete his work, he has to look for a formula, and he tried to find one: "I'm trying to find this function...and I don't know how to find it. Finding it, this is the problem. I had an idea and I am guessing [probably from the graph shape], I am guessing and trying..."

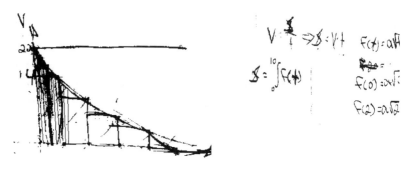

Despite the fact that mathematically it did make sense to him, he could not understand how the formula he arrived at could be valid for the given situation, and he gave up.

Points Are Fine and Averages Will Provide More Points

How can one find more information about the motion of the car in the Car Problem, in order to form a model? Clearly, because the problem gives data in intervals of 2 seconds (and not in each unit of 1 second), the natural operation is to look in between the given data. For most interviewees, the natural way to get more dense data was to calculate the arithmetic average of each pair of measures. Eido approached the problem by averaging the values and using the average value over the entire time interval. In describing what happens to the car, Eido began from the first 2 seconds: " I can say that this [the speed] should be between 14 and 20 meters... so it must be 17 meters. It starts at a speed of 20 and decreases its speed constantly so its average speed along the 2 first seconds must have been 17...And it [the distance] would be 34. 17 times 2."

Although Eido did not rush to find a formula and spent some time trying to complete the numerical data that was given in the problem, nothing in his answer suggests that he was thinking about the properties of the situation. For each given interval $\Delta(t)=2$, Eido replaced the speed curve by its linear approximation and computed the function value in the midpoint of each interval to be the arithmetic average of the values at both ends. This algorithm (as was also suggested by the sample of teachers Thompson reported in 1994) fits any linear phenomenon, and Eido did not evaluate whether it applies to the given situation.

Discussion

The previous examples above raise some serious issues regarding the value of the use of applications for the understanding of functions, and the use of functions to model the world. The gap between the mathematical and physical fields that dominates the process of problem solving, although commonly acknowledged (e.g. Kaput titled it as the gap between the island of formal mathematics and the mainland of real human experience, 1994), becomes sharper when exemplified by high achieving calculus and physics students. Formally, the mathematical signifiers are well known, but they are probably not meaningful and even when various representations of functions are employed, the models of mathematical concepts used by students (e.g., stairs, rectangles, and curves) do not become models for reasoning about the essential qualities of a temporal phenomena (e.g., average speed, nonlinear process), and their application cannot be extended. Different ways have been suggested to overcome this gap. One of them is to give the situated tasks a more central role and to use it as a motivational power for reinvention of mathematics by the students themselves. Gravemeijer and Doorman (1999) claim that if, "it would be possible to have the students invent distance-time and speed-time by themselves, the dichotomy between formal mathematics and authentic experience Kaput presupposes, would not arise" (p. 115). In order to support such processes they suggest that guided reinvention is an important distinction of the Realistic Mathematics Education (RME) approach. The guided inquiry approach to function based algebra that we will discuss below, combines elements of the guided reinvention environment in the RME approach and elements of the LS approach. Like RME, this guided inquiry approach provides opportunities for students to reinvent and construct mathematical notations and mathematical concepts that are connected to their intuitions and experience (Yerushalmy, 1997). The Learning System (LS) approach that this guided inquiry is based upon, suggests that the mathematical didactical models, which are cognitively mapped along the core of the conceptual knowledge to be acquired, should be predesigned for the learner. Before analyzing a few problem-solving episodes with students who experienced this approach, we take a short detour to describe the learning environment that supports this learning stance.

INTRODUCING A TOOL FOR EXPLORING PHENOMENA AND FOR EXPRESSING THE CONCEPT OF FUNCTION: A DETOUR

This study is based on an experimental curriculum that includes a full sequence of an algebra course supported by innovative software tools, especially designed materials for students and teachers, and workshops and materials for professional development. The experimental curriculum—*Visual Mathematics* (1995)—is a function-based approach to algebra and uses the idea that modeling (i.e., mathematizing authentic situations) is an important mathematical goal as well as a central pedagogical tool of the mathematics curriculum. Within this guided inquiry curriculum, students encounter situations that they need to mathematize from the first steps in their algebra course. The potential of the function concept to support a distinct set of models that are both familiar and mathematical models that can be used as didactical models and play a major role in modeling physical situations, is a main reason for the appropriateness of the concept of function as the major idea upon which the curriculum is built (Chazan, 2000; Schwartz & Yerushalmy, 1992). We are exploring these basic principles as a part of an algebra curriculum adopted by several schools (Grades 7 to 9) in Israel. Elsewhere, we describe in detail the rationale for a function approach to algebra that starts with qualitative modeling (Schwartz & Yerushalmy, 1995; Yerushalmy & Shternberg, 2001). Here we describe the design of the didactical models that promote and enable learning in this sequence.

"The Function Sketcher" software environment (Yerushalmy & Shternberg, 1993/1998) supports bringing reality into the language of graphs by an interaction between the mechanics of motion and the formulation of the graphical model. The situation of planar motion is simulated using freehand mouse drawing. The software provides graphs (appearing simultaneously with the drawing or upon request) of quantities involved in a planar motion as they change in time. In Fig. 26.1, the path is drawn on the X-Y plane (previously mentioned), and two graphs show the positions of X and Y in time (next).

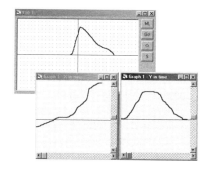

FIG. 26.1. The planar path (upper window) and the X and Y positions graphs in time.

The learner draws a path by activating time and moving the mouse along the mouse pad. This motion is expressed as a path drawn on the computer screen. The software stores the positions at any time and can reproduce the motion and the graphs that represent the X- and Y-components of the motion in time. One may try to invent a path according to given graphs and realize that different types of motion's graphs can create the same path. The two sets of different motions that are demonstrated in Fig. 26.2 describe the same paths (the path in the x–y plane appears in the square windows and the graphs in the rectangular windows—x(t) up and y(t) below).

26. MODELING-BASED LEARNING ENVIRONMENTS

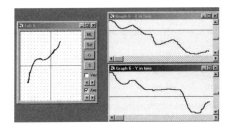
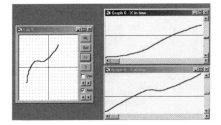

FIG. 26.2. Two different sets of graphs produce equal paths (Fig. 26.2a - left; Fig. 26.2b.right)

The general direction of the motion that created the path on the left (Fig. 26.2a) was from top right to bottom left, and therefore x(t) and y(t) graphs are descending from positive to negative values. The motion that created the exact same path that appears on the right (Fig. 26.2b) is described by two ascending graphs—x(t) and y(t) from bottom left to top right.

In order to analyze the major properties of the data plotted by the software and to preserve the essential physical actions of the learner, a didactical model which includes a set of seven graphical icons, and a limited verbal list of function properties (Fig. 26.3) is offered. There are synonymous relationships between the iconic and verbal elements of this linguistic representation. Although statements expressed using one lexicon can be restated using the other, the two distinct sets allow different degrees of precision. For example, describing a function in a region as "ascending" and "curved" or, even more accurately, as ascending at an ascending rate of change is far less limiting than is the display of a particular increasing curved function.

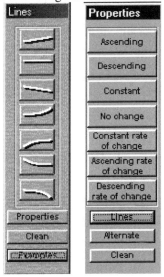

FIG. 26.3. The two lexicons used to represent the visual properties of functions.

In order to allow a finer taxonomy within a verbal description of a particular behavior, the software introduces stairs, a discrete companion to a continuous graph (Fig. 26.4).

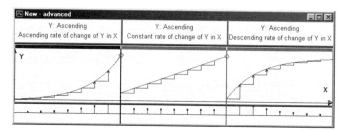

FIG. 26.4. Stair model of three ascending curves.

The different components of the lexical system (icons, verbs, and stairs) are eventually adopted as manipulable objects that support students when solving problems that are too complicated for them to describe symbolically. Using the grammar of objects and the operations on them, students are involved in constructing more complex mathematical models, and these are qualitatively analyzed by identifying the mode of variations using the stairs. This environment is designed to map the terrain of the concept of function in a simplified and abstract way. It provides mathematical terminology and an overview of the possible behavior of processes. The icons are didactical models of functions and are used for reasoning about functions. The icons are also mathematical models of situations and together with the stairs model are used for reasoning about temporal phenomena. Thus, the same model serves as the didactical model of the concept of function and as the mathematical model of a physical situation. Although mathematical modeling cannot be fully accomplished by this qualitative sign system, the issue for this study is to identify essential contributions of modeling using didactical models. In the following section we will make some distinctions for describing prealgebra students' modeling attempts while solving the Car Problem.

REASONING WITH DIDACTICAL MODELS: THE CASE OF PREALGEBRA STUDENTS

Interviews with pairs of 7th graders while solving the Car Problem are the core of this section. The students had recently spent a few weeks working within the Function Sketcher microworld on the qualitative formation of the concept of function and also gained a limited experience in connecting function expressions to graph shapes. Their activities included graphing and telling mathematical narratives using the terms of function variations to describe given situation stories (similar to the ones described by Nemirovsky, 1996 and Yerushalmy & Shternberg, 2001). In this setting the Car Problem raised three challenges: 1)

The students had not yet experienced modeling situations given by a list of discrete numerical measures; 2) They were used to analyzing variation by assuming stairs of one unit width. Here, measurements were given in intervals of two units, which invited exploration within the intervals while keeping in mind the monotonous behavior of the whole process; 3) In addition to descriptive modeling tasks (the sort of task they were used to), this problem required a numerical solution. This solution deals with computations of a distance traveled in a changing acceleration, and is designed to involve them in the construction of a model which is new to them: a model of quantity accumulation. From the work of a 7th grade student Sue, studied by Thompson (1994), we learn about the difficulties that might arise not so much in the attempt to reach a reasonable estimated answer, but in explaining the complexities and the actions taken in this situation. Thompson (1994) asked Sue to evaluate the distance an accelerating car traveled in a time unit. The Car Problem in our study was more constraining in the sense that it required the student to figure out the properties of the motion based on the given data and "zoom into" each of the two units intervals and describe the continuous motion in the same manner. Beyond the emergence of models, we were further interested in this problem because it provides discrete data for a continuous process. There is evidence (Bell, Breker, & Swan, 1987; Bergeron & Herscovics, 1982) that often, point plotting does not lead to an appropriate global view and to an analysis of the generality of the curve. This study attempts to explore how learning based on models of continuous functions is employed when discrete numerical values describe the phenomenon. The following is an analysis of major trends of modeling attempts taken by over 30 pairs[2] of 7th graders.[3]

Examining the given data table, all the students started by indicating that the speed should be described as a decreasing curve with a decreasing (in its absolute value) rate of change. In order to find the distance, some of them looked for an appropriate model to support the completion of the missing data within the intervals, and then looked for an answer to the distance question. Others decided that the search for the intermediate values was irrelevant, either because there are infinite number of such values or because they were not convinced that such computations would result a point that would fit the given description of the situation. Thus, their major concern was the appropriateness of the estimations of the distance. The most visual and global approach was probably taken by Gal and Alon. They chose from the three possible increasing lines the one with decreasing rate of change (in its absolute value), as a model of the traveled distance versus time. Using visual considerations and stairs as a model of variation, they explained the accumulation curve in conjunction with the speed curve.

---

[2] Students were used to work in pairs or other small size groups in the math class and it was natural to watch their work in this setting.
[3] The interviewees were students from several different 7th grade classes in 2 different years in the same school.

Gal: "The distance changes about the same way the speed is changing"
Alon: "This is not a constant rate. The speed decreased slowly."

The car starts from 0 and slowly arrives to the top. It starts by fast increasing and then slow down. The graph [distance with time] has to be inverted [relative to the speed] but is very similar.

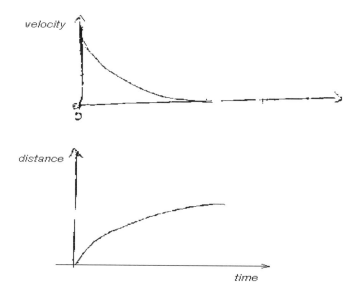

"We do the [distance] graph by the speed—it is hard to explain but ..I can learn from this one. Here is the speed graph: it decreased fast and then slower. The same way that the speed decreased here the distance has to increase there. So if I take the graphs they are the same but inverted. The speed is equal to the distance it moves ...[we went] up each time by what we measured."

The conversation ended here, and it suggests both the strengths and limitations of qualitative modeling. The students analyzed the given data and constructed mathematical models of the speed and the distance. They described the models in terms of properties of functions and treated the models as objects that can be transformed, with the transformation justified by their explanations of the situation. But do their profound constructions and explanations illuminate all nuances of the model? Locally (e.g. the shape of a quadratic and a part of a cubic can indeed be very similar) their analysis makes a lot of sense. Had Alon taken his explanation a step further in the last sentence and tried to draw the graph more accurately, he may have realized that the increase in distance had to be larger than the decrease of speed. But Gal and Alon did not continue and were satisfied with the qualitative model of the situation they had constructed.

Other students were looking for a numerical result, but quickly realized that they could not complete the computation. Once they took into consideration the

26. MODELING-BASED LEARNING ENVIRONMENTS

nonconstant rate of change of the speed, they faced problems. Completing the middle values became a difficult task for Nir, and he explains his hesitations:

> "It is hard to know, one can't know. If the decrease of speed as related to the seconds passed were constant, then the difference [corresponding to the odd values of time in the middle] would have been constant as well, but it is not."

Another pair of students drew a graphical model of the speed in time and hoped to get a better estimation of the distance, but soon realized that their only idea—computing average values in between—would not work.

> Marina: "I think that in the first second the car traveled more than the average and in the second less than the average. I know that the idea of average is not a good one, the graph is not straight, and the differences are changing, but I don't have a better way."

They continued to jot straight segments on the curve trying to estimate various values, but did not come up with a better idea for computing the distance.

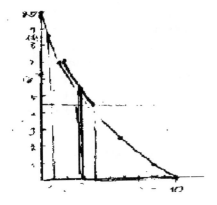

We indeed expected students to look for ways to explore the inner interval as they did in the two previous examples. We were surprised to find a different trend of modeling that treated the absence of information differently. Here are two examples.

Yaron and Omer did massive work with the graphical representation and came up with the idea of upper and lower boundaries. Initially Yaron assumed that the speed in each time interval is constant, but Omer explained why such assumption is invalid:

> "As you describes it, the car traveled at 20 and then suddenly at 14, a sudden fall to 14 [he points on the graph] no, it does not seem possible."

Yaron recognized that completing each time unit value by computing the middle value was a mistake, and he graphed a speed vs. time graph using linear approximation in each interval. From this graph and table he started to compute the distance.

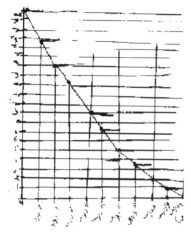

While doing it they refuted the idea that lead to the construction:

Yaron: "That is, assuming the car travels all the time at constant speed, as long as it doesn't say otherwise..."

Omer: "He is trying to use the intermediate values to complete this table, and I don't agree with him because I am not sure that every two minutes that he wrote here the car traveled at the same rate...It is possible that there was a decrease...That's why we can't complete it by the middle value... by the average of these two givens."

He then makes the distinction between the graph as a model of the motion and as a tool for computation:

"According to you, it traveled at 20, and then suddenly decreased to 14. You don't have stops on the way? You assume that the car travels in this way [points to the constant segments on the graphs]... I don't know, maybe it makes sense [to you], but doesn't make sense to me. Maybe it's like this, but I don't think we can base our answer on it." They worked hard on the table and tried to estimate the distance in each interval.

26. MODELING-BASED LEARNING ENVIRONMENTS

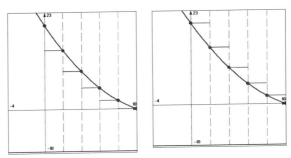

Time	Speed
0	20 — 37
1	17 — 51
2	14 — 62.5
3	11.5 — 71.5
4	9 — 78.5
5	7 — 83.5
6	5 — 87
7	3.5 — 89
8	2 — 90
9	1 — 90
10	0

Yaron: "Oh, so I know, it has to be larger than 90 and smaller than 100. I computed the maximum possible distance and now we will do the minimum."

Actually we can illustrate Yaron and Omer's approach as working in two rounds: The first involves finding the minimum in each interval (and summing up,) and the second involves the same action with the maximum in each interval:

They were not satisfied with merely taking the minimal speed and the maximal one in each interval, and tried to close the gap between them by improving the estimation. They made some technical mistakes and ended up with the wrong estimation $70 < x < 90$. Thus, the curve model of the function did fit to what they considered as the given situation, but not to the way they tried to compute the solution, and it lead to the upper–lower boundaries model.

In the next example, it was the contradiction between the monotonous continuous function as opposed to the given discrete values, that was the source of the complexity met by Guy and Tal, and they treated it using a similar boundaries strategy.

```
  40
  28
  18
 .10
   4
 ----
 100
```

Guy: "On this line it decreased from 20 to 14 …a non-constant decrease. The speed is lower by 6, so approximately the distance for the first 2 seconds is 40."
Tal: "No, it had decreased speed, so let's do the middle. Maybe 17?"
Guy: "But it will still not be accurate."
Tal: "Why?"
Guy: "It would only be accurate if we do it to [a part of] the seconds because the speed changes all the time and not each second…"

Guy and Tal argued about the appropriateness of taking "the middle", and they cross out the possibility of taking the average as a representing measure that could describe the curve. Guy made further efforts to convince his friend and he offered an argument about points on the connecting line between the two middles:

Guy: "If you look at the middle here [shows the horizontal segment of the stair] you will not necessarily approach the middle here [in the vertical segment]. We could find the maximal distance that he might have gone."

A formal interpretation of Guy's argument would read:

$$f(\frac{x_n - x_{n-1}}{2}) \neq \frac{f(x_n) - f(x_{n-1})}{2}:$$

This gave an alternative to Tal who was still thinking about the average, and he suggested:

Tal: "We can do the maximum, the minimum and the average."

And he suggested the following possible estimation for the computed distance [maximum on the right, minimum in the middle and the sum (twice the average) on the left].

26. MODELING-BASED LEARNING ENVIRONMENTS

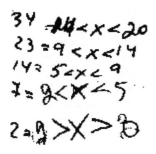

Practically, Guy did not see any way to calculate the distance in the situation that is modeled by a continuous speed curve and he was ready to make a general argument about that:

> Guy: "It decreases continuously, so it's decreasing all the time. We are missing more information. Here, 10 seconds had passed and it changed its speed and every time by a different difference. I don't think there is a way to find an accurate answer. Because the graph [the rate] is not constant and it is impossible to find accurate results.

Guy began to talk about a tool that would help him get infinite small measures and might help to find an answer:

> Guy: "Maybe only if I have a ruler that could measure 1/1,000 of 1/1,000. And then we can tell you what had happened in a 1/3 of a second but not otherwise.. I don't think so."

Discussion

Analyzing the qualities of the rate of change of a phenomena and sketching a graph according to the analysis became a regular activity for these pre-algebra students. This time, the students had to construct a model that went beyond qualitative analysis and to provide a method to get a numerical result. Using the story text and analyzing the pattern of the differences of the given data, they sketched, or more accurately, graphed a curve. The students' approach was to draw a continuous curve of the speed to calculate the distance. This method turned to be a difficult one. To reach a numerical solution for the distance, they had to go through multiple transitions: (a) To recognize the differences between the given discrete data and the continuous model that they chose to describe the speed with; (b) to depart from the continuous curve model and consider discrete models that would appropriately describe the curve (and are not necessarily the given points but rather slanted segments or tangents, etc.); (c) to distinguish qualitatively and quantitatively between the discrete description and the continuous curve; and (d) to use the discrete model to reach a numerical solution. Each transition requires applying a known set of representations of tools for modeling in an unknown state. Throughout the study, pairs of students argue about the correctness of the solution or of any of the intermediate variations of the original curve. In some cases, the students did not find any convincing alternative and ended up either with a qualitative model of the accumulated distance, or with an explanation why it is impossible to give an answer. In other cases, the students were more inventive and tried out one of two approaches: estimation of the accumulated magnitude by its upper and lower boundaries, or linear approximation of segments of the motion. In the former case they used numerical or graphical models or both and acknowledged that they can only provide an estimation within a range. In the latter they acknowledged that the linearity within intervals is a wrong idea, and argued about its deficiencies in terms of the given situation.

ATTEMPTING TO UNIFY THE MODEL OF THE CONCEPT AND THE MODEL OF PHENOMENON

The two groups that worked on the Car Problem—a group of high school students and a group of pre-algebra students—were quite unsimilar. They were used to different representations of different mathematical concepts. Thus we expected that they would take different approaches to the same task. Rather than making a comparative analysis of the groups, we wish to present cumulative observations on their modeling attempts. The calculus group used a wide range of mathematical concepts relevant to the problem: series of arithmetic differences, function expressions, slopes, equations, integrals, discrete sums, and approximations. They showed little concern for the situation itself or the appropriateness of the procedures and formulas for describing the situation. They were often upset when they could not find a numerical answer, and did not consider any non-numerical result to be a mathematical solution. Although they were fluent in symbolic descriptions of functions and operations on them (such as sums or integration), they did not devise any mathematical alternative for describing processes and operations other than a formal description. It seems that this absence of models of functions and of function variations left them in a state of being smart users of the familiar procedures and symbols, but disadvantaged when a new variation on familiar procedures had to be expressed.

The pre-algebra students modeled speed using the terms of the didactical model of variations they had experienced before. But they had to construct new representations and models (such as average and interpolation) when they tried to describe the distance's change and to reach a numerical solution. At each stage of the solution, they provided explanations and reasoning about the attempt. They were not eager to find a specific numerical solution, and were satisfied with an estimation of a range of possible answers or with a decision that a number cannot be computed. They were careful to check that their results did not contradict the model of the situation or the model of function. In terms of the qualities of their understanding of functions and the connection between the physical field and the mathematical field, the skills demonstrated by the two groups working on the same problem suggest an almost empty set. The didactical models used by the 7th graders enabled meaningful physical and

26. MODELING-BASED LEARNING ENVIRONMENTS

mathematical considerations. The role of the didactical model from solely being an object for constructing mathematical concept in the Learning System approach was extended to play a role in mathematical construction of a class of phenomena external to mathematics (in this case, temporal phenomena). The intermediate bridging language of the didactical model helped to form a mathematical construction with language that developed from acquaintance with physical scenarios. It supported the abstraction of everyday phenomena into a smaller set of mathematical signs that are manipulated with software as semi-concrete objects. The language used in their models is a bridging mathematical language, but definitely not yet the powerful formal symbolic language of functions that could illuminate additional aspects of the situation. In closure, we would like to review a few possible ways to use current available technology (software and hardware) to support the emergence of appreciation of symbolic formal language. This may help to further extend the contributions of modeling based learning environments.

Technology allows us to move part of the required expertise from the handmade into the machine-made arena. In the case of the Car Problem, numerical data were given. In other cases, students may run experiments using probes of different sorts and have the computer store and present the data. No matter where the numerical data comes from, one can use a curve fitting software (e.g., calculators, spreadsheets) that would represent the data by a graph and a symbolic expression. We believe that while such tool can appropriately serve those who are interested in working with symbolic models, shifting most responsibility of generating symbolic descriptions to a hidden algorithm is not a desired pedagogical idea, and we should further seek for meaningful ways to describe phenomena symbolically.

Technology provides ways to link situations and their symbolic models in a bi-directional mode. One is the direction from the situation to the model that has been mentioned earlier here. The other allows to make "the absent present" (Nemirovsky & Monk, 2000), and produce phenomena according to given mathematical models. Here the students input into the computer a mathematical model, either graphical (as in the SimCalc environment, Kaput & Roschelle, 1993) or symbolical (Schnepp & Nemirovsky, 2001) and get as an output an example of a phenomenon that behaves according to the model. The parameters of the mathematical model can be manipulated and the synthesized phenomenon changes accordingly. Another attempt to approach the construction of 'symbolic intuitions', one that we are exploring, offers systematic ways of using software components to support the transition from drawn pictures to graphic qualitative mathematical symbols, and through quantitative analysis of these symbols to approach a formal algebraic representation. Based on evidence from work with calculus students (Gravemeijer & Doorman, 1999), on the analysis of the development of algebraic representations (Dennis & Confrey, 1996), and on our own first experiments (Yerushalmy & Shternberg, 2001), we conjecture that symbolic constructions associated with visual properties of rate of change, constitute an important stage in the formation of the concept of function. We wonder whether establishing the didactical model as both a model of the concept of function and as a tool for modeling phenomena would bridge the two signified fields—the formal mathematical and the physical one—in meaningful ways.

Changes in the school algebra curriculum pose potentially fascinating challenges to curriculum developers, software designers and researchers interested in the impact of technology on student learning. The different designs mentioned here exemplify a collection of diversified goals: Teaching mathematical concepts at various levels through physical experimentation before teaching the formality of mathematics; motivating the use of mathematical representations in a meaningful way and helping learners to acquire modeling skills while further understanding functions and calculus in a less formal way. Although research of students' learning, of some of the innovative approaches, points to students appreciating symbolic models, a profound understanding of mathematical symbols and symbols manipulations seems to remain an epistemological obstacle. Whether constructions of mathematical symbolic models are necessary in order to access profound understanding and how to support such construction call for an exciting research agenda into student learning and for inventive uses of technology.

ACKNOWLEDGMENTS

We would like to thank our colleagues Ricardo Nemirovsky, Helen Doerr and Koeno Gravemeijer whose insightful comments on our drafts challenged our thinking and helped us to shape this final version.

VI
NEXT STEPS

Chapter 27

From Problem Solving to Modeling: The Evolution of Thinking About Research on Complex Mathematical Activity

Frank K. Lester, Jr.
Indiana University

Paul E. Kehle
Illinois Wesleyan University

This chapter focuses on the change in thinking over the past 30 years, within the mathematics education research community, about the nature of complex mathematical activity. We begin by looking at the roots of mathematics education research on problem solving, roots deeply planted in the European Gestalt tradition. Then, we offer a synthesis of the problem-solving research by mathematics educators. Finally, we discuss a shift in thinking about problem solving and its relation to other forms of mathematical activity. In brief, we discuss the shift from regarding the act of doing mathematics as consisting of performing low-level procedures (i.e., performing skills and routines), recalling facts and formulas, applying conceptual knowledge, and engaging in high-level problem solving (sometimes referred to as critical thinking or creative thinking) to a view exemplified by many of the other chapters in this volume.

IN THE BEGINNING WERE THE GESTALTISTS[1]

The early experimental work of the Gestaltists in Germany (e.g., Duncker, 1935; Katona, 1940) marks the beginning of interest among psychologists in studying problem solving, an interest that continued through the early 1970s. Psychological research on problem solving was typically conducted with relatively simple, laboratory tasks—for example, Duncker's X-ray problem, Ewert and Lambert's (1932) disk problem (later to be known as the Tower of

---

[1] The discussion in this section comes from the excellent overview of the history of problem solving as a research domain in psychology by psychologists Frensch and Funke (1995). The personal commentaries embedded in this discussion are, of course, our responsibility alone.

Hanoi Problem) and Katona's (1940) matchstick problems. Simple tasks were used because they had clearly defined optimal solutions, they were solvable within a relatively short time period, and subjects' problem-solving steps could be traced. The underlying assumption was that tasks such as the Tower of Hanoi Problem captured the main properties of realistic problems, and that the cognitive processes underlying subjects' solution attempts on simple problems were representative of the processes engaged in when solving realistic problems. Thus, simple problems were used for reasons of convenience, and generalizations to more complex problems were thought possible. In the United States at least, the efforts of the European Gestaltists eventually gave way to a new field of study—cognitive science. Perhaps the best known example of this line of research is the seminal work by Newell and Simon (1972). The ultimate goal of Newell and Simon and of other cognitive scientists of the time was to develop a global theory of problem solving, a theory that accounted for problem-solving performance independently of task domain.

> Comment: Without a doubt, mathematics education researchers of this era (i.e., the mid-1970s to late 1980s) were quick to adopt the cognitive science perspective toward mathematical activity, particularly among those mathematics educators interested in problem solving. Syntheses of mathematical problem-solving research published in the 1980s provide compelling evidence for this claim (e.g., Charles & Silver, 1988; Lester & Garofalo, 1982; Schoenfeld, 1987a; Silver, 1985)[2]

During the period from the 1930s to the early 1970s, the prevailing definitions of problem solving among experimental psychologists (and, later, cognitive scientists) contained three assumptions: (a) the theoretical goal was to understand the cognitive processes of a person solving a problem, (b) cognitive processes were guided by internal goals, and (c) the cognitive processes were essentially the same for all kinds of problems.[3] Problems typically were defined so that they represented situations for the person that could not be solved by the mere application of existing knowledge; as a result, more often than not, problems were domain-general. However, beginning in the 1970s, researchers became increasingly convinced that empirical findings and theoretical concepts derived from simple laboratory tasks could not be extended to more complex, real-life problems. Moreover, it became apparent that the processes underlying complex problem solving in one domain were different from those underlying another domain.

Initiated by the work of Simon on learning in semantically rich domains (e.g., Anzai & Simon, 1979; Bhaskar & Simon, 1977), researchers began to investigate problem solving in different natural knowledge domains (e.g.,

---

[2] At various places throughout this paper, we provide personal comments intended to situate traditional mathematics education research in the broader issues we are discussing.

[3] The third assumption was implicit in all definitions of problem solving, rather than explicitly stated.

physics, writing, and chess playing) and gave up on their attempt to develop a global theory of problem solving. These researchers frequently focused on the development of problem solving expertise within a certain domain (e.g., Anderson, Boyle, & Reiser, 1985; Chase & Simon, 1973; Chi, Feltovich, & Glaser, 1981). Indeed, until quite recently, psychologists were content to compare the cognitive processes of experts and novices engaged in solving tasks in natural knowledge domains; that is, tasks that could be considered problems for novices, but exercises for experts.

The relatively recent realization among cognitive scientists that problem-solving processes differ across knowledge domains and across levels of expertise and that findings obtained in the laboratory cannot necessarily be generalized to problem-solving situations outside the laboratory ultimately has led to the current emphasis on real-world problem solving (see, e.g., Frensch & Funke, 1995, for extended discussions of contemporary research on complex problem solving by experimental psychologists and cognitive scientists).

> Comment: Educational researchers, mathematics educators among them, did not immediately follow the lead of their cognitive science counterparts with respect to developing a domain-specific focus. It was not until the 1980s that a critical mass of mathematics education researchers began to call for studying problem solving in the context of learning and doing mathematics, rather than as a separate topic. Furthermore, the aforementioned recent realization is one consequence of the complex interaction between research agendas and research methodologies. What is studied both shapes and is shaped by the available tools. Although discussed only briefly here, this point is an important one to bear in mind while reading this chapter and this book.

Among the various topics studied by experimental psychologists and cognitive scientists, perhaps none has received more attention than the notion of expertise. Psychologists have been studying expertise at least since the beginning of the 20th century, but the modern study of expertise began in earnest about 35 years ago and continues to the present day. However, very few of the studies of expertise in problem solving have taken seriously the question of just what is an expert; indeed, according to Sternberg (1995), a leading researcher in the area of human intelligence, at least 10 views of expertise have been prominent at one time or another among these researchers.[4] In general, typical cognitive science work has compared experts to novices in the performance of tasks requiring high levels of professional expertise, and the study of differences between experts and novices in the natural sciences and mathematics has predominated in this research. As Hunt (1991) has pointed out,

---

[4] These views have ranged from those who regard experts as people who simply know more than non experts, to those who think experts solve problems using different processes and more efficiently than non experts, and from those who regard experts as people who organize their knowledge more effectively than non experts, to those who think of experts as having greater analytic ability than non experts.

the view of expertise that has emerged from this research is "one of a problem solver who has memorized a huge number of tricks of the trade, rather than one who has developed a single powerful strategy for reasoning. Local optimality acquired at the expense of global consistency" (p. 394). Briefly put, cognitive scientists have tended to study expertise in knowledge-rich domains, and have attributed expertise to greater knowledge about these domains.

Comment: Surely, after so much research cognitive science has more to offer than the rather obvious observation that experts (usually) know more than novices and non experts! Moreover, we wonder why there has been so much interest in studying expertise. We believe that the intensive study of extremely specialized pursuits holds little promise for informing us about the questions and issues that should be of interest to mathematics educators. In the first place, there is the issue of relevance: What percentage of Americans become mathematicians, or actually even ever study math beyond basic levels? Second, the underlying assumption of this emphasis is that what is learned about expertise in some specialized domain generalizes to other domains. There is ample reason to doubt this. Do everyday people think like mathematicians? Third, studying how experts think, make decisions, and solve problems does not guarantee that one is studying the experts at what actually makes them experts. For example, one can certainly give expert mathematicians textbook problems to solve, and compare their strategies and mental representations to those of novices. But expert mathematicians do not solve textbook problems for a living. Our point is that even if cognitive scientists have studied the right people, they may have studied them doing the wrong tasks. Fourth, a lot of attention has been given to differences between experts and novices, but relatively little attention has focused on individual differences among experts. Clearly, expert scientists, writers, or artists do not all succeed in the same way. It is not clear that there is any common thing that makes them experts. In short, without considering individual differences, it is difficult to fully understand expertise. Finally, even if there was a robust theory of expertise that identified what makes an expert an expert, it is doubtful that this theory would provide any information about how to make anyone else an expert. There has been little, if any, research on how experts become experts.

RESEARCH ON PROBLEM SOLVING BY MATHEMATICS EDUCATORS

An argument could be made that systematic research on problem solving in mathematics began in the early 1970s because in May 1975, a "Research Workshop on Problem Solving in Mathematics Education" was held at the University of Georgia (Hatfield & Bradbard, 1978). Three years later a conference was held on applied mathematical problem solving at Northwestern

27. FROM PROBLEM SOLVING TO MODELING

University (Lesh, Mierkiewicz, & Kantowski, 1979). The Georgia workshop brought together individuals who were already deeply involved in mathematical problem-solving research and, more than any other single event, stimulated a level of collaboration among mathematics education researchers that had not existed earlier. The Northwestern conference involved researchers who were beginning to recognize that little progress had been made "toward dealing with the issue of what it is beyond having an idea, that allows a normally intelligent person to use the idea to deal with the math related problems in everyday situations" (Lesh et al., 1979, p. 1). The very fact that these two meetings were held is evidence that during the mid 1970s a critical mass of researchers actively engaged in the study of problem solving in mathematics was forming.

Since the appearance of a review of the research on mathematical problem solving by Kilpatrick (1969) more than 30 years ago, extensive reviews of the research literature have appeared from time to time—for example, those by Lester (1980, 1983) and Schoenfeld (1992). There also have been several attempts to interpret problem-solving research for classroom practice (Driscoll, 1981, 1983; Hembree & Marsh, 1993; Kroll & Miller, 1993; Suydam, 1980; Wilson, Fernandez, & Hadaway, 1993). In addition, several comprehensive analyses of problem-solving research have been published, notably, in chronological order, by Lester and Garofalo (1982), Schoenfeld (1985), Silver (1985), and Charles and Silver (1988). (The Charles and Silver volume is particularly noteworthy because it resulted from one of four "Research Agenda Project" conferences under the sponsorship of NCTM held in 1987. The conference was a response to needs in schools related to the teaching and evaluation of problem solving as well as to the lack of research-based knowledge in these areas.) Thus, rather than offering yet another analysis, we provide brief discussions of those areas of problem-solving research that received the most attention over the past 30 years.

Research Emphases and Methodologies

In his seminal review, Kilpatrick (1969) characterized the preponderance of research literature on mathematical problem solving as atheoretical, unsystematic and uncoordinated, interested almost exclusively in standard textbook word problems, and restricted completely to quantitative measures of problem-solving behavior. It would be an overstatement to say that problem-solving research today is theory-based, systematic and coordinated, concerned with a wide range of problem types, and conducted by researchers using a wide range of methods—both quantitative and qualitative. Nevertheless, there is no doubt that the nature of the research on problem solving has matured tremendously in the past 30 years.

How the research has been conducted perhaps best illustrates the evolution of the field. At the 1975 Georgia Workshop, Kilpatrick (1978) noted: "A broader conception of research is needed, and an openness to new techniques, if studies of problem-solving processes and the teaching of heuristics are to have an impact" (p. 18). Ten years later he observed that it is the changes in our approach to the study of problem solving that are the most notable (Kilpatrick, 1985).

Table 27.1 is perhaps an oversimplified encapsulation of the past 30 years of research in the field. We include it to help frame the research and highlight the aspects of problem solving that have been primary foci of interest. Although it is not readily evident from looking at Table 27.1, our study of the literature uncovered four areas of inquiry in which noticeable progress has been made: (1) determinants of problem difficulty, (2) distinctions between good and poor problem solvers, (3) attention to problem-solving instruction, and (d) the study of metacognition in problem solving (cf. Kilpatrick, 1985; Schoenfeld, 1992; Silver, 1985). Each area is discussed briefly next.

TABLE 27.1
An Overview of Problem-Solving Research
Emphases and Methodologies: 1970–1995

Dates[a]	Problem solving Research Emphases	Research Methodologies Used
1970-1982	Isolation of key determinants of problem difficulty; identification of characteristics of successful problem solvers; heuristics training.	Statistical regression analysis; early "teaching experiments."
1978-1985	Comparison of successful & unsuccessful problem solvers (experts vs. novices); strategy training.	Case studies; "think aloud" protocol analysis.
1982-1990	Metacognition; relation of affects/beliefs to problem solving; metacognition training.	Case studies; "think aloud" protocol analysis.
1990-1995	Social influences; problem solving in context (situated problem solving).	Ethnographic methods.

[a] Of course, the dates shown are only approximate; however, the chronology is reasonably accurate.

Determinants of Problem Difficulty. From the early 1970s until the early 1980s, quite a lot of attention was devoted to the study of determinants of problem difficulty. These efforts focused almost exclusively on features of the types of problems students were asked to solve in school. In 1979, a landmark book was published that synthesized the research on what were then referred to as "task variables" in mathematical problem solving (Goldin & McClintock, 1979/1984). To briefly summarize this and other closely related research, four classes of variables were identified that contribute to problem difficulty: content and context variables, structure variables, syntax variables, and heuristic behavior variables. Initially, these classes were studied via linear regression models, later via information-processing techniques. This line of inquiry was replaced eventually by investigations of the interaction between task variables and the characteristics of the problem solver (Kilpatrick, 1985).

27. FROM PROBLEM SOLVING TO MODELING

Today there is general agreement that problem difficulty is not so much a function of various task variables, as it is of characteristics of the problem solver such as traits (e.g., spatial visualization ability, ability to attend to structural features of problems), dispositions (e.g., motivation and beliefs), and experiential background (e.g., instructional history, familiarity with types of problems). Perhaps what is most notable about the early attempts to identify what makes problems difficult for students was that they marked the beginning of efforts to make mathematical problem-solving research more systematic and analytic.

> Comment: This new understanding of problem difficulty is consistent with constructivists' urging that we, as researchers or teachers, shift from a task-centered perspective to a student-centered one. Such a perspective is also consistent with a semiotic approach to the study of problem-solving behavior: Within this approach models, representations, symbols, and skills are inert and meaningless when considered separately from a person using them.

Distinctions Between Good and Poor Problem Solvers. Beginning in the late 1970s and continuing until the mid-1980s, problem-solving research focused primarily on individual problem-solving competence and performance (Charles & Silver, 1988). More specifically, a substantial amount of attention was devoted to the study of the distinctions between so-called "good" and "poor" problem solvers. This emphasis probably stemmed, at least initially, from work being done in the rapidly developing fields of cognitive science and artificial intelligence. Schoenfeld (1985, 1987a, 1987b) conducted the most extensive work in this area. Briefly put, this research suggests that good problem solvers can be distinguished from poor problem solvers in at least five important respects:

1. Good problem solvers know more than poor problem solvers and what they know, they know differently—their knowledge is well connected and composed of rich schemas.
2. Good problem solvers tend to focus their attention on structural features of problems, poor problem solvers on surface features.
3. Good problem solvers are more aware than poor problem solvers of their strengths and weaknesses as problem solvers.
4. Good problem solvers are better than poor problem solvers at monitoring and regulating their problem-solving efforts.
5. Good problem solvers tend to be more concerned than poor problem solvers about obtaining elegant solutions to problems.

In addition to the concerns we posed earlier in this chapter about expert—novice research, Lesh (1985) cautioned that pinpointing the ways that experts solve problems and then trying to teach these ways to novices in a short amount of time may not result in desirable outcomes. He suggested that efforts to teach novices how to be experts must take into account the possibility that problem-solving processes and heuristics develop slowly over time in much the same way that other mathematical ideas are known to develop.

Problem Solving Instruction. About 15 years ago Lester observed that of the several, then new, problem-solving programs that had been created since 1975, none of them was firmly based in research (Lester, 1985). The situation is not so different today. What exists now are programs based largely on: (a) the folklore of mathematics teaching, particularly the sage advice of master teacher and problem solver George Pólya (1945/73), and (b) a vague understanding of constructivist principles. As good as some of these programs may be, few of them are firmly based on a solid foundation of what is known about how students learn and how they should be taught.

Does the problem-solving research literature have anything of value to guide instructional decision making? The answer is that, although the literature on problem-solving instruction presents ambiguous messages, five results stand out:

1. Students must solve many problems in order to improve their problem solving ability.
2. Problem solving ability develops slowly over a prolonged period of time.
3. Students must believe that their teacher thinks problem solving is important in order for them to benefit from instruction.
4. Most students benefit greatly from systematically planned problem-solving instruction.
5. Teaching students about problem-solving strategies and heuristics and phases of problem solving (e.g., Pólya's, 1945/1973, four-phase problem-solving model) does little to improve students' ability to solve general mathematics problem (cf. Zawojewski & Lesh, chap. 18 this volume).[5]

The Role of Metacognition in Problem Solving. Research on the role of metacognition in mathematical activity, especially mathematical problem solving, has been concerned with two related components: (1) knowledge of one's own thought processes, and (2) regulation and monitoring (also referred to as "control") of one's activity during problem solving.[6] Among the first references to the role of metacognition in mathematical problem solving, papers written in the same year by Lesh (1982), Schoenfeld (1982), and Silver (1982) were especially influential in drawing attention to the importance of this elusive construct. These three researchers regarded metacognitive actions as "driving

---

[5] Alternatives to "teaching about problem solving" are discussed elsewhere—in particular, "teaching via problem solving" (Lester, Maki, LeBlanc, & Kroll, 1992; Schroeder & Lester, 1989). However, these alternatives have not been subjected to enough research scrutiny to make any claims about their potential.

[6] Schoenfeld (1992) adds a third category to this list: beliefs and intuitions.

forces" in problem solving, influencing cognitive behavior at all phases of problem solving. A spate of research followed in which metacognition was a primary focus. By the end of the 1980s, metacognition not only was regarded as a force driving cognitive behaviors, but also was being linked to a wide range of non cognitive factors—in particular, beliefs and attitudes (Lester, Garofalo, & Kroll, 1989; Schoenfeld, 1987b). The degree to which metacognition influences problem-solving activity has not been resolved, but three results have come to be generally accepted:

1. Effective metacognitive activity during problem solving requires knowing not only what and when to monitor, but also *how* to monitor. Moreover, teaching students how to monitor their behavior is a difficult task.
2. Teaching students to be more aware of their cognitions and better monitors of their problem-solving actions should take place in the context of learning specific mathematics concepts and techniques (general metacognition instruction is likely to be less effective).
3. The development of healthy metacognitive skills is difficult and often requires unlearning inappropriate metacognitive behaviors developed through previous experience (Schoenfeld, 1992).

The Situation Today

There has been a noticeable drop off in the amount of research activity related to mathematical problem solving during the past 5 years or so. We can only speculate about the reasons for this change, but two possibilities come readily to mind.

First, by its very nature, problem solving is an extremely complex form of human endeavor that involves much more than the simple recall of facts or the application of well-learned procedures. Furthermore, as noted earlier, the ability to solve mathematics problems develops slowly over a very long period of time because success depends on much more than mathematical content knowledge. Problem-solving performance seems to be a function of several interdependent categories of factors including: knowledge acquisition and utilization, control, beliefs, affects, socio cultural contexts, implicit and explicit patterns of inference making, and facility with various representational modes (e.g., symbolic, visual, oral, and kinesthetic). These categories overlap (e.g., it is not possible to completely separate affects, beliefs, and socio-cultural contexts) and they interact in a variety of ways too numerous to discuss in these few pages (cf. Lesh, Lester, & Hjalmarson, chap. 21, this volume; Lester et al., 1992; Schoenfeld, 1992). There is general agreement that new perspectives are needed regarding the nature of problem solving and its role in school mathematics.

> Comment: Supplying and informing these perspectives is the focus of the remainder of this chapter, and the degree to which the categories of factors just mentioned overlap is crucial to our point of view.

A second possible reason for the decline of interest in studying problem solving is the opposite of the first one, that is, the mathematics education community thinks it knows all there is to know about problem solving. It is commonplace today for research articles, curriculum guides, and other reports to refer to the various *Standards* documents of the National Council of Teachers of Mathematics (NCTM) and related mathematics education reform materials as authoritative sources to develop a rationale for their activities (e.g., NCTM, 2000). In particular, all too often, researchers use the emphasis these documents give to problem solving as a rationale for their studies, instead of firmly positioning their work in research. Apparently they believe that there no longer is a need to refer to the body of relevant literature on problem solving to justify their work—instead, the *Standards* have become the authority. But, as the analysis in the previous section indicates, far too little is known about students' learning in mathematically—rich environments for research on problem solving to end. As we demonstrate in the next section, perhaps a different perspective on problem solving must be adopted in order to make problem solving the hot topic it was for the previous generation of researchers. We now shift our attention from discussing what was, to what might (should) be.

THINKING DIFFERENTLY ABOUT MATHEMATICAL PROBLEM SOLVING

Problem solving is an activity requiring the individual to engage in a variety of cognitive actions, each of which requires some knowledge and skill, and some of which are not routine.[7] Furthermore, these cognitive actions are influenced by a number of non cognitive factors. Although it is difficult to define problem solving, the following statement comes close to capturing our sense of what it involves:

> Successful problem solving involves coordinating previous experiences, knowledge, familiar representations and patterns of inference, and intuition in an effort to generate new representations and related patterns of inference that resolve the tension or ambiguity (i.e., lack of meaningful representations and supporting inferential moves) that prompted the original problem-solving activity.

It is probably safe to say most teachers agree that the development of students' problem-solving abilities is a primary objective of instruction. Although acceptance of the notion that problem solving should play a prominent role in the curriculum has been widespread, there has been anything but

---

[7] Actually, in this section we want to begin to purposely blur the distinction between problem solving and other types of mathematical activity because, we think the distinction is both unhelpful and unnecessary. We might well have opened this paragraph by stating "Doing mathematics is an activity. . . ."

27. FROM PROBLEM SOLVING TO MODELING 511

widespread acceptance of how to make it an integral part of the curriculum. To date, no school mathematics program has been developed in the United States that adequately addresses the issue of making problem solving the central focus of the curriculum. Instead of programs with coherence and direction, teachers have been given a well-intentioned mélange of story problems, lists of strategies to be taught, and suggestions for classroom activities. If problem solving is to become a more prominent goal of mathematics instruction, more serious and thoughtful attention must be given to what it means to make problem solving the focus of school mathematics. Thus, for students who are struggling to become better problem solvers, their difficulty due to the complexity of problem solving is compounded by the fact that most of them do not receive adequate instruction, in either quality or quantity.

Comment: By now, it should be apparent that we believe the roles played by representations and patterns of inference in problem-solving activity and in responses to the model-eliciting tasks described in this volume need more careful study. Although a few teachers turn to Polya's (1968) less well known volumes, *Mathematics and Plausible Reasoning*, for guidance, the interest in inference we urge must be broader in scope than his very narrow treatment of it wholly within rarefied mathematical contexts.

Our view is that the relative ineffectiveness of instruction to improve students' ability to solve problems stems in large part from the fact that problem solving has often been conceptualized in a simplistic way by a perspective like the one shown in Fig. 27.1. This perspective has two levels or "worlds": the everyday world of things, problems, and applications of mathematics; and the idealized, abstract world of mathematical symbols and operations. In this perspective, the problem-solving process has three steps. Beginning with a problem posed in terms of physical reality, the problem solver first translates (Arrow A) the problem into abstract mathematical terms, then operates (Arrow B) on the mathematical representation in order to come to a mathematical solution of the problem, which is then translated (arrow C) into the terms of the original problem. According to this view, mathematics may be and often is learned separately from its applications. Teachers who adhere to this perspective are very concerned about developing skillfulness in translating real world problems into mathematical representations and vice versa. However, they tend to deal with problems and applications of mathematics only after the mathematical concepts and skills have been introduced, developed, and practiced.

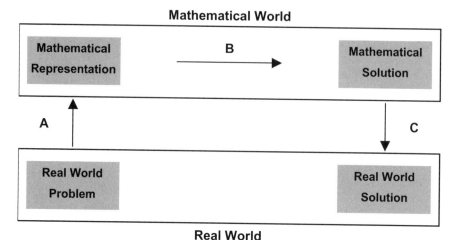

FIG. 27.1. A naïve perspective of mathematical problem solving.

The difficulty with this naive perspective is that it applies to quite routine "story" problems better than to non routine ones. Problems like those that have been classified as "translation problems" (Charles & Lester, 1982) often are solved exactly as the naive perspective indicates. But for more challenging, substantive problems, such as the model-eliciting tasks (METs) discussed throughout this book, the problem solver cannot simply apply a previously learned procedure to solve the problem. In addition to translation and interpretation, these METs also demand more complex processes such as planning, selecting strategies, identifying sub-goals, conjecturing, and verifying that a solution has been found. Put simply, for non routine tasks, a different type of perspective is required, one that emphasizes either the discovery and/or making of new meanings through construction of new representations and inferential moves.

Figure 27.2 shows a modification of the problem-solving perspective for translation problems that can be used to illustrate thinking processes when certain types of non routine problems are involved. Like the previous perspective, this one also contains two levels representing the everyday world of problems and the abstract world of mathematical symbols and operations. In this perspective, however, the mathematical processes in the upper level are "under construction" (i. e., being learned, as opposed to already learned; coming to be understood, as opposed to being understood) and the most important features are the relationships between steps in the mathematical process (in the upper level) and actions on particular elements in the problem (in the lower level). It is in the forging of these relationships that Fig. 27.2 implies the meaning making that is central to mathematical activity of all kinds.

27. FROM PROBLEM SOLVING TO MODELING

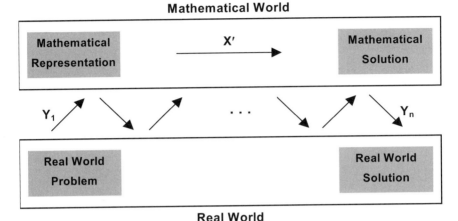

FIG. 27.2. A modified perspective of mathematical problem solving.

In the figure, some of the Y_i arrows point upward to indicate that the problem solver is learning to make abstract written records of the actions that are understood in a concrete setting. These upward-pointing arrows represent the processes of abstraction and generalization. And, some of the arrows point downward to show that the problem solver is able to connect a mathematical process to the real-world actions that the mathematical process represents. A downward-pointing arrow might also be used to suggest that a problem solver who had forgotten the details of a mathematical procedure would be able to reconstruct that process by imagining the corresponding concrete steps in the world in which the problem was posed. The collection of Y_i arrows taken together illustrate that typically the problem solver moves back and forth between the two worlds—the real and the mathematical—as the need arises. It may happen that, for a particular problem, the problem solver is able to move directly from the real world along Arrow Y_1 to the mathematical world and proceed along Arrow X' directly to a mathematical generalization and hence to a solution to the original real-world problem (In such a case the solution process is essentially the same as that modeled by Fig. 27.1).

Unfortunately, despite the fact that this perspective is an improvement over the original, it too falls short of what is needed because it does not account for many of the most important actions (both cognitive and non cognitive) involved during real problem solving. Figure 27.3 is yet another modification—a major modification—of the original perspective depicted in Fig. 27.1 (cf. Kehle & Lester, chap. 5, this volume). This figure subsumes problem solving within a much broader category—"mathematical activity"—and gives a prominent role to metacognitive activity engaged in by the individual.[8]

---

[8] In this discussion we refer to the "individual" as being engaged in some mathematical activity, but we recognize that much of the sort of mathematical activity discussed throughout this book involves groups, usually of students. The discussion applies to the mathematical behavior of groups working collaboratively as well.

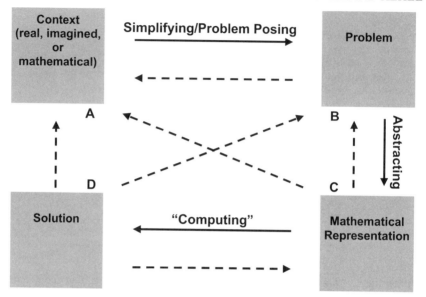

Note: Dashed arrows denote "comparing."

Fig. 27.3. An "ideal" model of mathematical activity.

We hasten to point out that this depiction is a representation of ideal, rather than typical, performance during an individual's work on some mathematical task. It is ideal in the sense that it denotes key actions in which the individual should engage in order to obtain acceptable results, multiphase process that begins when an individual, working in a complex realistic, imaginary, and mathematical context (Box A), poses a specific problem (the solid arrow between A and B). To start solving the problem, the individual simplifies the complex setting by identifying those concepts and processes that seem to bear most directly on the problem. This simplifying and problem posing phase involves making decisions about what can be ignored, developing a sense about how the essential concepts are connected, and results in a realistic representation of the original situation. This realistic representation is a model of the original context from which the problem was drawn because it is easier to examine, manipulate, and understand than the original situation.

Next comes the abstraction phase (solid arrow from B to C), which introduces mathematical concepts and notations (albeit perhaps idiosyncratic). This abstraction phase involves the selection of mathematical concepts to represent the essential features of the realistic model. Often the abstraction phase is guided by a sense of what a given representation makes possible in the subsequent computation phase. The explicit representation of the original setting and problem in mathematical symbolism constitutes a mathematical representation of both the setting and problem.

Once a problem solver has generated a mathematical representation of the original situation, the realistic problem now becomes a specific mathematical problem related to the representation. This mathematical problem acquires a meaning all its own, becoming an isolated, well-defined mathematical problem (Box C).

The third phase of the process (from C to D) involves manipulating the mathematical representation and deducing some mathematical conclusions—depicted in Fig. 27.3 by the computing arrow. During this phase, a person's store of mathematical facts, skills, mathematical reasoning abilities, and so forth, are relied on. For example, the problem might call for a solution of a system of equations and solving this system of equations does not depend on the original context of the initial problem. The final phase (from D to A, D to B, and D to C), then, should involve the individual in comparing the conclusions/results obtained with the original context and problem, as well as with the mathematical representation (refer to the dashed arrows between boxes). But, the act of comparing does not occur only after conclusions are drawn and a solution is obtained. Rather, it might take place at any time and at any point during the entire process. Indeed, research indicates that this regular and continual monitoring—metacognitive activity—of one's work is a key feature of success on complex mathematical tasks. In general, the act of comparing the current state of one's work, thinking, and decisions denotes how complex mathematical activity can be. The degree to which the individual chooses to compare her or his current state with earlier states can be considered a determinant of task complexity. For example, performing routine calculations using whole numbers typically requires little comparing, whereas work on the Linear Irrigation Problem (see Kehle & Lester, chap. 5, this volume) usually necessitates quite a lot of comparing throughout ones work on it. Consequently, the Linear Irrigation Problem is both more complex and involves more high-order thinking than a straightforward calculation.

FINAL THOUGHTS

The Ideal Model portrayed in Fig. 27.3 is more likely than previous models to be useful in researching students' mathematics learning in general and problem-solving behavior in particular. Kehle and Lester (chap. 5, this volume) provide one detailed examination of the cognitive movement (indicated by the arrows) between only Boxes B and C in Fig. 27.3. The semiotic approach[9] used in that work has potential for revealing more about other aspects of students' mathematical thinking framed by the Ideal Model.

As mentioned in footnote #6, we see a fruitful blurring of problem solving and other mathematical activity emerging from research on mathematical

---

[9] The rich nature of the cognitive activity portrayed by the Ideal Model and by our discussion of the evolution of thinking about problem solving demands that research in this area utilize multiple approaches. Although we see much value in semiotic approaches, this emphasis does not preclude other valuable alternatives.

problem solving and constructivist thinking about learning. In concluding, we would like to propose that this blurring be taken even further. Our work, and that of many of the authors represented in this volume, implies a blurring of task, person, mathematical activity, nonmathematical activity, learning, applying what has been learned, and other features of mathematical problem solving. As researchers have zeroed in on particular aspects of problem solving, they often have found salient issues within their narrow fields of view that have demanded expansion of those views.

Many of the distinctions that led historically to the isolation of mathematical problem solving as a research focus and subsequent distinctions that resulted from this isolation are due in part to strong traditions of disciplinary boundaries. One advantage of current approaches within cognitive science is the interdisciplinary collaboration (another type of blurring) researchers have been driven to in their quest for a fuller understanding of human cognition.

To some, our blurring metaphor might suggest less precision. We disagree. Instead, we are coming to see, just as sharply as past researchers have seen through their models, richness and complexity that was previously unseen, or at least not seen as important. Figure 27.4 shows how the blurring has resulted in an inversion in the way the Ideal Model frames complex mathematical activity. In the top oval, mathematical understanding was subsumed under problem solving. In the bottom oval, mathematical understanding subsumes problem posing and solving (along with which comes many of the top oval's concepts), and emphasis is placed on several other concepts absent in the top oval.

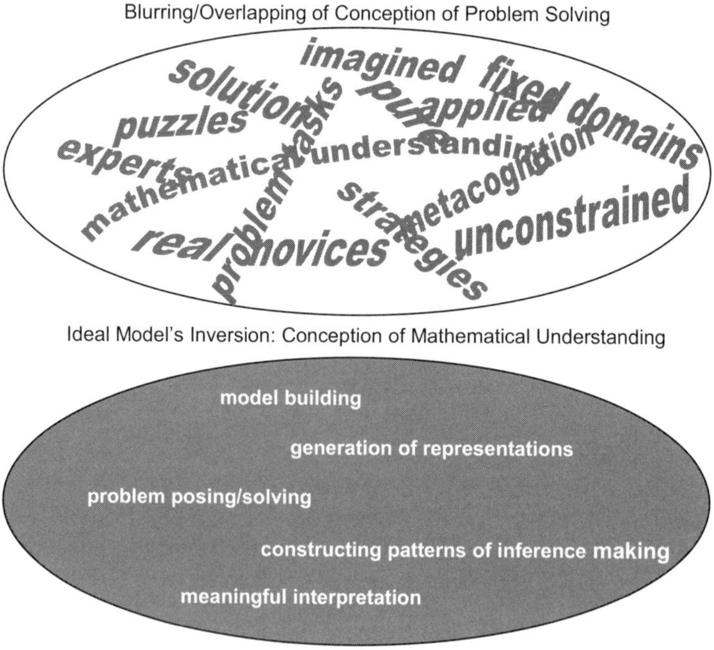

Blurring/Overlapping of Conception of Problem Solving

(oval containing overlapping words: solutions, imagined, fixed, puzzles, applied, experts, mathematical, understanding, domains, tasks, real, problems, novices, strategies, metacognition, unconstrained)

Ideal Model's Inversion: Conception of Mathematical Understanding

(oval containing: model building, generation of representations, problem posing/solving, constructing patterns of inference making, meaningful interpretation)

FIG. 27.4. Blurring and inversion of conceptions of mathematical activity.

27. FROM PROBLEM SOLVING TO MODELING

As one last example of productive blurring, consider the notion of model-eliciting tasks discussed in this volume. We see the use of the word *model* in this context as a unification of the traditional sense of mathematical model, and something more akin to cognitive scientists' mental models that codify understanding. This unification echoes one of our earlier comments regarding the inertness of models when considered independently of the person using them.

In conclusion, the blurring and inversion traced in Fig. 27.4 results in a more authentic view of students' cognitions as they exist in busy classrooms and in complex realistic settings. In fact, the richer view implied by the Ideal Model is truer to what teachers see in their classrooms as they seek to develop their students' ways of thinking.

Against the charge that all we have done is prescribe an enlarging of the spotlight research trains on mathematical problem solving, we note that new methodological illumination can now be provided, not simply broader illumination. Thus, the essence of one of our opening comments applies here as well.

> Comment: The complex interaction between methodology and subject—the way they mutually define each other—continues to shape our understanding of mathematical problem solving. We believe the perspectives shared in this book help enable researchers to generate richer, more authentic accounts of human sense making that possesses a strong mathematical dimension.

Chapter 28

In What Ways Does a Models and Modeling Perspective Move Beyond Constructivism?

Richard Lesh
Purdue University

Helen M. Doerr
Syracuse University

The first half of this chapter briefly describes some of the most significant philosophical foundations of models and modeling (M&M) perspectives. Then, in the second half, a two-column table format is used to describe a variety of ways that we believe a models and modeling perspective moves beyond constructivist foundations for thinking about the nature of children's developing mathematical knowledge and abilities–and about mathematics teaching, learning, and problem solving. The left hand column identifies significant contributions that constructivism (in many variations) has made to the field of mathematics education (Steffe & Thompson, 2000). The right hand column briefly describes some of the most important points that have been described in greater detail throughout the chapters of this book concerning: (a) the nature of children's developing mathematical knowledge, (b) the nature of real life situations (beyond school) where elementary-but-powerful mathematical constructs are useful, (c) the nature of mathematical understandings and abilities that contribute to success in the preceding problem solving situations, and (d) the nature of teaching and learning situations that contribute to the development of the preceding understandings and abilities. ... Overall, the table is organized around the following questions.

- What Does It Mean To Be a Constructivist?
  o What is Different About *Assembly-ists*, *Constructivists* and *CONSTRUCT-ivists*?
  o Do Constructs Need To Be Discovered Independently By Students?
  o What is the Relationship Between the Mastery of *Basics Skills* and the Development of *Deeper and Higher Order Understandings and Abilities*?
  o What is the Nature of "Real" Experience?
  o Can We Know What is in the Mind Of Another Person?

519

- In What Ways Do M&M Perspectives Go Beyond Piagetian Views About Students' Developing Mathematical Constructs?
  o In What Ways Do M&M Perspectives Go Beyond Cognitively Guided Instruction?
  o In What Ways Do M&M Perspectives Go Beyond Ladder-Like Stages of Development?
  o In What Ways Do M&M Perspectives Go Beyond Research on Situated Cognition?
- In What Ways Do M&M Perspectives Go Beyond Research On Vygotsky's Zones of Proximal Development?
  o In What Ways Do M&M Perspectives Go Beyond Social Constructivism?
  o In What Ways Do M&M Perspectives Go Beyond Current Research On Students' Learning Trajectories?
  o In What Ways Do M&M Perspectives Go Beyond Current Research on Representation and Semiotic Functioning?
- In What Ways Do M&M Perspectives Go Beyond Research on What is Needed for Success Beyond School of Education in a Technology-Based Age of Information?
  o In What Ways Do M&M Perspectives Go Beyond Traditional Research on Problem Solving—And Information Processing?
  o In What Ways Do M&M Perspectives Go Beyond Current Research on Metacognition and Higher Order Thinking?
  o In What Ways Do M&M Perspectives Go Beyond Constructivist Research on Teacher Development?

A BRIEF OVERVIEW OF ISSUES EMPHASIZED THROUGHOUT THIS BOOK

Part I of this book, titled *Introduction to a Models and Modeling Perspective*, emphasizes that thinking mathematically is about interpreting situations mathematically at least as much as it is about computing. Mathematical models are the kind of conceptual tools that mathematicians use to interpret situations mathematically; and, the processes of interpretation (description, explanation, or construction) involve imposing schemes that organize, systematize, dimensionalize, coordinatize, or in general mathematize the relevant problem solving situations. Therefore, some of the most important goals of mathematics instruction should focus on developing these mathematizing abilities, and on developing a collection of elementary-but-powerful mathematical models.

Part II, titled *A Models and Modeling Perspective on Teacher Development*, focuses on teachers' developing knowledge. As students develop models for making sense of mathematical decision-making situations, teachers develop models for making sense of students' model development behaviors, and researchers develop models for making sense of interactions between teachers and students during model development activities. In all cases, people (students, teachers, researchers) develop by expressing their current ways of thinking in forms that must be tested and refined or revised repeatedly. Therefore, teacher development programs can be based on principles similar to those that govern the tasks that promote student development.

Part III, titled *Models and Modeling as Viewed by Heavy Users of Mathematics*, emphasizes that, in a technology-based *age of information*, preparation for success beyond school includes the ability to construct, describe, explain, predict, or manipulate complex systems. Such mathematically interesting systems range from financial planning systems, to communication systems, to school systems, and to environmental and health-care decision making. Making sense of complexity—while working in teams of specialists, adapting to continually evolving technical tools, and communicating about the nature of complex artifacts that are produced as intermediate and final results—is at the heart of what is needed for success in the 21st century. In other words, we claim that mathematical models and modeling abilities should be among the most important goals of mathematics instruction

Part IV, titled *Models and Modeling in Problem Solving and Learning*, describes significant ways that conceptions of problems solving abilities need to change when attention shifts beyond traditional word problems toward *model-eliciting activities* in which the products that are produced are not simply short answers to artificially constrained questions about pre-mathematized situations. Instead, the goal is to develop conceptual tools that are sharable (with others) and re-useable (in a variety of structurally similar situations) in addition to being useful for specific purposes in a given situation.

Part V, titled *Models and Modeling Before and After Middle School*, describes how modeling activities that focus on middle school mathematics can be extended down to topics emphasized in primary school—as well as up to topics emphasized in high school or college. It also describes several important ways that new computer-based tools and representational media can influence students' learning and problem solving capabilities.

A meta-theme that applies to all of the preceding sections of the book reflects the fact that models and modeling perspectives trace their lineage not only to modern descendents of Piaget and Vygotsky but also (and perhaps even more significantly) to modern descendents of American Pragmatists such as:

- William James—who is widely regarded as the founder of modern psychology in America.
- Charles Sanders Peirce—the founder of *semiotics*, the study of symbolism, signs, sign-functions, signifiers and tokens.
- Oliver Wendell Holmes—the foremost legal architect of modern conceptions about *academic freedom, freedom of speech*, and other basic liberties that characterize the best traditions of American education, and
- John Dewey—the leading advocate of *instrumentalist* approaches to education, and the founder of the *University of Chicago's* famous *Laboratory School,* which he used as a "philosophy laboratory" to investigate pragmatist principles of teaching and learning.

PRAGMATIST PHILOSOPHICAL ROOTS OF MODELS AND MODELING PERSPECTIVES

Two of the most significant factors that influenced *pragmatist perspectives* were: (a) the Civil War in the United States, and (b) increasing ideological conflicts between religion and science (e.g., Christian Fundamentalists and Darwinists). The American Civil War stamped a traumatic lesson on the minds of survivors such as Oliver Wendell Holmes. They saw firsthand that certitude based on "grand philosophies" often led dogmatists to believe that the self-perceived correctness of their beliefs justifies imposing them on anyone who does not happen to agree (Russell, 1961). For this reason, *pragmatism* was partly a reaction against the notion that "grand theories" could or should govern most significant forms of human decision making. Pragmatists argued that, in peoples' daily lives, "grand theories" should not be expected to determine most choices about courses of action. For example:

- Oliver Wendell Holmes was especially adamant about his belief that to stake important human concerns on the truth of this or that philosophical theory was arrogant and absurd (Holmes, 1997). To emphasize his point, Holmes is famous for inviting his fellow justices to name any legal principle(s) that they believed to be relevant to a given case under consideration. Then, he would use their principle(s) to decide the case in either way. He joked that one of the primary virtues of common law, as it was evolving in the United States during his lifetime, was that judges and juries decided results first and only later figured out plausible idealized accounts of the arguments that they used to support their decisions (Hadot, 2002). Holmes believed that this approach was both proper and necessary because (for example) a case that seemed to begin as a legal/logical problem often evolved into an economic problem, which evolved into a social problem, which evolved into a political problem, which ultimately cycled back to influence the underlying legal logic of the situation. Therefore, trade-offs often need to be considered among competing principles; and, decisions often need to consider phenomena that modern "systems' theorists" refer to as feedback loops, second-order effects, and emergent characteristics that result from interactions among multiple agents with conflicting agendas. In other words, even though Holmes and other pragmatists such as Chauncey Wright (*The Winds and the Weather*, 1858), knew nothing about the modern field of *complexity theory* (Sterman, 2000), they were familiar with the fact that, even when a system consists of parts that obey simple lawful principles, interactions among these principles often lead to characteristics of the system-as-a-whole that are essentially unpredictable.

- Today, observations similar to those emphasized by the pragmatists are familiar in fields such as engineering or business management—where heavy users of mathematics, science, and technology investigate complex systems that are dynamic, interacting, self-organizing, and continually adapting (Sterman, 2000). Among scientists at the forefront of such fields, it is considered to be common knowledge that "real problems" seldom can be solved using only the conceptual tools associated with a single disciplinary topic or theory. Solutions to non-trivial problems typically involve trade-offs that must integrate ways of thinking from a variety of disciplines—as well as from a variety of topic areas within a given disciple. For example, relevant trade-offs may involve factors such as timeliness versus completeness, cost versus quality, or simplicity versus complexity; and, solutions that are most useful are not necessarily the ones that are most complex, most sophisticated, most costly, and most time-consuming to produce.
- Themes related to complex systems also apply to teachers and teaching. For example, a decision or an issue that appears to be resolvable using (only) some grand theory of psychology may evolve into a classroom management issue, which evolves into an administrative policy issue, which evolves into a political issue involving parents or school board members, which cycles back to influence the underlying psychological issues that were thought to be sufficient to determine actions in the original decision-making situation.

For William James and other founders of pragmatist perspectives, debates between religion and science (e.g., Darwinism) convinced them that one of the main goals of the *pragmatist* movement should be to develop a coherent framework for decision-making that would enable decision-makers to integrate useful ideas and beliefs from a variety of competing "grand theories"—while at the same time making appropriate use of time-tested wisdom from successful and experienced practitioners. For example, in his book *Principles of Psychology* (1918), James spoke of a "pluriverse"—by which he meant that people often draw on a variety of (perhaps partly conflicting) worldviews as bases for making judgments in everyday situations.

KNOWING IS NOT ABOUT MIRRORING REALITY

A second factor that strongly influenced the *pragmatist movement* was that, at the time when *pragmatist perspectives* were being formulated, Newton's clockwork view of the universe was developing exceedingly high levels of observational precision. Consequently, it became apparent that, when people observe complex systems in fields such as astronomy, our knowledge of the world is characterized by certain unavoidable amounts of uncertainty and subjectivity (Rorty, 1979). The world is filled with unique things, the same thing never happens twice, and certain errors in perception are unavoidable. This is why pragmatists argued that knowing cannot be a matter of an individual mind "mirroring" reality. Each mind reflects differently; a single mind reflects differently at different moments; and, reality does not stand still long enough to be accurately mirrored.

When James B. Conant critiqued the views of Newton, Maxwell, and other 19th century scientists concerning the absolute separation of mind and reality, he emphasized that assumptions about things like causality are "metaphysical beliefs" rather than being things that we actually observe in nature (Conant, 1951). Scientific laws about causality rely on the assumption that like causes produce like effects. But, in reality, every event tends to have a multiplicity of interdependent causes; and, nothing ever happens twice. What we see is one phenomenon followed by another. We cannot see causation. Causation is attributed when a causal model is developed that is sufficiently useful to explain and predict events. Yet, Peirce's book *Design and Chance* (1884) emphasized that we not only believe in the reality of constructs such as space, time, and causality; we cannot not believe in them.

For Peirce and other *pragmatists*, causality and other fundamental concepts of science are not considered to be realities of nature. They are tools of inquiry; and, they are thought of as being predictions more than conclusions. That is, ideas like *cause* and *effect*, or *uncertainty* and *chance*, are not thought of as naming actual properties of nature. Instead, *pragmatists* consider them to be mostly predictions about how things will behave most of the time—based on patterns and regularities that we infer but cannot observe directly.

KNOWING INVOLVES SOCIAL DIMENSIONS

Partly because of the perspectives described in the preceding section, *pragmatists* rejected the notion that science is independent of our own (or our society's) interests and preferences. In fact, they rejected the very notion of separating mind and reality into separate compartments—as 19th century psychological research was predisposed to do. For example, in William James' famous *Remarks on Spencer's Definition of Mind as Correspondence* (1878) as well as in Peirce's equally famous article on *How to make Our Ideas Clear* (1878), both men emphasized that it is a mistake to think of ideas as being true (or justified) only if they mirror "the way things really are." In fact, in James's book, *Principles of Psychology* (1918), he gives an extended answer to the question posed by Locke and Kant: *"How do we acquire our ideas about the world outside ourselves?"* Conceptual systems, James claimed, are not justified by their correspondence with reality; and, mirroring reality is not the purpose of such tools. For example, James argued that the reason we believe in causation is because experience shows that it pays to believe in causation. So, unlike positivist or empiricist perspectives, which emphasized distinctions between the mind and reality, *pragmatists* emphasized that both the mind and reality are abstractions from one indissociable milieu. Ideas, they claimed, are mainly tools—somewhat like power saws and food processors—that people devise to create (and make sense of) their experiences.

28. MOVING BEYOND CONSTRUCTIVISM

Dewey (1963) was especially emphatic about his conviction that it makes no more sense to talk about a "split" between the mind and reality. For Dewey, both the "mind" and "reality" are abstractions that are derived from a single and indivisible process of interaction and prediction. For example, in his famous paper, On *The Reflex Arc Concept in Psychology* (1896), Dewey wrote:

Things are what they are experienced as. ... Knowledge is not a copy of something that exists independently of its being known. It is an instrument or organ of successful action. pp. 224.

Pragmatists also emphasized that the development of knowledge is inherently a social enterprise. That is, they emphasized that ideas (or constructs) are not "out there" waiting to be discovered; nor do they develop entirely according to some inner logic of their own. Because they are, above all, conceptual tools that people create to accomplish specific goals, and because peoples' reasons for needing tools are always changing, the tools themselves always need to be changing. Therefore, the Darwin-style survival of conceptual tools depends on their adaptability as much as on their immutability; and, it depends on their social viability as much as on their abilities to predict the behaviors of physical systems. In fact, these facts are especially true because conceptual tools often produce results that involve fundamental change to the world in which the tools exist. ... In other words, using James' terminology, the universe is continually "in the making" pp. 117 (1997).

Social considerations tend to be especially obvious and straightforward for model development. One reason this is true is because models seldom are worth bothering to create unless they are intended to be sharable and re-useable—either by others, or by yourself in situations beyond the one in which the model was first needed. Furthermore, the purposes for which models are created usually cannot be completely determined by model developers themselves. Nor can the model developers completely control the criteria that will be used to assess the usefulness of models that are generated. Also, in many cases, the model developer is not simply an isolated individual, but instead is a team of specialists who have access to a variety of different conceptual technologies. So, for all of these reasons, modeling is inherently a social enterprise.

HOW ARE MODELS DIFFERENT FROM THEORIES?

Models and theories both involve holistic conceptual systems that are expressed using representational media for purposes such as generating purposeful descriptions or explanations of (allowable classes of) situations. Therefore, in some respects, models can be thought of as being similar to *small theories.*

- In mathematics, even something as small as an equation that describes a specific problem-solving situation may be referred to as being a model of the situation. On the other hand, even in these cases, the size of such models may be much larger and more complex than their immediate appearances might suggest. This is because the equations (or descriptions, or explanations) are like the tips of icebergs. They are simply the visible parts of much larger conceptual systems that support their meanings. Nonetheless, even if we recognize the hidden conceptual systems that underlie the visible part of a model, the relevant conceptual systems still tend to be much smaller than the "grand theories" in which these conceptual systems are embedded.

- Because models are created for specific purposes in specific situations, they tend to be thought of as being much smaller than "grand theories" which seem to be more like all-purpose languages or "world views." On the other hand, these "world views" are self-contained in the sense that they place strict limitations on problems, assumptions and arguments that are recognized as being legitimate. Whereas, *models and modeling perspectives* adopt the pragmatists' approach of emphasizing practical problems and "ends in view" that occur outside any specific philosophical systems that problem solvers might draw upon in order to make sense of them. Therefore, in the case of model development, or the development of other kinds of conceptual tools, the "ends in view" (or the practical goals that need to be achieved) provide criteria for making decisions about the relative usefulness of competing ways of thinking (Dewey, 1982). In other words, model development tends to be much more difficult to seal off from "reality" the way theory development so often is able to do (Popper, 1972).

For theories, the theory itself determines whether a problem (or a critic, or an assumption) is significant. Whereas, model developers usually do not have the luxury of being able to dismiss assumptions, difficulties, critics, or alternative ways of thinking that occur naturally in the course of dealing with real life problems and situations. As applied scientists emphasize, solutions to realistically complex problems seldom fall into neat and tidy single-discipline perspectives. Consequently, when the goal is to develop a model to describe (or construct, or explain) realistically complex systems, model development usually needs to draw upon more than a single theory. Therefore, because models often cannot be contained within the boundaries of any single theory. So, they often actually appear to be bigger than the individual theories that they draw upon.

Beyond the preceding similarities and differences between models and "grand theories", there are other ways that desirable characteristics of models often need to be quite different from those that apply to theories.

- Grand theories generally try to provide a single coherent conceptual system that does not need to be changed in order to provide reasonably complete descriptions or explanations for a large class of situations and purposes. But, even if immutability generally is thought of as being a desirable characteristic of grand theories, the usefulness of models often depends on the ease with which they can be modified and transformed for a variety of changing goals and situations.

- Grand theories generally try to express as much of their meanings as possible within a single representation media. Therefore, the meaning of the theory seems to be contained within these single representation media—which, in mathematics, usually is a written symbolic system. But, models typically are expressed using a variety of interacting media—which often range from spoken language, to written symbols, to constructed diagrams, to dynamic simulations, to experience-based metaphors—each emphasizing and de-emphasizing (ignoring or distorting) somewhat different aspects of the situation being described or explained. So, the meaning of a model tends to be distributed across a variety of representational media. It is not contained within any single one of them. In some ways, this sometimes seems to make models seem larger than corresponding "pure" theoretical constructs.
- In the time of Plato, when theories were expressed primarily in oral rather than written forms, the accepted way for philosophers to demonstrate the superiority of their ways of thinking was to engage in debate against formidable adversaries. In other words, philosophers demonstrated the superiority of their thinking in somewhat the same way that football teams demonstrate their superiority today (Hadot, 2002). ... Grand theories, on the other hand, shift attention away from (or beyond) carrying on arguments against formidable adversaries. Instead, most arguments are carried on as if they were directed at an idealized "opponent" who is assumed to embody the collective norms and wisdom of "the field" about (i) which assumptions and procedures that are considered to be acceptable, and (ii) which problems and issues are worthy of investigation. ... Of course, this latter approach has many virtues which are among the most important reasons why science theories have evolved as cumulatively as they have. Nonetheless, it is not the only approach that people use to demonstrate the relative power of complex systems. In fact, in some ways, the criteria that superior models must satisfy sometimes appear to be even more difficult to satisfy than those that apply to corresponding "pure" constructs. This is because successful models must serve purposes that do not lie within their own control.

WHAT FACTORS LED TO THE REJECTION OF PRAGMATISM IN AMERICAN EDUCATION?

Even though many of the most distinctive, significant, and positive characteristics of American schooling trace their origins to ideas introduced by pragmatists such as William James, Charles Sanders Peirce, Oliver Wendell Holmes, and John Dewey, it long ago became unfashionable to think of *pragmatism* (or Dewey's *instrumentalism*) as a driving force behind the ways of thinking that are emphasized in American schools. Yet, from time to time in recent history, themes emphasized by the *pragmatists* reappear explicitly in the work of modern educators and educational psychologists. For example, Bruner's *evolutionary instrumentalism* drew explicitly on the *instrumentalist* views of Dewey (Bruner, 1986); and, he also drew on the *semiotic perspectives* of Peirce (See chapter 5 in this book). Similarly, many more recent artificial intelligence models of teaching and learning draw heavily on these same themes.

Precisely because pragmatism is no longer fashionable, and precisely because so many modern ways of thinking are so closely related to basic ideas introduced by the *pragmatists*, it is productive for modern educators to ask: (i) *What factors led to the rejection of pragmatism in American education?* (ii) *What can be done to combat similar forces today?* ... To answer these questions, two clusters of issues are especially relevant to consider.

First, *pragmatist perspectives* were closely associated with principles of *academic freedom, freedom of speech*, and other basic liberties that continue to be sources of controversy today as much as when they were introduced. For example, during the period of United States' history from *World War II* through the *Cold War*, and especially during periods when *McCarthyism* flourished, concerns grew about perceived misuses of these liberties. Consequently, educational philosophies that supported such liberties became unfashionable.

Second, *pragmatists* advocated the notion of a *marketplace of ideas* in which the value of an idea was not thought to be determined by its correspondence to either a preexisting reality or to a metaphysical truth. Instead, it value was based mainly on the difference that it makes in the life of the relevant community. ... Unfortunately, these views often were misinterpreted as supporting a form of laissez-faire individualism in which anything goes—so that no standards exist for assessing the quality of the work that individuals produce. In particular, many people interpreted Dewey's project-based approach to learning as providing no clear way of judging whether the results students produced were worth pursuing. Many educators assumed (incorrectly) that the pragmatists were saying that just any activity is valuable as long as students are engaged. Consequently, pragmatism came to be perceived as a philosophy that leads to a directionless curriculum. Furthermore, because pragmatists emphasized the usefulness of knowledge, pragmatism came to be perceived (again incorrectly) as advocating a decreasing emphasis on "pure" mathematics or the pursuit of "knowledge for it's own sake" (King and Brownell, 1966).

The truth is that the creators of *pragmatist perspectives* were strong opponents of laissez-faire individualism. In general, they were Darwinists; and, as such, their 19th century commitment to free markets was quite different than the 20th century commitment to individual freedom. They did not advocate the free expression of ideas because they believed that any idea was as good as any other. They believed in free expression of ideas because the development of a community-as-a-whole needs the resources of the whole group to produce ways of thinking that will be optimally useful, sharable, powerful, and transportable. But, while emphasizing the importance of diversity, their Darwinian roots also led them to emphasize selection, communication, and preservation. For example, Dewey clearly intended his *philosophy of experience* to start with values of the community—as determined by teachers and other representatives

28. MOVING BEYOND CONSTRUCTIVISM

of relevant disciplines or communities (Dewey, 1916). So, *pragmatists* relied largely on traditions of the relevant disciplines to determine what ideas are worth learning. In fact, they were strong supporters of the "liberal arts" curriculum of their time.

Pragmatists generally believed that good mathematics is useful mathematics. But, they did not view utility and quality to be conflict; and, they also emphasized that thinking is inherently a social activity. For example, Dewey in particular was outspoken about rejecting hard distinctions between the development of the individual and the development of the communities in which they functioned. So, the content disciplines were viewed as playing important roles to provide values for judging the importance of various pursuits and results. On the other hand, they did not believe that "pure" mathematicians should be the only voices that should be heard concerning the value of mathematical ways of thinking; and they emphatically opposed the notion that logic alone should determine what should be taught and what achievements should be valued. Mainly, *pragmatism* is about *how* things are learned (or what it means to learn them) rather than being about *what* things should be learned.

HOW CAN MODELS AND MODELING PERSPECTIVES SURVIVE THE FATE THAT BEFELL PRAGMATISM?

Today, many of the criticisms that were leveled against *pragmatism* are being heard once again as criticisms against student centered curriculum innovations which often are seen as: (a) avoiding accountability, (b) pandering to the superficial interests of individuals at the expense of "hard nosed" scientific rigor, and (c) valuing inquiry and discourse for its own sake—apart from the quality of the results that inquiry and discourse produce.

Because *models and modeling perspectives* draw on many of the most basic ideals that were emphasized by *pragmatists*, while at the same time drawing on the work of modern descendants of Piaget and Vygotsky, it behooves us to ask about ways that future research on *models and modeling* should draw on resources that were not available to *pragmatists* such as James, Holmes, Peirce, and Dewey. Only if we do so is it reasonable to expect modern mathematics educators to overcome criticisms that led their early predecessors to abandon many of the promising practices advocated by pragmatists. ... Examples of such resources include the following that have been referred to in chapters throughout this book. Each should be strongly emphasized in future research on *models and modeling perspectives*.

- William James surely would urge advocates of *models and modeling perspectives* to continue to contribute to (and draw upon) the wealth of Piaget-inspired and Vygotsky-inspired research that mathematics educators have produced during the past thirty years about the nature of students' developing mathematical constructs and conceptual systems—and about social dimensions of development.

- John Dewey surely would urge advocates of *models and modeling perspectives* to contribute to (and draw upon) the rapidly growing body of knowledge about students' abilities to develop and use powerful new types of conceptual tools that put extraordinary achievements within the grasp of ordinary students. He also would be pleased to see how heavy users of mathematics are emerging in virtually every field of human endeavor—ranging from the arts to the sciences. So, he would encourage mathematics educators to emphasize the fact that mathematics is no longer the exclusive domain of professional mathematicians.
- Oliver Wendell Holmes surely would be pleased to see that the "case studies" methods of instruction and assessment that he pioneered in the field of law are now being used in virtually every field of applied science—from engineering to business management. So, he would encourage mathematics educators to investigate the kind of *case studies for kids* that have been described throughout this book.
- Charles Sanders Peirce surely would urge advocates of *models and modeling perspectives* to draw on the following four recent trends that have developed as byproducts to advances in computation and communication technologies.
    o First, the development of complexity theory (Sterman, 2000) as a science provides coherent mathematical models to explain the behaviors of complex systems (such as those that characterize the thinking of students, teachers, and learning communities) that are dynamic, interacting, self organizing, and continually adapting. So, "systems thinking" should continue to provide useful conceptual tools for making sense of students' thinking, teachers' thinking, and the thinking of parents, policy makers, and other relevant practitioners.
    o Second, in fields such as *geographic information systems*, rapid advances in graphic, dynamic, and interactive displays of information provide powerful tools to generate multi-dimensional progress reports to describe the development of complex systems that range from weather systems to economic systems. So, when conceptual systems are developed by students, teachers, or programs, mathematics educators should not restrict themselves to characterizations based on static linear equations (Lesh and Lamon, 1994).
    o Third, dynamically linked and highly interactive multi-media representation systems provide powerful new tools to create new types of learning environments in which students are able to iteratively express, test, and revise or refine their own ways of thinking—in order to produce powerful, sharable, and re-useable conceptual tools that are needed for success beyond the mathematics classroom. So, mathematics educators should

continue to investigate powerful conceptual tools that are not contained completely within the minds of students.

o Fourth, schools no longer have a monopoly as a source of mathematics learning—due to recent advances in communication and computation technologies. So, mathematics educators should continue to investigate new ways to use the Internet, and other increasingly ubiquitous pocket-sized communication devices, to democratize access to powerful conceptual tools.

One especially important factor that enhances the potential power of *models and modeling perspectives* results from the fact that future-oriented fields of applied science are now filled with heavy users of mathematics, science, and technology; and, people who are responsible for training leaders in these fields tell us that many of the most important characteristics that are most attractive during job interviews include:

- The ability to work and communicate effectively at a variety of roles within diverse teams of specialists.
- The ability to adapt to rapidly changing advanced technology tools; the ability to describe situations in forms so that relevant tools can be used; and the ability to interpret results that are generated by the tools.
- The ability to plan, monitor, and assess progress during projects that are aimed at designing complex systems.

Like the *pragmatists*, advocates of *models and modeling perspectives* believe that elementary mathematics is useful mathematics. But, when we ask *"What should be the most important goals of mathematics instruction?"* models and modeling perspectives emphasize that we should go beyond asking *"What kind of computations or constructions can they do?"* to also ask *"What kinds of situations can they describe?"* So, many of the most important goals of mathematics instruction focus on helping students develop powerful constructs and conceptual systems for making sense of their experiences. That is, models and modeling is at the heart of many purposes of mathematics instructions.

TABLE 28.1
In What Ways Do M&M Perspectives
Advance Beyond Constructivism?

Constructivist Perspectives	Modeling Perspectives
What Does It Mean To Be a Constructivist?	
What are the basic principles on which all constructivists agree? Clements and Battista give the following two-part answer—which has come to function as a sort of constructivist's creed. *Constructivists claim that knowledge is actively constructed by the child, not passively received from the environment.* (1990, p. 34) Using this conception of the essence of constructivism, virtually every modern theory of cognitive science can claim to be based on constructivist philosophy. That is, almost none would argue students passively receive knowledge from teachers. This is why, in the book, *Constructivist Views on the Teaching and Learning of Mathematics*, Davis, Maher, & Noddings state that: *A great many people now think and write about it [constructivism], and the people who do so do not entirely agree with one another* (1990, p. 187). To further emphasize this point, Goldin (1990, in press) found that different constructivists often advocate opposite courses of action for nearly any of the following kinds of issues. • What relationships exist between the development of basic skills and the development of deeper or higher-order constructs? • When and how should calculators, concrete materials, or "real life" problem solving situations be used in instruction? • How much and what kind of guidance should be provided by teachers? • What should students learn from discussions or interactions with peers?	M&M perspectives focus on the noun *construct* rather than the verb *construct*. We believe that helping children develop powerful constructs (or conceptual systems) for interpreting experience are among the most important goals of mathematics instruction. And, we agree with constructivists that constructs must be *developed* by students themselves. Nonetheless, we believe that it is misleading to suggest that all constructs must be constructed. According to M&M perspectives, both parts of the *constructivists' creed* are at best half-truths. Consider the following. • All goals of mathematics instruction do not need to be achieved through processes of personal construction (Kilpatrick, 1987). Construction processes apply mainly to constructs. But, in mathematics classrooms, many other important things that we want students to learn are not constructs at all; and, they do not need to be invented independently by students (Clements, 1997; Kamii, 1982). They are things like notational and procedural conventions that can be learned through demonstration, imitation, and practice with feedback. • Construction is only one among many processes that contribute to the development of constructs. For example, development also typically involves sorting out, differentiating, reorganizing, refining, adapting, or reflectively abstracting conceptual systems that already exist at some concrete or intuitive level in students' thinking. And, sorting our existing conceptual systems is not at all the same thing as constructing new ones.

28. MOVING BEYOND CONSTRUCTIVISM

To explain how two constructivists could hold such exceedingly different views, Noddings explains: *Constructivism can dictate only guidelines for good teaching. We cannot derive... specific teaching methods* (Davis, Maher & Noddings, 1990, p. 83).	• Construction processes do not necessarily lead to important constructs. For example, *construction* may lead to nothing more than the assembly of chains of low-level skills.

Following traditions established by the *American Pragmatists*, M&M perspectives are not intended to be "grand theories" developed by philosophers to address questions that are priorities to philosophers. Nor are they simply intended to provide after-the-fact cover stories to describe instructional decisions that already have been made. Instead, they are intended to provide "blue-collar" conceptual frameworks for integrating the best theories of the best theorists with the best practices of the best practitioners—and for developing sharable conceptual tools for dealing with day-to-day decision-making for teachers, curriculum developers, teacher educators and others practitioners.

As constructivists surely would agree, the key issue is not whether a theory is true or false, but rather whether it is useful. So, one implication of this policy is that, for researchers whose goals are to test and revise or refine theories, a philosophy that is accepted by nearly everyone is not useful. In other words, constructivism itself ceases to be useful to theory developers precisely when virtually every potentially competing theory claims to adopt this philosophy. ... Currently, we believe, commitments to construction processes are far too easy to reconcile with heavy-handed forms of teacher-directed instruction – and with views of students' knowledge that continue to be based on mechanical (hardware-based and/or software-based) views of knowledge in which:

- complex conceptual systems continue to be thought of as nothing more than the sums nor their parts,
- the parts may be assumed to be defined operationally using naive checklists of condition-action rules,
- each part may be thought of as being teachable and testable one-at-a-time, in isolation, and out of context.

In contrast to constructivist claims that all knowledge must be constructed, M&M perspectives recognize that there are at least four qualitatively distinct types of instructional objectives that are important in mathematics education – and that not all of these need to be constructed independently by students. They are: (a) *behavioral objectives* (BO's) such as basic facts and skills, (b) *process objectives* (PO's) such as habits of mind that are not connected to any particular mathematical constructs, (c) *affective objectives* (AO's) such as attitudes, beliefs, feelings, and (d) *cognitive objectives* (CO's) such as models and accompanying conceptual systems (constructs) for constructing, describing, explaining, manipulating, and controlling mathematically (or structurally) interesting systems. ... According to M&M perspectives, CO's, BO's, AO's, and PO's are all considered to be important. But, all are viewed as being parts of students' mathematical interpretations of their experiences; and, models are considered to be what students develop in order to create relevant interpretations.

What is Different About *Assembly-ists, Constructivists* and *CONSTRUCT-ivists*?

Many leading constructivists (Confrey, 1995; Davis, 1984; Steffe & Thompson, 2000) have been leaders helping math educators explain the nature of children's developing mathematical knowledge. In general, these researchers have challenged us to:

- Shift beyond hardware-based metaphors, where (a) conceptual systems are considered to be no more than the sum of their parts, (b) interactions among systems are assumed to involve no more than simple one-way cause-and-effect relationships, and (c) assembling knowledge in children's minds is thought of as being like assembling machine parts.
- Shift beyond software-based metaphors, where silicone-based electronic circuits may involve layers of recursive interactions which often lead to emergent phenomena at higher levels that are not derived from characteristics of phenomena at lower levels—but where assembling knowledge in children's minds continues to be thought of as being like programming a computer.
- Shift toward wetware-based metaphors where organic neurochemical interactions may involve "logics" that are fuzzy, partly redundant, partly inconsistent, and unstable. So, developing knowledge is thought of as being similar to developing communities of complex, dynamic, continually adapting biological systems.

M&M perspectives emphasize that that mathematics is about *structure* at least as much as it is about *algorithms*; and, they also emphasize that: (a) the noun *construct* is as significant as the verb *construct*. (b) the noun *design* is as significant as the verb *design*, and, (c) the noun *model* is as significant as the verb *model*. Therefore, because M&M perspectives emphasize the importance of constructs and conceptual systems it might seem appropriate to refer to M&M as a CONSTRUCT-ivist perspective. But, M&M is a "blue collar perspective." So, we avoid such terms mainly because constructivists already have been criticized for excessive use of jargon such as *RE-presentations* and *TAKEN-AS-SHARED* concepts.

Regardless whether attention focuses on the developing capabilities of students, groups of students, teachers, schools, or other learning communities, M&M recognizes that, for each of these problem solvers, developing knowledge is less like a machine or a computer program than it is like a community of living, adapting, and continually evolving biological systems. Conceptual systems are not simply inert and waiting to be stimulated. They are dynamic, interacting, and self-regulating systems that develop emergent characteristics and competencies that generally cannot be reduced to simpleminded checklists of condition-action rules (Lesh, 2001).

Many constructivists have been leaders clarifying the nature of students' developing constructs in topic areas ranging from early number concepts (Steffe, 1990), to geometry concepts (Clements, 1997), to statistics concepts (Cobb, 1999), to early algebra concepts (Davis, 1984). So, it is ironic that many of these same individuals have ended up supporting a philosophy that focuses on the verb *construct* rather than the noun *construct*. In the past, advocates of discovery learning made a similar strategic error when they advocated discovery (the process) even when nothing of substance was being discovered. Mathematics educators have a long history of treating discovery (or construction) as if it were an end-in-itself—rather than being a means-to-an-end (namely children's development of mathematical constructs).

28. MOVING BEYOND CONSTRUCTIVISM

Do Constructs Need To Be Discovered Independently By Students?	
Many constructivists are strong supporters of discovery learning that is similar to the type associated with the "new math" movement of the 1960s and 1970s (Goldin, 1990). On the other hand, the term *discovery* tends to focus attention on insights about patterns and systems that are outside students' minds, whereas constructivism emphasizes emergent phenomena associated with conceptual systems that develop inside students' minds.	M&M perspectives recognize that both discovery and construction are important in some cases and for some purposes. But, both construction and discovery are treated as means-to-ends rather than as ends-in-themselves; and, both are recognized as being only two among the many means by which students develop meaningful understandings of elementary-but-deep mathematical constructs (Ausubel, 1978).

When theory is translated into practice by curriculum developers, teacher educators, or teachers, many over-generalizations associated with constructivism are similar to those associated with discovery learning. For example, in the past, misuses and over-generalizations often occurred when "new math" zealots claimed that discovery learning was the (only) path for meaningful understanding to develop, and when students were engaged in discovery activities simply for the sake of discovery. M&M perspectives recognize that many student discoveries have little or no positive instructional value; many important things that need to be learned are not likely to come from students' independent discoveries; and, many things that need to be learned can be learned meaningfully and efficiently without being discovered (Ausubel, 1978). Similar statements apply to:

- concrete materials which are used for their own sake,
- classroom discourse which is engaging in for its own sake, or
- construction processes which are emphasized for their own sake.

What is the Relationship Between the Mastery of *Basics Skills* and the Development of *Deeper and Higher Order Understandings and Abilities*?	
Constructivism is partly a political movement that was created as a response to excesses associated with behaviorist philosophies. Therefore, many constructivists have been reluctant to recognize legitimate ways that the mastery of facts and skills interacts with the development of deeper and higher-order constructs (Goldin, in press; Kamii, 1982).	M&M perspectives consider skills, values, feelings, attitudes, & higher-order processes to be important components of the mathematical models that students develop to interpret their experiences. Ignoring skill-level fundamentals is as foolish in mathematics as in other fields where students develop complex conceptual/procedural systems.

In fields like basketball (or cooking, or carpentry), it is obvious that excellent coaches, teachers, and mentors usually emphasize both complex performance and drills and skills. What's needed is a sensible mix of complexity and fundamentals. Both must evolve in parallel; one does not come before (or without) the other; and, even coaches who stress fundamentals do not neglect scrimmages. They recognize that it is not necessary to master the names and skills associated with every item at Sears before students can begin to cook or to build things.

According to M&M perspectives, well coordinated and smoothly functioning conceptual systems (or models) are similar, in many respects, to well coordinated and smoothly functioning physical systems—such as those that are involved in hitting tennis balls, dancing, or creating clay pots. Coordination and smooth functioning are acquired gradually over time. Not all aspects of performance are mastered at once or at the same level of achievement. And, in given learning or problem-solving situations, the amounts (and kinds) of information that students notice are directly related to the degree of coordination of the relevant systems. The degree of coordination of a given system directly impacts a student's ability to go beyond thinking with the system to also think about it. In other words, the degree of coordination of lower level skills influences metacognitive abilities as much as metacognitive abilities influence the coordination of lower level skills (chapter 21).

Beyond recognizing important interactions that exist between skill-level and higher order abilities, M&M perspectives also emphasize that going "back to basics" is likely to be quite different than emphasizing "foundations for the future." What is needed for success beyond school in a technology-based *age of information* is not likely to be the same as minimum competencies needed to avoid failure in tests focusing on 19th century shopkeeper arithmetic. This is why M&M emphasizes many types of problems, skills, abilities, and constructs that have been ignored in traditional textbooks, tests, and teaching. For example:

- Thinking mathematically involves *interpretation* (description) as much as *computation* (see chapter 18 in this book).

- Thinking mathematically involves *multimedia representational fluency* as much as *proficiency at manipulating written symbols* (see chapter 15 in this book).

- Thinking mathematically involves *working within teams* of specialists and using (and adapting to) rapidly evolving and powerful conceptual tools, at least as much as it involves working in isolation using only a pencil and paper (see chapter 19 in this book).

According to M&M perspectives, one of the most important parts of thinking mathematically involves developing powerful constructs and conceptual tools for creating (and making sense of) complex systems.

What is the Nature of "Real" Experience?

Constructivism begins with an emphasis on the constructed world of the knower and the relationship of that world to reality. As von Glasersfeld observes:

It is necessary to keep in mind the most fundamental trait of constructivist epistemology, that is, that the world which is constructed ... makes no claim whatsoever about 'truth' in the sense of correspondence with an ontological reality (1984, p. 29).

Rather than beginning with the claim that the reality that you think you see isn't really out there, M&M perspectives begin with much more simple and straight forward assumptions. (a) people interpret their experiences using models, (b) these models consist of conceptual systems that are expressed using a variety of interacting media (concrete materials, written symbols, spoken language) for constructing, describing, explaining, manipulating or controlling systems that occur in the world.

28. MOVING BEYOND CONSTRUCTIVISM

On receiving an honorary doctorate from Montclair University, Lawrence Peter "Yogi" Berra is quoted as saying:

In theory there is no difference between theory and practice. In practice there is. (USA Today, May 24, 1996)

Among philosophers, Kant (1724-1804) usually is recognized as the main person to introduce the notion that the mind does not passively reflect experience but rather actively creates meanings by attributing patterns and regularities that cannot be perceived directly. So, two centuries after Kant's death, radical constructivists' preoccupations with naïve interpretations of reality seem quaint – especially in a field like mathematics where the notion that mathematics is about truth was abandoned hundreds of years ago—following the discovery of non-Euclidean geometries. In fact, since the discovery of Godel's Theorem, the Axiom of Choice, and other related issues in mathematics, mathematicians have recognized that they can not even guarantee the internal consistency of any system that is more complex than the integers (Hofstadter, 1999). Similarly, readers of the magazine *Scientific American* are likely to be familiar with many articles that have been published during the past 30 years showing that even visual perception involves complex interactions and interpretations that are nothing at all like photographic processes. Nonetheless, if we refer the reader's attention to the chair shown below, few people need to be reminded that what we are referring to is not a real chair. Most recognize that different people (ranging from farmers to furniture makers) will consider quite different chairs to be similar to the one that is shown here.

When aeronautical engineers or other modern applied scientists speak of model-reality mismatches, this does not imply that they are reverting back to a Platonic conception of reality. It simply means that their models fail to predict (or fit) certain aspects of the situations they are trying to describe or explain. As modern scientists increasingly move toward investigations of systems that cannot be observed directly, it is clear that many levels of theory and inference lie between the situations that we hope to understand and the interpretations that our models attribute to them.

Every model has some characteristic that the described system does not have, and every model does not have some characteristic that the described system does have. Otherwise the model and the described system would be the same. So, when our goals are to understand, and communicate about, complex systems, it is important to recognize that what we understand is based on patterns that we attribute to these systems, based on the models that we use. Or, a more accurate way to express this fact is to say that our interpretations result from interactions between our models and reality. Our interpretation is what we know about these systems, and our models are what we use to make interpretations. So, if we are talking to professional philosophers about issues of philosophy, rather than talking to practitioners about issues of practical importance, then it sometimes is worthwhile to acknowledge that model-reality mismatches implicitly involve within-model mismatches—and that the distinction between within-model mismatches and between-model mismatches is purely in the eye of the beholder. But, when we are talking to teachers who would like to emphasize the usefulness of mathematics beyond school, or when we are working with people who are trying to write effective model-eliciting activities, we've found that it is seldom useful to continually get distracted by philosophical nuances about the nature of reality.

Returning to the Question: What Kind of Theory Do We Intend to Develop?
Following the *pragmatists*, a point that was emphasized earlier in this table was that M&M does not concern itself with truth. Models are adopted or rejected because they are useful; and, useful models (or useful theories, or useful conceptual systems) are considered to be those that: (a) begin with "axioms" or claims that are simple and clearly understood, and (b) generate "theorems" or predictions that are powerful and not obvious. So, from M&M perspectives, shortcomings of radical constructivist assumptions have nothing to do with the truth or falsity claims. Instead: • Constructivist claims, such as those about the nature of reality, are far from obvious. So, they don't provide useful entry-level assumptions for practitioners. • The theorems that can be derived from constructivist claims simply have not provided useful information for the main decision-making issues that we confront in our work with teachers, curriculum developers, teacher educators, and researchers. To mathematics education researchers, a useful theory should generate predictions that are falsifiable. So, as long as virtually every modern theory of cognitive science can claim to be based on constructivist principles, constructivism won't be useful to researchers for testing hypotheses that discriminate among alternative theories.

Can We Know What is in the Mind Of Another Person?	
Radical constructivists (Steffe, 1990; von Glasersfeld, 1995) emphasize that one person cannot know what is in the mind of another. *One constructs general as well as specific models of students. Although realists tend to think that their models should, and to some extent do, reflect the students as they really are, constructivists must remain aware of the fact that models cannot reflect anything but the model builders' own conceptual constructs that they have externalized and kept constant by a continual process of assimilation and accommodation* (von Glasersfeld, 1990b, p. 33).	M&M perspectives recognize that: (a) the only way that anybody can have access to another person's internal conceptual systems (or even their own internal conceptual systems) is when they are expressed using some external media (Lesh & Doerr, chap. 1, this volume), and (b) external media can never reveal everything about the conceptual systems that produced them. Yet, it is not necessary for these facts to lead to philosophical solipsism - where each person is assumed to exist in his or her own little world which is totally inaccessible to others. In fact, in model-eliciting activities, the products that students produce typically involve conceptual tools that are designed explicitly to be sharable and reuseable (by other people, or by ourselves at some later time or in other situations).
According to M&M perspectives, models are based on conceptual systems that can be thought of as residing inside the minds of students. But, these conceptual systems seldom have much power unless they are expressed using some collection of external media; and, humans are continually projecting their conceptual schemes into the world. That is, as soon as humans develop a conceptual system for making sense of some domain of their experiences, they typically use these conceptual systems to mold and shape their world of experience. As a result, the world is filled with coordinate systems	

28. MOVING BEYOND CONSTRUCTIVISM

communication systems, information processing systems, economic and accounting systems, and other kinds of complex systems that impact the lives of ordinary people. In other words, the universe is still (using William James' term) "in the making."

Of course, when a person expresses internal constructs and conceptual systems using external media, or when internal constructs and conceptual systems are used to construct external system, there is no reason to assume that what one person creates (or draws, or says) is identical to what another person sees (or hears). In fact, the very act of externalizing an internal conceptual system often induces significant changes in the conceptual systems that were used. For example, when a person (or a team) is solving a problem, one reason why problem solvers express their current ways of thinking using external media is to help clarify (or modify, refine, and revise) each individual's ways of thinking. Therefore, the conceptual system that is used to produce a drawing (or artifact) cannot be assumed to be the same as the conceptual system that is "seen" in the result that is produced. So, when a person first draws and then examines a diagram, his or her thinking may change in fundamental ways. Thus, there is a sense in which the only ways we can know what is in our own minds is to externalize our ways of thinking, and the very act of externalizing induces change in the ways of thinking that produced them. So, M&M perspectives go one step further than the constructivist claim that you can't know what's in the mind of another person. You can't even know what's in your own mind.

In What Ways Do M&M Perspectives Go Beyond Piagetian Views About Students' Developing Mathematical Constructs?

Whereas Piaget studied the natural development of conceptual systems that emerge as a result of ordinary experiences during periods of major conceptual reorganization (at about 2, 6, and 12 years of age), many constructivist researchers in mathematics education have extended Piaget's work by studying the induced development of conceptual systems that (a) seldom develop beyond primitive levels unless artificially rich mathematical experiences are provided, and (b) emerge during intermediate stages between and after Piaget's periods of concrete operational reasoning and his period of formal operational reasoning.	When students develop conceptual tools (or other mathematical artifacts) during model development activities, the modeling cycles they go through often are remarkably similar to stages of development that children go through in Piagetian investigations about the development of the relevant conceptual systems. Furthermore, processes that promote model development often are remarkably similar to those that Piagetians have identified that promote general conceptual development. So, we sometimes refer to model development activities as *local conceptual development activities* (Lesh & Harel, in press).

Like Piaget, M&M perspectives focus on the development of elementary-but-deep cognitive structures with holistic characteristics that are not derived from characteristics of constituent parts. But, according to M&M perspectives:

- Conceptual development is far more piecemeal and situated than Piagetians generally suggest (diSessa, 1988). In particular, at early stages of development, students' knowledge tends to be organized around experiences far more than it is organized around abstractions.

- Knowledge develops along a variety of dimensions beyond those emphasized by Piaget who focused on the gradually increasing structural complexity of children's constructs and on development from concrete to abstract understandings. In addition to Piaget's emphasis on concrete abstract and simple complex, from a M&M perspective, we would also include dimensions such as: situated decontextualized, specific general, external internal, intuitive formal, and unstable stable. (note: The models that are most useful are not always the ones that are most abstract, most complex, most general, and/or most formal, we point out that the right side of these pairs is not necessarily better than the left.) For example, because M&M perspectives emphasize both internal conceptual systems and the external media in which they are expressed, they emphasize (for example) the role of language development and increasing representational fluency

- Whereas Piaget's theory tends to be remembered (inappropriately, we believe) as being pessimistic about the possibility of artificially encouraging the development of elementary-but-deep mathematical constructs, one of the most obvious facts resulting from research on model-eliciting activities is that even middle school children who have been classified as being below average routinely invent (or modify, or refine) significant mathematical constructs that typically have been thought to be beyond their capabilities. So, even though there was more than a bit of hyperbole in Bruner's famous claim that "The foundations of any subject can be taught to any child at any time, in some intellectually honest way" (1960, p.15), the overall message that comes from research on M&M is quite optimistic.

In spite of differences between Piagetian perspectives and M&M perspectives, strong similarities result from the fact that both perspectives emphasize the development of holistic cognitive structures—and their impacts on children's mathematical thinking. For example, by interpreting model eliciting activities as local conceptual development activities, it often is possible to apply the basic principles of Piaget's theories to help explain children's problem solving processes. During local conceptual development sessions, it is often possible to directly observe many of the processes that enable children to develop beyond their own initial conceptualizations of problem solving situations. In some ways, then, M&M is a situated version of Piagetian perspectives.

In What Ways Do M&M Perspectives Go Beyond Cognitively Guided Instruction?

Cognitively Guided Instruction (Carpenter & Fennema, 1992; Fennema, Carpenter, Franke, & Carey, 1993) is a theory emphasizing that one of the best ways to positively influence how teachers teach is to help them develop more sophisticated understandings about what CGI researchers believe to be the nature of students' developing mathematical knowledge. When CGI has been effective, one reason is because expertise in teaching is reflected not only in what teachers can do but also what they see in teaching, learning, and problem-solving

- M&M perspectives enable CGI to be extended beyond primary school grades toward courses in which more complex conceptual systems are emphasized.

- Rather than explaining the nature of students' thinking, M&M perspectives use model-eliciting activities (and accompanying tools for teachers) so that teachers can make direct observations about the nature of their own students' ways of thinking—even in mathematical topic areas that CGI researchers have not investigated.

28. MOVING BEYOND CONSTRUCTIVISM

situations. Unfortunately, so far, the success of CGI has been restricted mainly to the teaching of whole number arithmetic concepts in the primary grades. At later grade levels that emphasize more complex and interacting domains of concepts, it has been far more difficult for CGI researchers to explain the nature of students' models to teachers.

- Development is encouraged, for both students and teachers, because both are engaged in activities where they continually express their current ways of thinking in forms that must be examined and tested and revised— based on formative feedback and the shared use of tools and interpretations.

- While students are developing models (and conceptual systems) for making sense of mathematical problem solving experiences, teachers are developing models (and conceptual systems) for making sense of students' modeling activities. When using model-eliciting activities, teachers' daily teaching activities provide the basis for classroom-based teacher development.

M&M perspectives emphasize the principle of teaching teachers using strategies that we want them to use to teach their students. In particular, we believe that it rarely makes sense to lecture teachers about being constructivist teachers. Or, if we believe that collaborative problem solving experiences are important for students, then we also assume that effective teacher development activities may involve learning through problem solving - while working in collaborative groups to develop useful, powerful, transportable, and sharable conceptual tools such as:

- tools for giving students feedback about strengths and weaknesses of their work on thought-revealing modeling tasks.
- tools for observing the roles that students play, and the ideas and processes that they use, during their work on thought-revealing modeling tasks.

In What Ways Do M&M Perspectives Go Beyond Ladder-Like Stages of Development?

Like Piaget, mathematics educators have tended to describe the development of mathematical constructs using ladder-like sequences of stages that involve increasingly complex conceptual systems. According to this point of view, when the claim is made that a given child is "at stage N" for a given construct or conceptual system, this implies that the child's behavior is expected to be invariant across most tasks that are characterized by the relevant construct or conceptual system.

Piaget used the term *decalage* to refer to the fact that a given child's apparent level of performance may be different

M&M perspectives recognize that a *decalage* is not simply an unexplained phenomenon that occurs infrequently. It is a common phenomenon that is to be expected for the following reasons.

- Development occurs along a variety of dimensions: concrete-abstract, simple-complex, situated-decontextualized, external-internal, specific-general, intuitive-formal, and unstable-stable. So, if tasks differ along any of these dimensions, then different levels of performance are predicted.

- In general, knowledge and abilities (and models) are only organized around

for tasks that Piaget considered to be characterized by the same underlying conceptual structure.	abstractions at relatively late stages of development. At earlier stages, knowledge tends to be organized around experience. That is, two ideas are considered to be related, not because they are logically connected, but because they've been used together in familiar problem solving situations. • In applied sciences, serious problems seldom lend themselves to descriptions using only a single disciplinary perspective. Similarly, even for children, the development of models for making sense of real experiences generally involves the parallel and interactive development of clusters of constructs and conceptual systems. • During early stages of development, the preceding conceptual systems tend to be characterized by fuzzy, piecemeal, poorly differentiated, and poorly coordinated ways of thinking that are gradually sorted out so that similarities and differences are clear. Therefore, M&M perspectives begin with the assumption that development cannot be collapsed to a single dimension—and progress is not like taking steps on a ladder.

An important component of M&M research seeks to describe the multidimensional terrains in which conceptual development occurs. As mathematics educators have investigated the development of increasing numbers of logically related constructs and conceptual systems, it has become obvious that it is far too simplistic to describe conceptual development using the metaphor of a single ladder-like sequence. For example, the following points have become clear as colleagues have gone beyond investigating the development of whole number arithmetic concepts to also focus on the interacting development of basic constructs related to fractions, ratios, rates, proportional reasoning, or other fundamental concepts involving measurement, geometry, algebra, calculus, probability, or statistics.

• Primitive understandings of both elementary and advanced constructs often begin quite early and continue over time periods of several years. For example, at the same time that many whole number arithmetic concepts are still at intermediate stages of development, constructs and conceptual systems also are beginning to develop that are related to fractions, ratios, rates, proportions, linear relationships, and a variety of different types of measures (length, area, volume, time, and speed).

- At any given point in time, most children are at intermediate stages of development for nearly any construct or conceptual system that teachers want their students to develop further. That is, the concepts are not completely mastered, nor are they completely new. For example, when a teacher wants to introduce a new concept, a wide variety of concrete, intuitive, or situated understandings can be expected to exist already. So, the challenge for teachers is not simply to introduce new ideas. The goal is to put students in situations where they express their current ways of thinking in forms that will be tested and revised in directions of increasing power.

- For related concepts, such as those involving fractions, ratios, rates, quotients, proportions, and linear relations, a student that someone might consider to be "at stage N" for one concept may be at a very different stage for another logically related concept. So, it is not true that development of the second concept necessarily precedes or follows the development of the first concept on a predictable basis.

- Even for several tasks that apparently involve the same construct, a child whose performance on one task appears to be "at stage N" frequently appears to be at a completely different stage for a slightly different task—or (for example) if the same task is expressed using different representation media.

- Even within a single learning or problem-solving situation, performance on model-eliciting activities shows that children often develop (locally) through a series of stages during the solution of a single problem. In other words, students often do not operate at a single stage even throughout the course of a 60-minute activity.

For the preceding kinds of reasons, a given student's apparent stage of reasoning often varies significantly across constructs, across contexts, across tasks, and across representations, as well as from moment to moment even within a given learning or problem solving situation. Yet, the phenomenon of *local-conceptual development* implies that Piagetian theories in particular, and cognitive development theories in general, often have direct applicability to learning and problem solving activities.

In What Ways Do M&M Perspectives Go Beyond Research on Situated Cognition?

Situated cognition refers to the fact that, knowledge is organized around experience at least as much as it is organized around abstractions—especially during early stages of development. Furthermore, even for experts in a field, their conceptual systems often continue to be shaped significantly by the situations that led to their development (Greeno, 1991).	Models nearly always are developed in specific contexts for specific purposes and for specific clients. So, their forms and functions usually reflect these facts. However, in general, models also need to be sharable and reuseable by other people in other situations beyond the specific instance that led to their development. Models are seldom simply embodiments of isolated pieces of ungeneralizable ways of thinking.

Some models are easier to modify than others; and, some are more sharable and reuseable than others. By requiring students to develop sharable, easily modifiable, and reuseable models and conceptual tools, M&M perspectives often provide straightforward ways for teachers to deal with some important aspects of generalizability—while recognizing that a student who has produced sharable, transportable, and reuseable models may not be able to use them in situations that appear the same to an expert in the area.

In What Ways Do M&M Perspectives Go Beyond Research On Vygotsky's Zones of Proximal Development?	
Vygotsky (1962, 1978) emphasized the importance of language in the development of higher cognitive functioning. He also emphasized the process of *internalization*—whereby cognitive functions are expected to occur first on an inter-individual plane before later occurring on an intra-individual plane. In particular, a given function is not expected to go from un-mastered to mastered at a single point on a developmental number line. Instead, there exists *zones of proximal development* in which a child's ability to function can be influenced greatly by a teacher (or more able others) who provides appropriate guidance, language, and other social supports.	M&M perspectives recognize the important role that language plays in conceptual development and the usefulness of the notion of a *zone of proximal development*. At the same time, however, M&M perspectives emphasize that language is only one among many different kinds of representational media that are important in mathematical thinking, teaching, learning and problem solving. In particular, the availability of dynamic, computational media extends the notion of the role that language plays in the development of higher order thinking. M&M perspectives suggest that *zones of proximal development* are multi-dimensional regions where interactions include not only teachers or other adults who are more capable but also learners interacting with peers, with themselves, and with powerful representational tools and media.

Vygotsky's *zones of proximal development* provide ways for teachers to help students extend conceptual systems that they already have developed at some level. They do not say how students could develop conceptual systems that are completely new. However, this is not terribly problematic in practice, because for most conceptual systems that teachers want students to develop, most students can be expected to be at intermediate stages of development.

Similar points have been emphasized by theorists such as Bruner (1960) and Piaget (1968). That is, constructs develop, and for most constructs that students are expected to master, the relevant conceptual systems are not completely new—nor are they completely mastered. As we have argued earlier, the development of significant conceptual systems proceeds simultaneously along several dimensions and hence for any given learner the *zones of proximal development* are multidimensional. The kind of support that is needed to promote this development varies across contexts and settings and includes not only the language of the teacher and more able peers, but also many kinds of representational media. The critically important role of language (and representational media) is that of empowering the learner to express ideas that he or she already has in some stage of development.

In What Ways Do M&M Perspectives Go Beyond Social Constructivism?	
Social constructivists (e.g., Cobb, 1994) go beyond purely cognitive views of conceptual development by emphasizing *the mind in society*. That is, the conceptual development of an individual	M&M emphasizes *the society in the mind* as well as *the mind in society*. • Concerning the society in the mind: M&M perspectives assume that (a)

28. MOVING BEYOND CONSTRUCTIVISM 545

does not occur in a vacuum. Students are part of learning communities; and, the shared constructs and social norms of these communities strongly influence the conceptual development of individuals.	conceptual systems are complex, living, interacting, self organizing, and continually adapting systems, and (b) regardless of whether we focus on the development of individuals or groups, their ways of thinking are characterized by communities of complex and interacting conceptual systems. Consequently, for development to occur, it is necessary to encourage diversity, selection, communication, and preservation. • Concerning *the mind in society*: The kind of model-development sequences that were described in chapter 2 typically involve teams of individuals working cooperatively to produce sharable and reuseable conceptual tools. Products they produce include descriptions, explanations, & justifycations. Follow-up discussions often involve procedures to help students internalize social norms that encourage powerful forms of explanation, argumentation, & justification.

Whereas social constructivists extend purely cognitive perspectives by viewing the development of individuals through the lens of social theories, M&M perspectives extend social constructivist perspectives by also viewing communities through the lens of cognitive theories. Also, whereas social constructivism recognizes that one way that ideas are validated depends on social norms and beliefs and approved procedures, M&M perspectives recognize that there are at least four distinct way that models (and ways of thinking) come to be adopted by individuals or by groups:

• Based on consistency (pure math criteria): By checking whether the new construct is consistent with conceptual systems that already have been accepted.
• Based on usefulness (applied math criteria): By checking whether the new construct is useful for the purposes it was developed to address.
• Based on peer review (social correctness criteria): By checking whether the construct is judged to be acceptable (or exemplary) by peers.
• Based on judgments by an authority (power criteria): By checking whether the construct is judged to be acceptable (or exemplary) by teachers or other authorities.

Kuhn (1970), Lakatos (1978), and other philosophers of science have shown that the development of scientific knowledge is not the kind of purely logical and purely objective enterprise that it often is portrayed as being. For example, social norms determine answers to the following kinds of questions.

• Which constructs will be judged to be acceptable (or exemplary) by peers?
• Which constructs construct is judged to be acceptable (or exemplary) by teachers or other authorities?
• What kind of assumptions will be accepted as being obvious?

- What kind of problems or issues will be considered to be priorities to address?
- What kind of evidence and arguments will be judged to be persuasive?
- What kinds of tools will problem solvers be allowed to use?

Similar questions also arise in classrooms. For example, if a teacher tries to introduce new modes of instruction (such as those that involve problem-based learning, small group collaborations, or other student-centered approaches to instruction that break with tradition), new classroom norms need to be established before productive interactions are likely to occur. Nonetheless, these observations about the influences of social norms do not imply that usefulness and consistency are unimportant in the conceptual development of individuals or societies. In fact, in the development of knowledge by either individuals or groups, some of the most interesting events occur precisely at times when the beliefs of authorities and communities are rejected or revised substantially in the case of some construct or conceptual system that is especially powerful or fundamental. Piaget believed that, during these periods of major conceptual reorganization, *cognitive conflict* is a primary driving force that promotes qualitative changes in thinking. Therefore, in the design of instruction that is aimed at encouraging the development of a small number of big ideas for either individuals or groups, model-eliciting activities are designed to put problem solvers in situations where:

- they clearly recognize the need for the targeted construct,
- they are able to judge for themselves whether their current constructs are sufficiently useful

In the same way that major conceptual reorganizations don't occur often in the history of science, model-eliciting activities represent only one small part of instruction that's aimed at helping students develop a small number of "big ideas" in school mathematics. Other kinds of activities may be more appropriate for the large number of "little ideas," facts, and skills that also need to be learned in mathematics instruction. Furthermore, even in the case of model-eliciting activities that are aimed at "big ideas" in the school mathematics curriculum, M&M perspectives suggest that it often is productive to follow these with *model-exploration activities* that are intended to introduce powerful language and notation systems that depend heavily on social conventions (chapter 2) for specified purposes and situations.

In What Ways Do M&M Perspectives Go Beyond Research on Students' Learning Trajectories?

When the goal of instruction is for teachers to guide the thinking of a class toward targeted instructional goals, social constructivists (e.g., Simon, 1995) emphasize how important it is for teachers to be familiar with learning trajectories that correspond to typical developmental sequences that students often go through while developing relevant conceptual systems. They also emphasize the importance of adapting these trajectories to fit the emerging thinking of the class. Such trajectories are not inflexible linear sequences; they

M&M perspectives recognize there are times when teachers should assume the role of guides and authorities about the correctness of students' thinking. But:

- Because modeling activities are designed to reveal an explosion of diversity in thinking, M&M perspectives are especially sensitive to the diversity of thinking that exists even in the most homogeneously grouped classrooms or problem-solving groups—or even within the mind of an individual student. Modeling tasks (as well as discussions sessions that

28. MOVING BEYOND CONSTRUCTIVISM 547

are more like branching tree diagrams in which specific trajectories are only visible in retrospect.	surround them) assume that the class is a complex and heterogeneous problem-solving organism. • In model-eliciting activities, and in discussion sessions that accompany them, attention focuses on encouraging students to express, test, and refine or revise their own ways of thinking—much more than on guiding students toward teachers' ways of thinking.

For model-eliciting activities, the products that problem solvers produce involve descriptions, explanations, justifications (for actions or predictions), or other complex artifacts for which trade-offs nearly always occur that involve factors such as simplicity versus complexity of expression, or alternative modes of expression. Therefore, a variety of products often are potentially useful. Yet, it is not true that anything that students produce is satisfactory, nor is it true that all products are equally useful. In fact, in well-designed model-eliciting activities, the goals should be stated in ways that implicitly or explicitly include criteria that students themselves can use to test the usefulness of the products that they develop. Therefore, without consulting their teachers, students themselves should be able to identify strengths and weaknesses of alternative products that they produce. This selection of viable products is necessarily set in a social context. For example, in model development sequences, students generally work in teams on multi-step problems in which communicating, planning, monitoring, and assessing activities are highlighted. Because the products that these teams produce need to be sharable and re-useable by others, follow-up presentations, classroom discussions, and assessments emphasize consensus building among learners.

In What Ways Do M&M Perspectives Go Beyond Research on Representation and Semiotic Functioning?

In spite of the *reality phobia* that preoccupies many radical constructivists, they often have been explicit about the following facts.	In addition to the points about representation that are identified in the left-hand column, M&M perspectives also emphasize the following facts.
• Representational tools are among the most important artifacts that students both: (a) project into the world, and (b) encounter in the world (Steffe & Thompson, 2000). • In addition to adopting special mathematical language and symbol systems introduced by teachers and textbooks, students also create their own languages, notation systems, diagrams, experience-based analogies, and other media to express their thoughts (Lehrer & Schauble, 2000).	• Representational media that are significant in mathematics learning and problem solving include more than just systems that students look at. They also include systems that students look through to construct, describe, explain, or make sense of other systems. • When students develop responses to model-eliciting tasks, or when they think mathematically about other kinds of learning or problem solving situations, they often make simultaneous use of a variety of representational media—each of which clarifies (or ignores, or

- Several powerful types of expressive media involve technology-based tools; and, these often are multiple linked representations (e.g., tables, graphs, and equations); (Kaput, 1991).
- Unless special care is taken in instruction, the preceding systems may do little more than represent one another without serving as: (a) embodiments for students internal constructs and conceptual systems, or (b) descriptions or explanations of other systems that occur in students lives (Sfard & Thompson, 1994).

distorts) different aspects of the underlying conceptual scheme—or the situation that these schemes are intended to describe.

As an example of this, consider the fact that, when students are writing symbolic descriptions or drawing diagrams of problem solving situations, they also may be talking about what they are doing as well as imagining (or referring to) some experience-based metaphor that is considered to be similar to the situation at hand. In other words, students often use several media in parallel and interactively. An important part of developing a final way of thinking about many learning or problem solving situations often involves integrating meanings that are partly expressed using a variety of media.

In applied science activities beyond schools, "problem solvers" often consist of teams of specialists—each working on different parts of the problem, or different aspects of the solution process. Or, even when isolated students work on modeling tasks:

- the problems themselves often inherently involve multiple media
- the solution processes often involve the parallel use of a variety of different representational media
- the products that are produced involve multiple media.

One reason why communication capabilities and representational fluency are emphasized in M&M activities is because purposeful descriptions, explanations, and justifications are not just part of the processes used to produce the conceptual tools that need to be developed; they are significant parts of the conceptual tools that need to be developed. Another reason is because technological tools often are available in real life problem solving situations, and because what these tools tend to do best involves information processing. Therefore, some of the most important problem solving abilities that successful problem solvers need to develop include:

- describing problem solving situations in forms that can be used with the technological tools that are available.
- partitioning complex problem solving situations into pieces that can be addressed by different specialists.
- describing progress (and intermediate results for subproblems) in forms that can be communicated to, and assessed by, other specialists or nonspecialists.
- integrating results from subproblems into forms that address complex problems.

In What Ways Do M&M Perspectives Go Beyond Traditional Research on Problem Solving—And Information Processing?

Problem solving often is characterized as *getting from givens to goals when the path is not obvious*. Also, it is

Whereas what is problematic about traditional word problems is that students must develop *meanings for*

28. MOVING BEYOND CONSTRUCTIVISM

commonly assumed that problem solving means solving traditional textbook word problems – where the central thing that is problematic is that *students must make meaning of symbolically described situations*. Yet:

- Most traditional word problems do not require construct development.
- Many assembly-ist activities in solving such problems do not involve the creation of any tools or artifacts that the students consider to be useful.
- Many discussion sessions aimed at establishing negotiated meanings for language, symbols, or routines do not involve the resolution of any issues that students consider to be problematic.

symbolically described situations, model-eliciting tasks emphasize almost exactly the opposite kinds of processes. That is, what is problematic is that *students must develop symbolic descriptions for meaningful situations*. Consequently, the understandings and abilities that are needed for modeling tasks often are quite different than those emphasized in traditional word problems. For example:

- Because modeling often requires students to work in teams to develop models (or conceptual tools) that are sharable and reuseable, sociocultural factors are emphasized in addition to purely cognitive factors.
- Because the goals of modeling tasks involve developing complex artifacts (or conceptual tools), development processes generally involve a series of design cycles (or modeling cycles) that involve systematically rethinking the nature of givens, goals, and relevant solutions steps—or patterns & relationships that are attributed to surface-level data. Therefore, problem solvers go beyond being information processors to become construct processors and pattern matchers. And, productive problem solving strategies often go beyond helping students process external data to also help them assess, modify or transform their own ways of thinking.

- Because traditional word problems tend to involve only a single cycle of getting from givens to goals when the path is not obvious, the kinds of strategies and heuristics that are emphasized generally are intended to provide answers to the question *What should I do when I'm stuck?* But, for modeling activities, students are rarely stuck! That is, at every stage of problem solving, they usually have a variety of relevant (and some irrelevant) ideas and procedures. What is problematic is that these ideas and procedures generally need to be tested and revised repeatedly and iteratively before they are sufficiently useful to address the problem at hand. Therefore, the kind of strategies and heuristics that are most needed tend to be those that help students express, test, revise, and refine ideas that they do have—much more than learning to behave cleverly in situations where they have none.

- According to M&M perspectives, problem solving strategies and heuristics are expected to develop in a manner similar to other cognitive functions. For example, one of the most important dimensions along which understanding appears to develop involves Vygotsky's notions about the internalization of external functions: "The greatest change in children's capacity to use language as a problem-solving tool takes place somewhat later in their development, when socialized speech (which has previously been used to address an adult) *is turned inward.* Instead of appealing to the adult, children appeal to themselves; language thus takes on an *intrapersonal function* in addition to its *interpersonal use.*" (Vygotsky, 1978, p. 27)

- Important parts of what it means to understand a given strategy, heuristic, or function involves knowing when to use it. So, understanding involves seeing at least as much as it involves doing. For example, as the diagram below illustrates, if we envision a function as a nail-like condition-action rule (where the head of the nail corresponds to the condition, and the body of the nail is the action), then heuristics and strategies look more like thumbtacks than they do like finishing nails. That is, the condition component is relatively large compared with the action component.

Information Processing Theories: A "Finishing Nail" Metaphor.	Models and Modeling Theories: A "Thumbtack" Metaphor

In What Ways Do M&M Perspectives Go Beyond Research on Attitudes, Interests, Affect, Identity, and Beliefs?

Motivation, interest, and engagement often are among the main reasons cited by teachers for emphasizing instructional activities that they associate with constructivist philosophies.	According to M&M perspectives, a student's feelings about an experience is determined by how it is interpreted; and, the interpretation depends on the implicit or explicit models that are brought to bear. In other words, attitudes, values, interests, and beliefs are all considered to be parts of the models that students develop to make sense of their experiences. That is: - They vary from one situation to another, and from one interpretation to another for a single situation. - They are learned in the context of specific interpretations (models); they are not learned in the abstract. - As models develop, so do accompanying attitudes, interests, values, beliefs—and conceptions of personal roles, abilities and resources. Although attitudes and interests influence students' behaviors, students are not simply victims of these attitudes and interests. Students can manipulate their own profiles to suit circumstances.

One way models impact attitudes and interests can be seen by considering Csiksentmihalyi concept of flow (1990). "Flow" is a metaphor for a type of experience that is highly enjoyable—*"like being carried away by a current, everything moving smoothly without effort"* (p. xiii). On the one hand, time passes rapidly. Yet, because you are in control of the situation, everything seems to run in slow motion. Csikszentmihalyi found that flow experiences are more likely to occur:

- when people are involved in difficult and complex activities that stretch their physical or mental abilities, rather than in moments of leisure or during what we commonly consider to be entertainment.
- during activities where immediate feedback is available about progress toward challenging goals, rather than in no risk situations where no feedback is available.
- especially during moments when we rise above the welter of details to see the patterns and regularities that underlie them.

Well-designed model-eliciting activities sometimes elicit flow experiences in children. They should be extrinsically motivating because the goals are to develop powerful conceptual tools that are useful in real life situations. But, even more importantly, they should be intrinsically motivating because the whole purpose of a mathematical model is to see patterns and regularities beneath the surface of things.

In What Ways Do M&M Perspectives Go Beyond Research on Metacognition and Higher Order Thinking?

Because one of the most important original goals of the constructivist movement was to encourage the development of deeper and higher-order understanding, many constructivists are strong advocates of:

- Process objectives.
- Metacognitive abilities.
- Argumentation and justification
- Habits of mind.

M&M perspectives also emphasize process objectives, argumentation and justification, as well as the development of productive work habits. However, M&M research suggests that:

- Processes students use are determined by the way they interpret learning and problem solving situations. That is, they are influenced by the models students use. So, the development of process objectives does not occur in the abstract. To be most productive, learning occurs in the context of model development.

- According to M&M perspectives, arguments, explanations and justifications are considered to be activities that don't just occur before or after "results" are given. In M&M activities, they are important parts of the activities themselves - and also important parts of the products that are produced. Modeling tasks should make clear who needs the explanation—and when and for what purpose. Therefore, the criteria are available so that students can assess the quality of their own work. If this were not the case, then there would be no reason for students to go beyond their first primitive interpretations of problem solving situations.

- Metacognitive abilities generally involve thinking about thinking, thinking about the products of thought, or thinking about one's own roles during learning or problem-solving experiences. Modeling activities demystify such abilities, and facilitate their development, because students continually express their thinking in forms that can be examined, tested, compared, and revised or refined.

- Habits (like smoking) conjure up counterproductive metaphors for the kinds of attitudes beliefs and behaviors that we want students to develop. Habits tend to refer to dispositions and actions that function unconsciously. But, from a M&M perspectives, we want to foster the development of problem solving profiles (identities, personae) that can be monitored and controlled. For example, it sometimes is productive for students to be:

    o Impulsive (nonreflective)—such as during "brainstorming" stages of dealing with complex problems.
    o off-task—such as when a brief break is needed (often just prior to a breakthrough), or when a new perspective is needed (perhaps stepping back from the details of what you have been doing).

During multiple-cycle processes needed to develop complex conceptual tools, different roles often need to be emphasized at different stages in development. Therefore, productive individuals (and groups) and those who flexibly adapt their roles and personae to fit a variety of circumstances. In general, it seems unlikely that productive strategies for helping students develop such adaptable personae will be similar to those typically used to make and break unconscious habits.

When students work in groups to solve complex problems that were designed to encourage them to express their thinking in visible forms, teachers often think that it may be productive for them (the teachers) to assume the role of "the hint giver"—or (if they have taken a recent cognitive science course) "the metacognitive agent." However, during modeling activities, it may or may not be productive for teachers to play these roles—depending on how well the teacher recognizes that:

- The student's task has more to do with seeing than doing.
- The fundamental challenge is to develop new ways of thinking about the situation and sort out the variety of ways of thinking about the situation.
- Several modeling cycles are probably needed to accomplish the preceding goal.

Early in a problem solving session, if a hint has the effect of leading students to imagine that they have been given "the right ways to do it," then the long-range effect of this hint may be negative.

In What Ways Do M&M Perspectives Go Beyond Research on What is Needed for Success Beyond School in a Technology-Based Age of Information?

Constructivists have had strong influences on a variety of curriculum standards documents that have been produced by organizations ranging from the National Council of Teachers of Mathematics (1989, 2000), to the American Association for the Advancement of Science (1993), to many state Departments of Education. Furthermore, constructivist researchers have conducted naturalistic observations in order to investigate what kind of mathematical understandings and abilities are needed for success beyond school.	When M&M-based research investigates what's needed for success beyond school, we do not trust our own abilities to make unbiased judgments about: • where to observe? (grocery stores? engineering firms? Internet cafés?) • whom to observe? (farmers? cooks? shoppers? baseball fans?). • when to observe? (only when they are calculating with numbers?) • what to count as mathematical or scientific thinking in videotapes or naturalistic observations.

28. MOVING BEYOND CONSTRUCTIVISM

Preconceived answers to the preceding questions often expose questionable assumptions about what it means to think mathematically or scientifically and about the nature of real life situations in which mathematical or scientific thinking is useful. For this reason, M&M-based research often emphasizes research designs that are analogous to the procedures that applied scientists use to design aircraft or other complex systems. In particular, we often use *multi-tier design experiments* in which:

- Level #1: Students develop conceptual tools for making sense of mathematical problem solving situations.
- Level #2: Teachers and other participants develop conceptual tools for making sense of students' modeling activities.
- Level #3: Researchers develop conceptual tools for making sense of students' and teachers' modeling activities.

At level #2 of the preceding multi-tier design studies, teachers, parents, policy makers, professors, researchers, and curriculum designers are all considered to be evolving experts. Each has important views that should be considered about what's needed for success beyond school in the 21st century. Yet, different experts are expected to hold conflicting views; none have exclusive insights about truth; and, all tend to evolve significantly during design experiments where current ways of thinking are expressed in forms that must go through a series of testing-and-revising cycles in which formative feedback and consensus-building influence final conclusions that are reached about:

- the kinds of mathematical or scientific problem solving situations that (they believe) will be especially important to master as preparation for success in a technology-based age of information.
- the most important levels and types of elementary-but-deep understandings and abilities that (they believe) are likely to be needed for success in the preceding kinds of situations.

Results from multitier *evolving expert studies* are showing that the kind of mathematical understandings and abilities that are emerging are often quite insightful, quite supportive of focusing on deep treatments of a small number of big ideas, and yet quite different than those that emphasized by school people in curriculum standards documents whose main concern is to make incremental improvements in the traditional curriculum. In general, the kind of understandings and abilities that are emphasized are similar to those typically emphasized in job interviews following graduation from professional interviews following graduation from professional schools in engineering, business, and other applied sciences where the kind of people that are sought need to be proficient at: (a) representational fluency needed to generate multimedia descriptions and explanations of complex systems, (b) working in teams of diverse specialists, and (c) adapting to rapidly evolving conceptual tools. These points are especially significant to recognize at a time when a back to basics backlash is threatening to reverse progress made by standards-based curriculum innovations—and when one reason for this backlash is because curriculum standards documents generally have made only inadequate attempts to enlist the understanding and support of parents, policy makers, or professionals or professors in fields that are heavy users of mathematics, science, and technology.

In What Ways Do M&M Perspectives Go Beyond Constructivist Research on Teacher Development?

At the extreme, one could argue that from a constructivist perspective, all teaching is constructivist, because students must necessarily construct their own knowledge. However, such a point of view is hardly useful for supporting the meaningful decisions of teachers or providing principles for the design of instructional materials. Indeed, constructivists themselves make clear the limits of constructivism for understanding the development of teaching. "But the cognitive premises of constructivism can dictate only guidelines for good teaching. We cannot derive from them, any more than we can from any other cognitive position, specific teaching methods" (Noddings, 1990, p. 15)

The norms of the classroom learning community become the criteria by which constructs are validated. Establishing these norms in ways that are aligned with broadly accepted mathematical thinking is a central role for the teacher.

Teachers play a central role in guiding the students' collective thinking along a learning trajectory so they can construct particular mathematical understandings. In practice, this guidance is often in the form of carefully posed questions by the teacher accompanied by the selection of responses that further the movement along the path that the teacher has chosen towards a convergence of mathematical meaning.

Even though teachers' constructs are expected to be more complex than children's, principles that govern the development of useful models by students are expected to be similar to those that govern the development of useful models by teachers. M&M perspectives assume that teaching is about seeing and interpreting situations as much as it is about "doing." However, compared to what is known about the nature of children's mathematical knowledge, relatively little is known about the constructs of teachers to make sense of situations that involve mathematical problem solving, learning, or teaching.

In effective teaching and learning situations, it sometimes is important to guide students efficiently toward the experts' "correct" ways of thinking. But, at other times, it is important to encourage students to express, test, and revise their own current ways of thinking – so that development occurs in directions of increasing conceptual power. ... John Dewey's emphasize this distinction when he claimed that "making mathematics more practical" is often quite different than "making practice more mathematical" – and that "mathematizing reality" is quite different than "realizing mathematics" (Steffe & Thompson, 2000). M&M perspectives focus on the roles of teachers in both kinds of situations.

M&M perspectives have powerful implications for how we understand the role of the teacher and for how we support the professional development of teachers. For example if the teacher's role is not simply to guide students through careful sequences of questions along preplanned learning trajectories, then teacher's knowledge needs to include an understanding of the multiplicity of children's models as those models develop along multiple dimensions – and as the classroom learning community goes through processes of selection, refinement, and revision of children's existing ways of thinking.

M&M perspectives on teacher development focuses on designing effective and sophisticated ways of helping teachers see and interpret children's thinking and support the development of that thinking. Rather than telling teachers what to do or recommending courses of action, M&M perspectives shift attention toward finding

28. MOVING BEYOND CONSTRUCTIVISM

> ways to support teachers in their efforts to see and interpret the complexities of teaching and learning. This emphasis on interpretation includes the ways in which students might learn, the mathematics itself, and its pedagogical and logical development, the relevant curricular materials, and possible ways of proceeding with a sequence of learning activities. The variability that teachers bring in their perceptions and interpretations of rich contexts for teaching and learning provide diversity in understanding the complex and ill-structured domain of practice.

SUMMARY

Speaking on behalf of all 31 authors who contributed to this book, we want to invite readers to join us in future work of the models and modeling research groups. These groups are continuing to function in subgroups that are focusing on problem solving, teacher development, curriculum development, and student development involving a variety of grade levels and mathematical topic areas. Perhaps the best way to contact these groups is by sending an e-mail message to us, the editors of this book. We intend to remain in contact with these research-working groups.

To be a research community, it's important to establish sufficiently shared theoretical and practical perspectives - as well as shared research sites and tools. But, to participate productively in *models and modeling research groups*, it is not necessary to adopt any rigid orthodoxy. The nature of a *models and modeling perspective* is that there is a clear need for diversity in thinking; and, even collaboration-minded curmudgeons often play productive roles. In fact, like the *pragmatist perspectives* that provide so many philosophical foundations for the research reported in this book, a *models and modeling perspective* is intended to provide a framework for integrating productive ideas from a variety of theoretical and practical perspectives. On the other hand, as the chapters throughout this book illustrate, *models and modeling perspectives* certainly are not theoretically neutral. For example, as the table in this chapter illustrates, we believe that the language of models and modeling pushes mathematics educators to advance their thinking considerably beyond current constructivist ways of thinking about mathematical problem solving, learning, and teaching—while at the same time integrating and simplifying many the most productive ideas and procedures that have been developed by modern descendents of Piaget, Vygotsky, Dewey, Peirce, and James. *Models and modeling perspectives* also are ideally suited to make use of a wealth of recent advances in fields of mathematics ranging from complexity theory to game theory—where a variety of different types of "systems thinking" tends to be highlighted.

Systemic curriculum reform recognizes that the development of new programs and curriculum materials is not likely to have a lasting impact on student development. For example, in order for innovative standards-based curriculum materials to succeed, there are perhaps no issues that are higher priorities to address than those related to: (i) assessment, (ii) teacher development, and (iii) enlisting the understanding and support of parents, policy makers, and community leaders. So, the dilemma that confronts most systemic curriculum reform initiatives is: *How can you neglect nothing without failing to focus sufficiently on anything?* ... The answer, of course, is that it is necessary to focus on a small number of "leverage points" that significantly impact the most important systemic characteristics of the interacting systems we're attempting to understand and influence.

An important characteristic of the *models and modeling perspectives* emphasized in this book is that students, teachers, researchers, curriculum developers, program developers, and teacher developers—all are considered to be developers of powerful, sharable, and reusable conceptual tools in which models for making sense of experience are among the most important components. Therefore, modeling activities provide just the kind of "leverage point" referred to in the preceding paragraph. For example, for many especially illuminating kinds of learning and problem solving situations, model development activities for students often also provide ideal contexts for productive model development activities for teachers, researchers, and others. As students develop mathematical models for making sense of problem solving situations, teachers develop models of students' modeling behaviors; and, researchers, curriculum developers, and teacher developers develop models of both teachers' and students' modeling behaviors. Consequently, as chapter throughout this book have illustrated, it often is possible for researchers to share resources (problem sets, video taped problem solving episodes, and research tools) so that productive collaborations can be forged among researchers who are focusing on student development, teacher development, program development, or curriculum materials development.

For a description of the multi-tier design research that has been emphasized in the models and modeling working groups, see the chapter about design experiments (Lesh, 2002) in the *International Handbook of Research Design in Mathematics Education*. These multi-tier research designs were developed especially so that multiple researchers, at multiple sites, and representing multiple perspectives can coordinate their work so that realistically complex problems can be more effectively addressed. We hope the research reported in this book will stimulate readers to contribute to these efforts.

REFERENCES

Aliprantis, C. D. (1999). *Games and decision making.* Cambridge, England: Oxford University Press.
American Association for the Advancement of Science. (1993). *Benchmarks for science literacy.* Washington, DC: American Association for the Advancement of Science.
Amit, M. (1988). Career choice, gender and attribution patterns of success and failure in mathematics. In A. Bourbas (Ed.), *Proceedings of the 12th annual conference of the International Group for the Psychology of Mathematics Education* (Vol. 1, (pp. 125–130). Veszprem, Hungary: Hungarian National Centre for Educational Technology.
Anderson, J. R., Boyle, C. B., & Reiser, B. J. (1985). Intelligent tutoring systems. *Science, 228,* 456–462.
Anzai, K., & Simon, H. A. (1979). The theory of learning by doing. *Psychological Review, 86,* 124–140.
Archambault, R. D. (1964). (Ed.). *John Dewey on education: Selected writings.* Chicago: University of Chicago Press.
Asch, S. E. (1952). *Social psychology.* New Jersey: Prentice Hall.
Ausubel, D. (1978). In defense of advance organizers: A reply to the critics. *Review of Educational Research, 48,* 251–257.
Baker, E. L. (1990). Developing comprehensive assessments of higher-order thinking. In G. Kulm (Ed.), *Assessing higher-order thinking in mathematics* (pp. 7–20). Washington D C: American Association for the Advancement of Science.
Ball, D. (1993). With an eye on the mathematical horizon: Dilemmas of teaching elementary school mathematics. *Elementary School Journal, 93*(4), 373–397.
Baratta-Lorton, M. (1976). *Mathematics their way.* Menlo Park, CA: Addison Wesley.
Behr, M. Lesh, R., & Post, T. (1987). Theoretical analyses: Structure and hierarchy, missing value proportion problems. In J. Bergeron, N. Herscovics, & C. Kieren (Eds.), *Psychology of mathematics education.* Montreal, Canada: University of Montreal.
Behr, M., Wachsmuth, I., Post, T., & Lesh, R., (1984). Order and equivalence of rational number: A clinical teaching experiment. *Journal for Research in Mathematics Education, 15*(5), 323–341.
Bell, A., Breke, G., & Swan, M. (1987). Diagnostic teaching: Four graphical interpretations. *Mathematics Teaching, 119,* 56–59
Bell, A., Swan, M., & Taylor, G. (1981). Choice of operation in verbal problems with decimal numbers. *Educational Studies in Mathematics, 12,* 399–420.
Bell, M., Fuson, K., & Lesh, R. (1976). *Algebraic and arithmetic structures: A concrete approach for elementary school teachers.* New York: The Free Press.
Bergeron, J. C., & Herscovics, N. (1982). Levels in the understanding of the function concept. In G. vanBarneveld & H. Kabbendam (Eds.) *Proceedings of the conference on functions* (pp. 39–46). Enschede, The Netherlands: National Institute for Curriculum.

Berman, E., Bound, J., & Machin, S. (1997). *Implications of skill-biased technological change: International evidence.* Cambridge, MA: National Bureau of Economic Research.

Beth, E., & Piaget, J. (1966). *Mathematical epistemology and psychology.* Dordrecht: Reidel.

Bezuk, N., & Cramer, K. (1989). Teaching about fractions: What, when and how? In P. Trafton (Ed.), *National council of teachers of mathematics 1989 yearbook: New directions for elementary school mathematics* (pp. 156–167). Reston, VA: National Council of Teachers of Mathematics.

Bhaskar, R., & Simon, H. A. (1977). Problem solving in semantically-rich domains: An example from engineering thermodynamics. *Cognitive Science, 1,* 193–215.

Bishop, A. J. (1988). *Mathematical enculturation: A cultural perspective on mathematics education.* Dordrecht,The Netherlands: Kluwer.

Black, P., & William, D. (1998, October). Inside the back box. Raising standards through classroom assessment. *Phi Delta Kappan, Vol. 80*(2) pp. 139–148.

Black, S., & Lynch, L. (2000). *What's driving the new economy: The benefits of workplace innovation.* Cambridge, MA: National Bureau of Economic Research.

Boero, P., Dapueto, C., Ferrari, P.L., Ferrero, E. Garuti, R., Lemut, E., C. Parenti, L. & Scali, E. (1995). Aspects of mathematics–culture relationship in mathematics teaching–learning in compulsory school. *Proceedings of the 19th International Conference of the PME,* Recife (Vol. 1, pp. 151–166).

Borko, H., Mayfield, V., Marion, S., Flexer, R., & Cumbo, K. (1997). Teachers' developing ideas and practices about mathematics performance assessment: Successes, stumbling blocks, and implications for professional development. *Teaching and Teacher Education, 13,* 259–278.

Breshnahan, T., Brynjolfsson, E., & Hitt, L. (1999). *Information technology, workplace organization, and the demand for skilled-labor: Firm-level evidence.* Cambridge, MA: National Bureau of Economic Research.

Brouseau, G. (1977). *Theory of didactical situations in mathematics.* Dordrecht, The Netherlands: Kluwer.

Brown, A. L. (1987). Metacognition, executive control, self-regulation, and other more mysterious mechanisms. In F. E. Weinert & R. H. Kluwe (Eds.), *Metacognition, motivation, and understanding* (pp. 65–116). Hillsdale, NJ: Lawrence Erlbaum Associates.

Brown, A. L. (1990). Domain-specific principles affect learning and transfer in children. *Cognitive Science, 14,* 107–134.

Brown, S. I., & Walter, M. I. (1993). *Problem posing: Reflections and applications.* Hillsdale, NJ: Lawrence Erlbaum Associates.

Bruner, J. (1960). *The process of education.* Cambridge, MA: Harvard University Press.

Bruner, J. (1962). *On knowing: Essays for the left hand.* Cambridge, MA: Harvard University Press.

Bruner, J. (1963). *The process of education.* New York: Vantage Books.

Bruner, J. (1966). *Toward a theory of instruction.* Cambridge, MA: Harvard University Press.

Bruner, J. (1986) *Actual minds, possible worlds.* Cambridge, MA: Harvard University Press.

Buchler, J. (Ed.). (1955). *Philosophical writings of Peirce.* New York: Dover.

Burner, J. S. (1973). Going beyond the information given. In J. M. Anglin (Ed.), *Beyond the information given: Studies in the psychology of knowing.* New York, NY: Norton. (pp. 218–238). (Original work published in 1957 *Contemporary approaches to cognition.* Cambridge MA: Harvard University Press).

REFERENCES

Burton, G., & Maletsky, E. M. (1998). *Mathematics advantage: Teacher's edition*. Orlando, FL: Harcourt Brace.

Byrne, B. M. (1984). The general/academic self-concept nomological network: A review of construct validation research. *Review of Educational Research, 54*, 427–456.

Campione, J. C., Brown, A. L, & Connell, M. L. (1989). Metacognition: On the importance of understanding what you are doing. In R. I. Charles & E. A. Silver (Eds.), *The teaching and assessing of mathematical problem solving* (Vol. 3, pp. 93–114). Reston, VA: National Council of Teachers of Mathematics.

Carlson, M. (1998). A cross-sectional investigation of the development of the function concept. In E. Dubinsky, A. H. Schoenfeld, & J. J. Kaput (Eds.), *Research in collegiate mathematics education, 1*. (Vol. 7, pp. 115–162). Providence, RI: American Mathematical Society.

Carlson, M. (1998). *Notation and language: Obstacles for undergraduate students' concept development*. Psychology of Mathematics Education: North America, Conference Proceedings, Columbus, OH: ERIC Clearinghouse for Science, Mathematics, and Environmental Education.

Carlson, M., Jacobs, S., Coe, E., & Hsu, E. (2001). *Applying covariational reasoning while modeling dynamic events: A framework and a study*. Manuscript under review.

Carlson, M., Jacobs, S., & Larsen, S. (2001). An investigation of covariational reasoning and its role in learning the concepts of limit and accumulation. *Proceedings of the Twenty Third Annual Meeting of the North American Chapter of the International Group for the Psychology of Mathematics Education*; Columbus, OH: Eric Clearinghouse.

Carpenter, T. P., Corbitt, M. K., Kepner, H. S., Lindquist, M. M., & Reys, R. E. (1981). *Results from the second mathematics assessment of the national assessment of educational progress*. Reston, VA: National Council of Teachers of Mathematics.

Carpenter, T. P., & Fennema, E. (1991). Research and cognitively guided instruction. In E. Fennema, T. P. Carpenter, & S. J. Lamon (Eds.), *Integrating research on teaching and learning mathematics* (pp. 1–16). Albany: State University of New York Press.

Carpenter, T. P., & Fennema, E. (1992). Cognitively guided instruction: Building on the knowledge of students and teachers. In W. Secada (Ed.), *Curriculum reform: The case of mathematics in the United States* (Special issue of *International Journal of Educational Research* (pp. 457–470). Elmsford, NY: Pergamon.

Carpenter, T., Fennema, E., & Lamon, S. J. (Eds). (1988). *Integrating research on teaching and learning mathematics*. Madison, WI: Wisconsin Center for Education Research.

Carpenter, T. P., Fennema, E., Peterson, P. L., Chiang, C., & Loef, M. (1989). Using knowledge of children's mathematical thinking in classroom teaching: An experimental study. *American Educational Research Journal, 26*, 499–532.

Carpenter T. P., & Lehrer, R. (1999). Teaching and learning mathematics with understanding. In E. Fennema & T. Romberg (Eds.), *Mathematics classrooms that promote understanding* (pp. 19–32). Mahwah, NJ: Lawrence Erlbaum Associates.

Carraher, T. N., Carraher, D. W., & Schliemann, A. D. (1985). Mathematics in the streets and in schools. *British Journal of Developmental Psychology, 3*(1), 21–29.

Charles, R. I., & Lester, F. K. (1982). *Teaching problem solving: What, why, and how*. Palo Alto, CA: Dale Seymour Publications.

Charles, R. I., & Silver, E. A. (1988). (Eds.). *The teaching and assessing of mathematical problem solving*. Reston, VA: National Council of Teachers of Mathematics.
Charles, R. I., Mason, R. P., Nofsinger, J. M., & White, C. A. (1985). *Problem solving experiences in mathematics*. Menlo Park, CA: Addison–Wesley.
Chase, W. G., & Simon, H. A. (1973). Perception in chess. *Cognitive Psychology, 4*, 55–81.
Chazan, D. (2000). *Beyond formulas in mathematics and teaching: Dynamics of the high school algebra classroom*. New York: Teachers College Press.
Chi, M. T. H., Feltovich, P. J., & Glaser, R. (1981). Categorization and representation of physics problems by experts and novices. *Cognitive Science, 5*, 121–152. Chicago: Encyclopedia Britannica.
Chomsky, N. (1965). *Aspects of the theory of syntax*. Cambridge, MA: M.I.T. Press.
Cohen, E. G. (1996 April). *A sociologist looks at talking and working together in the mathematics classroom*. Paper presented at the Annual Meeting of the American Education Research Association, New York.
Clements, D. H. (1997). (Mis?) Constructing constructivism. *Teaching children mathematics, 4*(4), 198–200.
Clements, D. H. & Battista, M. T. (1990). Constructivist learning and teaching. *Arithmetic Teacher, 38(1)*, 34–35.
Cobb, P. (1994). Where is the mind? Constructivist and sociocultural perspectives on mathematical development. Educational Researcher, 23(7), 13-20. [Reprinted in Fosnot, C. (Ed.) (1996). Constructivism: Theory, perspectives, and practice (pp. 34-52). New York: Teachers College Press, and in P. Murphy (Ed.) (in press), Learners' learning and assessment. Milton Keynes, England: Open University Press].
Cobb, P. (1995). Cultural tools and mathematical learning: A case study. *Journal for Research in Mathematics Education, 24*(4), 362–385.
Cobb, P. (1995). Mathematical learning and small-group interaction: Four case studies. In P. Cobb & H. Bauersfeld (Eds.), *The emergence of mathematical meaning: Interaction in classroom cultures* (pp. 25–129). Hillsdale, NJ: Lawrence Erlbaum Associates.
Cobb, P. (1999). Individual and collective mathematical learning: The case of statistical data analysis. *Mathematical Thinking and Learning, 1*, 5–44.
Cobb, P., Gravemeijer, K., Yackel, E., McClain, K., & Whitenack, J. (1997). Mathematizing and symbolizing: The emergence of chains of signification in one first-grade classroom. In D. Kirshner & J. A. Whitson (Eds.), *Situated cognition: Social, semiotic, and psychological perspectives* (pp. 151–233), Mahwah, NJ: Lawrence Erlbaum Associates.
Cobb, P., McClain, K., & Gravemeijer, K. (April, 2000). *Learning about statistical covariation*. Paper presented at the annual meeting of the American Educational Research Association, New Orleans.
Cobb, P., Stephan, M., McClain, K., & Gravemeijer, K. (2001). Participating in classroom mathematical practices. *Journal for the Learning Sciences, 10*(1&2), 113-164.
Cobb, P., Wood, T., & Yackel, E. & McNeal, B. (1993). Mathematics as procedural instructions and mathematics as meaningful activity: The reality of teaching for understanding. In R. B. Davis & C. A. Maher (Eds.), *Schools, mathematics and the world of reality* (pp. 119–134). Needham Heights, MA: Allyn and Bacon.
Cobb, P., & Yackel, E. (1995, October). *Constructivist, emergent, and sociocultural perspectives in the context of developmental research*. Paper presented at the Seventeenth Annual Meeting of the North American Chapter of the

REFERENCES

International Group for the Psychology of Mathematics Education, Columbus, OH.

Cobb, P., & Yackel, E. (1996). Sociomathematical norms, argumentation, and autonomy in mathematics. *Journal for Research in Mathematics Education, Vol. 27*(4) pp.458–474.

Cobb, P., Yackel, E., & Wood, T. (1991). Curriculum and teacher development: Psychological and anthropological perspectives. In E. Fennema, T. P. Carpenter, & S. J. Lamon (Eds.), *Integrating research on teaching and learning mathematics* (pp. 83–120). Albany, NY: State University of New York Press.

Cobb, P., Yackel, E., & Wood, T. (1992). Interaction and learning in mathematics classroom situations. *Educational Studies in Mathematics, 23*, 99–122.

Cohen, D. K. (1988). Teaching practice: Plus que ça change. In P. W. Jackson (Ed.), *Contributing to educational change: Perspectives on research and practice* (pp. 27–84). Berkeley, CA: McCutchan.

Cohen, D. K. & Ball, D. L. (1990). Policy and practice: An overview. *Educational Evaluation and Policy Analysis, 12*, 247–353.

Cohen, D. K., & Barnes, C. A. (1993). Pedagogy and policy. In D. K. Cohen, M. W. McLaughlin, & J. E. Talbert (Eds.), *Teaching for understanding: Challenges for policy and practice* (pp. 207–239). San Francisco, CA: Jossey-Bass.

Cohen, E. G. (1994). Restructuring the classroom: Conditions for productive small groups. *Review of Educational Research, 64*, 1–35.

Cohen, E. G. (1996 April). *A sociologist looks at talking and working together in the mathematics classroom*. Paper presented at the Annual Meeting of the American Education Research Association, New York.

Conant, J. B. (1951). *Science and common sense*. New Haven, CT: Yale University Press.

Confrey, J. (1995). A theory of intellectual development. *For the Learning of Mathematics, 15*(1), 38–48.

Confrey, J., & Smith, E. (1995). Splitting, covariation, and their role in the development of exponential functions. *Journal for Research in Mathematics Education, 26*, 66-86.

Cooney, T. J. (1999). Conceptualizing teachers' ways of knowing. *Educational Studies in Mathematics, 38*(1–3), 163–87.

Cramer, K. (2001). Using concrete models to build middle-grade students understanding of functions. *Mathematics Teaching in the Middle School, 6*, 310–319.

Cramer, K., Behr, M., Lesh, R., & Post, T. (1997a). *Rational number project fraction lessons for the middle grades: Level 1*. Dubuque, IA: Kendall-Hunt.

Cramer, K., Behr, M., Lesh, R., & Post, T. (1997b). *Rational number project fraction lessons for the middle grades: Level 2*. Dubuque, IA: Kendall-Hunt.

Cramer, K., & Bezuk, N. (1991). Multiplication of fractions: Teaching for understanding. *Arithmetic Teacher, 39*(3), 34–37.

Cramer, K., & Henry, A. (in press). Using manipulative models to build number sense for addition of fractions. In B. Litwiller (Ed.), *Making sense of fractions, ratio and proportions: NCTM 2002 yearbook*. Reston, VA: National Council of Teachers of Mathematics.

Cramer,K., Post, T., & delMas, R. (in press). Initial fraction learning by fourth- and fifth-gradestudents: A comparison of the effects of using commercial curricula or the Rational Number Project curriculum. *Journal for Research in Mathematics Education.*

Cramer, K., Post, T., Lesh, R. & Behr, M. (1998). *Rational Number Project. Level 1and Level 2*. Dubuque, IA: Kendall–Hunt.

Csiksentmihalyi, M. (1990). *Flow: The psychology of optimal experience*. New York: Harper & Row, Publishers.

Cunningham, D. (1992). Beyond educational psychology: Steps toward an educational semiotic. *Educational Psychology Review, 4*, 165–194.

Cuoco, A., Goldenberg, E. P., & Mark, J. (1996). Habits of mind: An organizing principle for mathematics curricula. *Journal of Mathematical Behavior, 15*, 375–402.

Danesi, M. (1994). *Messages and meanings: An introduction to semiotics.* Toronto: Canadian Scholars' Press.

Dapueto, C., & Parenti, L. (1999). Contributions and obstacles of contexts in the development of mathematical knowledge. *Educational Studies in Mathematics,39*, 1–21.

Davidson, N. (1985). Small-group learning and teaching in mathematics: A selective review of the literature. In R. Slavin, S. Sharan, S. Kagan, R. Lazarowitz, C. Webb, & R. Schmuck (Eds.), *Learning to cooperate, cooperating to learn* (pp. 211-230). New York: Plenum.

Davis, R. B. (1984). *Learning mathematics: The cognitive science approach to mathematics education.* Norwood, NJ: Ablex.

Davis, R. B., Maher, C. A., & Noddings, N. (1990). Suggestions for the improvement of mathematics education. In R.B. Davis, C.A. Maher, & N. Noddings (Eds.), *Constructivist views on the teaching and learning of mathematics (Journal for Research in Mathematics Education* Monograph No. 4) (pp. 187–191). Reston, VA: National Council for Teachers of Mathematics.

Davis, R., Maher, C. A. & Noddings, N. (1990). *Constructivist views on the teaching and learning of mathematics.* National Council of Teachers of Mathematics. Reston, VA: Author.

Debellis, V. A., & Goldin, G. A. (1998, July). *Aspects of affect: Mathematical intimacy, mathematical integrity.* Paper presented at the 22nd conference of the International Group for the Psychology of Mathematics Education. Stellenbosch, South Africa.

DeCorte, E., Verschaffel, L., Op 't Eynde P. (2000). Self-regulation: A characteristic and a goal of mathematics education. In P. Pintrich, M. Boekaerts, & M. Zeidner (Eds.), *Self-regulation: Theory, research, and applications* (pp. 687–726). Mahwah, NJ: Lawrence Erlbaum Associates.

Deely, J. (1986). The coalescence of semiotic consciousness. In J. Deely, B. Williams, & F. Kruse (Eds.), *Frontiers of semiotics* (pp. 5–34). Bloomington, IN: Indiana University Press.

Deely, J. (1990). *Basics of semiotics.* Bloomington, IN: Indiana University Press.

deLoache, J. S. (1989). The development of representation in young children. In H. W. Reese (Ed.), *Advances in child development and behavior,* Vol. 22, (pp. 1–39). New York: Academic Press.

Dennis, D., & Confrey, J. (1996). The creation of continuous exponents: A study of the methods and epistemology of John Wallis. *CBMS Issues in Mathematics Education, 6,* 33–56.

Dewey, J. (1896). The reflex arc concept in psychology. Psychological Review. 3, 357-370. Washington: American Psychological Association.

Dewey, J. (1916) *Democracy and education.* New York: Macmillian.

Dewey, J. (1963) *Experience and education.* London, England: Collier Books.

Dewey, J. (1982). In H.S. Thayer (Ed.) *Pragmatism: The classic writings* (pp. 253-334). Indianapolis, IN: Hackett Publishing Company.

Dewey, J. (1990). John Dewey's pragmatic technology. In D. Ihde (Ed.), *The Indiana Series in the Philosophy of Technology.* Bloomington and Indianapolis, IN: Indiana University Press.

Dewey, J. (1997). In L. Menand (Ed.) *Pragmatism: A reader.* New York: NY. Random House.

REFERENCES

Dienes, Z. (1960). *Building up mathematics*. London: Hutchinson Educational Ltd.

DiSessa, A. A. (1988). Knowledge in pieces. In G. Forman & P. B. Pufall (Eds.), *Constructivism in the computer age* (pp. 49–70). Hillsdale, NJ: Lawrence Erlbaum Associates.

DiSessa, A. A., Hammer, D., Sherin, B., & Kolpakowski, T. (1991). Inventing graphing: Meta–representational expertise in children. *Journal of Mathematical Behavior, 10* (2), 117-160.

Doerr, H. M., & Bowers, J. S. (1999). Revealing preservice teachers' thinking about functions through concept mapping. In F. Hitt & M. Santos (Eds.), *Proceedings of the twenty-first annual meeting: Psychology of mathematics education*. Columbus, OH: ERIC.

Doerr, H. M., Masingila, J. & Teich, V. (2000, April). *Linking the emerging practices of pre-service teachers to an experienced teacher's practice through the use of multimedia case studies*. Paper presented at the Research Pre–session of the Annual Meeting of the National Council of Teacher of Mathematics, April, 2000, Chicago, IL.

Doerr, H. M., & Tinto, P. (2000). Paradigms for teacher-centered classroom-based research. In A. Kelly, & R. Lesh (Eds.) *Handbook of research design in mathematics and science education*. Mahwah, NJ: Lawrence Erlbaum Associates.

Driscoll, M. (1983). *Research within reach: Secondary school mathematics*. Reston, VA: National Council of Teachers of Mathematics.

Driscoll, M. (1981). *Research within reach: Elementary school mathematics*. Reston, VA: National Council of Teachers of Mathematics.

Duncker, K. (1935). *The psychology of productive thinking*. Berlin: Springer.

Dweck, C. S. (1986). Motivational processes affecting learning. *American Psychologist, 41*, 1040–1048.

Eccles, J., Wigfield, A., & Reuman, D. (1987, April). *Changes in self-perceptions and values at early adolescence*. Paper presented at the annual meeting of the American Educational Research Association, San Francisco.

Educational Testing Service. (1995). *Packets project*: Princeton, NJ: Author.

English, L. D. (1997). Children's reasoning processes in classifying and solving computational word problems. In L. D. English (Ed.), *Mathematical reasoning: Analogies, metaphors, and images* (pp. 191–220). Mahwah, NJ: Lawrence Erlbaum Associates.

English, L. D. (1998). Children's problem posing within formal and informal contexts. *Journal for Research in Mathematics Education, 29*(1), 83–106.

English, L. D., Charles, K., & Cudmore, D. (2000). Students' statistical reasoning during a data modeling program. In T. Nakahara & M. Koyama (Eds.), *Proceedings of the 24th Conference of the International Group for the Psychology of Mathematics Education* (Vol. 2, pp. 265–272). Hiroshima, Japan: Hiroshima University.

Lesh, R. (2002) Research Design in Mathematics Education: Focusing on Design Experiments. In L. English (Ed.) *International Handbook of Research Design in Mathematics Education*. Hillsdale, NJ: Lawrence Erlbaum Associates.

English, L. D., & Cudmore, D. (2000). Using extranets in fostering international communities of mathematical inquiry. In M. J. Burke (Ed.), *Learning mathematics for a new century* (2000 NCTM Yearbook, pp. 82–95). Reston, VA: National Council of Teachers of Mathematics.

Ewert, P. H., & Lambert, J. F. (1932). Part II. The effect of verbal instructions upon the formation of a concept. *Journal of General Psychology, 6,* 400–411.

Feltovich, P. J., Spiro, R. J., & Coulson, R. L. (1997). Issues of expert flexibility in contexts characterized by complexity and change. In P. J. Feltovich, K. M. Ford, & R. R. Hoffman (Eds.), *Expertise in context: Human and machine* (pp. 125–146). Cambridge, MA: AAAI/MIT Press.

Fennema, E., Carpenter, T. P., Franke, M. L., & Carey, D. A. (1993). Learning to use children's mathematics thinking: A case study. In R. B. Davis & C. A. Maher (Eds.), *Schools, mathematics and the world of reality*. Needham Heights, MA: Allyn and Bacon.

Fennema, E., Carpenter, T. P., Franke, M. L., Levi, L. Jacobs, V. R., & Empson, S. B. (1996). A longitudinal study of learning to use children's thinking in mathematics instruction. *Journal of Research in Mathematics Education, 27*(4), 403–434.

Flavell, J. H. (1976). Metacognitive aspects of problem solving. In L. Resnick (Ed.) *The nature of intelligence* (pp. 231–236). Hillsdale, NJ: Lawrence Erlbaum Associates.

Frensch, P. A., & Funke, J. (Eds.) (1995). *Complex problem solving: The European perspective*. Mahwah, NJ: Lawrence Erlbaum Associates.

Fuson, K.C. (1992). Research on whole number addition and subtraction. In D. A. Grouws (Ed.), *Handbook of research on mathematics teaching and learning*, (pp 243-275). New York: Macmillan.

Gagne (1965). *The conditions of learning*. New York: Holt, Rinehart Winston.

Gardner, H. (1983). *Frames of Mind*. New York: Basic Books.

Garofalo, J., & Lester, F. K. (1985). Metacognition, cognitive monitoring, and mathematical performance. *Journal for Research in Mathematics Education, 16*, 163–175.

Gee, J. P. (1997). Thinking, learning, and reading: The situated sociocultural mind. In D. Kirshner & J. A. Whitson (Eds.), *Situated cognition: Social, semiotic, and psychological perspectives* (pp. 235–259). Mahwah, NJ: Lawrence Erlbaum Associates.

Gentner, D., & Toupin, C. (1986). Systematicity and similarity in the development of analogy. *Cognitive Science, 10,* 277–300.

Giere, R. N. (1992). *Cognitive models of science*. Minneapolis, MN: University of Minnesota Press.

Ginsburg, H. P. (1998). *Entering the child's mind: The clinical interview in psychological research and practice.* New York: Cambridge University Press.

Glenn, J. (2000). *Before it's too late: A report to the nation from the National Commission on Mathematics and Science Teaching for the 21st Century.* Washington, DC: U.S. Department of Education.

Goldin, G. A. (1990). Epistemology, constructivism, and discovery learning in mathematics. In R. B. Davis, C. A. Maher, & N. Noddings (Eds.), *Constructivist views on the teaching and learning of mathematics (Journal for Research in Mathematics Education* Monograph No. 4) (pp. 31–47). Reston, VA: National Council of Teachers of Mathematics.

Goldin, G. A. (in press). Representation in mathematical learning and problem solving. In L. English, (Ed.), *International handbook of research design in mathematics education.* Mahwah, NJ: Lawrence Erlbaum Associates.

Goldin, G. A., & McClintock, C. E. (Eds.). (1984*). Task variables in mathematical problem solving.* Hillsdale, NJ: Lawrence Erlbaum Associates. (Original work published 1979 by the Franklin Institute Press).

Goleman, D. (1995). *Emotional intelligence: Why it can matter more than IQ.* New York: Bantam Books.

Good, T. L., & Biddle, B. (1988). Research and the improvement of mathematics instruction: The need for observational resources. In D. Grouws & T. Cooney

REFERENCES

(Eds.), *Perspectives on research on effective mathematics teaching* (pp. 114–142). Hillsdale, NJ: Lawrence Erlbaum Associates.

Good, T. L., Mulryan, C., & McCaslin, M. (1992). Grouping for instruction in mathematics: A call for programmatic research on small-group processes. In D. Grouws (Ed.), *Handbook of research on mathematics teaching and learning*. New York: McMillan.

Good, T. L., Reys, B., Grouws, D., & Mulryan, C. (1989–1990). Using work groups in mathematics instruction. *Educational Leadership, 47,* 56–62.

Graduate Record Examination Board. (1997). *The GRE fairness, access, multiculturalism, and equity (FAME)—Report Series* (Vol. 1–3). Princeton, NJ: Educational Testing Service.

Gravemeijer, K., & Doorman, M. (1999). Context problems in realistic mathematics education: A calculus course as an example. *Educational Studies in Mathematics, 39,* 111–129

Greeno, J. (1991). Number sense as situated knowing in a conceptual domain. *Journal for Research in Mathematics Education, 22*(3), 170–218.

Greeno, J. (1997). The middle-school mathematics through applications project group. Theories and practices of thinking and learning to think. *American Journal of Education, 106,* 85-126.

Greeno, J. (1998). The situativity of knowing, learning, and research. *American Psychologist, 53,* 5–26.

Griffiths, J., & Hassan, E. (1978). Increasing the shipping capacity of the Suez Canal. *Journal of Navigation, 31,* 219–231.

Grosslight, L., Unger, C., Jay, E., & Smith, C. (1991). Understanding models and their use in science: Conceptions of middle and high school students and experts. *Journal of Research in Science Teaching, 28,* (pp. 799–822).

Grouws, D. A. (Ed.). (1992). *Handbook of research on mathematics teaching and learning*. New York: Macmillan.

Guilford, J. P. (1967). *The nature of human intelligence.* New York: McGraw–Hill.

Hadot, P. (2002). *What is ancient philosophy?* Cambridge, MA: Harvard University Press.

Harel, G. (2001). The development of mathematical induction as a proof scheme: A model for DNR-based instruction. In S. Campbell & Dr. Zargkis (Eds.). *Learning and teaching of number theory*. Norwood, NJ: Ablex Publishing Corporation.

Harel, G. (in press). Students' proof schemes revisited: Historical and epistemological considerations. In P. Boero (Ed.) *Theorems in School*. Dordrecht, The Netherlands: Kluwer.

Harel, G. & Sowder, J. (1998). Students' proof schemes. In E. Dubinsky, A. Schoenfeld, & J. Kaput (Eds.), Research on collegiate mathematics education. (Vol. III, pp. 234–283). Providence, Rhode Island: American Mathematical Society.

Harre, R., & Gillett, G. (1994). *The discursive mind.* Thousand Oaks, CA: Sage.

Harter, S. (1986). Processes underlying the construction, maintenance, and enhancement of the self-concept in children. In J. Suls & A. C. Greenwald (Eds.), *Psychological perspectives of the self* (pp. 55–121). New York: Academic Press.

Hatfield, L. L., & Bradbard, D. (Eds.). (1978). *Mathematical problem solving: Papers from a research workshop.* Columbus, OH: ERIC Clearinghouse for Science, Mathematics, and Environmental Education.

Hembree, R., & Marsh, H. (1993). Problem solving in early childhood: Building foundations. In R. J. Jensen (Ed.), *Research ideas for the classroom: Early*

childhood mathematics (pp. 151–170). Reston, VA: National Council of Teachers of Mathematics.
Hestenes, D. (1992). Modeling Games in the Newtonian World. *American Journal of Physics, 60*, 732–748.
Hidi, S., & Anderson, V. (1992). Situational interest and its impact on reading and expository writing. In K.A. Renninger, S. Hidi, & A. Krapp (Eds.), *The role of interest in learning and development* Hillsdale, NJ: Lawrence Erlbaum Associates.
Hiebert, J., & Behr, M. (1988). Introduction: capturing the major themes. In J. Hiebert & M. Behr (Eds.), *Research agenda for mathematics education: Number concepts and operations in the middle grades* (Vol. 2, pp. 1–18). Reston, VA: The National Council of Teachers of Mathematics.
Hilton, P. J. (1981). Avoiding math avoidance. In L. A. Steen (Ed.), *Mathematics tomorrow*, (pp. 73–82). New York: Springer–Verlag.
Hofstadter, D. R. (1999). *Gödel, Escher, Bach: An eternal golden braid.* New York: Basic Books.
Holmes, O. W., (1982). *Pragmatism: The classic writings.* In H.S. Thayer. Indianapolis, IN: Hackett Publishing Company.
Holmes, O. W., (1997). In L. Menand (Ed.) *Pragmatism: A reader.* New York: NY. Random House.
Horton, P. B., McConney, A., Gallo, M., Woods, A. L., Senn, G. J., & Hamelin, P. B. (1993). An investigation of the effectiveness of concept mapping as an instructional tool. *Science Education, 77*(1), 95–111.
Houser, N. (1987). Toward a Peircean semiotic theory of learning. *The American Journal of Semiotics, 5*, 251–274.
Hoyles, C. (1982). The pupil's view of mathematics learning. *Educational Studies in Mathematics, 13*, 349–372.
Hoyles, C., Noss, R., & Pozzi, S. (2001). Proportional reasoning in nursing practice. *Journal for Research in Mathematics Education, 32*, 4–27.
Hunt, E. (1991). Some comments on the study of complexity. In R. J. Sterneberg & P. A. Frensch (Eds.), *Complex problem solving: Principles and mechanisms* (pp. 317–340). Hillsdale, NJ: Lawrence Erlbaum Associates.
Indiana State Department of Education. (2000). *Indiana standards.* Indianapolis: Indiana Department of Education Library.
Inhelder, B. & Piaget, J. (1958). *The growth of logical thinking from childhood to adolescence.* New York: Basic Books.
James, W. (1878). Remarks on Spencer's definition of mind as correspondence. *Journal of Speculative Philosophy, XII* (1878), 1-18. Reprinted in Essays in Philosophy. Cambridge, MA: Harvard University Press, 1978.
James, W. (1918). *The principles of psychology.* (Reprint, August 1982). Harvard University.
James, W. (1982). In H. S. Thayer (Ed.) *Pragmatism: The classic writings* (pp. 123-250). Indianapolis, IN: Hackett Publishing Company.
James, W. (1987). The varieties of religious experience / Pragmatism /A Pluralistic Universe / The Meaning of Truth /Some Problems of Philosophy / Essays (Library of America) by William James. In B. Kuklick (Ed.) *William James (Writings 1902-1910).* New York, NY: Literary Classics of the United States.
James, W. (1997). Pragmatism: A reader. In L. Menand. New York: NY. Random House.
Jonassen, D. H., & Reeves, T. C. (1996). Learning with technology: using computers as cognitive tools. In D. H. Jonassen (Ed.), *Handbook of research on educational communications and technology,* (pp. 693–719). New York: Macmillan.

REFERENCES

Johnson, D., & Johnson, R. (1974). Instructional goal structure: Cooperative, competitive, or individualistic. *Review of Educational Research, 44,* 213–240.
Judson, T. W. (1999). E. A. Japan: A different model of mathematics education. In E. A. Gavosto, S. G. Krantz, & W. McCallum (Eds.), *Contemporary issues in mathematics education* (pp. 75–81). Cambridge, MA: Cambridge University Press.
Kamii, C. (1982). *Number in preschool and kindergarten.* Washington, DC: National Association for the Education of Young Children.
Kaput, J. (1987). Representation systems and mathematics. In C. Janvier (Ed.), *Problems of representation in the teaching and learning of mathematics* (pp. 19–26). Hillsdale, NJ: Lawrence Erlbaum Associates.
Kaput, J. (1989). Linking representations in the symbol system of algebra. In C. Kieren & S. Wagner (Eds.), *A research agenda for the teaching and learning of algebra.* Reston, VA: National Council of Teachers of Mathematics.
Kaput, J. (1991). Notations and representations as mediators of constructive processes. In E. v. Glasersfeld (Ed.), *Constructivism in mathematics education* (pp. 53–74). Dordrecht, The Netherlands: Kluwer.
Kaput, J. (1992). Patterns in students' formalization of quantitative patterns: The concept of function aspects of epistemology and pedagogy. *MAA Notes, 25,* 290–318.
Kaput J. (1994). Democratizing access to calculus: new routes to old roots. In A.H. Schoenfeld (Ed.), *Mathematical thinking and problem solving.* Hillsdale, NJ: Lawrence Erlbaum Associates.
Kaput, J. (1994). The representational roles of technology in connecting mathematics with authentic experience. In R. Bieler, R. W. Scholz, R. Strasser & B. Winkelman (Eds.), *Mathematics didactics as a scientific discipline.* Dordrecht, The Netherlands: Kluwer.
Kaput, J. (in press). Overcoming physicality and the eternal present: Cybernetic manipulatives. In R. Sutherland & J. Mason (Eds.), *Visualization and technology in mathematics education.* New York: Springer-Verlag.
Kaput, J., & Roschelle, J. (1993). SimCalc: Simulations for calculus learning [Computer software].
Kaput, J., & Thompson, P. (1994). The first 25 years of research on technology use in mathematics education. *Journal for Research in Mathematics Education, 25th Anniversary Edition,* (December).
Karplus, R., & Peterson, (1970). In R. Lesh & M. Landau (Eds.), *Acquisition of mathematics concepts and processes.* New York: Academic Press.
Karplus, R., Pulos, S., & Stage, E. (1983). Proportional reasoning in early adolescents. In R. Lesh & M. Landau (Eds.), *Acquisition of mathematics concepts and processes* (pp. 45–90). Orlando, FL: Academic Press.
Katona, G. (1940). *Organizing and memorizing.* New York: Columbia University Press.
Kauffman, S. (1993). *The origins of order: Self organization and selection in evolution.* New York: Oxford University Press.
Kauffman, S. (1995). *At home in the universe: The search for laws of self-organization and complexity.* New York: Oxford University Press.
Kehle, P. (1998). An empirical semiotic analysis of abstraction in mathematical modeling. *Dissertation Abstracts International, 60*(06A), 1949. (University Microfilms No. AAG9932664).
Kelly, E., & Lesh, R. (Eds.) (2000). *Handbook of research design in mathematics and science education.* Mahwah, NJ: Lawrence Erlbaum Associates
Kelly, E., & Lesh, R. (2000). Trends and shifts in research methods. In E. Kelly & R. Lesh (Eds.), *Handbook of research design in mathematics and science education.* Mahwah, NJ: Lawrence Erlbaum Associates.

REFERENCES

Kemme, S. (1990). *Uitleggen van Wiskunde* [Explaining mathematics]. Doctoral dissertation, Utrecht: Onderzoek Wiskunde Onderwijs en Onderwijs Computer Centrum.

Kilpatrick, J. (1967). *Analyzing the solution of word problems in mathematics: An exploratory study.* Unpublished doctoral dissertation, Stanford University.

Kilpatrick, J. (1969). Problem solving and creative behavior in mathematics. In J. W. Wilson & L. R. Carey (Eds.), *Reviews of recent research in mathematics education. Studies in Mathematics Series*, (Vol. 19, pp. 153–187). Stanford, CA: School Mathematics Study Group.

Kilpatrick, J. (1978). Variables and methodologies in research on problem solving. In L. L. Hatfield & D. A. Bradbard (Eds.), *Mathematical problem solving: Papers from a research workshop* (pp. 7–20). Columbus, OH: ERIC/SMEAC.

Kilpatrick, J. (1985). Reflection and recursion. *Educational Studies in Mathematics, 16*, 1–26.

Kilpatrick, J. (1985). A retrospective account of the past 25 years of research on teaching mathematical problem solving. In E. A. Silver (Ed.), *Teaching and learning mathematical problem solving: Multiple research perspectives* (1–15). Hillsdale, NJ: Lawrence Erlbaum Associates.

Kilpatrick, J. (1987, July). *What constructivism might mean in mathematics education.* Paper presented at the Eleventh International Conference on the Psychology of Mathematics Education, Montreal, Canada.

Kilpatrick, J., & Silver, E. (2000). Unfinished business: Challenges for mathematics educators in the next decades. In M. Burke & F. Curcio (Eds.), *Learning mathematics for a new century* (pp. 223–235). Reston, VA: National Council of Teachers of Mathematics.

King, A. & Brownell, W. (1966). *The curriculum and the disciplines of knowledge.* New York: John Wiley & Sons.

Klein, R., & Tirosh, D. (2000). Does a research based teacher development program affect teachers' lesson plans? In T. Nakahara & M. Koyama (Eds.) *Proceedings of the 24th Conference of the International Group for the Psychology of Mathematics Education.* Hiroshima, Japan.

Kloosterman, P., & Gorman, J. (1990). Building motivation in the elementary mathematics classroom. *School Science and Mathematics, (90)*5, 375–382.

Koedinger, K. (1998). *Intelligent cognitive tutors as modeling tool and instructional model.* Paper prepared for the NCTM Standards 2000 Technology Conference. Pittsburgh, PA: Carnegie Mellon University.

Kozak, M., & Robb, J. (1991). Education about technology. In *Technological Literacy.* Macmillan/McGraw-Hill.

Kroll, D. L., & Miller, T. (1993). Insights from research on mathematical problem solving in the middle grades. In D. Owens (Ed.), *Research ideas for the classroom: Middle grades mathematics* (pp. 58–77). Reston, VA: National Council of Teachers of Mathematics.

Kuhn, D. (1999). A developmental model of critical thinking. *Educational Researcher, 28*, 16–25.

Kuhn, T. (1970). *The structure of scientific revolutions.* Chicago: University of Chicago Press. (Original work published 1962).

Kuhn, T. (1970). Logic of discovery or psychology of research? In I. Lakatos & A. Musgrave (Eds.) *Criticism and the growth of knowledge.* Cambridge, MA: Cambridge University Press.

Kuhn, T. (1971). *Concepts of cause in the development of physics.* Etudes depistemologie genetique. Universitaires de Frances.

Kuhn, T. (1973). *Objectivity, value judgment, and theory choice.* Chicago, IL: University of Chicago Press.

REFERENCES

Lakatos, I. (1978). *The methodology of scientific research programmes* (Vol. 1). Cambridge, England: Cambridge University Press.

Lambdin, D. V. (1993). Monitoring moves and roles in cooperative mathematical problem solving. *Focus on learning problems in mathematics, 15*(2–3), 48–64.

Lamon, S. J. (1999). *Teaching fractions and ratios for understanding: Essential content knowledge and instructional strategies for teachers.* Mahwah, NJ: Lawrence Erlbaum Associates.

Lamon, S. J. (2001a). Presentations and re-presentations: From fractions to rational numbers. In A. Cuoco & F. R. Curcio (Eds.), *The roles of representation in school mathematics. 2001Yearbook* (pp. 146–165). Reston, VA: National Council of Teachers of Mathematics.

Lamon, S. J. (2001b). A Over B: Meanings over applications. In R. Speiser, C. A. Maher, & C. N. Walter (Eds.), *Proceedings of the 23rd annual meeting of the North American Chapter of the International Group for the Psychology of Mathematics Education* (pp. 235–243). Columbus, OH: ERIC Clearinghouse for Science, Mathematics, and Environmental Education.

Lamon, S. J. (in press). Part-whole comparisons with unitizing. In B. Litwiller (Ed.), *Making sense of fractions, ratios, and proportions. 2002 Yearbook.* Reston, VA: National Council of Teachers of Mathematics.

Lamon, S. J., & Lesh, R. (1994). Interpreting responses to problems with several levels and types of correct answers. In R. Lesh & S. Lamon (Eds), *Assessment of authentic performance in school mathematics.* Hillsdale, NJ: Lawrence Erlbaum Associates.

Lampert, M. & Ball, D. (1998). *Teaching, multimedia, and mathematics: Investigations of real practice.* New York: Teachers College Press.

Lange, J.de (1999). *Framework for classroom assessment in mathematics.* Utrecht, The Netherlands: Freudenthal Institute.

Latour, B. (1990). Drawing things together. In M. Lynch & S. Woolgar (Eds.), *Representation in scientific practice* (pp. 19–68). Cambridge, MA: MIT Press.

Lave, J. (1988). *Cognition in practice: Mind, mathematics and culture in everyday life.* Cambridge, MA: Cambridge University Press.

Lave, J., & Wenger, E. (1991). *Situated Learning: Legitimate Peripheral Participation.* Cambridge, UK: Cambridge University Press.

Lehrer, R., Carpenter, S., Schauble, L., & Putz, A. (2000). Designing classrooms that support inquiry. In J. Minstrell & E. H. van Zee (Eds.), *Inquiring into inquiry learning and teaching in science* (pp. 80–99). American Association for the Advancement of Science.

Lehrer, R., Jacobson, C., Kemeny, V., & Strom, D. (1999). Building on children's intuitions to develop mathematical understanding of space. In E. Fennema & T. A. Romberg (Eds.), *Mathematics classrooms that promote understanding* (pp. 63–87). Mahwah, NJ: Lawrence Erlbaum Associates.

Lehrer, R., & Pritchard, C. (in press). Symbolizing space into being. In K. Gravemeijer, L. Verschaffel, B. Van Oers, & R. Lehrer (Eds.), *Symbolizing, modeling, and tool use in mathematics education.* The Netherlands: Kluwer.

Lehrer, R., & Schauble, L. (2000). Inventing data structures for representational purposes: Elementary grade students' classification models. *Mathematical Thinking and Learning, 2,* 51–74.

Lehrer, R., & Schauble, L. (2000). Model-based reasoning in mathematics and science. In R. Glaser (Ed.), *Advances in instructional psychology (Vol. 5).* Mahwah, NJ: Lawrence Erlbaum Associates.

Lehrer, R., & Schauble, L. (in press-a). Symbolic communication in mathematics and science: Co-constituting inscription and thought. In J. Byrnes, J. & E. Amsel, (Eds.), *Language, literacy, and cognitive development: The development and consequences of symbolic communication.* Mahwah, NJ: Lawrence Erlbaum Associates.

Lehrer, R., & Schauble, L. (in press-b). Similarity of form and substance: Modeling material kind. In D. Klahr & S. Carver (Eds.), *Cognition and instruction: 25 years of progress.* Mahwah, NJ: Lawrence Erlbaum Associates.

Leher, R., Schauble, L,. Carpenter, S., & Penner, D. (2000). The inter-related development of inscriptions and conceptual understanding. In P. Cobb, E. Yackel, & K. McClain (Eds.), *Symbolizing and communicating in mathematics classrooms* (pp. 325–360). Mahwah, NJ: Lawrence Erlbaum Associates.

Leinhardt, G. (1990). Capturing craft knowledge in teaching. *Educational Researcher,* 19(2), 18–25.

Lemke, J. (1997). Cognition, context, and learning: A social semiotic perspective. In D. Kirshner & J. Whitson (Eds.), *Situated cognition: Social, semiotic, and psychological perspectives* (pp. 37–55). Mahwah, NJ: Lawrence Erlbaum Associates.

Lesh, R. (1979). Mathematical learning disabilities: Considerations for identification, diagnosis and remediation. In R. Lesh, D. Meierkiewicz, & M. G. Kantowski (Eds.), *Applied mathematical problem solving.* Columbus, OH: ERIC Clearinghouse for Science, Mathematics, and Environmental Education.

Lesh, R. (1982). *Metacognition in mathematical problem solving.* Unpublished manuscript.

Lesh, R. (1983). Conceptual analyses of problem solving performance. In E. Silver (Ed.), *Teaching and learning mathematical problem solving* (pp. 309–329). Hillsdale, NJ: Lawrence Erlbaum Associates.

Lesh, R. (1985). Conceptual analysis of problem-solving performance. In E. A. Silver (Ed.), *Teaching and learning mathematical problem solving: Multiple research perspectives* (pp. 309-329). Hillsdale, NJ: Lawrence Erlbaum Associates.

Lesh, R. (1986b). Review of Mathematical Problem Solving, by Alan Schoenfeld. *Teaching, thinking, and problem solving* 8(1), 39-48.

Lesh, R. (1987). The evolution of problem representations in the presence of powerful conceptual amplifiers. In C. Janvier (Ed.), *Problems of representation in teaching and learning mathematics.* Hillsdale, NJ: Lawrence Erlbaum Associates.

Lesh, R. (1996). Mathematizing: The "real" need for representational fluency. In L. Puig & A. Gutierrez (Eds.), *Research Forum Plenary Address. Proceedings of the 20th Conference of the International Group for the Psychology of Mathematics Education.* (pp. 3–13), Universitat do Valencia: Valencia, Spain.

Lesh, R. (1998). The development of representational abilities in middle school mathematics. In I. Sigel (Eds.), *Representations and student learning.* Mahwah, NJ: Lawrence Erlbaum Associates.

Lesh, R. (2000). *Dealing with complexity: New paradigms for research in mathematics education.* Plenary session in Fernandez, M. (Ed.) Proceedings of the Twenty Third Annual Meeting of the North American Chapter of the International Group for the Psychology of Mathematics Education. Columbus, OH: ERIC Clearinghouse for Science, Mathematics, and Environmental Education.

Lesh, R. (2001). What mathematical abilities are needed for success beyond school in a technology-based age of information? In M. Thomas (Ed.), *Technology in mathematics education.* Auckland, New Zealand: University of Auckland Press.

Lesh, R. (2001) Beyond constructivism: A new paradigm for identifying mathematical abilities that are most needed for success beyond school in a technology Based age of information. In M. Mitchelmore (Ed.) *Technology in mathematics*

REFERENCES

learning and teaching: Cognitive considerations: A special issue of the mathematics education research journal. Australia Mathematics Education Research Group. Melbourne Australia.

Lesh. R. (in press). Models and modeling in mathematics education. Monograph for *International Journal for Mathematical Thinking & Learning.* Hillsdale, NJ: Lawrence Erlbaum Associates.

Lesh, R., & Akerstrom, M. (1982). Applied problem solving: Priorities for mathematics education research. In F. K. Lester & J. Garofalo (Eds.), *Mathematical problem solving: Issues in research* (pp. 117–129). Philadelphia, PA: Franklin Institute Press.

Lesh, R., Amit, M. & Schorr, R. Y. (1997). Using "real-life" problems to prompt students to construct conceptual models for statistical reasoning. In I. Gal & J. Garfield (Eds.), *The assessment challenge in statistics education.* [A publication for the International Statistical Institute, IOS Press]. Amsterdam, The Netherlands: IOS Press.

Lesh, R., Behr, M. & Post, T. (1987). Rational number relations and proportions. In C. Janvier (Ed.), *Problems of representation in teaching and learning mathematics.* Hillsdale, NJ: Lawrence Erlbaum Associates.

Lesh, R., Behr, M., & Post, T. (1987). The role of representational translations in proportional reasoning and rational number concepts. In C. Janvier (Ed.), *Problems of representation in mathematics learning and problem solving.* Hillsdale, NJ: Lawrence Erlbaum Associates.

Lesh, R., & Clarke, D. (2000). Formulating operational definitions of desired outcomes of instruction in mathematics and science education. In A. Kelly & R. Lesh (Eds.), *Handbook of research design in mathematics and science education.* Mahwah, NJ: Lawrence Erlbaum Associates.

Lesh, R., Crider, J., & Gummer, E. (2001). Emerging possibilities for collaborating doctoral programs. In R. Reys (Ed.) *The state of doctoral research in mathematics education.* Reston, VA: National Council of Teachers of Mathematics.

Lesh, R., & Doerr, H. M. (1998). Symbolizing, communicating, and mathematizing: Key components of models and modeling. In P. Cobb & E. Yackel (Eds.). *Symbolizing and communicating in mathematics classrooms.* Mahwah, NJ: Lawrence Erlbaum Associates.

Lesh, R. & Harel, G. (in press). *Problem solving, modeling and local conceptual development.* Monograph for *International Journal for Mathematical Thinking and Learning.* Hillsdale, NJ: Lawrence Erlbaum Associates.

Lesh, R., Hoover, M., & Kelly, A. (1993). Equity, assessment, and thinking mathematically: Principles for the design of model-eliciting activities. In I. Wirszup & R. Streit (Eds.), *Developments in school mathematics education around the world, Vol. 3,* (pp. 104–130). Reston, VA: National Council of Teachers of Mathematics.

Lesh, R., Hoover, M., & Kelly, A. (1993). Equity, technology and teacher development. In I. Wirszup & R. Streit (Eds.), *Developments in school mathematics education around the world: Vol. 3.* Reston, VA: National Council of Teachers of Mathematics.

Lesh, R., Hoover, M. Hole, B., Kelly, A., & Post, T. (2000). Principles for developing thought-revealing activities for students and teachers. In A. Kelly & R. Lesh (Eds.). *Handbook of research design in mathematics and science education* (pp. 591–646). Mahwah, NJ: Lawrence Erlbaum Associates.

Lesh, R., & Kaput, J. (1988). Interpreting modeling as local conceptual development. In J. DeLange & M. Doorman (Eds.), *Senior secondary mathematics education.* Utrecht, The Netherlands: OW & OC.

Lesh, R., & Kelly, A. (2000). Multitiered teaching experiments. In E. Kelly, & R. Lesh (Eds.), *Handbook of research design in mathematics and science education.* Mahwah, NJ: Lawrence Erlbaum Associates.

Lesh, R., & Lamon, S. (Eds.). (1992). *Assessment of authentic performance in school mathematics.* Washington, DC: American Association for the Advancement of Science.

Lesh, R., & Lamon, S. (Eds.) (1994). *Assessment of authentic performance in school mathematics.* Hillsdale, NJ: Lawrence Erlbaum Associates.

Lesh, R., Landau, M., & Hamilton, E. (1983). Conceptual models in applied mathematical problem solving research. In R. Lesh & M. Landau (Eds.), *Acquisition of mathematics concepts and processes* (pp. 263–343). New York: Academic Press.

Lesh, R., & Lehrer, R. (2000). Iterative refinement cycles for videotape analyses of conceptual change. In A. Kelly & R. Lesh (Eds.), *Handbook of research design in mathematics and science education.* Mahwah, NJ: Lawrence Erlbaum Associates.

Lesh, R., Mierkiewicz, D., & Kantowski, M. (1979). *Applied mathematical problem solving.* Columbus, OH: ERIC Clearinghouse for Science, Mathematics, and Environmental Education.

Lesh, R., Post, T., & Behr, M. (1987). Representations and translations among representations in mathematics learning and problem solving. In C. Janvier (Ed.), *Problems of representation in teaching and learning mathematics.* Hillsdale, NJ: Lawrence Erlbaum Associates.

Lesh, R., Post, T., & Behr, M. (1987). Dienes revisited: Multiple embodiments in computer environments. In I. Wirszup & R. Streit (Eds.), *Developments in school mathematics education around the world.* Reston, VA: National Council of Teachers of Mathematics.

Lesh, R., Post, T., & Behr, M. (1989). Proportional reasoning. In J. Hiebert & M. Behr (Eds.), *Number concepts and operations in the middle grades* (pp. 93–118). Reston, VA: National Council of Teachers of Mathematics.

Lesh, R., & Zawojewski, J. S. (1987). Problem solving. In T. Post (Ed.), *Teaching mathematics in grades K–8: Research-based methods.* Boston, MA: Allyn & Bacon.

Lesh, R., & Zawojewski, J. S. (1992). Problem solving. In T. Post (Ed.), *Teaching mathematics in grades K-8: Research-based methods* (2nd Edition, pp. 49-88). Newton, MA: Allyn & Bacon.

Leslie, A. M. (1987). Pretense and representation: The origins of "theory of mind." *Psychological Review, 94,* 412-426.

Lester, F. K. (1978). Mathematical problem solving in the elementary school: Some educational and psychological considerations. In L. L. Hatfield (Ed.), *Mathematical problem solving: Papers from a research workshop.* Columbus, OH: ERIC Clearinghouse for Science, Mathematics, and Environmental Education.

Lester, F. K. (1980). Mathematical problem solving research. In R. J. Shumway (Ed.), *Research in mathematics education* (pp. 286–323). Reston, VA: National Council of Teachers of Mathematics.

Lester, F. K. (1983). Trends and issues in mathematical problem-solving research. In R. Lesh & M. Landau (Eds.), *Acquisition of mathematics concepts and processes.* New York: Academic Press.

Lester, F. K. (1985). Methodological considerations in research on mathematical problem-solving instruction. In E. A. Silver (Ed.), *Teaching and learning mathematical problem solving: Multiple research perspectives* (pp. 41–69). Hillsdale, NJ: Lawrence Erlbaum Associates.

REFERENCES

Lester, F. K., & Garofalo, J. (Eds.). (1982). *Mathematical problem solving: Issues in research*. Philadelphia, PA: Franklin Institute Press.

Lester, F. K., Garofalo, J., & Kroll, D. L. (1989). Self-confidence, interest, beliefs, and metacognition: Key influences on problem-solving behavior. In D. B. McLeod & V. M. Adams (Eds.), *Affect and mathematical problem solving: A new perspective* (pp. 75–88). New York: Springer–Verlag.

Lester, F. K., Maki, D., LeBlanc, J. F., & Kroll, D. L. (Eds.). (1992). Content component. Vol. II, final report to the National Science Foundation: *Preparing elementary teachers to teach mathematics: A problem-solving approach* (Grant Number TEI 8751478). Bloomington: Mathematics Education Development Center, Indiana University.

Library of America. (1987). William James writings 1902-1910. In B. Kuklick (Ed.). *Literary classics of the United States*. New York: Penguin Books.

Lucas, J. (1972). *An exploratory study of the diagnostic teaching of heuristic problem-solving strategies in calculus*. Unpublished doctoral dissertation, University of Wisconsin.

Maher, C. A. (1987). The teacher as designer, implementer, and evaluator of children's mathematical learning environments. *Journal of Mathematical Behavior*, 6, 295–303.

Maki, D., & Thompson, M. (1973). *Mathematical models and applications, with emphasis on the social, life, and management sciences* (2nd ed.). Englewood Cliffs, NJ: Prentice-Hall.

Manufacturing Skill Standards Council. (2000). *Manufacturing skill standards*. Washington, DC: National Skill Standards Board.

Mason, J., Burton, L., & Stacey, K. (1982). *Thinking mathematically*. New York: Addison Wesley.

Mayer, R. E. (1985). Implications of cognitive psychology for instruction in mathematical problem solving. In E. A. Silver (Ed.), *Teaching and learning mathematical problem solving: Multiple research perspectives*. Hillsdale, NJ: Lawrence Erlbaum Associates.

McClain, K., Cobb, P. & Gravemeijer, K. (2000). Supporting students' ways of reasoning abut data. In M. Burke (Ed.), *Learning mathematics for a new century* (2001 Yearbook of the National Council of Teachers of Mathematics). Reston, VA: National Council of Teachers of Mathematics.

McClain, K., McGatha, M., & Hodge, L. (2000). The importance of argumentation in supporting students' mathematical development. *Mathematics Teaching in the Middle Grades*, 5(8), 548–553.

McLeod, D. B. (1992). Research on affect in mathematics education: A reconceptualization. In D. A. Grouws (Ed.), *Handbook of research on mathematics teaching and learning* (pp. 575–596). New York: Macmillan.

Mead, G. H. (1910). Social consciousness and the consciousness of meaning. *Psychological Bulletin*, 7, 397–405.

Mead, G. H. (1982). In H.S. Thayer (Ed.) Pragmatism: The classic writings (pp. 337-358). Indianapolis, IN: Hackett Publishing Company.

Meece, J. L., Wigfield, A., & Eccles, J. S. (1990). Predictors of math anxiety and its influence on young adolescents' course enrollment intentions and performance in mathematics. *Journal of Educational Psychology*, 82(1), 60–70.

Menand, L. (Ed.). (1997). *Pragmatism: A reader*. New York: Random House.

Menand, L. (Ed.). (2001). *The metaphysical club: A story of ideas in America*. New York: Farrar, Straus and Giroux.

Merrell, F. (1997). *Peirce, signs, and meaning*. Toronto: University of Toronto Press.

Meyer, M. R., & Fennema, E. (1985). Predicting mathematics achievement for females and males from causal attributions. In S. K. Damarin & M. Shelton (Eds.),

REFERENCES

Procedings of the seventh annual conference of the North American Chapter for the Psychology of Mathematics Education (pp. 201–206). Columbus, OH: Authors.

Middleton, J. A. (1999). Curricular influences on the motivational beliefs and practice of two middle school mathematics teachers: A follow-up study. *Journal for Research in Mathematics Education, 30*(3), 349–358.

Middleton, J. A., Leavy, A., & Leader, L. (1998). Student attitudes and motivation towards mathematics: What happens when you change the rules? Paper presented at the annual meeting of the American Educational Research Association, San Diego, CA.

Middleton, J. A., Littlefield, J., & Lehrer, R. (1992). Gifted children's conceptions of academic fun: An examination of a critical construct in gifted education. *The Gifted Child Quarterly, 36*(1), 38–44.

Middleton, J. A., & Spanias, P. (1999). Motivation for achievement in mathematics: Findings, generalizations, and criticisms of the recent research. *Journal for Research in Mathematics Education, 30*(1), 65–88.

Middleton, J. A., & Toluk, Z. (1999). First steps in the development of an adaptive, decision-making theory of motivation. *Educational Psychologist, 34*(2), 99–112.

Midgley, C., Feldlaufer, H., & Eccles, J. S. (1989). Student/teacher relations and attitudes toward mathematics before and after transition to junior high school. *Child Development, 60*, 981–992.

Minsky, M. (1987). *The society of mind.* New York: Simon & Schuster.

Monk, S. (1992). Students' understanding of a function given by a physical model: The concept of function, aspects of epistemology and pedagogy. *MAA Notes, 25*, 175–194.

Monk, S., & Nemirovsky, R. (1994). The case of Dan: Student construction of a functional situation through visual attributes. *CBMS Issues in Mathematics Education, 4*, 139–168.

National Council of Teachers of Mathematics. (1989). *Curriculum and evaluation standards for school mathematics.* Reston, VA: Author.

National Council of Teachers of Mathematics [Swetz, F., & Hartzler, J. (Eds.)]. (1991). *Mathematical modeling in the secondary school curriculum.* Reston, VA: National Council of Teachers of Mathematics.

National Council of Teachers of Mathematics. (1995). *Assessment standards for school mathematics.* Reston, VA: Author.

National Council of Teachers in Mathematics (2000). *Principles and standards for school mathematics: An overview.* National Council of Teachers in Mathematics. Reston, VA: Author.

National Research Council (1999a). *Being fluent with information technology.* Washington DC: National Academy Press.

National Research Council (1999b). *The changing nature of work; Implications for occupational analysis.* Washington, DC: National Academy Press.

National Science and Technology Council and the Office of Science and Technology Policy. (1996). *Technology in the national interest.* (online). Available: http://www.ta.doc.gov/Reports/TechNI/

Nemirovsky, R. (1996). Mathematical narratives, modeling and algebra. In N. Bednarz, C. Kieran, & L. Lee (eds.), *Approaches to algebra: Perspectives for research and teaching* (pp. 197–220). Dordrecht: Holland, Kluwer.

Nemirovsky, R., & Monk, S. (2000). If you look at it the other way. An exploration into the nature of symbolizing. In P. Cobb, E. Yackel & K. McClain (Eds.) *Symbolizing and communicating in mathematics classrooms: Perspectives on*

REFERENCES

discourse, tools, and instructional design. Mahwah, NJ: Lawrence Erlbaum Associates.

Nesher, P. (1989). Microworlds in mathematical education: A pedagogical realism. In L.B. Resnick (Ed.), *Knowing, learning and instruction.* Hillsdale, NJ: Lawrence Erlbaum Associates.

Newell, A., & Simon, H. A. (1972). *Human problem solving.* Englewood Cliffs, NJ: Prentice Hall.

Noddings, N. (1989). Theoretical and practical concerns about small groups in mathematics. *Elementary School Journal, 89,* 607–623.

Noddings, N. (1990). Constructivism in mathematics education. In R. B. Davis, C. A. Maher, & N. Noddings (Eds.), *Constructivist views on the teaching and learning of mathematics (Journal for Research in Mathematics Education* Monograph No. 4) (pp. 7–18). Reston, VA: National Council of Teachers of Mathematics.

Norman, D. A. (1993). *Things that make us smart.* Reading, MA: Addison-Wesley.

Novak, J. D., & Gowin, D. B. (1984). *Learning how to learn.* New York: Cambridge University Press.

Nunes, T. (1992). Ethnomathematics and everyday cognition. In D. A. Grouws (Ed.), *Handbook of research on mathematical teaching and learning* (pp. 557–574). New York: Macmillan.

Nunes, T., Schliemann, A. D. & Carraher, D. W. (1993). *Street mathematics and school mathematics.* Cambridge, England: Cambridge University Press

Papert, S. (1991). Situating constructionism. In I. Harel & S. Papert (Eds.), *Constructionism.* Norwood, NJ: Ablex.

Pearlman, K. (1997). Twenty-first century measures for twenty-first century work. In *Transitions in work and learning; Implications for Assessment.* Washington, DC: National Academy Press.

Peirce, C. (1878). How to Make our Ideas Clear. *Popular Science Monthly, 12,* 286-302.

Peirce, C. S. (1884). Design and Chance. In N. P. A. Lynn A. Ziegler, Edward C. Moore, Max H. Fisch, Christian J. Kloesel, Don D. Roberts (Ed.) (1982), *Writings of Charles S. Peirce: A Chronological Edition* (Vol. 4, pp. 544-554). Bloomington, IN: Indiana University Press.

Penner, D. E., Giles, N. D., Lehrer, R., & Schauble, L. (1977). Building functional models: Designing an elbow. *Journal of Research in Science Teaching, 34,* 125–143.

Perkins, D. N., Jay, E., & Tishman, S. (1993). Beyond abilities: A dispositional theory of thinking. *Merrill-Palmer Quarterly, 39*(1), 1-21.

Perkins, D. N., & Tishman, S. (1998). *Dispositional aspects of intelligence.* Paper presented at the Winter Colloquium Series, Arizona State University, Tempe.

Piaget, J. (1960). *The psychology of intelligence.* Littlefield, NJ: Adams.

Piaget, J. (1964). Development and learning. In R. E. Ripple & V. N. Rockcastle (Eds.), *Piaget rediscovered: Report of the conference on cognitive studies and curriculum development.* Ithaca, NY: Cornell University Press.

Piaget, J. (1967). *The child's conception of number.* New York: Norton.

Piaget, J. (1968). *Genetic epistemology.* New York Columbia University Press.

Piaget, J. (1970). *Genetic epistemology.* E. Duckworth (Trans.). New York: Columbia University Press.

Piaget, J. (1983). Piaget's theory. In P. H. Mussen (Ed.), Handbook of child psychology, (4th ed., pp. 103-128). Wiley.

Piaget, J., & Beth, E. (1966). *Mathematical epistemology and psychology.* Dordrecht, Netherlands: D. Reidel.

REFERENCES

Piaget, J., & Inhelder, B. (1958). *The growth of logical thinking from childhood to adolescence.* London: Routledge & Kegan Paul.

Piaget, J., Inhelder, B., & Szeminska, A. (1960). *The child's conception of geometry.* London: Butler and Tanner.

Pimm, D. (1995). *Symbols and meanings in school mathematics.* London, England: Routledge.

Pinker, S. (1997). *How the mind works.* New York: W. W. Norton.

Polya, G. (1945). *How to solve it.* Princeton, NJ: Princeton University Press.

Polya, G. (1968). *Mathematics and plausible reasoning* (2nd ed., Vols. 1–2). Princeton, NJ: Princeton University Press.

Polya, G. (1973). *How to solve it.* Princeton, NJ: Princeton University Press. (Original work published 1945)

Popper, C. (1972). *Objective knowledge: An evolutionary approach.* New York: Oxford: Clarendon Press.

Post, T. (1988). Some notes on the nature of mathematics learning. In T. Post (Ed.), *Teaching mathematics in grades K–8* (pp. 1–19). Boston: Allyn and Bacon.

Post, T., Behr, M., Lesh, R., & Harel, G. (1988). Intermediate teachers' knowledge of rational number concepts. In E. Fennema, T. Carpenter, & S. J. Lamon (Eds.), *Integrating research on teaching and learning mathematics.* Madison, WI: National Center for Research in the Mathematical Sciences Education.

Post, T., & Cramer, K. (1987). Children's strategies when ordering rational numbers. *Arithmetic Teacher, 35* (2), 33–35.

Post, T., Wachsmuth, I., Lesh, R., & Behr, M. (1985). Order and equivalence of rational number: A cognitive analysis. *Journal for Research in Mathematics Education, 16* (1), 18–36.

Pozzi, S., Noss, R., & Hoyles, C. (1998). Tools in practice, mathematics in use. *Educational Studies in Mathematics, 36*, 105–122.

Resnick, M. (1994). *Turtles, termites, and traffic jams.* Cambridge, MA: MIT Press.

Rissing, S. W., Pollock, G. B., Higgins, M. R., Hagen, R. H., & Smith, D. R. (1989). Foraging specialization without relatedness or dominance among co-founding ant queens. *Nature, 338*, 420–422.

Rissland, E. L. (1985). Artificial intelligence and the learning of mathematics: A tutorial sampling. In E. A. Silver (Ed.), *Teaching and learning mathematical problem solving: Multiple research perspectives.* Hillsdale, NJ: Lawrence Erlbaum Associates.

Romberg, T. A (1997). (Ed.) *Mathematics in context: A connected curriculum for grades 5–8.* Chicago, IL: Encyclopedia Britannica Educational Corporation.

Rorty, R., (1979). *Philosophy and the mirror of nature.* Princeton, NJ: Princeton University Press.

Russell, B. (1961). *Religion and science.* New York, NY: Oxford University Press.

Russell, R. S., & Taylor III, R. B. (2000). *Operations management.* Upper Saddle River, NJ: Prentice Hall.

Saldanha, L., & Thompson, P. W. (1998). Re-thinking co-variation from a quantitative perspective: Simultaneous continuous variation. In S. B. Berensen & W. N. Coulombe (Eds.), *Proceedings of the Annual Meeting of the Psychology of Mathematics Education North America.* Raleigh: North Carolina State University.

Saxe, G. B. (1988). Candy selling and math learning. *Educational Researcher, 16*(6), 14–21.

Saxe, G. B. (1991). *Culture and cognitive development: Studies in mathematical understanding.* Hillsdale, NJ: Lawrence Erlbaum Associates.

Schank, R. C. (1990). *Tell me a story: Narrative and intelligence.* Evanston, IL: Northwestern University Press.

REFERENCES

Shannon, A., & Zawojewski, J. S. (1995). Performance assessment in school mathematics: New rules of the game for students. *Mathematics Teacher, 88*(9), 752-757.

Schifter, D., & Fosnot, C. T. (1993). *Reconstructing mathematics education: Stories of teachers meeting the challenges of reform.* New York: Teachers College Press.

Schmuck, R. A., & Schmuck, P. A. (1983). *Group processes in the classroom.* Dubuque, IA: William C. Brown.

Schnepp, M. J., & Nemirovsky, R. (2001). Constructing a foundation for the fundamental theorem of calculus. In A. Cuoco & F. Curcio (Eds.), *The Roles of Representations in School Mathematics.* Reston, VA: National Council of Teachers of Mathematics Yearbook.

Schoenfeld, A. H. (1982). Some thoughts on problem-solving research and mathematics education. In F. K. Lester & J. Garofalo (Eds.), *Mathematical problem solving: Issues in research* (pp. 27–37). Philadelphia, PA: Franklin Institute Press.

Schoenfeld, A. H. (1985). Metacognitive and epistemological issues in mathematical understanding. In E. A. Silver (Ed.), *Teaching and learning mathematical problem solving: Multiple research perspectives.* Hillsdale, NJ: Lawrence Erlbaum Associates.

Schoenfeld, A. H. (1985). *Mathematical problem solving.* San Diego, CA: Academic Press.

Schoenfeld, A. H. (1985a). Artificial intelligence and mathematics education: A discussion of Rissland's paper. In E. A. Silver (Ed.), *Teaching and learning mathematical problem solving: Multiple research perspectives.* Hillsdale, NJ: Lawrence Erlbaum Associates.

Schoenfeld, A. H. (Ed.). (1987a). *Cognitive science and mathematics education.* Hillsdale, NJ: Lawrence Erlbaum Associates.

Schoenfeld, A. H. (1987b). What's all the fuss about metacognition? In A. H. Schoenfeld (Ed.), *Cognitive science and mathematics education* (pp. 189–215). Hillsdale, NJ: Lawrence Erlbaum Associates.

Schoenfeld, A. (1989). Problem solving in context(s). In R. Charles & E. Silver (Eds.), *The teaching and assessing of mathematical problem solving.* Reston, VA: National Council of Teachers of Mathematics.

Schoenfeld, A. H. (1992). Learning to think mathematically: Problem solving, metacognition, and sense making in mathematics. In D. Grows (Ed.), *Handbook of research on mathematics teaching and learning* (pp. 334-370). New York: Macmillan.

Schorr, R. Y. (2000). Impact at the student level. *Journal of Mathematical Behavior 19* 209–231.

Schorr, R. Y., & Alston, A. S. (1999). Teachers evolving ways of thinking about student work. In Fernando Hitt and Manuel Santos, *Proceedings of the 21st Conference of the North American Chapter of the International Group for the Psychology of Mathematics Education* (pp. 169–177). Cuernavaca: Mexico.

Schorr, R. Y., & Firestone, W.A. (2001). Changing mathematics teaching in response to a state testing program: A fine-grained analysis. Paper presented at the Annual Meeting of the American Educational Research Association, Seattle, WA.

Schorr, R. Y., Ginsburg, H. P. (2000). Using clinical interviews to promote pre-service teachers' understanding of children's mathematical thinking. *Proceedings of the 22nd Annual Meeting of the North American Chapter of the International Group for the Psychology of Mathematics Education,* (pp. 599–605). Tucson, AZ.

Schorr, R. Y. & Lesh, R. (1998). Using thought-revealing activities to stimulate new instructional models for teachers. In S. Berenson, K. Dawkins, M. Blanton, W. Coulcombe, J. Kolb, K. Norwood, & L. Stiff (Eds.), *Proceedings of the 20th*

REFERENCES

Annual Meeting of the North American Chapter of the International Group for the Psychology of Mathematics Education. (pp. 723–731). Raleigh, NC.

Schroeder, T. L., & Lester, F. K. (1989). Developing understanding in mathematics via problem solving. In P. R. Trafton (Ed.), *New directions for elementary school mathematics* (pp. 31–56). Reston, VA: National Council of Teachers of Mathematics.

Schwartz, J. L., & Yerushalmy, M. (1992). Getting students to function in and with algebra. In G. Harel & E. Dubinsky (Eds.), *The concept of function: Aspects of epistemology and pedagogy* (pp. 261–289). Washington: Mathematical Association of America.

Schwartz, J. L., Yerushalmy, M. (1995). On the need for a bridging language for mathematical modeling. *For the Learning of Mathematics* 15 (2), 29-35

Sfard, A. (1998). On two metaphors for learning and the dangers of choosing just one. *Educational Researcher, 27*(2), 4–13.

Sfard, A., & Thompson, P. (1994). Problems of reification: Representations and mathematical objects. In D. Kirshner (Ed.), *Proceedings of the Sixteenth Annual Meeting of the North American Chapter of the International Group for the Psychology of Mathematics Education:* (Vol. 1, pp. 3–34). Baton Rouge: Louisiana State University.

Shafer, M., & Romberg, T. (1999). Assessment in classrooms that promote understanding. In E. Fennema, E. and T. Romberg (Eds.), *Mathematics classrooms that promote understanding.* (pp. 159–184) Mahwah, NJ: Lawrence Erlbaum Associates.

Shank, G. (1988). Three into two will go: Juxtapositional strategies for empirical research in semiotics. In J. Deely (Ed.), *Semiotics 1987* (pp. 123–127). New York: University Press of America.

Shank, G., & Cunningham, D. (1996). *Modeling the six modes of Peircean abduction for educational purposes.* Available at http://www.indiana.edu/~cunningh/

Sharan, R. (1980). Cooperative learning in small groups: Recent methods and effects on achievement, attitudes, and ethnic relations. *Review of Educational Research, 50,* 241–271.

Shaughnessy, J. M., Garfield, J., & Greer, B. (1996). Data handling in the media. In I. Gal & J. B. Garfield, (Eds.), *The assessment challenge in statistics education* (pp. 107–121). Amsterdam, The Netherlands: IOS Press.

Shell Centre for Mathematical Education. (1984). *Problems with patterns and numbers.* Nottingham, England: Shell Centre.

Shulman, L. S. (1986). Those who understand: Knowledge growth in teaching. *Educational Researcher, 15*(2), 4–14.

Sierpinska, A. (1994). *Understanding in mathematics.* London: The Falmer Press.

Silver, E. A. (1982, January*). Thinking about problem solving: Toward an understanding of metacognitive aspects of mathematical problem solving.* Paper presented at the Conference on Thinking, Suva, Fiji.

Silver, E. A. (1985). Research on teaching mathematical problem solving: Some underrepresented themes and needed directions. In E. Silver (Ed.), *Teaching and learning mathematical problem solving: Multiple research perspectives* (pp. 247–266). Hillsdale, NJ: Lawrence Erlbaum Associates.

Silver, E. A. (Ed.). (1985). *Teaching and learning mathematical problem solving: Multiple research perspectives.* Hillsdale, NJ: Lawrence Erlbaum Associates.

Silver, E. A. (1988). Teaching and assessing mathematical problem solving: Toward a research agenda. In R. I. Charles & E. A. Silver (Eds.), *The teaching and assessing of mathematical problem solving* (pp. 273-282). Reston, VA: National Council of Teachers of Mathematics.

REFERENCES

Silver, E. A., Mamona–Downs, J., Leung, S. S., & Kenny P. A. (1996). Posing mathematical problems: An exploratory study. *Journal for Research in Mathematics Education, 27*(3), 293–309.

Simmt, E., & Kieran, T. (2000). On being embodied in the body of mathematics: Mathematics knowing in action. In Fernandez (Ed.), *Proceedings of the 22nd Annual meeting of the North American Chapter of the International Group for the Psychology of Mathematics Education.* (Vol. 1, pp. 205–211). Columbus, OH: ERIC Clearinghouse for Science, Mathematics, and Environmental Education.

Simon, M. A. (1995). Reconstructing mathematics pedagogy from a constructivist perspective. *Journal for Research in Mathematics Education, 26*(2), 114–145.

Simon, M. A., & Schifter, D. (1991). Towards a constructivist perspective: An intervention study of mathematics teacher development. *Educational Studies in Mathematics, 22,* 309–331.

Simon, M.A., Tzur, R., Heinz, K., Kinzel, M., & Smith, M. S. (2000). Characterizing a perspective underlying the practice of mathematics teachers in transition. *Journal of Research in Mathematics Education, 31*(5), 579–601.

Slavin, R. (Ed.). (1989). *School and classroom organization.* Hillsdale, NJ: Lawrence Erlbaum Associates.

Slavin, R. (1989–1990). Research on cooperative learning: Consensus and controversy. *Educational Leadership, 47,* 52–54.

Slavin, R., Sharan, S., Kagan, S., Lazarowitz, R., Webb, C., & Schmuck, R. (Eds.). (1985). *Learning to cooperate, cooperating to learn.* New York: Plenum.

Smith, J. (1999). Tracking the mathematics of automobile production: Are schools failing to prepare students for work? *American Educational Research Journal, 36*(4), 835–878.

Sowder, L. K. (1985). Cognitive psychology and mathematical problem solving: A discussion of Mayer's paper. In E. A. Silver (Ed.), *Teaching and learning mathematical problem solving: Multiple research perspectives.* Hillsdale, NJ: Lawrence Erlbaum Associates.

Spiegel, M. R., & Stephens, L. J. (1999). *Schaum's outlines in statistics.* (3rd Ed.). Columbus, OH: McGraw-Hill.

Spillane, J. P., & Zeuli, J. S. (1999). Reform and teaching: Exploring patterns of practice in the context of national and state mathematics reforms. *Educational Evaluation and Policy Analysis, 21*(1) 1–27.

Spiro. R. J., Coulson, R. L., Feltovich, P. J., & Anderson, D. K. (1988). Cognitive flexibility theory: Advanced knowledge acquisition in ill-structured domains. In *Tenth Annual Conference of the Cognitive Science Society* (pp. 375–383). Hillsdale, NJ: Lawrence Erlbaum Associates.

Stacey, K. (1992). Mathematical problem solving in groups: Are two heads better than one? *Journal of Mathematical Behavior, 11,* 261–275.

Steen, L. A. (1987). Mathematics education: A predictor of scientific competitiveness. Science, 237, 251–252, 302.

Steen, L. A. (1988). *Calculus for a new century: A pump, not a filter.* (MAA Notes Number 8). Washington, D C: Mathematical Association of America.

Steen, L. A. (1990). Pattern. In L. A. Steen (Ed.), *On the shoulders of giants. New approaches to numeracy.* (pp. 1–10). Washington, DC: National Academy Press.

Steffe, L. P. (1990). Mathematics curriculum design: A constructivist's perspective. In L. P. Steffe & T. Wood (Eds.), *Transforming children's mathematics education* (pp. 389–398). Hillsdale, NJ: Lawrence Erlbaum Associates.

REFERENCES

Steffe, L. P. (1994). Children's multiplying schemes. In G. Harel & J. Confrey (Eds.), The development of multiplicative reasoning in the learning of mathematics (pp. 3–39). Albany, NY: SUNY Press.

Steffe, L. P., Cobb, P., & von Glasersfeld, E. (1988). *Construction of arithmetical meanings and strategies.* New York: Springer–Verlag.

Steffe, L. P., & Thompson, P. (2000). Teaching experiment methodology: Underlying principles and essential elements. In E. Kelly, & R. Lesh (Eds.), Handbook of research design in mathematics and science Education (pp. 267-306). Mahwah, NJ: Lawrence Erlbaum Associates.

Stein, M. K., Grover, B. W., & Henningsen, M. (1996). Building student capacity for mathematical thinking and reasoning: An analysis of mathematical tasks used in reform classrooms. *American Educational Research Journal, 33,* 455–488.

Stein, M. K., & Smith, M. S (1998). Mathematical tasks as a framework for reflection: From research to practice. *Mathematics Teaching in the Middle School, 3*(4), 268–275.

Sterman, J. D. (Ed.) (2000). *Business Dynamics: Systems thinking and modeling for a complex world.* New York, NY: McGraw-Hill.

Sternberg, R. J. (1985). *Beyond IQ.* New York: Cambridge University Press.

Sternberg, R. J. (1995). Expertise in complex problem solving: A comparison of alternative conceptions. In P. A. Frensch & J. Funke (Eds.), Complex problem solving: The European perspective (pp. 295-321). Hillsdale, NJ: Lawrence Erlbaum Associates.

Stewart, I., & Golubitsky, M. (1992). *Fearful symmetry: Is God a geometer?* London: Penguin Books.

Stigler, J., & Hiebert, J. (1999). *The teaching gap: Best ideas from the world's teachers for improving education in the classroom.* New York, NY: Free Press.

Sutherland, R., & Balacheff, N. (1999). Didactical complexity of computational environments for learning of mathematics. *International Journal of Computers for Mathematical Learning, 4*(1), 1–26.

Suydam, M. N. (1980). Untangling clues from research on problem solving. In S. Krulik (Ed.), *Problem solving in school mathematics* (pp. 34–50). Reston, VA: National Council of Teachers of Mathematics.

Tall, D. (1992). The transition to advanced mathematical thinking: Function, limits, infinity, and proof. In D. A. Grouws (Ed.), *Handbook of research on mathematics teaching and learning.* New York: Macmillan.

Tall, D., & Vinner, S. (1981). Concept images and concept definitions in mathematics with particular reference to limits and continuity. *Educational Studies in Mathematics, 12,* 151–169.

Thompson, A. G. (1992). Teachers' beliefs and conceptions: A synthesis of the research. In D. Grouws (Ed.), *Handbook of research on mathematics teaching and learning* (pp. 127–146). New York: Macmillan.

Thompson P. W. (1994). Images of rate and operational understanding of the fundamental theorem of calculus. *Educational Studies in Mathematics, 26,* 229–274.

Thompson, P. W. (1994b). Students, functions, and the undergraduate curriculum. In E. Dubinsky, A. H. Schoenfeld, & J. Kaput (Eds.), *Research in collegiate mathematics education, 1.* (Vol. 4, pp. 21–44). Providence, RI: American Mathematical Society.

Thompson, P. W. & Steffe, L. (Ed.) (2000). *Radical constructivism in action: Building on the pioneering work of Ernst Von Glasersfeld.* Manhattan, NY: Taylor and Francis.

Thorndike, E. L. (1920). Intelligence and its uses. *Harper's Magazine, 140,* 227–235.

REFERENCES

Tierney, C., Nemirovsky, R., & Noble, T. (1996). *Patterns of change.* Palo Alto, CA: Dale Seymour Publications.

Tzou, C.T. (2000). Supporting students' understanding of the relationships between data creation and data analysis. In P. Cobb (Chair), Supporting the learning of classroom communities: A case from middle school statistics. Symposium conducted at the meeting of the American Educational Research Association, New Orleans, LA.

U. S. National Research Center. (1996). *Third International Mathematics and Science Study.* Report no. 7. Washington, DC: Author.

USA Today *(*May 24, 1996) "Yogi Berra Receives Honorary Doctorate."

Van Hiele, P. M. (1986). *Structure and insight.* Orlando, FL: Academic Press.

Visual Mathematics (1995). *Algebra and functions* [Curriculum In Hebrew] Center for Educational Technology, Tel-Aviv.

von Glasersfeld, E. (1984). An introduction to radical constructivism. In P. Watzlawick (Ed.), *The invented reality* (pp. 17–40). New York: Norton.

von Glasersfeld, E. (1990a). An exposition of constructivism: Why some like it radical. In R. B. Davis, C. A. Maher, & N. Noddings (Eds.), *Constructivist views on the teaching and learning of mathematics* (*Journal for Research in Mathematics Education* Monograph No. 4) (pp. 19–29). Reston, VA: National Council of Teachers of Mathematics.

von Glasersfeld, E. (1990b). Environment and communication. In L. P. Steffe & T. Wood (Eds.), *Transforming children's mathematics education* (pp. 30–38). Hillsdale, NJ: Lawrence Erlbaum Associates.

von Glasersfeld, E. (1995). *Radical constructivism: A way of knowing and learning.* London: Taylor & Francis.

Vuyk, R. (1981). Overview and critique of Piaget's genetic epistemology 1965–1980, Vol. 1. London: Academic Press.

Vygotsky, L. S. (1962). *Thinking and speaking.* New York: Cambridge University Press.

Vygotsky, L. S. (1962). *Thought and language.* Cambridge, MA: MIT Press.

Vygotsky, L. S. (1978). *Mind in society: The development of higher psychological processes.* Cambridge, MA: Harvard University Press.

Watson, J. M., & Chick, H. (in press). Factors influencing the outcomes of collaborative mathematical problem solving—An introduction. *Mathematical Thinking and Learning.*

Webb, N. M. (1991). Task-related verbal interaction and mathematics learning in small groups. *Journal for Research in Mathematics Education, 22,* 366–389.

Webb, N. M., Troper, J. D., & Fall, R. (1995). Constructive activity and learning in small groups. *Journal of Educational Psychology, 87,* 406–423.

Weinstein, H., Rosenblum, L., Frugoli, P., Moeller, L., Osburn, J., Peirce, B., Veneri, C., & Wash, P. (1999). The many facets of skills. In *Report on the American workforce.* U.S. Department of Labor. (On-line). Available: http://www.bls.gov/opub/rtaw

Wenger, E. (1998) *Communities-of-practice: Learning, meaning, and identity.* Cambridge: Cambridge University Press.

Wertsch, J. (1985). *Vygotsky and the social formation of mind.* Cambridge, MA: Harvard University Press.

Wertsch, J. (1991). *Voices of the mind: A sociocultural approach of mediated action.* London: Harvester.

Wertsch, J., & Toma, C. (1995). Social and cultural dimensions of knowledge. In L. Steffe & J. Gale (Eds.), *Constructivism in education* (pp. 175–184) Hillsdale, NJ: Lawrence Erlbaum Associates.

Wickelgren, W. A. (1974). *How to solve problems.* San Francisco: W. H. Freeman.

REFERENCES

Wigfield, A., & Eccles, J. (1992). The development of achievement task values: A theoretical analysis. *Developmental Review, 12*, 265–310.

Williams, C. G. (1998). Using concept maps to assess conceptual knowledge of function. *Journal for Research in Mathematics Education, 29*(4), 414–421.

Wilson, J., Fernandez, M., & Hadaway, N. (1993). Mathematical problem solving. In P. S. Wilson (Ed.), *Research ideas for the classroom: High school mathematics.* (pp. 57-78). Reston, VA: Macmillan/NCTM.

Wisconsin Center for Educational Research and Freudenthal Institute. (Eds.). (1998). *Mathematics in context: A connected curriculum for grades 5–8.*

Womack, J. P., & Jones, D. T. (1996). *Lean thinking.* New York: Simon & Schuster.

Wright, C. (1858). The winds and the weather. Atlantic Monthly, 1, 273.

Yackel, E., & Cobb, P. (1996). Socio-mathematical norms, argumentation, and autonomy in mathematics. *Journal for Research in Mathematics Education, 27*(4), 458–77.

Yackel, E., Cobb, P., & Wood, T. (1991). Small-group interaction as a source of learning opportunities in second-grade mathematics. *Journal for Research in Mathematics Education, 22,* 390–408.

Yackel, E., Cobb, P., Wood, T., Wheatley, G., & Merkel, G. (1990). The importance of social interaction in children's construction of mathematical knowledge. In T. J. Cooney & C. R. Hirsch (Eds.), *Teaching and learning mathematics in the 1990s* (pp. 12–21). Reston, VA: National Council of Teachers.

Yerushalmy, M. (1997). Mathematizing qualitative verbal descriptions of Situations: A language to support modeling. *Cognition and Instruction 15*(2), 207–264.

Yerushalmy, M., & Shternberg, B. (1993/1998). The function sketcher [Computer software]. Tel-Aviv, Israel: Center for Educational Technology.

Yerushalmy, M., & Shternberg, B. (2001). A visual course to the concept of function. In A. Cuoco, & F. Curcio (Eds.), *The roles of representations in school mathematics. 2001 National Council of Teachers of Mathematics Yearbook.*

Young, R. M. (1983). Surrogates and mappings: Two kinds of conceptual models for interactive devices. In D. Gentner & A. L. Stevens (Eds.), *Mental models.* (pp. 35–52). Hillsdale, NJ: Lawrence Erlbaum Associates.

AUTHOR INDEX

A

Akerstrom, M., 21, 361
Aliprantis, C. D., 214, 256
Alston, A. S., 143
Amit, M., 143, 407, 416
Anderson, D. K., 128
Anderson, J. R., 503
Anderson, V., 428
Anzai, K., 502
Archambault, R. D., 298
Asch, S. E., 421
Ausubel, D., 535

B

Baker, E. L., 389
Balacheff, N., 481
Ball, D., 125, 131, 188
Ball, D. L., 143
Baratta-Lorton, M., 453
Barnes, C. A., 141
Battista, 532,
Behr, M., 13, 19, 25, 37, 39, 46, 51, 72, 129, 135, 275, 361, 453, 454
Bell, A., 325, 326, 489
Bell, M., 37, 39,
Bergeron, N., 489
Berman, E., 286
Beth, E., 36, 55, 83, 274, 350
Bezuk, N., 453, 463
Bhaskar, R., 502
Biddle, B., 422
Bishop, A. J., 445
Black, P., 191
Black, S., 279, 280
Boero, P., 481
Borko, H., 127
Bound, J., 286
Bowers, J. S., 161
Boyle, C. B., 503
Bradbard, D., 504
Breker, G., 489
Bresnahan, T., 279
Brouseau, G., 363
Brown, A. L., 60, 386
Brown, S. I., 304
Brownell, W., 528
Bruner, J., 44, 206, 256, 274, 347, 438, 450, 451, 453, 462, 528, 540, 544
Brynjolfsson, E., 279
Buchler, J., 100, 101, 111
Burton, L 325
Burton, G., 462
Byrne, B. M., 415

C

Campione, J. C., 386
Carey, 540
Carlson, M., 39, 465, 466, 467
Carmona, L., 214, 364, 366
Carpenter, S., 62, 66, 70, 176
Carpenter, T. P., 9, 32, 61, 66, 72, 128, 131, 143, 188, 192, 407, 540
Carraher, D. W., 129, 397
Carraher, T. N., 129
Charles, K., 297,
Charles, R. I., 319, 502, 505, 507, 512
Chase, W. G., 503
Chazan, D., 486
Chi, M. T. H., 503
Chiang, C., 127
Chick, H., 341, 342
Christiansen, D., 210
Clarke, D., 213
Clements, D. H., 532, 534
Cobb, P., 51, 130, 143, 175, 179, 181, 188, 192, 346, 364, 365, 385, 411, 412, 421, 445, 534, 544
Cohen, D. K., 141, 143
Cohen, E. G., 341, 343
Conant, J. B., 524
Confrey, J., 497, 534
Connell, M. L., 386
Cooney, T. J., 130
Corbitt, M. K., 407
Coulson, R. L., 128
Cramer, K., 25, 51, 135, 338, 342, 347, 350, 453, 454, 455, 462, 463
Crider, J., 268
Csiksentmihalyi, M., 411, 415, 551
Cuban, 281
Cudmore, D., 297, 307
Cumbo, K., 128
Cunningham, D., 103, 104, 105, 106, 120
Cuoco, A., 385, 391

D

Danesi, M., 101
Dapueto, C., 483
Dark, M. J., 389
Davidson, N., 422
Davis, R. B., 445, 532, 533, 534, 535
Debellis, V. A., 408
de Corte, E., 406
Deely, J., 101, 104
delMas, R., 454
DeLoache, J. S., 60

583

AUTHOR INDEX

Dennis, D., 497
Dewey, J., 521, 525, 526, 527, 529, 530, 555, 557
Dienes, Z., 36, 275, 451, 452, 453, 462
diSessa, A. A., 29, 128, 392, 394, 539, 540
Doerr, H. M., ix, 10, 21, 36, 39, 42, 99, 138, 161, 256, 301, 318, 338, 342, 347, 350, 355, 360, 470, 479, 498, 538
Doorman, M., 485, 497
Driscoll, M., 505
Dubinsky, E., 365
Duncker, K., 501
Dweck, C. S., 431

E

Eccles, J. S., 407, 410, 411
Empson, S. B., 131, 143
English, L. D., 53, 54, 55, 297, 300, 304, 306, 307, 308, 319, 387, 558
Ewert, P. H., 502

F

Feldlaufer, H., 410
Feltovich, P. J., 128, 503
Fennema, E., 9, 32, 61, 72, 128, 131, 143, 188, 416, 540
Fernandez, M., 505
Firestone, W. A., 142
Flavell, J. H., 386
Flexer, R., 128
Fosnot, C. T., 128, 131
Franke, M. L., 131, 143, 540
Frensch, P. A., 501, 503
Funke, J., 501, 503
Fuson, K., 37, 39, 72

G

Gagne, 320
Gallo, M., 161
Gardner, H., 419, 422, 430
Garfield, J. B., 305
Garofalo, J., 383, 386, 502, 505, 509
Gee, J. P., 417
Gentner, D., 60, 63
Giere, R. N., 59
Giles, N. D., 61
Gillette, G., 420
Ginsburg, H. P., 144
Glaser, R., 503
Glenn, J., 208
Goldenberg, E. P., 385, 391
Goldin, G. A., 256, 259, 408, 506, 532, 535, 536

Goleman, D., 422
Golubitsky, M., 59
Good, T. L., 341, 422, 424, 425
Gorman, J., 410
Gowin, D. B., 161
Gravemeijer, K., 175, 179, 181, 412, 485, 497, 498
Greeno, J., 29, 397, 543
Greer, B., 305
Griffiths, J., 97
Grosslight, L., 61
Grouws, D. A., 405, 422
Grover, B. W., 388
Guilford, J. P., 430
Gummer, E., 268
Gundlach, L., 70

H

Hadaway, N., 505
Hadot, 522, 527
Hagen, R. H., 422
Hamelin, P. B., 161
Hamilton, E., ix, 21, 211, 361
Hammer, D., 392
Harel, G., 135, 359, 360, 361, 362, 366, 379, 380, 425, 539
Harre, R., 420
Hart, 20
Harter, S., 415
Hassan, E., 97
Hatch, 422
Hatfield, L. L., 504
Heger, M., 302, 303, 342, 343, 346, 383, 396
Heinz, K., 143, 147
Hembree, R., 505
Henningsen, M., 388
Henry, A., 455
Herscovics, N., 489
Hestenes, D., 59, 408
Hidi, S., 428
Hiebert, J., 137, 275
Higgins, M. R., 422
Hilton, P. J., 410
Hitt, L., 279
Hjalmarson, M., 51, 350, 509
Hodge, L., 184
Hofstadter, D. R., 421, 537
Hole, B., 14, 35, 44, 207, 212, 257, 338, 381, 427, 465, 470
Holmes, O., 521, 522, 527, 528, 530
Hoover, M., ix, 5, 14, 24, 35, 41, 44, 207, 212, 215, 257, 338, 381, 427, 465, 470
Horton, P. B., 161
Houser, N., 101
Hoyles, C., 397, 411
Hunt, E., 503

AUTHOR INDEX

I

Inhelder, B., 359, 361, 379, 382

J

Jacobs, S., 131, 143, 465
Jacobson, C., 63
James, W., 521, 523, 524, 525, 527, 529, 525, 530, 540, 555
Jay, E., 61, 412, 430
Johnson, D., 422
Johnson, R., 422
Johnson, T., 266, 402, 451
Jonassen, D. H., 266
Jones, D. T., 287

K

Kagan, S., 422
Kamii, C., 532, 535, 536
Kant, 537
Kantowski, M. G., 505
Kaput, J. J., 13, 21, 72, 207, 267, 359, 360, 379, 466, 485, 497, 548
Kardos, G., 214
Karplus, R., 443
Katona, G., 501
Kauffman, S., 421
Kehle, P., 100, 122, 112, 513, 515
Kelly, A., ix, 5, 14, 22, 24, 35, 41, 44, 207, 212, 213, 215, 257, 338, 381, 385, 427, 465, 470
Kemeny, V., 63
Kemme, S., 192
Kenny, P. A., 304
Kepner, H. S., 407
Kieren, C., 28, 408
Kilpatrick, J., 100, 386, 435, 505, 506, 532
King, A., 528
Kinzel, M., 143, 147
Klein, R., 143
Kline, 59
Kloosterman, D., 410
Koedinger, K., 426
Kolpakowski, T., 392
Kozak, M., 285
Kroll, D. L., 383, 505, 508, 509
Kuhn, D., 386
Kuhn, T., 545

L

Lakatos, I., 545

Lambdin, D. V., 394
Lambert, J. F., 502
Lamon, S. J. ix, 9, 12, 31, 208, 436, 530
Lampert, M., 125
Landau, M., ix, 21, 211, 361
Larsen, S., 465
Latour, B., 61, 62, 66
Lave, J., 28, 128, 129, 397, 427
Lazarowitz, R., 422
Leader, L., 416
Leavy, A., 416
LeBlanc, J. F., 508
Lehrer, R., 25, 61, 62, 63, 64, 65, 66, 176, 192, 411, 547
Leinhardt, G., 128
Lemke, J., 100, 407, 411, 416, 426
Lesh, R., ix, 5, 10 13, 14, 19, 21, 22, 23, 25, 30, 31, 35, 36, 37, 39, 41, 42, 43, 44, 46, 51, 53, 54, 55, 69, 72, 99, 129, 135, 143, 144, 160, 192, 207, 208, 211, 212, 213, 215, 256, 257, 268, 275, 308, 318, 319, 322, 325, 327, 338, 340, 341, 342, 343, 346, 347, 350, 359, 360, 361, 364, 366, 379, 381, 383, 384, 385, 387, 395, 401, 402, 427, 449, 451, 453, 454, 465, 466, 470, 472, 479, 505, 507, 508, 509, 530, 534, 538, 556
Leslie, A. M., 60
Lester, F. K., 51, 100, 318, 350, 383, 386, 502, 505, 508, 509, 512, 513, 515
Leung, S. S., 304
Levi, L., 131, 143
Lindquist, M. M., 407
Littlefield, J., 411
Loef, M., 128
Lucas, J., 100
Lynch, L., 279, 280

M

Machin, S., 286
Maher, C. A., 188, 445, 532, 533
Maki, D., 98, 508
Maletsky, E. M., 462
Mamona-Downs, J., 304
Marion, S., 128
Mark, J., 385, 391
Marsh, H., 505
Martin, 380
Masingila, J., 138
Mason, J., 325
Mason, R. P., 319
Mayer, R. E., 318
Mayfield, V., 128
McCaslin, M., 341, 424, 425
McClain, K., 50, 51, 175, 179, 181, 184, 412
McClintock, C. E., 506
McConney, A., 161

McGatha, M., 184
McLeod, D. B., 416
McNeal, B., 143
Mead, G. H., 412
Meece, J. L., 407
Merrell, F., 100
Meyer, M. R., 416
Middleton, J. A., 342, 343, 346, 383, 396, 405, 409, 410, 411, 412, 414, 416, 426, 428
Midgley, C., 410
Mierkiewicz, D., 505
Miller, T., 505
Minsky, M., 53
Monfils, 142
Monk, S., 466, 497
Mulryan, C., 341, 422, 424, 425

N

Nemirovsky, R., 461, 466, 488, 497, 498
Nesher, P., 480, 481
Newell, A., 502
Noble, T., 461
Noddings, N., 341, 422, 445, 532, 533, 554
Nofsinger, J. M., 319
Norman, D. A., 61
Noss, R., 397
Novak, J. D., 161
Nunes, T., 28, 397

O

Oakes, W., 214, 421
Op'T Enyde, P., 406

P

Papert, S., 30
Parenti, L., 483
Pearlman, K., 281, 288, 289
Peirce, B., 100, 101
Peirce, C. S., 362, 521, 524, 527, 530, 531, 555, 557
Penner, D. E., 61, 63, 66, 176
Perkins, D. N., 406, 412, 430
Peterson, P. L., 19, 128
Piaget, J., 19, 21, 36, 55, 83, 274, 350, 359, 361, 362, 379, 382, 451, 453, 529, 541, 544, 555
Pimm, D., 100, 481
Pinker, S., 442, 445
Pollock, G. B., 422
Polya, G., 100, 317, 325, 508, 511
Popper, C., 523
Post, T., 13, 14, 19, 25, 35, 37, 39, 44, 46, 51, 72, 129, 135, 207, 212, 257, 275, 338, 342, 347, 350, 361, 381, 427, 451, 452, 453, 454, 465, 470
Pozzi, S., 397
Pritchard, C., 65
Pulos, S., 443
Putz, A., 62

R

Reeves, T. C., 266
Reiser, B. J., 503
Resnick, M., 68, 70
Reuman, D., 410
Reys, B., 422
Reys, R. E., 407
Rissing, S., 422
Rissland, E. L., 318
Robb, J., 285
Romberg, T. A., 51, 191
Rorty, R., 523
Roschelle, J., 497
Rud, A. G., 214, 421
Russell, B., 522, 524
Russell, R. S., 285

S

Saldana, L., 466
Saxe, G. B., 28, 397
Schank, R. C., 397
Schauble, L., 25, 61, 62, 64, 66, 176, 547
Schifter, D., 127, 131, 188
Schliemann, A. D., 129, 397
Schmuck, P. A., 422
Schmuck, R. A, 422
Schnepp, M. J., 497
Schoenfeld, A. H., 100, 318, 324, 325, 385, 386, 502, 505, 506, 507, 508, 509
Schorr, R. Y., 25, 142, 143, 144, 160
Schroeder, T. L., 508
Schwartz, J. L., 486
Senn, G. J., 161
Sfard, A., 130, 548
Shafer, M., 191
Shank, G., 104, 105
Shannon, A., 357
Sharan, R., 422
Sharan, S., 422
Shaughnessy, J. M., 305, 306
Sherin, B., 392
Shternberg, B., 486, 488, 497
Shulman, L. S., 53, 130, 159
Sierpinska, A., 100

AUTHOR INDEX

Silver, E. A., 304, 435, 502, 505, 506, 507, 508
Simmt, E., 408
Simon, H. A., 72, 127, 131, 143, 147, 188, 502, 503, 546, 548
Slavin, R., 422
Smith, C., 61
Smith ,D. R., 422
Smith, E., 466
Smith , J., 408
Smith , M. S., 143, 147, 342
Sowder, J., 359, 361, 362, 379, 380
Sowder, L. K., 318
Spanias, P., 405, 414, 416
Spiegel, M. R., 50
Spillane, J. P., 142
Spiro, R. J., 128
Stacey, K., 325
Stage, E., 443
Steen, L. A., 36
Steffe, L. P., 72, 364, 365, 519, 534, 538, 547, 554
Stein, M. K., 342, 388
Stephan, M., 179
Stephens, L. J., 50
Sterman, 522, 530
Sternberg, R. J., 419, 430, 503
Stewart, I., 59
Stigler, J., 137
Strom, D., 63
Sutherland, R., 481
Suydam, M. N., 505
Swan, M., 325, 326, 489
Szeminska, A., 359

T

Tall, D., 466, 480
Taylor III, R. B., 285
Taylor, G., 325, 326
Teich, V., 138
Thompson, A. G., 188
Thompson, M., 98
Thompson, P. W., 466, 483, 489, 519, 534, 538, 547, 548, 554
Thorndike, E. L., 419
Tierney, C., 461
Tinto, P., 39
Tirosh, D., 143
Tishman, S., 406, 412, 430
Toluk, Z., 409, 412, 414, 426, 428
Toma, C., 189
Toupin, C., 60, 63
Tzou, C. T., 176
Tzur, R., 143, 147

U

Unger, C., 61
USA Today, 537

V

Van Hiele, P. M., 371, 382
van Reeuwijk, M., 192
Verschafel, L., 406
Vinner, S., 480
von Glasersfeld, E., 364, 365, 536, 537, 538, 539
Vuyk, R., 365
Vygotsky, L. S., 56, 275, 405, 431, 529, 544, 550, 555

W

Wachsmuth, I., 453
Walter, M. I., 304
Watson, J. M., 341, 342
Webb, C., 422
Weinstein, H., 286, 288
Wenger, E., 128, 411, 427
Wertsch, J., 51, 56, 189, 275
White, C. A., 319
Whitenack, J., 412
Wickelgren, W. A., 317, 318
Wigfield, A., 407, 410, 411
Wijers, M., 192
Wilensky, U., 70
William, D., 191
Williams, C. G., 161
Wilson, J., 505
Womack, J. P., 287
Wood, T., 143, 188, 346
Woods, A. L., 161
Wright, C., 522

Y

Yackel, E., 51, 130, 143, 188, 192, 346, 412
Yerushalmy, M., 480, 485, 486, 488, 497
Young, R. M., 61

Z

Zawojewski, J. S., 22, 53, 54, 55, 308, 327, 338, 340, 341, 342, 347, 350, 357, 384, 395, 396, 401, 424, 425, 508
Zeuli, J. S., 142

SUBJECT INDEX

A

absent present, 497
abstraction phase, 98
abstraction, 359
academic development, 401
academic ostracism, 407
accountability, 209
activities for teachers, 32
adult-level conservation tasks, 75
aeronautical engineering, 7
affective objectives (AOs), 215, 533
affects, 509
age of information, 15
agent-centered explanations, 70
algorithms, 440
alternative perspectives, 54
American pragmatists, 533
answer-producing routines, 322
application, 252
applied math criteria, 545
applied mathematical problem solving, 504
applying, 59
arbitrary rules, 407
armchair case, 250
artifacts, 71
ascertaining, 362
assembly-ists, xi, 519
assessing intermediate results, 42
assessment
    instruments, 302
    issues, 150
    items, 463
ATI (Aptitude-treatment-interaction), 435
attitudes, 383
authoritarian proof scheme, 362
awarenesses, 383

B

back to basics, 208
basic skills, 519
becoming aware stage, 452
behavioral objectives (BOs), 9, 141, 215, 533
beliefs, 383, 509
beyond constructivism, 517
big ideas in mathematics, 438
Bigfoot transcript, 424
by-products of TCCT, 206

C

calculus students, 481
Candid Camera, 12
case method, 241
cases
    and the laboratory, 249
    as history, 249
    as problems, 247
    as projects, 246
case studies, 530
center of gravity, 82
centering, 52
changed their perceptions, 157
changed their view, 157
classroom
    analysis, 182
    discourse, 534
    norms, 201
cognitive, 408
    conflict, 26, 546
    development, x, 346
    guided instruction, 520, 540
    intelligences, 428
    objectives (COs), 9, 215, 518, 533
    processes, 502
    social dimensions, 51
    structures, 19, 72, 364
coherent sign system, 121
Cold War, 528
common
    definitions, 386
    knowledge, 42
    sense, 74
communicating and modeling, 267
communicating, 42, 288
community of practice, 412, 427
comparable units, 94
competency-based education, 279
complex artifacts, 303
complex mathematical activity, 501
complex mathematical domain, 437
complex
    situations, 402
    systemic nature, 29
    systems, 214, 288
complexity, 128, 493, 521, 522
composers, 438
computer-based tools, 521
computer tool, 180
Conant, James, 524
concatenators, 438

589

SUBJECT INDEX

concept map, 161
concept of area, 81
conceptual
  egocentrism, 53
  evolution, 28
  models, 13
  reorganization, 72
  schemes, 53
  systems, 19, 21, 383
  tools, ix, 3, 9,
  understanding, 142, 143
concrete
  manipulatable materials, 39
  manipulatives, 141
  models, 457
concrete-to-abstract, 36
condition-action rules, 282
conflicts, 252
conformity, 437
conservation, 27
  tasks, 73
constrained result, 439
construct
  development problems, 219
  eliciting activities, 207
construct generalization principle, 473
constructing, 59
construction processes, 534
constructions, 16
constructivist, 519
constructivist perspectives, 445, 532
constructs, 21
constructs
  conceptual systems, 71
consumer guide book, 41
context parsing, 107
context protocols, 107
context-and-context-dependent, 397
contextual, 129
continual, 130
control, 508
controlled transformation, 371
conventionalizing, 62
coordination between transformation, 374
coordination within transformation, 373
corporations, 224
correct conception, 80
counting, 78
covariation task, 465
covariational reasoning, 465
critique of the arguments, 176
cross-functionalizaton, 287
curriculum design
  construct documentation principle, 473
  construction principle, 37, 135
  constructivist foundations, 519
  constructive principle, 452
  Dienes' instructional principles, 37
  dynamic principle, 38, 135, 452
  four principles, 37

Lesh translation model, 449
mathematics variability principle, 452
model-externalization principle, 43
model generalization principle, 43
multiple embodiment
  principle, 38
  perceptual
  variability principles, 135
multilevel principle, 133
perceptual variability principle, 38, 452
personal meaningfulness principle, 43
reality principle, 133, 472
self-evaluation principle, 43, 134, 472
self-management, 289
self-modeling, 411
semiosis and inference, 104
sharing principle, 134
simple prototype principle, 43
simplicity principle, 473
situated cognitive structures, 8
six principles, 472
six principles of instructional design, 43
curricular development, 133, 279
curriculum material
  Alaskan Pipeline Problem, 111
  Assessing Myself, 154
  Big Foot Problem, 5, 338, 418
  Bottle Problem, 467
  Car Problem, 481
  Carpentry Problem, 322
  CD Deals, 162
  Classic Man, 155
  CBRs (Computer Based Rangers), 473
  Committee Scheduling Problem, 111
  Community Outreach Program, 226
  Constructed Wetlands, 231
  CROWS, 438
  cuisenaire rods, 480
  exercise problem, 298
  Fast Plants, 67
  Flow Activities, 415
  Gravity Rules, 85
  Greek Ideal Athlete, 155
  Historic Hotels Problem, 257
  How To Toolkit, 50, 268, 338
  Linear Irrigation Problem, 109, 515
  Lion Problem, 156
  Musical Madness, 314
  O'Hare Problem, 347
  Olympic Proportions Problem, 154
  Paper Airplane Problem, 7, 393
  Penny Pitch, 163
  Quilt Problem, 85, 303
  Reallotment, 194
  River Watch, 171
  Running Speed Problem, 27
  Sears Catalog Problem, 27
  Shadow Box Problem, 302
  Smart Shadows Problem, 360, 366, 367
  Sneaker Problem, 301

SUBJECT INDEX

591

Tower of Hanoi Problem, 502
Suez Canal, 97
Summer Camp Problem, 301
Summer Jobs Problem, 270, 289
Summer Reading Program Problem, 218
Toy-Train, 303
Urba and Cursa Problem, 195
Volleyball Problem, 6
Wallpaper Problem, 343
customer, 230
cyclic process, 178

D

data handling, 305
debriefing tools, 385, 401
deduction, 104
deductive proof scheme, 362
Descartes, 221
description, 17
Design and Chance, 524
design
    experience, 227
    principles, 341
    research, 179
developmental landscapes, 63
Dewey, John, 521
diagnostic prompts, 427
didactical models, 479
different modeling cycles, 22
digital appendices, x
digitization, 286
dimensions of mathematical knowledge, 441
discourses, 417
dispositional construct, 413
distancing and brainstorming, 332
distinctions between good and poor
    problem solving, 507
distribution, 70
distributional thinking, 70
diversity, 27, 435
domain-general, 502
dominating sign systems, 117
dynamic events, 466
dynamics, 427

E

EC (engineering criteria), 224
economic problem, 255
egocentrism, 52
Einstein's Theory of Relativity, 80
elementary-but-deep, 539
elements, 96
embodiment, 37
eMedia-electronic media, 265
empirical, 362

enactive mode, 453
ends-in-itself, 324, 534
ends-in-view-problems, 297
engineering education, 223
equalizing areas, 120
equity principle, 435
equivalence relation, 83
estimation, 95
Euclid, 221
European Gestalt tradition, 501
evaluating models, 59
evolving
    capacity levels, 277
    expert activities, 214
    expert studies, 553
    examples of communicating, 266
expectancies, 407
expertise in teaching, 127
expert-novice, 165
experts, 503
explain your answer, 193
explanations and assessment problems, 191
extending, 150
external conviction, 362
external-to-internal
    dimensions, 51
    processes, 350
evolutionary instrumentalism, 528

F

factory-based models, 30
failure, 407
Flow, 551
follow-up activities, 50, 56
formal mathematics, 485
formulation, 376
foundations for mathematical thinking, 25
Four Rs, 205
free response evaluation, 237
function concept, 486
functional relationships, 70
functions and proportionality, 456
fundamentals, 209
funneling, 165

G

geographic information systems, 530
geometry, 95
Gestalt tradition, 501
Gestaltist, 501
given-find, 248
givens, 319
global age, 281
goals, 319
grand theories, 523

grounded, 65
group functioning, 289, 422, 383
guided inquiry, 485

H

habits of mind, 386
hands-on activities, 142, 436
heuristics, 22, 317
high performance in global, technological age, 284
higher order objectives, 390
higher-order thinking, 383
historical dimension, 170
historical schema, 160
holistic systems, 96
Holmes, Wendall, 521
horizontal integration, 287
how cases are used, 244
How to Make Our Ideas Clear, 524
human cognition, 516
human performance, 279
hypothetical-deductive modeling, 67, 70

I

ideal model, 515
identify, 383, 401
implementation
    designs, 160
    issues, 150
implementing small group modeling activities, 353
implications of model and modeling perspective, 409
implicit and explicit patterns of inference, 509
indexical, 111
induction, 104
inductive proof scheme, 362
inertia, 82
inference, 104
inference types, 122
information gathering, 42
information processing-based definitions, 318
information processing theories, 387
input variable, 476
instruction, 511
instructional
    development, 265, 381
    dimension, 172
    practices, 141
    sequence, 179
instructionist theory, 444
instrumentalism, 528
integral domain, 83
integration with research and practice, 465
intellectual development, 451

intelligent behavior, 402
interacting, 268
interdependent, 150
interest tool, 428
interest, 401
internalization, 544
internal and external components, 71
internal constructs, 52
internal goals, 502
Internet-based resources, x
international survey, 306
interpersonal intelligence, 419
interplay of sign, 122
interpretant, 87, 101
interpreting, 59
intervention tools, 401
intervention, 150
investigating, 59
investigations curriculum, 461
isolated problem-solving activities, 43

J

James, William, 521

K

knowledge
    acquisition, 509
    tailoring, 325
knowledge-rich domains, 504

L

ladder-like stages of development, 520, 541
Lafayette, 238
learned knowledge, 410
learning systems, 479
leverage points, 556
limited control of transformation, 370
lists of data, 217
local competence, 28
local conceptual development, 19, 21, 359, 543
logical
    dimension, 170
    inference, 362
lower-order thinking, 384

M

M&M perspective, xi,
machine-based metaphors, 30
making sense of complex systems, 30

SUBJECT INDEX

manipulable objects, 488
manipulation, 17
manipulative, 120, 450, 454
MANOVA, 455
marketplace of ideas, 528
mathematical
    abilities, 24, 205
    abstraction, 102
    connections, 142
    constructs, 24
    eye, 308
    field, 480
    idea, ix
    knowledge, 255
    models, 10, 479, 521
    modeling, 411
    power, 341
    representation, 98, 512
    results, 98
    world, 512
mathematicians' formal logical systems, 82
mathematics standards, 435
mathematization, 61
mathematizing, 42
means-to-ends, 324, 534
measurement, 95
mental
    actions (MA), 467
    models, 61
    systems, 439
metacognition
    abilities, 394
    functioning, 383
    higher order thinking, 551
    problem solving, 508
    processes, 385
    semiosis, 103
metaphor finishing nail metaphor, 550
metaphysical beliefs, 524
metacognitive prompts, 426
methodologies, 505
methods, 150
middle labeling, 328
mind in society, 544
minitool, 181
mirroring reality, 523
mock job interviews, 212
model
    adaptation activity, 45
    application activities, 49
    based practices, 59
    based reasoning, 63
    construction principle, 43, 338, 472
    development, 134
    development cycles, 423
    development sequences, 35
    eliciting activities, 3
    exploration activity, 45
    extension activity, 49

    phenomenon, 496
modeling, 60
modeling
    abilities, 521
    cycles, 17, 85, 165, 361, 410
    environment, 408
    game, 59
    processes, 275
    session, 355
    task, 418
modeling-based learning environments, 479
models and modeling
    perspective, 4
    research group, 555
    theories, 387
models in physics, 483
models of concept, 496
models, 10
model-world relations, 64
modules, 35
monitoring, 42
monotonous continuous function, 493
multi-cycle processes, 552
multidimensional, 129
multiple contexts, 138
multiple media, 275
multi-media representational fluency, 536
multiple, 268
    disciplinary team tasks, 227
    modeling cycles, 25
    representational mappings, 275
    representations, 142
    teachers, 138
    transitions, 495
multiplicative proportional reasoning, 21
multitier
    design experiments, 553
    design research, 556
    professional development, 159, 173
    program development, 160
    teaching experiments, 160
    teaching study, 149
multivariate analysis, 455

N

natural knowledge domain, 503
negotiation, 446
new math, 535
newspaper article, 5, 258
Newton, 221, 524
non-quantitative symbolic proof scheme, 362
nontrivial problems, 17
norms for argumentation, 184
novice, 504
number conservation task, 79

SUBJECT INDEX

O

objective case study, 253
objectives of instruction, 141
objects/products of thought, 291
observation form, 32, 212
observation tool, 152
occupational analysis, 279
one-to-one correspondences, 75
on-the-job classroom-based teacher development activities, 33
on-the-job teacher development, 157
operational thought, 362
operational-relational-transformational systems, 82
operations, 96
organizations
    ABET (Accreditation Board of Engineering and Technology), 224
    American Association for the Advancement of Science, 207, 552
    American Society for Engineering Education, 224
    AT&T Foundation, 205
    Boeing Company, 224
    Child Development and Family Studies Department, 228
    Corporation for National Service, 238
    Department of Education, 207
    Educational Research and Development Center, x
    Educational Testing Services, 155, 162, 163
    Graduate Record Examination Board, 24
    Habitat for Humanity, 237
    Imagination Station, 227
    Microsoft Corporation, 238
    National Center for Improving Student Learning and Achievement, x
    National Council of Teachers of Mathematics, 128, 207, 263, 510, 552
    National Science Foundation, ix, 155, 162, 163, 189, 224, 238, 449, 456
    Office of Educational Research and Improvement, x
    Orbital Engine Corporation, 303
    Salvation Army, 237
    School of Electrical and Computer Engineering, 226
    U. S. Department of Education, 228
    YWCA, 237
overview, x
own construction, 198
own production, 198

P

paleontology, 5
parallel, 268
parallel processing, 396
part-part comparisons, 94
part-part relationships, 91
part-whole comparisons, 94
pattern
    of change, 461
    recognition, 20
    recognition and replication, 20
    regularities, 96
Peano's Postulates, 83
Pierce, Charles Sanders, 521
Peirce's conception of signs, 111
Peircean semiotic framework, 100
perceptual proof scheme, 362
perceptual reasoning, 379
personalities, 252
perspective taking, 350
persuading, 362
phenomena, 479
philosophy of experience, 529
philosophical foundations, 519
physical microcosms, 64
physics, 482
Piagetian Conceptual Systems, 71
Piagetian perspectives, 29
pictures, 450
pieces of information, 217
places rated almanac, 41
planning, 42
play stages, 452
pluralistic, 128
pocketful of models, 69
power of the group, 342
powerful
    conceptual tools, ix, 466
    mathematical ideas, 255
    mathematical tools, 143
    sense-making systems, 26
practical development, 133
practice-based situations, 136
pragmatism in American Education, 527
pragmatist movement, 523
preparation, 356
prerequisite ideas, 4
primitive multiplicative reasoning, 20
Principles of Psychology, 523
problem gallery, 308
problem posing, 42
problem solving
    instruction, 508
    processes, 22
    strategies, 317
    teams, 357
process objectives (POs), 9, 141, 215, 533
processual rush of semiosis, 102
product, 152
productivity of metacognitive functions, 399
professional development, 125
pure math criteria, 545
project

SUBJECT INDEX

partner, 232
proposal, 231
team , 231
titles, 231
projects
  Applied Problem Solving Project, 322
  Center for 21st Century Concepatual Tools (TCCT), xi, 205
  Cognitively Guided Instruction, 131
  Connected Math Project, 191
  EPICS (Engineering Projects in Community Service), 223
  Frontiers in Education Conference, 224
  Function Sketcher, 486
  Georgia Workshop, 505
  Harvard Calculus Project, 451
  HPN (Homelessness Prevention Network), 225
  Logo™, 68
  Math in Context, 51
  Mathematical Modeling in the Secondary School Curriculum (NCTM, 1991), 109
  Mathematics and Plausible Reasoning, 511
  Mathematics in Context, 191
  MathScape, 191
  MathThematics, 191
  Mathematics Their Way, 453
  MiC unit, 194
  Models and Modeling Working Group, x
  Nesher's Learning System (LS), 481
  PACKETS , 154, 162
  PriMath, 456
  Rational Number Project (RNP Fraction Lesson), 449, 453, 454
  Rational Number Project: Fraction Lessons for Middle Grades, 51
  rational mumbers, 79
  Realistic Mathematics Education (RME), 485
  Research Agenda Project, 505
  Research on Cognitively Guided Instruction, 32
  Research Workshop on Problem Solving in Mathematics Education, 504
  Simcalc, 466
  StarLogo™, 68
  Third International Mathematics and Science Study (TIMSS), 303
  Urban Systemic Initiative, 462
  Visual Mathematics, 485
  Wabash Center Children's Services, 228
prompts, 194
proof scheme, 359
propagation, 27
proportional reasoning, 19, 135
prototypes, 341
proximal development, 520
prove, 78
pseudo empirical abstraction, 365
psychological
  development, 133
  dimension, 170
psychometrics, 435
Purdue's Director of Admissions, 209
pure mathematics, 36, 351, 528

Q

qualitative
  information, 42
  reasoning, 19
  relationships, 19
quality assessment
  guides, 212
  procedures, 32
quality ratings, 41
quantitative relationships, 94

R

radical constructivist, 537
reasoning, 379
rate of change, 465
real experience, 536
real problems, 225, 523
realistic model, 98
realistic problem, 98
real-life
  contexts, 450
  problems, 142
  situation, ix, 255
real-world mathematics, 408
reasoning 20, 205, 466
  abilities, 477
  capabilities, 22
recommendations, 242
recurrence, 118
refining, 150
reflected abstraction, 365
reflection, 151, 205
reflection and debriefing activities, 50
reflection tools, 55, 385, 425
The Reflex Arc Concept in Psychology, 525
relations, 96
relationships between metacognition and cognition, 384
Remarks on Spencer's Definition of Mind as Correspondence, 524
representation system, 48
representational
  activities, 62
  conventions, 62
  development, 133
  fluency, 12, 205, 288, 449
  media, 11, 521
  modes, 509
  systems, 64, 261

SUBJECT INDEX

representations, 8, 62, 547
research methodologies, 211
researcher field notes, 151
researchers, 150
responsibility, 205
reversibility, 20
revising, 150
rich experiences, 255
ritual proof scheme, 362
Robin and Ellen's modeling activity, 110
Rube Goldberg, 439
rule of four, 451
rule of three, 451

S

Schaum Outlines, 50
schemata, 8
selection, 27
semiosis, 101
semiotic
    framework, 99
    functioning, 547
sequence of events, 252
service-learning program, 229
setting, 252
shareable products, 338
shared experiences, 340
sharing observations, 148
short stories, 242
show your work, 192
sign systems, 114
signified field, 480
signifier, 480
situations, 73
situated cognitions, 520
situation-specific interest, 428
skill standards, 279
skillfulness, 282
skills for high performance workplace, 279
small group
    learning activity, 337
    modeling activities, 346
social
    constructivism, 520
    correctness criteria, 545
social and development perspectives, 325
social
    aspects, 338
    dimensions, 55, 524
    functioning, 401, 413, 419
    intelligence, 419
    norms, 194
socially accepted language, 58
societies of mind, 52
society in the mind, 544
socio cultural contexts, 509
sociocultural perspectives, 445

soft skills, 223
solution
    cycles, 154
    development, 321
    patterns, 262
solutions, 242
South Bend, 238
spatial constructions, 302
stage of reasoning, 28
stages of development, 19
Stair model, 480
standards, 510
standards-based reform, 142
stimulation and control, 428
strategies, 22
structural similarity, 49
structure, 73
structured stage, 452
student
    centered, 507
    collaboration, 141
    knowledge, 125
    products, 152, 477
    understanding, 142
students'
    abilities, 27
    data models, 188
    learning trajectories, 520, 546
    solutions, 475
    work, 151
subjective case study, 253
suggested strategies, 427
symbolic, 111, 497
symbolic mode, 453
syntactic modeling, 66
system design, 233
system-based intuitions, 84
systemic reform, 226
systems-as-a-whole, 73

T

taken-as-shared, 421, 534
task
    design principles, 338
    domain, 502
    variables, 506
    analysis cycles, 175
    centered, 507
tautology, 78
TCCT researchers, 222
teacher
    criteria, 151
    development activities, 148, 520, 554
    education design, 149
    knowledge, 125, 159, 127
    learning, 143
    level tasks, 137

SUBJECT INDEX

models, 132
professional development, 126
responses, 151
role, 355
teaching problem-solving, 142
teammates , 230
technological innovation, 285
technology-based
    age of information, 216
    representational media, 265,
template, 92, 364
template theory, 444
term model, 450
textbook problems, 247
theoretical
    framework, 465
    goal, 502
    grounding, 139
    implications, 173
    perspective, 139, 266, 360
theory development, 381
thought processes, 508
thought-revealing activities, 32, 145, 474
threaded conversations, 412
three dimensions, 170
thumbtack metaphor, 550
three Rs, 205
time-on-task versus time-off-task, 424
tool as a product, 301
tool building, 320
traditional conceptions, 15, 207
traditional
    constructivist philosophy, 140
    media, 267
    perspectives, 31
    research on problem solving, 548
    views versus modeling views, 31
transcripts, 24
transformation, 78, 362
transformational
    proof scheme, 362
    reasoning, 379
transient signs, 118
transition periods, 72
translation problem, 512

translation, 17
treatments, 437
trends, 20
trial tools, 213

U

under construction, 512
unit, 74
unstable models, 349
USA Today, 216
utilization, 509

V

vaccinations and growth of mathematical
    knowledge, 443
values, 407
variability, 128
verification, 17
vertical integration, 227
visual design, 227
Vygotsky's zones, 520, 544

W

warm-up activities, 56
ways of thinking sheets, 25, 32, 212
well-organized structures, 73
What is a Case?, 243
whole-class discussion, 176
why questions, 142
windows, 54
workplace definition, 282
World War II, 528
writing cases, 242
written
    post, 455
    symbols, 450